Nissan Sentra & 200SX Automotive Repair Manual

by Larry Warren, Tim Imhoff and John H Haynes
Member of the Guild of Motoring Writers

Models covered:
All Nissan Sentra and 200SX models
1995 through 1999

(3C2 - 72051)

ABCDE
FGHIJ
KLMNO
PQ

Haynes Publishing Group
Sparkford Nr Yeovil
Somerset BA22 7JJ England

Haynes North America, Inc
861 Lawrence Drive
Newbury Park
California 91320 USA

About this manual

Its purpose

The purpose of this manual is to help you get the best value from your vehicle. It can do so in several ways. It can help you decide what work must be done, even if you choose to have it done by a dealer service department or a repair shop; it provides information and procedures for routine maintenance and servicing; and it offers diagnostic and repair procedures to follow when trouble occurs.

We hope you use the manual to tackle the work yourself. For many simpler jobs, doing it yourself may be quicker than arranging an appointment to get the vehicle into a shop and making the trips to leave it and pick it up. More importantly, a lot of money can be saved by avoiding the expense the shop must pass on to you to cover its labor and overhead costs. An added benefit is the sense of satisfaction and accomplishment that you feel after doing the job yourself.

Using the manual

The manual is divided into Chapters. Each Chapter is divided into numbered Sections, which are headed in bold type between horizontal lines. Each Section consists of consecutively numbered paragraphs.

At the beginning of each numbered Section you will be referred to any illustrations which apply to the procedures in that Section. The reference numbers used in illustration captions pinpoint the pertinent Section and the Step within that Section. That is, illustration 3.2 means the illustration refers to Section 3 and Step (or paragraph) 2 within that Section.

Procedures, once described in the text, are not normally repeated. When it's necessary to refer to another Chapter, the reference will be given as Chapter and Section number. Cross references given without use of the word "Chapter" apply to Sections and/or paragraphs in the same Chapter. For example, "see Section 8" means in the same Chapter.

References to the left or right side of the vehicle assume you are sitting in the driver's seat, facing forward.

Even though we have prepared this manual with extreme care, neither the publisher nor the author can accept responsibility for any errors in, or omissions from, the information given.

NOTE

A **Note** provides information necessary to properly complete a procedure or information which will make the procedure easier to understand.

CAUTION

A **Caution** provides a special procedure or special steps which must be taken while completing the procedure where the Caution is found. Not heeding a Caution can result in damage to the assembly being worked on.

WARNING

A **Warning** provides a special procedure or special steps which must be taken while completing the procedure where the Warning is found. Not heeding a Warning can result in personal injury.

Acknowledgements

Wiring diagrams were originated exclusively for Haynes North America, Inc. by Valley Forge Technical Communications.

© Haynes North America, Inc. 1998, 2000

With permission from J.H. Haynes & Co. Ltd.

A book in the Haynes Automotive Repair Manual Series

Printed in the U.S.A.

All rights reserved. No part of this book may be reproduced or transmitted in any form or by any means, electronic or mechanical, including photocopying, recording or by any information storage or retrieval system, without permission in writing from the copyright holder.

ISBN 1 56392 377 7

Library of Congress Catalog Card Number 00-102336

While every attempt is made to ensure that the information in this manual is correct, no liability can be accepted by the authors or publishers for loss, damage or injury caused by any errors in, or omissions from, the information given.

Contents

Introductory pages

About this manual	0-2
Introduction to the Nissan Sentra and 200SX	0-4
Vehicle identification numbers	0-5
Buying parts	0-6
Maintenance techniques, tools and working facilities	0-6
Jacking and towing	0-12
Booster battery (jump) starting	0-12
Automotive chemicals and lubricants	0-13
Conversion factors	0-14
Safety first!	0-15
Troubleshooting	0-16

Chapter 1
Tune-up and routine maintenance — 1-1

Chapter 2   Part A
Engine — 2A-1

Chapter 2   Part B
General engine overhaul procedures — 2B-1

Chapter 3
Cooling, heating and air conditioning systems — 3-1

Chapter 4
Fuel and exhaust systems — 4-1

Chapter 5
Engine electrical systems — 5-1

Chapter 6
Emissions and engine control systems — 6-1

Chapter 7   Part A
Manual transaxle — 7A-1

Chapter 7   Part B
Automatic transaxle — 7B-1

Chapter 8
Clutch and driveaxles — 8-1

Chapter 9
Brakes — 9-1

Chapter 10
Suspension and steering systems — 10-1

Chapter 11
Body — 11-1

Chapter 12
Chassis electrical system — 12-1

Wiring diagrams — 12-14

Index — IND-1

Haynes mechanic, author and photographer with 1995 Nissan Sentra

Introduction to the Nissan Sentra and 200SX

The Nissan Sentra is available in four-door sedan body styles while the 200SX is a two-door coupe.

The transversely mounted inline four-cylinder engines used in these models are equipped with electronic fuel injection. Two different engines of similar design are used on these models; the 1.6L GA16DE and 2.0L SR20DE.

The engine drives the front wheels through either a five-speed manual or a four-speed automatic transaxle via independent driveaxles.

Independent suspension, featuring coil spring/strut damper units, is used on the front wheels, while a beam axle located by trailing arms and a lateral link and using coil spring/shock absorber units is used at the rear. The rack-and-pinion steering gear is mounted behind the engine with power-assist available as an option.

The brakes are disc at the front and drums or disc at the rear, with power assist standard. An Anti-lock Brake System (ABS) is used on some models.

0-5

Vehicle identification numbers

Modifications are a continuing and unpublicized process in vehicle manufacturing. Since spare parts manuals and lists are compiled on a numerical basis, the individual vehicle numbers are essential to correctly identify the component required.

Vehicle Identification Number (VIN)

This very important identification number is stamped on the firewall in the engine compartment and on a plate attached to the dashboard inside the windshield on the driver's side of the vehicle (see illustration). The VIN also appears on the Vehicle Certificate of Title and Registration. It contains information such as where and when the vehicle was manufactured, the model year and the body style.

VIN engine code

One particularly important piece of information is found in the VIN is the engine code. Counting from the left, the engine code letter designation is the 4th digit.

On the models covered by this manual the engine codes are:

A ....GA16DE 1.6L DOHC
B ....SR20DE 2.0L DOHC

Manufacturer's Certification Regulation label

The manufacturer's Certification Regulation label is attached to the A-pillar on the driver's side door. The plate contains the name of the manufacturer, the month and year of production, the Gross Vehicle Weight Rating (GVWR), the Gross Axle Weight Rating (GAWR) and the certification statement.

Engine number

The engine code number can be found on a pad on the front (radiator) side of the cylinder block, near the transaxle (see illustration).

The Vehicle Identification Number (VIN) is visible through the driver's side of the windshield

The engine number (arrow) is stamped onto a machined pad on the front side of the engine block, near the transaxle

Buying parts

Replacement parts are available from many sources, which generally fall into one of two categories - authorized dealer parts departments and independent retail auto parts stores. Our advice concerning these parts is as follows:

Retail auto parts stores: Good auto parts stores will stock frequently needed components which wear out relatively fast, such as clutch components, exhaust systems, brake parts, tune-up parts, etc. These stores often supply new or reconditioned parts on an exchange basis, which can save a considerable amount of money. Discount auto parts stores are often very good places to buy materials and parts needed for general vehicle maintenance such as oil, grease, filters, spark plugs, belts, touch-up paint, bulbs, etc. They also usually sell tools and general accessories, have convenient hours, charge lower prices and can often be found not far from home.

Authorized dealer parts department: This is the best source for parts which are unique to the vehicle and not generally available elsewhere (such as major engine parts, transmission parts, trim pieces, etc.).

Warranty information: If the vehicle is still covered under warranty, be sure that any replacement parts purchased - regardless of the source - do not invalidate the warranty!

To be sure of obtaining the correct parts, have engine and chassis numbers available and, if possible, take the old parts along for positive identification.

Maintenance techniques, tools and working facilities

Maintenance techniques

There are a number of techniques involved in maintenance and repair that will be referred to throughout this manual. Application of these techniques will enable the home mechanic to be more efficient, better organized and capable of performing the various tasks properly, which will ensure that the repair job is thorough and complete.

Fasteners

Fasteners are nuts, bolts, studs and screws used to hold two or more parts together. There are a few things to keep in mind when working with fasteners. Almost all of them use a locking device of some type, either a lockwasher, locknut, locking tab or thread adhesive. All threaded fasteners should be clean and straight, with undamaged threads and undamaged corners on the hex head where the wrench fits. Develop the habit of replacing all damaged nuts and bolts with new ones. Special locknuts with nylon or fiber inserts can only be used once. If they are removed, they lose their locking ability and must be replaced with new ones.

Rusted nuts and bolts should be treated with a penetrating fluid to ease removal and prevent breakage. Some mechanics use turpentine in a spout-type oil can, which works quite well. After applying the rust penetrant, let it work for a few minutes before trying to loosen the nut or bolt. Badly rusted fasteners may have to be chiseled or sawed off or removed with a special nut breaker, available at tool stores.

If a bolt or stud breaks off in an assembly, it can be drilled and removed with a special tool commonly available for this purpose. Most automotive machine shops can perform this task, as well as other repair procedures, such as the repair of threaded holes that have been stripped out.

Flat washers and lockwashers, when removed from an assembly, should always be replaced exactly as removed. Replace any damaged washers with new ones. Never use a lockwasher on any soft metal surface (such as aluminum), thin sheet metal or plastic.

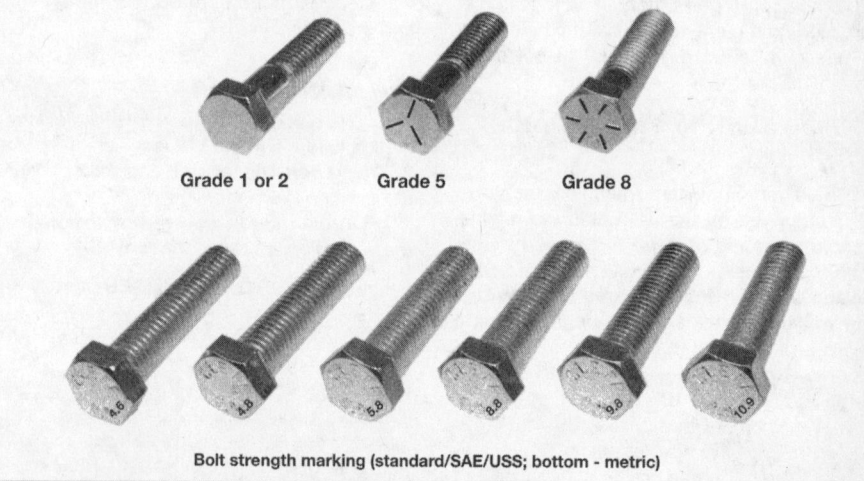

Bolt strength marking (standard/SAE/USS; bottom - metric)

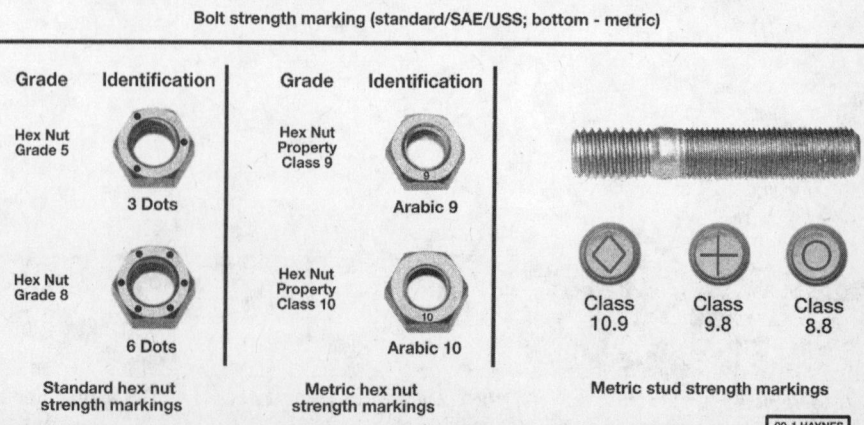

Standard hex nut strength markings

Metric hex nut strength markings

Metric stud strength markings

Maintenance techniques, tools and working facilities

Fastener sizes

For a number of reasons, automobile manufacturers are making wider and wider use of metric fasteners. Therefore, it is important to be able to tell the difference between standard (sometimes called U.S. or SAE) and metric hardware, since they cannot be interchanged.

All bolts, whether standard or metric, are sized according to diameter, thread pitch and length. For example, a standard 1/2 - 13 x 1 bolt is 1/2 inch in diameter, has 13 threads per inch and is 1 inch long. An M12 - 1.75 x 25 metric bolt is 12 mm in diameter, has a thread pitch of 1.75 mm (the distance between threads) and is 25 mm long. The two bolts are nearly identical, and easily confused, but they are not interchangeable.

In addition to the differences in diameter, thread pitch and length, metric and standard bolts can also be distinguished by examining the bolt heads. To begin with, the distance across the flats on a standard bolt head is measured in inches, while the same dimension on a metric bolt is sized in millimeters (the same is true for nuts). As a result, a standard wrench should not be used on a metric bolt and a metric wrench should not be used on a standard bolt. Also, most standard bolts have slashes radiating out from the center of the head to denote the grade or strength of the bolt, which is an indication of the amount of torque that can be applied to it. The greater the number of slashes, the greater the strength of the bolt. Grades 0 through 5 are commonly used on automobiles. Metric bolts have a property class (grade) number, rather than a slash, molded into their heads to indicate bolt strength. In this case, the higher the number, the stronger the bolt. Property class numbers 8.8, 9.8 and 10.9 are commonly used on automobiles.

Strength markings can also be used to distinguish standard hex nuts from metric hex nuts. Many standard nuts have dots stamped into one side, while metric nuts are marked with a number. The greater the number of dots, or the higher the number, the greater the strength of the nut.

Metric studs are also marked on their ends according to property class (grade). Larger studs are numbered (the same as metric bolts), while smaller studs carry a geometric code to denote grade.

It should be noted that many fasteners, especially Grades 0 through 2, have no distinguishing marks on them. When such is the case, the only way to determine whether it is standard or metric is to measure the thread pitch or compare it to a known fastener of the same size.

Standard fasteners are often referred to as SAE, as opposed to metric. However, it should be noted that SAE technically refers to a non-metric fine thread fastener only. Coarse thread non-metric fasteners are referred to as USS sizes.

Since fasteners of the same size (both standard and metric) may have different strength ratings, be sure to reinstall any bolts, studs or nuts removed from your vehicle in their original locations. Also, when replacing a fastener with a new one, make sure that the new one has a strength rating equal to or greater than the original.

Metric thread sizes	Ft-lbs	Nm
M-6	6 to 9	9 to 12
M-8	14 to 21	19 to 28
M-10	28 to 40	38 to 54
M-12	50 to 71	68 to 96
M-14	80 to 140	109 to 154

Pipe thread sizes	Ft-lbs	Nm
1/8	5 to 8	7 to 10
1/4	12 to 18	17 to 24
3/8	22 to 33	30 to 44
1/2	25 to 35	34 to 47

U.S. thread sizes	Ft-lbs	Nm
1/4 - 20	6 to 9	9 to 12
5/16 - 18	12 to 18	17 to 24
5/16 - 24	14 to 20	19 to 27
3/8 - 16	22 to 32	30 to 43
3/8 - 24	27 to 38	37 to 51
7/16 - 14	40 to 55	55 to 74
7/16 - 20	40 to 60	55 to 81
1/2 - 13	55 to 80	75 to 108

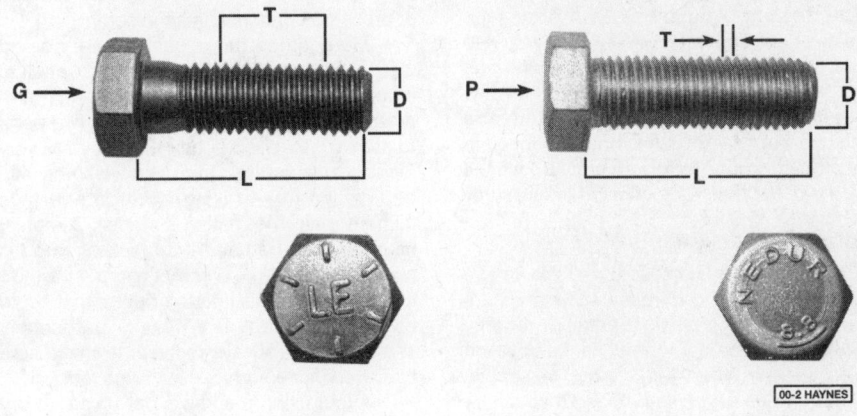

Standard (SAE and USS) bolt dimensions/grade marks

- G Grade marks (bolt strength)
- L Length (in inches)
- T Thread pitch (number of threads per inch)
- D Nominal diameter (in inches)

Metric bolt dimensions/grade marks

- P Property class (bolt strength)
- L Length (in millimeters)
- T Thread pitch (distance between threads in millimeters)
- D Diameter

Tightening sequences and procedures

Most threaded fasteners should be tightened to a specific torque value (torque is the twisting force applied to a threaded component such as a nut or bolt). Overtightening the fastener can weaken it and cause it to break, while undertightening can cause it to eventually come loose. Bolts, screws and studs, depending on the material they are made of and their thread diameters, have specific torque values, many of which are noted in the Specifications at the beginning of each Chapter. Be sure to follow the torque recommendations closely. For fasteners not assigned a specific torque, a general torque value chart is presented here as a guide. These torque values are for dry (unlubricated) fasteners threaded into steel or cast iron (not aluminum). As was previously mentioned, the size and grade of a fastener determine the amount of torque that can safely be applied to it. The figures listed here are approximate for Grade 2 and Grade 3 fasteners. Higher grades can tolerate higher torque values.

Fasteners laid out in a pattern, such as cylinder head bolts, oil pan bolts, differential cover bolts, etc., must be loosened or tight-

Maintenance techniques, tools and working facilities

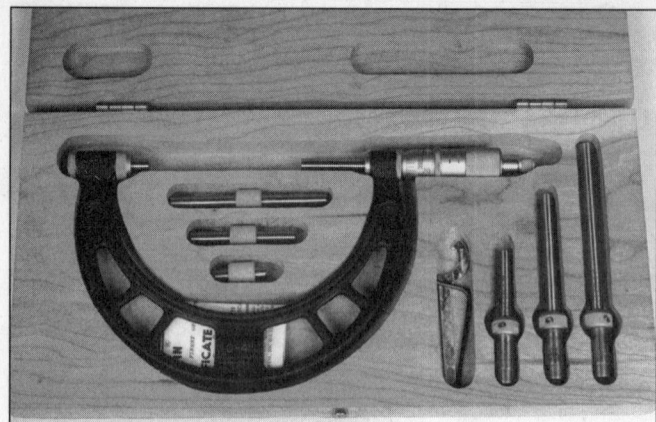

Micrometer set

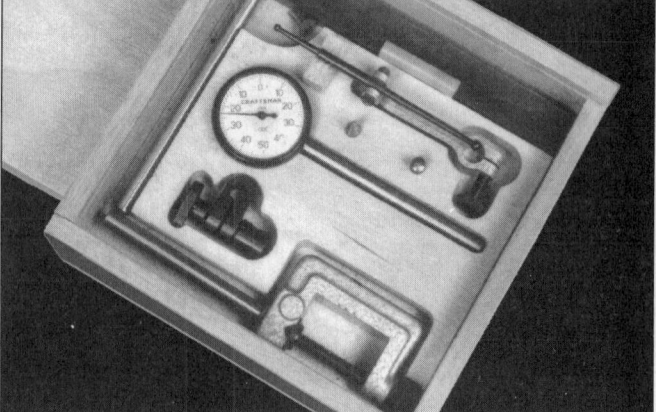

Dial indicator set

ened in sequence to avoid warping the component. This sequence will normally be shown in the appropriate Chapter. If a specific pattern is not given, the following procedures can be used to prevent warping.

Initially, the bolts or nuts should be assembled finger-tight only. Next, they should be tightened one full turn each, in a criss-cross or diagonal pattern. After each one has been tightened one full turn, return to the first one and tighten them all one-half turn, following the same pattern. Finally, tighten each of them one-quarter turn at a time until each fastener has been tightened to the proper torque. To loosen and remove the fasteners, the procedure would be reversed.

Component disassembly

Component disassembly should be done with care and purpose to help ensure that the parts go back together properly. Always keep track of the sequence in which parts are removed. Make note of special characteristics or marks on parts that can be installed more than one way, such as a grooved thrust washer on a shaft. It is a good idea to lay the disassembled parts out on a clean surface in the order that they were removed. It may also be helpful to make sketches or take instant photos of components before removal.

When removing fasteners from a component, keep track of their locations. Sometimes threading a bolt back in a part, or putting the washers and nut back on a stud, can prevent mix-ups later. If nuts and bolts cannot be returned to their original locations, they should be kept in a compartmented box or a series of small boxes. A cupcake or muffin tin is ideal for this purpose, since each cavity can hold the bolts and nuts from a particular area (i.e. oil pan bolts, valve cover bolts, engine mount bolts, etc.). A pan of this type is especially helpful when working on assemblies with very small parts, such as the carburetor, alternator, valve train or interior dash and trim pieces. The cavities can be marked with paint or tape to identify the contents.

Whenever wiring looms, harnesses or connectors are separated, it is a good idea to identify the two halves with numbered pieces of masking tape so they can be easily reconnected.

Gasket sealing surfaces

Throughout any vehicle, gaskets are used to seal the mating surfaces between two parts and keep lubricants, fluids, vacuum or pressure contained in an assembly.

Many times these gaskets are coated with a liquid or paste-type gasket sealing compound before assembly. Age, heat and pressure can sometimes cause the two parts to stick together so tightly that they are very difficult to separate. Often, the assembly can be loosened by striking it with a soft-face hammer near the mating surfaces. A regular hammer can be used if a block of wood is placed between the hammer and the part. Do not hammer on cast parts or parts that could be easily damaged. With any particularly stubborn part, always recheck to make sure that every fastener has been removed.

Avoid using a screwdriver or bar to pry apart an assembly, as they can easily mar the gasket sealing surfaces of the parts, which must remain smooth. If prying is absolutely necessary, use an old broom handle, but keep in mind that extra clean up will be necessary if the wood splinters.

After the parts are separated, the old gasket must be carefully scraped off and the gasket surfaces cleaned. Stubborn gasket material can be soaked with rust penetrant or treated with a special chemical to soften it so it can be easily scraped off. A scraper can be fashioned from a piece of copper tubing by flattening and sharpening one end. Copper is recommended because it is usually softer than the surfaces to be scraped, which reduces the chance of gouging the part. Some gaskets can be removed with a wire brush, but regardless of the method used, the mating surfaces must be left clean and smooth. If for some reason the gasket surface is gouged, then a gasket sealer thick enough to fill scratches will have to be used during reassembly of the components. For most applications, a non-drying (or semi-drying) gasket sealer should be used.

Hose removal tips

Warning: *If the vehicle is equipped with air conditioning, do not disconnect any of the A/C hoses without first having the system depressurized by a dealer service department or a service station.*

Hose removal precautions closely parallel gasket removal precautions. Avoid scratching or gouging the surface that the hose mates against or the connection may leak. This is especially true for radiator hoses. Because of various chemical reactions, the rubber in hoses can bond itself to the metal spigot that the hose fits over. To remove a hose, first loosen the hose clamps that secure it to the spigot. Then, with slip-joint pliers, grab the hose at the clamp and rotate it around the spigot. Work it back and forth until it is completely free, then pull it off. Silicone or other lubricants will ease removal if they can be applied between the hose and the outside of the spigot. Apply the same lubricant to the inside of the hose and the outside of the spigot to simplify installation.

As a last resort (and if the hose is to be replaced with a new one anyway), the rubber can be slit with a knife and the hose peeled from the spigot. If this must be done, be careful that the metal connection is not damaged.

If a hose clamp is broken or damaged, do not reuse it. Wire-type clamps usually weaken with age, so it is a good idea to replace them with screw-type clamps whenever a hose is removed.

Tools

A selection of good tools is a basic requirement for anyone who plans to maintain and repair his or her own vehicle. For the owner who has few tools, the initial investment might seem high, but when compared to the spiraling costs of professional auto maintenance and repair, it is a wise one.

To help the owner decide which tools are needed to perform the tasks detailed in this manual, the following tool lists are offered: *Maintenance and minor repair, Repair/overhaul* and *Special*.

The newcomer to practical mechanics

Maintenance techniques, tools and working facilities 0-9

Dial caliper

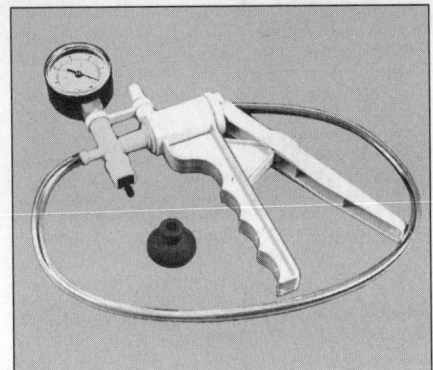
Hand-operated vacuum pump

Timing light

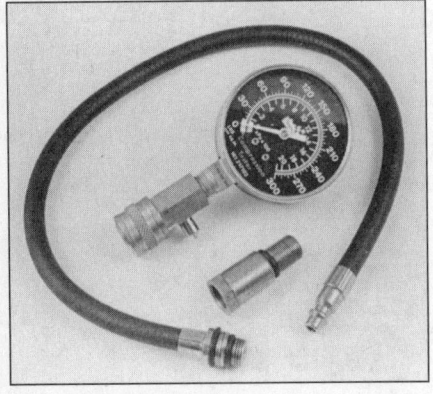
Compression gauge with spark plug hole adapter

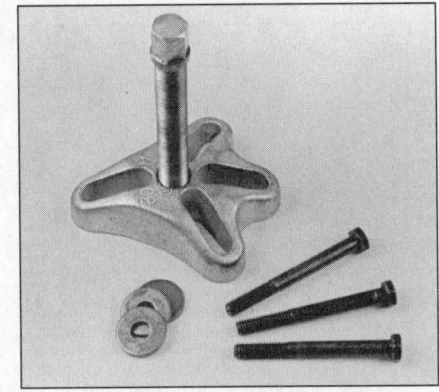
Damper/steering wheel puller

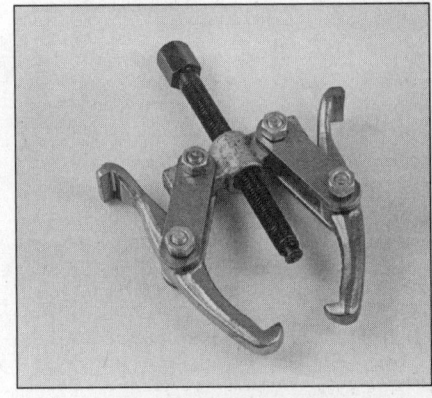
General purpose puller

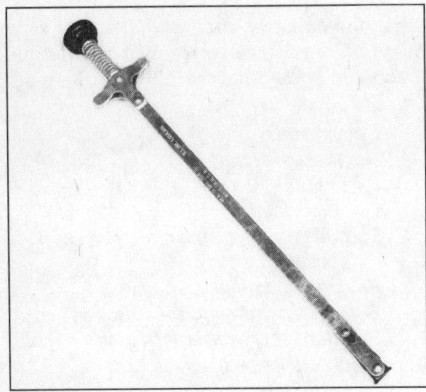
Hydraulic lifter removal tool

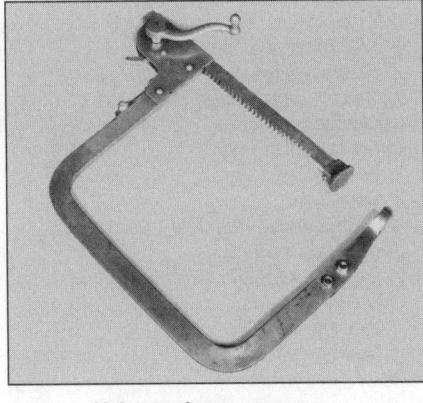
Valve spring compressor

Valve spring compressor

Ridge reamer

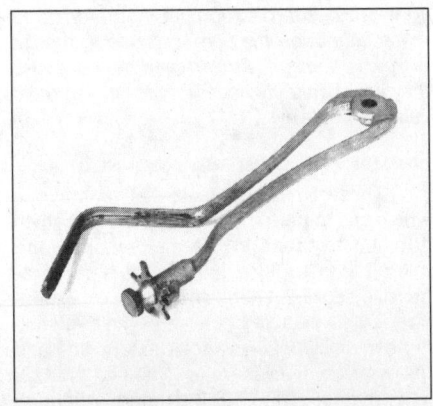
Piston ring groove cleaning tool

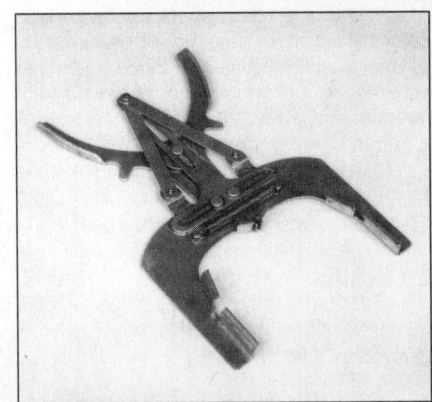
Ring removal/installation tool

0-10 Maintenance techniques, tools and working facilities

Ring compressor

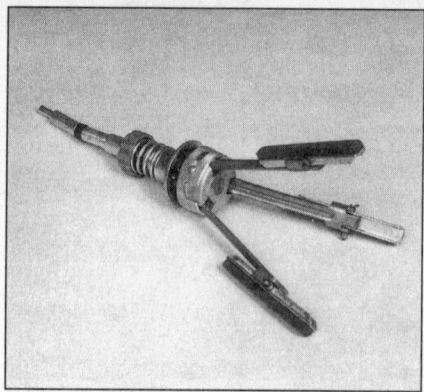
Cylinder hone

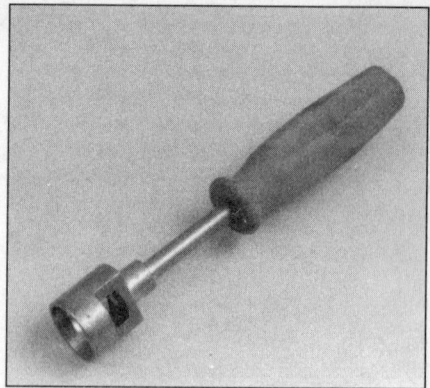
Brake hold-down spring tool

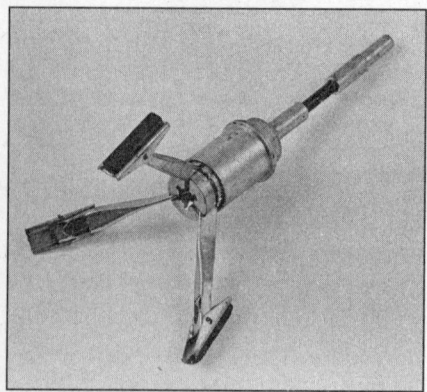
Brake cylinder hone

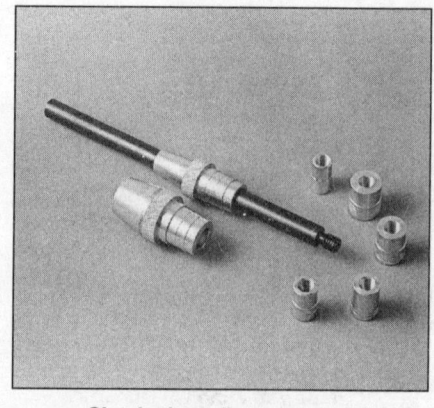
Clutch plate alignment tool

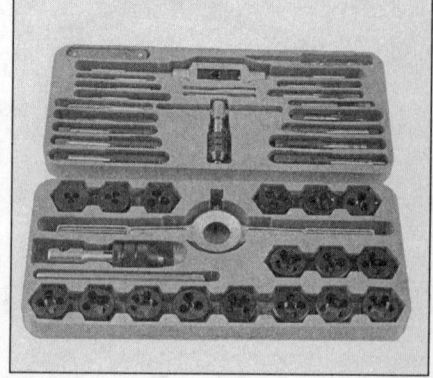
Tap and die set

should start off with the *maintenance and minor repair* tool kit, which is adequate for the simpler jobs performed on a vehicle. Then, as confidence and experience grow, the owner can tackle more difficult tasks, buying additional tools as they are needed. Eventually the basic kit will be expanded into the *repair and overhaul* tool set. Over a period of time, the experienced do-it-yourselfer will assemble a tool set complete enough for most repair and overhaul procedures and will add tools from the special category when it is felt that the expense is justified by the frequency of use.

Maintenance and minor repair tool kit

The tools in this list should be considered the minimum required for performance of routine maintenance, servicing and minor repair work. We recommend the purchase of combination wrenches (box-end and open-end combined in one wrench). While more expensive than open end wrenches, they offer the advantages of both types of wrench.

Combination wrench set (1/4-inch to 1 inch or 6 mm to 19 mm)
Adjustable wrench, 8 inch
Spark plug wrench with rubber insert
Spark plug gap adjusting tool
Feeler gauge set
Brake bleeder wrench
Standard screwdriver (5/16-inch x 6 inch)

Phillips screwdriver (No. 2 x 6 inch)
Combination pliers - 6 inch
Hacksaw and assortment of blades
Tire pressure gauge
Grease gun
Oil can
Fine emery cloth
Wire brush
Battery post and cable cleaning tool
Oil filter wrench
Funnel (medium size)
Safety goggles
Jackstands (2)
Drain pan

Note: *If basic tune-ups are going to be part of routine maintenance, it will be necessary to purchase a good quality stroboscopic timing light and combination tachometer/dwell meter. Although they are included in the list of special tools, it is mentioned here because they are absolutely necessary for tuning most vehicles properly.*

Repair and overhaul tool set

These tools are essential for anyone who plans to perform major repairs and are in addition to those in the maintenance and minor repair tool kit. Included is a comprehensive set of sockets which, though expensive, are invaluable because of their versatility, especially when various extensions and drives are available. We recommend the 1/2-inch drive over the 3/8-inch drive. Although the larger drive is bulky and more expensive, it has the capacity of accepting a very wide range of large sockets. Ideally, however, the mechanic should have a 3/8-inch drive set and a 1/2-inch drive set.

Socket set(s)
Reversible ratchet
Extension - 10 inch
Universal joint
Torque wrench (same size drive as sockets)
Ball peen hammer - 8 ounce
Soft-face hammer (plastic/rubber)
Standard screwdriver (1/4-inch x 6 inch)
Standard screwdriver (stubby - 5/16-inch)
Phillips screwdriver (No. 3 x 8 inch)
Phillips screwdriver (stubby - No. 2)
Pliers - vise grip
Pliers - lineman's
Pliers - needle nose
Pliers - snap-ring (internal and external)
Cold chisel - 1/2-inch
Scribe
Scraper (made from flattened copper tubing)
Centerpunch
Pin punches (1/16, 1/8, 3/16-inch)
Steel rule/straightedge - 12 inch
Allen wrench set (1/8 to 3/8-inch or 4 mm to 10 mm)
A selection of files
Wire brush (large)
Jackstands (second set)
Jack (scissor or hydraulic type)

Maintenance techniques, tools and working facilities

Note: *Another tool which is often useful is an electric drill with a chuck capacity of 3/8-inch and a set of good quality drill bits.*

Special tools

The tools in this list include those which are not used regularly, are expensive to buy, or which need to be used in accordance with their manufacturer's instructions. Unless these tools will be used frequently, it is not very economical to purchase many of them. A consideration would be to split the cost and use between yourself and a friend or friends. In addition, most of these tools can be obtained from a tool rental shop on a temporary basis.

This list primarily contains only those tools and instruments widely available to the public, and not those special tools produced by the vehicle manufacturer for distribution to dealer service departments. Occasionally, references to the manufacturer's special tools are included in the text of this manual. Generally, an alternative method of doing the job without the special tool is offered. However, sometimes there is no alternative to their use. Where this is the case, and the tool cannot be purchased or borrowed, the work should be turned over to the dealer service department or an automotive repair shop.

Valve spring compressor
Piston ring groove cleaning tool
Piston ring compressor
Piston ring installation tool
Cylinder compression gauge
Cylinder ridge reamer
Cylinder surfacing hone
Cylinder bore gauge
Micrometers and/or dial calipers
Hydraulic lifter removal tool
Balljoint separator
Universal-type puller
Impact screwdriver
Dial indicator set
Stroboscopic timing light (inductive pick-up)
Hand operated vacuum/pressure pump
Tachometer/dwell meter
Universal electrical multimeter
Cable hoist
Brake spring removal and installation tools
Floor jack

Buying tools

For the do-it-yourselfer who is just starting to get involved in vehicle maintenance and repair, there are a number of options available when purchasing tools. If maintenance and minor repair is the extent of the work to be done, the purchase of individual tools is satisfactory. If, on the other hand, extensive work is planned, it would be a good idea to purchase a modest tool set from one of the large retail chain stores. A set can usually be bought at a substantial savings over the individual tool prices, and they often come with a tool box. As additional tools are needed, add-on sets, individual tools and a larger tool box can be purchased to expand the tool selection. Building a tool set gradually allows the cost of the tools to be spread over a longer period of time and gives the mechanic the freedom to choose only those tools that will actually be used.

Tool stores will often be the only source of some of the special tools that are needed, but regardless of where tools are bought, try to avoid cheap ones, especially when buying screwdrivers and sockets, because they won't last very long. The expense involved in replacing cheap tools will eventually be greater than the initial cost of quality tools.

Care and maintenance of tools

Good tools are expensive, so it makes sense to treat them with respect. Keep them clean and in usable condition and store them properly when not in use. Always wipe off any dirt, grease or metal chips before putting them away. Never leave tools lying around in the work area. Upon completion of a job, always check closely under the hood for tools that may have been left there so they won't get lost during a test drive.

Some tools, such as screwdrivers, pliers, wrenches and sockets, can be hung on a panel mounted on the garage or workshop wall, while others should be kept in a tool box or tray. Measuring instruments, gauges, meters, etc. must be carefully stored where they cannot be damaged by weather or impact from other tools.

When tools are used with care and stored properly, they will last a very long time. Even with the best of care, though, tools will wear out if used frequently. When a tool is damaged or worn out, replace it. Subsequent jobs will be safer and more enjoyable if you do.

How to repair damaged threads

Sometimes, the internal threads of a nut or bolt hole can become stripped, usually from overtightening. Stripping threads is an all-too-common occurrence, especially when working with aluminum parts, because aluminum is so soft that it easily strips out.

Usually, external or internal threads are only partially stripped. After they've been cleaned up with a tap or die, they'll still work. Sometimes, however, threads are badly damaged. When this happens, you've got three choices:

1) *Drill and tap the hole to the next suitable oversize and install a larger diameter bolt, screw or stud.*
2) *Drill and tap the hole to accept a threaded plug, then drill and tap the plug to the original screw size. You can also buy a plug already threaded to the original size. Then you simply drill a hole to the specified size, then run the threaded plug into the hole with a bolt and jam nut. Once the plug is fully seated, remove the jam nut and bolt.*
3) *The third method uses a patented thread repair kit like Heli-Coil or Slimsert. These easy-to-use kits are designed to repair damaged threads in straight-through holes and blind holes. Both are available as kits which can handle a variety of sizes and thread patterns. Drill the hole, then tap it with the special included tap. Install the Heli-Coil and the hole is back to its original diameter and thread pitch.*

Regardless of which method you use, be sure to proceed calmly and carefully. A little impatience or carelessness during one of these relatively simple procedures can ruin your whole day's work and cost you a bundle if you wreck an expensive part.

Working facilities

Not to be overlooked when discussing tools is the workshop. If anything more than routine maintenance is to be carried out, some sort of suitable work area is essential.

It is understood, and appreciated, that many home mechanics do not have a good workshop or garage available, and end up removing an engine or doing major repairs outside. It is recommended, however, that the overhaul or repair be completed under the cover of a roof.

A clean, flat workbench or table of comfortable working height is an absolute necessity. The workbench should be equipped with a vise that has a jaw opening of at least four inches.

As mentioned previously, some clean, dry storage space is also required for tools, as well as the lubricants, fluids, cleaning solvents, etc. which soon become necessary.

Sometimes waste oil and fluids, drained from the engine or cooling system during normal maintenance or repairs, present a disposal problem. To avoid pouring them on the ground or into a sewage system, pour the used fluids into large containers, seal them with caps and take them to an authorized disposal site or recycling center. Plastic jugs, such as old antifreeze containers, are ideal for this purpose.

Always keep a supply of old newspapers and clean rags available. Old towels are excellent for mopping up spills. Many mechanics use rolls of paper towels for most work because they are readily available and disposable. To help keep the area under the vehicle clean, a large cardboard box can be cut open and flattened to protect the garage or shop floor.

Whenever working over a painted surface, such as when leaning over a fender to service something under the hood, always cover it with an old blanket or bedspread to protect the finish. Vinyl covered pads, made especially for this purpose, are available at auto parts stores.

Jacking and towing

Jacking

Warning: *The jack supplied with the vehicle should only be used for changing a tire or placing jackstands under the frame. Never work under the vehicle or start the engine while this jack is being used as the only means of support.*

The vehicle should be on level ground. Place the shift lever in Park, if you have an automatic, or Reverse if you have a manual

The jack fits over the rocker panel flange (there are two jacking points on each side of the vehicle, indicated by a notch in the rocker panel flange)

transaxle. Block the wheel diagonally opposite the wheel being changed. Set the parking brake.

Remove the spare tire and jack from stowage. Remove the wheel cover and trim ring (if so equipped) with the tapered end of the lug nut wrench by inserting and twisting the handle and then prying against the back of the wheel cover. Loosen the wheel lug nuts about 1/4-to-1/2 turn each.

Place the scissors-type jack under the side of the vehicle and adjust the jack height until it fits in the notch in the vertical rocker panel flange nearest the wheel to be changed. There is a front and rear jacking point on each side of the vehicle **(see illustration)**.

Turn the jack handle clockwise until the tire clears the ground. Remove the lug nuts and pull the wheel off. Replace it with the spare.

Install the lug nuts with the beveled edges facing in. Tighten them snugly. Don't attempt to tighten them completely until the vehicle is lowered or it could slip off the jack. Turn the jack handle counterclockwise to lower the vehicle. Remove the jack and tighten the lug nuts in a diagonal pattern.

Install the cover (and trim ring, if used) and be sure it's snapped into place all the way around.

Stow the tire, jack and wrench. Unblock the wheels.

Towing

As a general rule, the vehicle should be towed with the front (drive) wheels off the ground. If they can't be raised, place them on a dolly. The ignition key must be in the ACC position, since the steering lock mechanism isn't strong enough to hold the front wheels straight while towing. To avoid possible serious transaxle damage, never tow an automatic transaxle equipped vehicle backwards with all four wheels on the ground

If a vehicle equipped with an automatic transaxle must be towed with all four wheels on the ground, it should be towed from the front only, at speeds that don't exceed 30 mph and for a distance that is not over 40 miles. Before towing, check the transmission fluid level (see Chapter 1). If the level is below the HOT line on the dipstick, add fluid or use a towing dolly.

When towing a vehicle equipped with a manual transaxle with all four wheels on the ground, be sure to place the shift lever in neutral and release the parking brake.

Equipment specifically designed for towing should be used. It should be attached to the main structural members of the vehicle, not the bumpers or brackets.

Safety is a major consideration when towing and all applicable state and local laws must be obeyed. A safety chain system must be used at all times.

Booster battery (jump) starting

Observe these precautions when using a booster battery to start a vehicle:

a) *Before connecting the booster battery, make sure the ignition switch is in the Off position.*
b) *Turn off the lights, heater and other electrical loads.*
c) *Your eyes should be shielded. Safety goggles are a good idea.*
d) *Make sure the booster battery is the same voltage as the dead one in the vehicle.*
e) *The two vehicles MUST NOT TOUCH each other!*
f) *Make sure the transaxle is in Neutral (manual) or Park (automatic).*
g) *If the booster battery is not a maintenance-free type, remove the vent caps and lay a cloth over the vent holes.*

Connect the red jumper cable to the positive (+) terminals of each battery **(see illustration)**.

Connect one end of the black jumper cable to the negative (-) terminal of the booster battery. The other end of this cable should be connected to a good ground on the vehicle to be started, such as a bolt or bracket on the body.

Start the engine using the booster battery, then, with the engine running at idle speed, disconnect the jumper cables in the reverse order of connection.

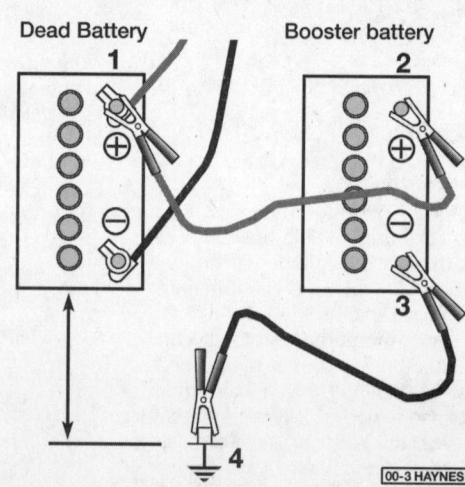

Make the booster battery cable connections in the numerical order shown (note that the negative cable of the booster battery is NOT attached to the negative terminal of the dead battery)

Automotive chemicals and lubricants

A number of automotive chemicals and lubricants are available for use during vehicle maintenance and repair. They include a wide variety of products ranging from cleaning solvents and degreasers to lubricants and protective sprays for rubber, plastic and vinyl.

Cleaners

Carburetor cleaner and choke cleaner is a strong solvent for gum, varnish and carbon. Most carburetor cleaners leave a dry-type lubricant film which will not harden or gum up. Because of this film it is not recommended for use on electrical components.

Brake system cleaner is used to remove grease and brake fluid from the brake system, where clean surfaces are absolutely necessary. It leaves no residue and often eliminates brake squeal caused by contaminants.

Electrical cleaner removes oxidation, corrosion and carbon deposits from electrical contacts, restoring full current flow. It can also be used to clean spark plugs, carburetor jets, voltage regulators and other parts where an oil-free surface is desired.

Demoisturants remove water and moisture from electrical components such as alternators, voltage regulators, electrical connectors and fuse blocks. They are non-conductive, non-corrosive and non-flammable.

Degreasers are heavy-duty solvents used to remove grease from the outside of the engine and from chassis components. They can be sprayed or brushed on and, depending on the type, are rinsed off either with water or solvent.

Lubricants

Motor oil is the lubricant formulated for use in engines. It normally contains a wide variety of additives to prevent corrosion and reduce foaming and wear. Motor oil comes in various weights (viscosity ratings) from 0 to 50. The recommended weight of the oil depends on the season, temperature and the demands on the engine. Light oil is used in cold climates and under light load conditions. Heavy oil is used in hot climates and where high loads are encountered. Multi-viscosity oils are designed to have characteristics of both light and heavy oils and are available in a number of weights from 5W-20 to 20W-50.

Gear oil is designed to be used in differentials, manual transmissions and other areas where high-temperature lubrication is required.

Chassis and wheel bearing grease is a heavy grease used where increased loads and friction are encountered, such as for wheel bearings, balljoints, tie-rod ends and universal joints.

High-temperature wheel bearing grease is designed to withstand the extreme temperatures encountered by wheel bearings in disc brake equipped vehicles. It usually contains molybdenum disulfide (moly), which is a dry-type lubricant.

White grease is a heavy grease for metal-to-metal applications where water is a problem. White grease stays soft under both low and high temperatures (usually from -100 to +190-degrees F), and will not wash off or dilute in the presence of water.

Assembly lube is a special extreme pressure lubricant, usually containing moly, used to lubricate high-load parts (such as main and rod bearings and cam lobes) for initial start-up of a new engine. The assembly lube lubricates the parts without being squeezed out or washed away until the engine oiling system begins to function.

Silicone lubricants are used to protect rubber, plastic, vinyl and nylon parts.

Graphite lubricants are used where oils cannot be used due to contamination problems, such as in locks. The dry graphite will lubricate metal parts while remaining uncontaminated by dirt, water, oil or acids. It is electrically conductive and will not foul electrical contacts in locks such as the ignition switch.

Moly penetrants loosen and lubricate frozen, rusted and corroded fasteners and prevent future rusting or freezing.

Heat-sink grease is a special electrically non-conductive grease that is used for mounting electronic ignition modules where it is essential that heat is transferred away from the module.

Sealants

RTV sealant is one of the most widely used gasket compounds. Made from silicone, RTV is air curing, it seals, bonds, waterproofs, fills surface irregularities, remains flexible, doesn't shrink, is relatively easy to remove, and is used as a supplementary sealer with almost all low and medium temperature gaskets.

Anaerobic sealant is much like RTV in that it can be used either to seal gaskets or to form gaskets by itself. It remains flexible, is solvent resistant and fills surface imperfections. The difference between an anaerobic sealant and an RTV-type sealant is in the curing. RTV cures when exposed to air, while an anaerobic sealant cures only in the absence of air. This means that an anaerobic sealant cures only after the assembly of parts, sealing them together.

Thread and pipe sealant is used for sealing hydraulic and pneumatic fittings and vacuum lines. It is usually made from a Teflon compound, and comes in a spray, a paint-on liquid and as a wrap-around tape.

Chemicals

Anti-seize compound prevents seizing, galling, cold welding, rust and corrosion in fasteners. High-temperature ant-seize, usually made with copper and graphite lubricants, is used for exhaust system and exhaust manifold bolts.

Anaerobic locking compounds are used to keep fasteners from vibrating or working loose and cure only after installation, in the absence of air. Medium strength locking compound is used for small nuts, bolts and screws that may be removed later. High-strength locking compound is for large nuts, bolts and studs which aren't removed on a regular basis.

Oil additives range from viscosity index improvers to chemical treatments that claim to reduce internal engine friction. It should be noted that most oil manufacturers caution against using additives with their oils.

Gas additives perform several functions, depending on their chemical makeup. They usually contain solvents that help dissolve gum and varnish that build up on carburetor, fuel injection and intake parts. They also serve to break down carbon deposits that form on the inside surfaces of the combustion chambers. Some additives contain upper cylinder lubricants for valves and piston rings, and others contain chemicals to remove condensation from the gas tank.

Miscellaneous

Brake fluid is specially formulated hydraulic fluid that can withstand the heat and pressure encountered in brake systems. Care must be taken so this fluid does not come in contact with painted surfaces or plastics. An opened container should always be resealed to prevent contamination by water or dirt.

Weatherstrip adhesive is used to bond weatherstripping around doors, windows and trunk lids. It is sometimes used to attach trim pieces.

Undercoating is a petroleum-based, tar-like substance that is designed to protect metal surfaces on the underside of the vehicle from corrosion. It also acts as a sound-deadening agent by insulating the bottom of the vehicle.

Waxes and polishes are used to help protect painted and plated surfaces from the weather. Different types of paint may require the use of different types of wax and polish. Some polishes utilize a chemical or abrasive cleaner to help remove the top layer of oxidized (dull) paint on older vehicles. In recent years many non-wax polishes that contain a wide variety of chemicals such as polymers and silicones have been introduced. These non-wax polishes are usually easier to apply and last longer than conventional waxes and polishes.

Conversion factors

Length (distance)

Inches (in)	X	25.4 = Millimetres (mm)	X	0.0394	= Inches (in)
Feet (ft)	X	0.305 = Metres (m)	X	3.281	= Feet (ft)
Miles	X	1.609 = Kilometres (km)	X	0.621	= Miles

Volume (capacity)

Cubic inches (cu in; in^3)	X	16.387 = Cubic centimetres (cc; cm^3)	X	0.061	= Cubic inches (cu in; in^3)
Imperial pints (Imp pt)	X	0.568 = Litres (l)	X	1.76	= Imperial pints (Imp pt)
Imperial quarts (Imp qt)	X	1.137 = Litres (l)	X	0.88	= Imperial quarts (Imp qt)
Imperial quarts (Imp qt)	X	1.201 = US quarts (US qt)	X	0.833	= Imperial quarts (Imp qt)
US quarts (US qt)	X	0.946 = Litres (l)	X	1.057	= US quarts (US qt)
Imperial gallons (Imp gal)	X	4.546 = Litres (l)	X	0.22	= Imperial gallons (Imp gal)
Imperial gallons (Imp gal)	X	1.201 = US gallons (US gal)	X	0.833	= Imperial gallons (Imp gal)
US gallons (US gal)	X	3.785 = Litres (l)	X	0.264	= US gallons (US gal)

Mass (weight)

Ounces (oz)	X	28.35 = Grams (g)	X	0.035	= Ounces (oz)
Pounds (lb)	X	0.454 = Kilograms (kg)	X	2.205	= Pounds (lb)

Force

Ounces-force (ozf; oz)	X	0.278 = Newtons (N)	X	3.6	= Ounces-force (ozf; oz)
Pounds-force (lbf; lb)	X	4.448 = Newtons (N)	X	0.225	= Pounds-force (lbf; lb)
Newtons (N)	X	0.1 = Kilograms-force (kgf; kg)	X	9.81	= Newtons (N)

Pressure

Pounds-force per square inch (psi; lbf/in^2; lb/in^2)	X	0.070 = Kilograms-force per square centimetre (kgf/cm^2; kg/cm^2)	X	14.223	= Pounds-force per square inch (psi; lbf/in^2; lb/in^2)
Pounds-force per square inch (psi; lbf/in^2; lb/in^2)	X	0.068 = Atmospheres (atm)	X	14.696	= Pounds-force per square inch (psi; lbf/in^2; lb/in^2)
Pounds-force per square inch (psi; lbf/in^2; lb/in^2)	X	0.069 = Bars	X	14.5	= Pounds-force per square inch (psi; lbf/in^2; lb/in^2)
Pounds-force per square inch (psi; lbf/in^2; lb/in^2)	X	6.895 = Kilopascals (kPa)	X	0.145	= Pounds-force per square inch (psi; lbf/in^2; lb/in^2)
Kilopascals (kPa)	X	0.01 = Kilograms-force per square centimetre (kgf/cm^2; kg/cm^2)	X	98.1	= Kilopascals (kPa)

Torque (moment of force)

Pounds-force inches (lbf in; lb in)	X	1.152 = Kilograms-force centimetre (kgf cm; kg cm)	X	0.868	= Pounds-force inches (lbf in; lb in)
Pounds-force inches (lbf in; lb in)	X	0.113 = Newton metres (Nm)	X	8.85	= Pounds-force inches (lbf in; lb in)
Pounds-force inches (lbf in; lb in)	X	0.083 = Pounds-force feet (lbf ft; lb ft)	X	12	= Pounds-force inches (lbf in; lb in)
Pounds-force feet (lbf ft; lb ft)	X	0.138 = Kilograms-force metres (kgf m; kg m)	X	7.233	= Pounds-force feet (lbf ft; lb ft)
Pounds-force feet (lbf ft; lb ft)	X	1.356 = Newton metres (Nm)	X	0.738	= Pounds-force feet (lbf ft; lb ft)
Newton metres (Nm)	X	0.102 = Kilograms-force metres (kgf m; kg m)	X	9.804	= Newton metres (Nm)

Vacuum

Inches mercury (in. Hg)	X	3.377 = Kilopascals (kPa)	X	0.2961	= Inches mercury
Inches mercury (in. Hg)	X	25.4 = Millimeters mercury (mm Hg)	X	0.0394	= Inches mercury

Power

Horsepower (hp)	X	745.7 = Watts (W)	X	0.0013	= Horsepower (hp)

Velocity (speed)

Miles per hour (miles/hr; mph)	X	1.609 = Kilometres per hour (km/hr; kph)	X	0.621	= Miles per hour (miles/hr; mph)

*Fuel consumption**

Miles per gallon, Imperial (mpg)	X	0.354 = Kilometres per litre (km/l)	X	2.825	= Miles per gallon, Imperial (mpg)
Miles per gallon, US (mpg)	X	0.425 = Kilometres per litre (km/l)	X	2.352	= Miles per gallon, US (mpg)

Temperature

Degrees Fahrenheit = (°C x 1.8) + 32   Degrees Celsius (Degrees Centigrade; °C) = (°F - 32) x 0.56

*It is common practice to convert from miles per gallon (mpg) to litres/100 kilometres (l/100km), where mpg (Imperial) x l/100 km = 282 and mpg (US) x l/100 km = 235

Safety first!

Regardless of how enthusiastic you may be about getting on with the job at hand, take the time to ensure that your safety is not jeopardized. A moment's lack of attention can result in an accident, as can failure to observe certain simple safety precautions. The possibility of an accident will always exist, and the following points should not be considered a comprehensive list of all dangers. Rather, they are intended to make you aware of the risks and to encourage a safety conscious approach to all work you carry out on your vehicle.

Essential DOs and DON'Ts

DON'T rely on a jack when working under the vehicle. Always use approved jackstands to support the weight of the vehicle and place them under the recommended lift or support points.
DON'T attempt to loosen extremely tight fasteners (i.e. wheel lug nuts) while the vehicle is on a jack - it may fall.
DON'T start the engine without first making sure that the transmission is in Neutral (or Park where applicable) and the parking brake is set.
DON'T remove the radiator cap from a hot cooling system - let it cool or cover it with a cloth and release the pressure gradually.
DON'T attempt to drain the engine oil until you are sure it has cooled to the point that it will not burn you.
DON'T touch any part of the engine or exhaust system until it has cooled sufficiently to avoid burns.
DON'T siphon toxic liquids such as gasoline, antifreeze and brake fluid by mouth, or allow them to remain on your skin.
DON'T inhale brake lining dust - it is potentially hazardous (see *Asbestos* below).
DON'T allow spilled oil or grease to remain on the floor - wipe it up before someone slips on it.
DON'T use loose fitting wrenches or other tools which may slip and cause injury.
DON'T push on wrenches when loosening or tightening nuts or bolts. Always try to pull the wrench toward you. If the situation calls for pushing the wrench away, push with an open hand to avoid scraped knuckles if the wrench should slip.
DON'T attempt to lift a heavy component alone - get someone to help you.
DON'T rush or take unsafe shortcuts to finish a job.
DON'T allow children or animals in or around the vehicle while you are working on it.
DO wear eye protection when using power tools such as a drill, sander, bench grinder, etc. and when working under a vehicle.
DO keep loose clothing and long hair well out of the way of moving parts.
DO make sure that any hoist used has a safe working load rating adequate for the job.
DO get someone to check on you periodically when working alone on a vehicle.
DO carry out work in a logical sequence and make sure that everything is correctly assembled and tightened.
DO keep chemicals and fluids tightly capped and out of the reach of children and pets.
DO remember that your vehicle's safety affects that of yourself and others. If in doubt on any point, get professional advice.

Asbestos

Certain friction, insulating, sealing, and other products - such as brake linings, brake bands, clutch linings, torque converters, gaskets, etc. - may contain asbestos. Extreme care must be taken to avoid inhalation of dust from such products, since it is hazardous to health. If in doubt, assume that they do contain asbestos.

Fire

Remember at all times that gasoline is highly flammable. Never smoke or have any kind of open flame around when working on a vehicle. But the risk does not end there. A spark caused by an electrical short circuit, by two metal surfaces contacting each other, or even by static electricity built up in your body under certain conditions, can ignite gasoline vapors, which in a confined space are highly explosive. Do not, under any circumstances, use gasoline for cleaning parts. Use an approved safety solvent.

Always disconnect the battery ground (-) cable at the battery before working on any part of the fuel system or electrical system. Never risk spilling fuel on a hot engine or exhaust component. It is strongly recommended that a fire extinguisher suitable for use on fuel and electrical fires be kept handy in the garage or workshop at all times. Never try to extinguish a fuel or electrical fire with water.

Fumes

Certain fumes are highly toxic and can quickly cause unconsciousness and even death if inhaled to any extent. Gasoline vapor falls into this category, as do the vapors from some cleaning solvents. Any draining or pouring of such volatile fluids should be done in a well ventilated area.

When using cleaning fluids and solvents, read the instructions on the container carefully. Never use materials from unmarked containers.

Never run the engine in an enclosed space, such as a garage. Exhaust fumes contain carbon monoxide, which is extremely poisonous. If you need to run the engine, always do so in the open air, or at least have the rear of the vehicle outside the work area.

If you are fortunate enough to have the use of an inspection pit, never drain or pour gasoline and never run the engine while the vehicle is over the pit. The fumes, being heavier than air, will concentrate in the pit with possibly lethal results.

The battery

Never create a spark or allow a bare light bulb near a battery. They normally give off a certain amount of hydrogen gas, which is highly explosive.

Always disconnect the battery ground (-) cable at the battery before working on the fuel or electrical systems.

If possible, loosen the filler caps or cover when charging the battery from an external source (this does not apply to sealed or maintenance-free batteries). Do not charge at an excessive rate or the battery may burst.

Take care when adding water to a non maintenance-free battery and when carrying a battery. The electrolyte, even when diluted, is very corrosive and should not be allowed to contact clothing or skin.

Always wear eye protection when cleaning the battery to prevent the caustic deposits from entering your eyes.

Household current

When using an electric power tool, inspection light, etc., which operates on household current, always make sure that the tool is correctly connected to its plug and that, where necessary, it is properly grounded. Do not use such items in damp conditions and, again, do not create a spark or apply excessive heat in the vicinity of fuel or fuel vapor.

Secondary ignition system voltage

A severe electric shock can result from touching certain parts of the ignition system (such as the spark plug wires) when the engine is running or being cranked, particularly if components are damp or the insulation is defective. In the case of an electronic ignition system, the secondary system voltage is much higher and could prove fatal.

Troubleshooting

Contents

Symptom	Section
Engine	
Engine backfires	15
Engine diesels (continues to run) after switching off	18
Engine hard to start when cold	3
Engine hard to start when hot	4
Engine lacks power	14
Engine lopes while idling or idles erratically	8
Engine misses at idle speed	9
Engine misses throughout driving speed range	10
Engine rotates but will not start	2
Engine runs with oil pressure light on	17
Engine stalls	13
Engine starts but stops immediately	6
Engine stumbles on acceleration	11
Engine surges while holding accelerator steady	12
Engine will not rotate when attempting to start	1
Oil puddle under engine	7
Pinging or knocking engine sounds during acceleration or uphill	16
Starter motor noisy or excessively rough in engagement	5
Engine electrical system	
Alternator light fails to go out	20
Battery will not hold a charge	19
Alternator light fails to come on when key is turned on	21
Fuel system	
Excessive fuel consumption	22
Fuel leakage and/or fuel odor	23
Cooling system	
Coolant loss	28
External coolant leakage	26
Internal coolant leakage	27
Overcooling	25
Overheating	24
Poor coolant circulation	29
Clutch	
Clutch pedal stays on floor	36
Clutch slips (engine speed increases with no increase in vehicle speed)	32
Grabbing (chattering) as clutch is engaged	33
High pedal effort	37
Noise in clutch area	35
Pedal travels to floor - no pressure or very little resistance	30
Transaxle rattling (clicking)	34
Unable to select gears	31
Manual transaxle	
Clicking noise in turns	41
Clunk on acceleration or deceleration	40

Symptom	Section
Knocking noise at low speeds	38
Leaks lubricant	47
Locked in second gear	48
Noise most pronounced when turning	39
Noisy in all gears	45
Noisy in neutral with engine running	43
Noisy in one particular gear	44
Slips out of gear	46
Vibration	42
Automatic transaxle	
Engine will start in gears other than Park or Neutral	52
Fluid leakage	49
General shift mechanism problems	51
Transaxle fluid brown or has burned smell	50
Transaxle slips, shifts roughly, is noisy or has no drive in forward or reverse gears	53
Driveaxles	
Clicking noise in turns	54
Shudder or vibration during acceleration	55
Vibration at highway speeds	56
Brakes	
Brake pedal feels spongy when depressed	64
Brake pedal travels to the floor with little resistance	65
Brake roughness or chatter (pedal pulsates)	59
Dragging brakes	62
Excessive brake pedal travel	61
Excessive pedal effort required to stop vehicle	60
Grabbing or uneven braking action	63
Noise (high-pitched squeal when the brakes are applied)	58
Parking brake does not hold	66
Vehicle pulls to one side during braking	57
Suspension and steering systems	
Abnormal or excessive tire wear	68
Abnormal noise at the front end	73
Cupped tires	78
Erratic steering when braking	75
Excessive pitching and/or rolling around corners or during braking	76
Excessive play or looseness in steering system	82
Excessive tire wear on inside edge	80
Excessive tire wear on outside edge	79
Hard steering	71
Poor returnability of steering to center	72
Rattling or clicking noise in steering gear	83
Shimmy, shake or vibration	70
Suspension bottoms	77
Tire tread worn in one place	81
Vehicle pulls to one side	67
Wander or poor steering stability	74
Wheel makes a thumping noise	69

Troubleshooting 0-17

This section provides an easy reference guide to the more common problems which may occur during the operation of your vehicle. These problems and their possible causes are grouped under headings denoting various components or systems, such as Engine, Cooling system, etc. They also refer you to the chapter and/or section which deals with the problem.

Remember that successful troubleshooting is not a mysterious "black art" practiced only by professional mechanics. It is simply the result of the right knowledge combined with an intelligent, systematic approach to the problem. Always work by a process of elimination, starting with the simplest solution and working through to the most complex - and never overlook the obvious. Anyone can run the gas tank dry or leave the lights on overnight, so don't assume that you are exempt from such oversights.

Finally, always establish a clear idea of why a problem has occurred and take steps to ensure that it doesn't happen again. If the electrical system fails because of a poor connection, check the other connections in the system to make sure that they don't fail as well. If a particular fuse continues to blow, find out why - don't just replace one fuse after another. Remember, failure of a small component can often be indicative of potential failure or incorrect functioning of a more important component or system.

Engine

1 Engine will not rotate when attempting to start

1   Battery terminal connections loose or corroded (Chapter 1).
2   Battery discharged or faulty (Chapter 1).
3   Automatic transaxle not completely engaged in Park (Chapter 7) or clutch pedal not completely depressed (Chapter 8).
4   Broken, loose or disconnected wiring in the starting circuit (Chapters 5 and 12).
5   Starter motor pinion jammed in flywheel ring gear (Chapter 5).
6   Starter solenoid faulty (Chapter 5).
7   Starter motor faulty (Chapter 5).
8   Ignition switch faulty (Chapter 12).
9   Starter pinion or flywheel teeth worn or broken (Chapter 5).

2 Engine rotates but will not start

1   Fuel tank empty.
2   Battery discharged (engine rotates slowly) (Chapter 5).
3   Battery terminal connections loose or corroded (Chapter 1).
4   Leaking fuel injector(s), faulty fuel pump, pressure regulator, etc. (Chapter 4).

5   Broken timing chain (Chapter 2).
6   Ignition components damp or damaged (Chapter 5).
7   Worn, faulty or incorrectly gapped spark plugs (Chapter 1).
8   Broken, loose or disconnected wiring in the starting circuit (Chapter 5).
9   Loose distributor is changing ignition timing (Chapter 5).
10 Broken, loose or disconnected wires at the ignition coil or faulty coil (Chapter 5).

3 Engine hard to start when cold

1   Battery discharged or low (Chapter 1).
2   Malfunctioning fuel system (Chapter 4).
3   Faulty coolant temperature sensor or intake air temperature sensor (Chapter 6).
4   Injector(s) leaking (Chapter 4).
5   Faulty ignition system (Chapter 5).

4 Engine hard to start when hot

1   Air filter clogged (Chapter 1).
2   Fuel not reaching the fuel injection system (Chapter 4).
3   Corroded battery connections, especially ground (Chapter 1).
4   Faulty coolant temperature sensor or intake air temperature sensor (Chapter 6).

5 Starter motor noisy or excessively rough in engagement

1   Pinion or flywheel gear teeth worn or broken (Chapter 5).
2   Starter motor mounting bolts loose or missing (Chapter 5).

6 Engine starts but stops immediately

1   Loose or faulty electrical connections at distributor, coil or alternator (Chapter 5).
2   Insufficient fuel reaching the fuel injector(s) (Chapters 1 and 4).
3   Vacuum leak at the gasket between the intake manifold/plenum and throttle body (Chapters 1 and 4).
4   Idle speed incorrect (Chapter 1).

7 Oil puddle under engine

1   Oil pan gasket and/or oil pan drain bolt washer leaking (Chapter 2).
2   Oil pressure sending unit leaking (Chapter 2).
3   Valve cover leaking (Chapter 2).
4   Engine oil seals leaking (Chapter 2).
5   Oil pump housing leaking (Chapter 2).

8 Engine lopes while idling or idles erratically

1   Vacuum leakage (Chapters 2 and 4).
2   Leaking EGR valve (Chapter 6).
3   Air filter clogged (Chapter 1).
4   Fuel pump not delivering sufficient fuel to the fuel injection system (Chapter 4).
5   Leaking head gasket (Chapter 2).
6   Timing chain worn (Chapter 2).
7   Camshaft lobes worn (Chapter 2).

9 Engine misses at idle speed

1   Spark plugs worn or not gapped properly (Chapter 1).
2   Faulty spark plug wires (Chapter 1).
3   Vacuum leaks (Chapter 1).
4   Incorrect ignition timing (Chapter 1).
5   Uneven or low compression (Chapter 2).
6   Problem with the fuel injection system (Chapter 4).

10 Engine misses throughout driving speed range

1   Fuel filter clogged and/or impurities in the fuel system (Chapter 1).
2   Low fuel output at the injector(s) (Chapter 4).
3   Faulty or incorrectly gapped spark plugs (Chapter 1).
4   Incorrect ignition timing (Chapter 5).
5   Cracked distributor cap, disconnected distributor wires or damaged distributor components (Chapters 1 and 5).
6   Leaking spark plug wires (Chapters 1 or 5).
7   Faulty emission system components (Chapter 6).
8   Low or uneven cylinder compression pressures (Chapter 2).
9   Weak or faulty ignition system (Chapter 5).
10 Vacuum leak in fuel injection system, air control valve or vacuum hoses (Chapters 2 and 4).

11 Engine stumbles on acceleration

1   Spark plugs fouled (Chapter 1).
2   Problem with the fuel injection system (Chapter 4).
3   Fuel filter clogged (Chapters 1 and 4).
4   Incorrect ignition timing (Chapter 5).
5   Intake manifold air leak (Chapters 2 and 4).
6   Problem with the emissions control system (Chapter 6).

12 Engine surges while holding accelerator steady

1   Intake air leak (Chapters 2 and 4).

Troubleshooting

2   Fuel pump or fuel pressure regulator faulty (Chapter 4).
3   Problem with the fuel injection system (Chapter 4).
4   Problem with the emissions control system (Chapter 6).

13 Engine stalls

1   Idle speed incorrect (Chapter 1).
2   Fuel filter clogged and/or water and impurities in the fuel system (Chapters 1 and 4).
3   Distributor components damp or damaged (Chapter 5).
4   Faulty emissions system components (Chapter 6).
5   Faulty or incorrectly gapped spark plugs (Chapter 1).
6   Faulty spark plug wires (Chapter 1).
7   Vacuum leak in the fuel injection system, intake manifold or vacuum hoses (Chapters 2 and 4).
8   Valve clearances incorrectly set (Chapter 1).

14 Engine lacks power

1   Incorrect ignition timing (Chapter 5).
2   Excessive play in distributor shaft (Chapter 5).
3   Worn rotor, distributor cap, spark plug wires or faulty coil (Chapters 1 and 5).
4   Faulty or incorrectly gapped spark plugs (Chapter 1).
5   Problem with the fuel injection system (Chapter 4).
6   Plugged air filter (Chapter 1).
7   Brakes binding (Chapter 9).
8   Automatic transaxle fluid level incorrect (Chapter 1).
9   Clutch slipping (Chapter 8).
10 Fuel filter clogged and/or impurities in the fuel system (Chapters 1 and 4).
11 Emission control system not functioning properly (Chapter 6).
12 Low or uneven cylinder compression pressures (Chapter 2).
13 Obstructed exhaust system (Chapter 4).

15 Engine backfires

1   Emission control system not functioning properly (Chapter 6).
2   Ignition timing incorrect (Chapter 5).
3   Faulty secondary ignition system (cracked spark plug insulator, faulty plug wires, distributor cap and/or rotor) (Chapters 1 and 5).
4   Problem with the fuel injection system (Chapter 4).
5   Vacuum leak at fuel injector(s), air control valve or vacuum hoses (Chapters 2 and 4).
6   Valve clearances incorrectly set and/or valves sticking (Chapter 1).

16 Pinging or knocking engine sounds during acceleration or uphill

1   Incorrect grade of fuel.
2   Ignition timing incorrect (Chapter 5).
3   Fuel injection system faulty (Chapter 4).
4   Improper or damaged spark plugs or wires (Chapter 1).
5   Worn or damaged distributor components (Chapter 5).
6   EGR valve not functioning (Chapter 6).
7   Vacuum leak (Chapters 2 and 4).

17 Engine runs with oil pressure light on

1   Low oil level (Chapter 1).
2   Idle rpm below specification (Chapter 1).
3   Short in wiring circuit (Chapter 12).
4   Faulty oil pressure sender (Chapter 2).
5   Worn engine bearings and/or oil pump (Chapter 2).

18 Engine diesels (continues to run) after switching off

1   Idle speed too high (Chapter 1).
2   Excessive engine operating temperature (Chapter 3).
3   Ignition timing in need of adjustment (Chapter 5).

Engine electrical system

19 Battery will not hold a charge

1   Alternator drivebelt defective or not adjusted properly (Chapter 1).
2   Battery electrolyte level low (Chapter 1).
3   Battery terminals loose or corroded (Chapter 1).
4   Alternator not charging properly (Chapter 5).
5   Loose, broken or faulty wiring in the charging circuit (Chapter 5).
6   Short in vehicle wiring (Chapter 12).
7   Internally defective battery (Chapters 1 and 5).

20 Alternator light fails to go out

1   Faulty alternator or charging circuit (Chapter 5).
2   Alternator drivebelt defective or out of adjustment (Chapter 1).
3   Alternator voltage regulator inoperative (Chapter 5).

21 Alternator light fails to come on when key is turned on

1   Warning light bulb defective (Chapter 12).
2   Fault in the printed circuit, dash wiring, bulb or bulb holder (Chapter 12).

Fuel system

22 Excessive fuel consumption

1   Dirty or clogged air filter element (Chapter 1).
2   Incorrectly set ignition timing (Chapter 5).
3   Emissions system not functioning properly (Chapter 6).
4   Fuel injection system not functioning properly (Chapter 4).
5   Low tire pressure or incorrect tire size (Chapter 1).

23 Fuel leakage and/or fuel odor

1   Leaking fuel feed or return line (Chapters 1 and 4).
2   Tank overfilled.
3   Evaporative canister filter clogged (Chapters 1 and 6).
4   Problem with the fuel injection system (Chapter 4).

Cooling system

24 Overheating

1   Insufficient coolant in system (Chapter 1).
2   Water pump drivebelt defective or out of adjustment (Chapter 1).
3   Radiator core blocked or grille restricted (Chapter 3).
4   Thermostat faulty (Chapter 3).
5   Electric coolant fan inoperative or blades broken (Chapter 3).
6   Radiator cap not maintaining proper pressure (Chapter 3).
7   Ignition timing incorrect (Chapter 5).

25 Overcooling

1   Faulty thermostat (Chapter 3).
2   Inaccurate temperature gauge sending unit (Chapter 3).

26 External coolant leakage

1   Deteriorated/damaged hoses; loose clamps (Chapters 1 and 3).
2   Water pump defective (Chapter 3).
3   Leakage from radiator core or coolant reservoir bottle (Chapter 3).
4   Engine drain or water jacket core plugs leaking (Chapter 2).

Troubleshooting 0-19

27 Internal coolant leakage

1 Leaking cylinder head gasket (Chapter 2).
2 Cracked cylinder bore or cylinder head (Chapter 2).

28 Coolant loss

1 Too much coolant in system (Chapter 1).
2 Coolant boiling away because of overheating (Chapter 3).
3 Internal or external leakage (Chapter 3).
4 Faulty radiator cap (Chapter 3).

29 Poor coolant circulation

1 Inoperative water pump (Chapter 3).
2 Restriction in cooling system (Chapters 1 and 3).
3 Water pump drivebelt defective/out of adjustment (Chapter 1).
4 Thermostat sticking (Chapter 3).

Clutch

30 Pedal travels to floor - no pressure or very little resistance

1 Damaged cable (Chapter 8).
2 Broken release bearing or fork (Chapter 8).

31 Unable to select gears

1 Faulty transaxle (Chapter 7).
2 Faulty clutch disc or pressure plate (Chapter 8).
3 Faulty release lever or release bearing (Chapter 8).
4 Faulty shift lever assembly or rods (Chapter 8).

32 Clutch slips (engine speed increases with no increase in vehicle speed)

1 Clutch plate worn (Chapter 8).
2 Clutch plate is oil soaked by leaking rear main seal (Chapter 8).
3 Clutch plate not seated (Chapter 8).
4 Warped pressure plate or flywheel (Chapter 8).
5 Weak diaphragm springs (Chapter 8).
6 Clutch plate overheated. Allow to cool.

33 Grabbing (chattering) as clutch is engaged

1 Oil on clutch plate lining, burned or glazed facings (Chapter 8).
2 Worn or loose engine or transaxle mounts (Chapters 2 and 7).
3 Worn splines on clutch plate hub (Chapter 8).
4 Warped pressure plate or flywheel (Chapter 8).
5 Burned or smeared resin on flywheel or pressure plate (Chapter 8).

34 Transaxle rattling (clicking)

1 Release lever loose (Chapter 8).
2 Clutch plate damper spring failure (Chapter 8).
3 Low engine idle speed (Chapter 1).

35 Noise in clutch area

1 Fork shaft improperly installed (Chapter 8).
2 Faulty bearing (Chapter 8).

36 Clutch pedal stays on floor

Broken cable, release bearing or fork (Chapter 8).

37 High pedal effort

1 Damaged cable.

Manual transaxle

38 Knocking noise at low speeds

1 Worn driveaxle constant velocity (CV) joints (Chapter 8).
2 Worn side gear shaft counterbore in differential case (Chapter 7A).*

39 Noise most pronounced when turning

Differential gear noise (Chapter 7A).*

40 Clunk on acceleration or deceleration

1 Loose engine or transaxle mounts (Chapters 2 and 7A).
2 Worn differential pinion shaft in case.*
3 Worn side gear shaft counterbore in differential case (Chapter 7A).*
4 Worn or damaged driveaxle inboard CV joints (Chapter 8).

41 Clicking noise in turns

Worn or damaged outboard CV joint (Chapter 8).

42 Vibration

1 Rough wheel bearing (Chapter 10).
2 Damaged driveaxle (Chapter 8).
3 Out of round tires (Chapter 1).
4 Tire out of balance (Chapters 1 and 10).
5 Worn CV joint (Chapter 8).

43 Noisy in neutral with engine running

1 Damaged input gear bearing (Chapter 7A).*
2 Damaged clutch release bearing (Chapter 8).

44 Noisy in one particular gear

1 Damaged or worn constant mesh gears (Chapter 7A).*
2 Damaged or worn synchronizers (Chapter 7A).*
3 Bent reverse fork (Chapter 7A).*
4 Damaged fourth speed gear or output gear (Chapter 7A).*
5 Worn or damaged reverse idler gear or idler bushing (Chapter 7A).*

45 Noisy in all gears

1 Insufficient lubricant (Chapter 7A).
2 Damaged or worn bearings (Chapter 7A).*
3 Worn or damaged input gear shaft and/or output gear shaft (Chapter 7A).*

46 Slips out of gear

1 Worn or improperly adjusted linkage (Chapter 7A).
2 Transaxle loose on engine (Chapter 7A).
3 Shift linkage does not work freely, binds (Chapter 7A).
4 Input gear bearing retainer broken or loose (Chapter 7A).*
5 Dirt between clutch cover and engine housing (Chapter 7A).
6 Worn shift fork (Chapter 7A).*

47 Leaks lubricant

1 Side gear shaft seals worn (Chapter 7).
2 Excessive amount of lubricant in transaxle (Chapters 1 and 7A).
3 Loose or broken input gear shaft bear-

0-20　Troubleshooting

ing retainer (Chapter 7A).*
4　Input gear bearing retainer O-ring and/or lip seal damaged (Chapter 7A).*

48　Locked in gear

Lock pin or interlock pin missing (Chapter 7A).*

* Although the corrective action necessary to remedy the symptoms described is beyond the scope of this manual, the above information should be helpful in isolating the cause of the condition so that the owner can communicate clearly with a professional mechanic.

Automatic transaxle

Note: *Due to the complexity of the automatic transaxle, it is difficult for the home mechanic to properly diagnose and service this component. For problems other than the following, the vehicle should be taken to a dealer or transaxle shop.*

49　Fluid leakage

1　Automatic transaxle fluid is a deep red color. Fluid leaks should not be confused with engine oil, which can easily be blown onto the transaxle by air flow.
2　To pinpoint a leak, first remove all built-up dirt and grime from the transaxle housing with degreasing agents and/or steam cleaning. Then drive the vehicle at low speeds so air flow will not blow the leak far from its source. Raise the vehicle and determine where the leak is coming from. Common areas of leakage are:
　a) Pan (Chapters 1 and 7).
　b) Dipstick tube (Chapters 1 and 7).
　c) Transaxle oil lines (Chapter 7).
　d) Speed sensor (Chapter 7).
　e) Driveaxle oil seals (Chapter 7).

50　Transaxle fluid brown or has a burned smell

Transaxle fluid overheated (Chapter 1).

51　General shift mechanism problems

1　Chapter 7, Part B, deals with checking and adjusting the shift linkage on automatic transaxles. Common problems which may be attributed to poorly adjusted linkage are:
　a) Engine starting in gears other than Park or Neutral.
　b) Indicator on shifter pointing to a gear other than the one actually being used.
　c) Vehicle moves when in Park.
2　Refer to Chapter 7B for the shift linkage adjustment procedure.

52　Engine will start in gears other than Park or Neutral

Neutral start switch malfunctioning (Chapter 7B).

53　Transaxle slips, shifts roughly, is noisy or has no drive in forward or reverse gears

There are many probable causes for the above problems, but the home mechanic should be concerned with only one possibility - fluid level. Before taking the vehicle to a repair shop, check the level and condition of the fluid as described in Chapter 1. Correct the fluid level as necessary or change the fluid and filter if needed. If the problem persists, have a professional diagnose the cause.

Driveaxles

54　Clicking noise in turns

Worn or damaged outboard CV joint (Chapter 8).

55　Shudder or vibration during acceleration

1　Excessive toe-in (Chapter 10).
2　Incorrect spring heights (Chapter 10).
3　Worn or damaged inboard or outboard CV joints (Chapter 8).
4　Sticking inboard CV joint assembly (Chapter 8).

56　Vibration at highway speeds

1　Out of balance front wheels and/or tires (Chapters 1 and 10).
2　Out of round front tires (Chapters 1 and 10).
3　Worn CV joint(s) (Chapter 8).

Brakes

Note: *Before assuming that a brake problem exists, make sure that:*
　a) *The tires are in good condition and properly inflated (Chapter 1).*
　b) *The front end alignment is correct (Chapter 10).*
　c) *The vehicle is not loaded with weight in an unequal manner.*

57　Vehicle pulls to one side during braking

1　Incorrect tire pressures (Chapter 1).
2　Front end out of alignment (have the front end aligned).
3　Front, or rear, tire sizes not matched to one another.
4　Restricted brake lines or hoses (Chapter 9).
5　Malfunctioning drum brake or caliper assembly (Chapter 9).
6　Loose suspension parts (Chapter 10).
7　Loose calipers (Chapter 9).
8　Excessive wear of brake shoe or pad material or disc/drum on one side.

58　Noise (high-pitched squeal when the brakes are applied)

Front and/or rear disc brake pads worn out. The noise comes from the wear sensor rubbing against the disc (does not apply to all vehicles). Replace pads with new ones immediately (Chapter 9).

59　Brake roughness or chatter (pedal pulsates)

1　Excessive lateral runout (Chapter 9).
2　Uneven pad wear (Chapter 9).
3　Defective disc (Chapter 9).

60　Excessive brake pedal effort required to stop vehicle

1　Malfunctioning power brake booster (Chapter 9).
2　Partial system failure (Chapter 9).
3　Excessively worn pads or shoes (Chapter 9).
4　Piston in caliper or wheel cylinder stuck or sluggish (Chapter 9).
5　Brake pads or shoes contaminated with oil or grease (Chapter 9).
6　Brake disc grooved and/or glazed (Chapter 1).
7　New pads or shoes installed and not yet seated. It will take a while for the new material to seat against the disc or drum.

61　Excessive brake pedal travel

1　Partial brake system failure (Chapter 9).
2　Insufficient fluid in master cylinder (Chapters 1 and 9).
3　Air trapped in system (Chapters 1 and 9).

62　Dragging brakes

1　Incorrect adjustment of brake light switch (Chapter 9).
2　Master cylinder pistons not returning correctly (Chapter 9).
3　Restricted brakes lines or hoses (Chap-

Troubleshooting

ters 1 and 9).
4 Incorrect parking brake adjustment (Chapter 9).

63 Grabbing or uneven braking action

1 Malfunction of proportioning valve (Chapter 9).
2 Malfunction of power brake booster unit (Chapter 9).
3 Binding brake pedal mechanism (Chapter 9).

64 Brake pedal feels spongy when depressed

1 Air in hydraulic lines (Chapter 9).
2 Master cylinder mounting bolts loose (Chapter 9).
3 Master cylinder defective (Chapter 9).

65 Brake pedal travels to the floor with little resistance

1 Little or no fluid in the master cylinder reservoir caused by leaking caliper piston(s) (Chapter 9).
2 Loose, damaged or disconnected brake lines (Chapter 9).

66 Parking brake does not hold

Parking brake linkage improperly adjusted (Chapters 1 and 9).

Suspension and steering systems

Note: *Before attempting to diagnose the suspension and steering systems, perform the following preliminary checks:*

a) *Tires for wrong pressure and uneven wear.*
b) *Steering universal joints from the column to the rack-and-pinion for loose connectors or wear.*
c) *Front and rear suspension and the rack-and-pinion assembly for loose or damaged parts.*
d) *Out-of-round or out-of-balance tires, bent rims and loose and/or rough wheel bearings.*

67 Vehicle pulls to one side

1 Mismatched or uneven tires (Chapter 10).
2 Broken or sagging springs (Chapter 10).
3 Wheel alignment out-of-specifications (Chapter 10).
4 Front brake dragging (Chapter 9).

68 Abnormal or excessive tire wear

1 Wheel alignment out-of-specifications (Chapter 10).
2 Sagging or broken springs (Chapter 10).
3 Tire out-of-balance (Chapter 10).
4 Worn strut damper (Chapter 10).
5 Overloaded vehicle.
6 Tires not rotated regularly.

69 Wheel makes a thumping noise

1 Blister or bump on tire (Chapter 10).
2 Improper strut damper or shock absorber action (Chapter 10).

70 Shimmy, shake or vibration

1 Tire or wheel out-of-balance or out-of-round (Chapter 10).
2 Loose or worn wheel bearings (Chapters 1, 8 and 10).
3 Worn tie-rod ends (Chapter 10).
4 Worn lower balljoints (Chapters 1 and 10).
5 Excessive wheel runout (Chapter 10).
6 Blister or bump on tire (Chapter 10).

71 Hard steering

1 Lack of lubrication at balljoints, tie-rod ends and rack-and-pinion assembly (Chapter 10).
2 Front wheel alignment out-of-specifications (Chapter 10).
3 Low tire pressure(s) (Chapters 1 and 10).

72 Poor returnability of steering to center

1 Lack of lubrication at balljoints and tie-rod ends (Chapter 10).
2 Binding in balljoints (Chapter 10).
3 Binding in steering column (Chapter 10).
4 Lack of lubricant in steering gear assembly (Chapter 10).
5 Front wheel alignment out-of-specifications (Chapter 10).

73 Abnormal noise at the front end

1 Lack of lubrication at balljoints and tie-rod ends (Chapters 1 and 10).
2 Damaged strut mounting (Chapter 10).
3 Worn control arm bushings or tie-rod ends (Chapter 10).
4 Loose stabilizer bar (Chapter 10).
5 Loose wheel nuts (Chapters 1 and 10).
6 Loose suspension bolts (Chapter 10).

74 Wander or poor steering stability

1 Mismatched or uneven tires (Chapter 10).
2 Lack of lubrication at balljoints and tie-rod ends (Chapters 1 and 10).
3 Worn strut assemblies (Chapter 10).
4 Loose stabilizer bar (Chapter 10).
5 Broken or sagging springs (Chapter 10).
6 Wheels out of alignment (Chapter 10).

75 Erratic steering when braking

1 Wheel bearings worn (Chapter 10).
2 Broken or sagging springs (Chapter 10).
3 Leaking wheel cylinder or caliper (Chapter 10).
4 Warped discs or drums (Chapter 10).

76 Excessive pitching and/or rolling around corners or during braking

1 Loose stabilizer bar (Chapter 10).
2 Worn struts, shock absorbers or mountings (Chapter 10).
3 Broken or sagging springs (Chapter 10).
4 Overloaded vehicle.

77 Suspension bottoms

1 Overloaded vehicle.
2 Worn strut dampers (Chapter 10).
3 Incorrect, broken or sagging springs (Chapter 10).

78 Cupped tires

1 Front wheel or rear wheel alignment out-of-specifications (Chapter 10).
2 Worn strut dampers (Chapter 10).
3 Wheel bearings worn (Chapter 10).
4 Excessive tire or wheel runout (Chapter 10).
5 Worn balljoints (Chapter 10).

79 Excessive tire wear on outside edge

1 Inflation pressures incorrect (Chapter 1).
2 Excessive speed in turns.
3 Front end alignment incorrect (excessive toe-in). Have professionally aligned.
4 Suspension arm bent or twisted (Chapter 10).

80 Excessive tire wear on inside edge

1 Inflation pressures incorrect (Chapter 1).
2 Front end alignment incorrect (toe-out).

Have professionally aligned.
3   Loose or damaged steering components (Chapter 10).

81 Tire tread worn in one place

1   Tires out-of-balance.
2   Damaged or buckled wheel. Inspect and replace if necessary.
3   Defective tire (Chapter 1).

82 Excessive play or looseness in steering system

1   Wheel bearing(s) worn (Chapter 10).
2   Tie-rod end loose (Chapter 10).
3   Steering gear loose (Chapter 10).
4   Worn or loose steering intermediate shaft (Chapter 10).

83 Rattling or clicking noise in steering gear

1   Steering gear loose (Chapter 10).
2   Steering gear defective.

Chapter 1
Tune-up and routine maintenance

Contents

Section		Section
Air filter replacement	17	Ignition timing check and adjustment	26
Automatic transaxle fluid change	30	Introduction	1
Automatic transaxle fluid level check	7	Maintenance schedule	2
Battery check, maintenance and charging	11	Manual transaxle lubricant change	31
Brake check	16	Manual transaxle lubricant level check	19
CHECK ENGINE light	See Chapter 6	Positive Crankcase Ventilation (PCV) valve and hose check	
Clutch pedal height, freeplay and release lever freeplay check		and replacement	32
and adjustment	10	Power steering fluid level check	6
Cooling system check	14	Spark plug check and replacement	23
Cooling system servicing (draining, flushing and refilling)	25	Spark plug wire, distributor cap and rotor check	
Driveaxle boot check	21	and replacement	24
Drivebelt check, adjustment and replacement	12	Steering and suspension check	20
Engine idle speed check and adjustment	27	Tire and tire pressure checks	5
Engine oil and oil filter change	8	Tire rotation	15
Evaporative emissions control (EVAP) system check	28	Tune-up general information	3
Exhaust system check	29	Underhood hose check and replacement	13
Fluid level checks	4	Valve clearance check and adjustment (GA16DE engine only)	33
Fuel filter replacement	22	Windshield wiper blade inspection and replacement	9
Fuel system check	18		

Specifications

Recommended lubricants and fluids

Note: *Listed here are manufacturer recommendations at the time this manual was written. Manufacturers occasionally upgrade their fluid and lubricant specifications, so check with your local auto parts store for current recommendations.*

Engine oil
  Type .... API grade SG or SG/CD multigrade and fuel efficient oil
  Viscosity .... See accompanying chart
Fuel .... Unleaded gasoline, 87 octane or higher

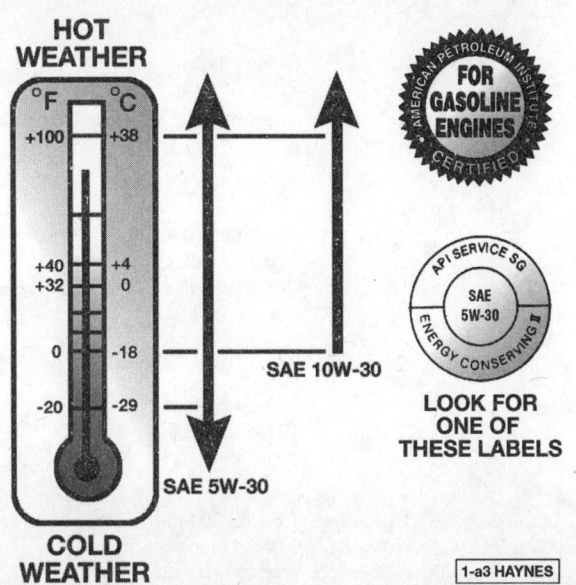

Recommended engine oil viscosity

Recommended lubricants and fluids (continued)

Automatic transaxle fluid	DEXRON III automatic transmission fluid
Manual transaxle lubricant	API GL-4 80W-90 gear oil
Brake fluid	DOT 3 brake fluid
Power steering system	DEXRON II or III automatic transmission fluid
Wheel bearings	NLGI no. 2 lithium-base grease

Capacities*

Engine oil (including filter)
- GA16DE engine .... 3-3/8 qts
- SR20DE engine .... 3-5/8 qts

Coolant .... 6 qts
Automatic transaxle (drain and refill) .... 2.6 qts

Manual transaxle
- GA16DE engine .... fill to within 2-1/2 inches of speedometer drive pinion opening (approximately 6-3/4 pints capacity)
- SR20DE engine .... fill to within 1-1/2 inches of speedometer drive pinion opening (approximately 8-1/4 pints capacity)

All capacities approximate. Add as necessary to bring up to appropriate level.

Ignition system

Spark plug type and gap
- SR20DE engine
  - Type .... NGK PFR5B-11 or equivalent platinum type
  - Gap .... 0.035 inch
- GA16DE engine
  - Type .... NGK BKR5E-11 or equivalent
  - Gap .... 0.043 inch

Spark plug wire resistance .... 10,000 to 25,000 ohms
Engine firing order 1-3-4-2

Ignition timing
- SR20DE engine
  - Manual transaxle .... 13 to 17-degrees BTDC*
  - Automatic transaxle (in Neutral) .... 13 to 17-degrees BTDC*
- GA16DE engine
  - Manual transaxle .... 6 to 10-degrees BTDC*
  - Automatic transaxle (in Neutral) .... 6 to 10-degrees BTDC*

Refer to the Emission Control Information label in the engine compartment and follow the specifications there if they differ.

GA16DE engine

SR20DE engine

Cylinder location and distributor rotation

Idle speed

SR20DE engine
- Manual transaxle .... 700 to 800 rpm*
- Automatic transaxle (in Neutral) .... 700 to 800 rpm*

GA16DE engine
- Manual transaxle .... 625 ± 50 rpm*
- Automatic transaxle in Neutral) .... 725 ± 50 rpm*

Refer to the Emission Control Information label in the engine compartment and follow the specifications there if they differ.

Valve clearances (engine hot)

GA16DE engine
- Intake valve .... 0.008 to 0.019 inch
- Exhaust valve .... 0.012 to 0.023 inch

SR20DE engine .... Not adjustable

Cooling system

Thermostat rating
- Starts to open .... 170-degrees F
- Fully open .... 194-degrees F

Clutch pedal

Freeplay	1/8 to 5/32 inch
Height	6 to 6-1/4 inches
Release lever freeplay	7/16 to 5/8 inch
Clutch interlock switch freeplay	1/32 to 1/16 inch

Chapter 1 Tune-up and routine maintenance

Brakes
Disc brake pad lining thickness (minimum)	1/16 inch
Drum brake shoe lining thickness (minimum)	1/16 inch
Parking brake adjustment	
Drum brake	7 to 8 clicks
Disc brake	8 to 9 clicks
Brake pedal	
Freeplay	3/64 to 1/8 inch
Depressed height	
Manual transaxle	3 inches
Automatic transaxle	3-1/3 inches
Free height	
Manual transaxle	5-3/4 to 6-1/4 inch
Automatic transaxle	6-1/4 to 6-1/2 inch

Suspension and steering
Steering wheel freeplay limit	1-3/16 inch
Balljoint allowable movement	0 inch

Torque specifications
	Ft-lbs
Automatic transaxle drain plug	22 to 29
Engine oil drain plug	22 to 29
Manual transaxle drain plug	18 to 25
Spark plugs	14 to 22
Wheel lug nuts	87

Engine compartment components (1995 model with GA16DE engine shown)

1. PCV valve (Section 32)
2. Spark plugs (Section 23)
3. Fuel filter (Section 18)
4. Brake fluid reservoir (Section 4)
5. Air filter housing (Section 17)
6. Battery (Section 11)
7. Fusible link and relay block (Chapter 12)
8. Fuse and fusible link block (Chapter 12)
9. Automatic transaxle fluid dipstick (Section 7)
10. Distributor (Section 24)
11. Engine oil dipstick (Section 4)
12. Upper radiator hose (Section 14)
13. Radiator cap (Section 14)
14. Coolant reservoir (Section 4)
15. Windshield washer fluid reservoir (Section 4)
16. Oil filler cap (Section 4)
17. Relay block (Chapter 12)
18. Power steering fluid reservoir (Section 6)

1-4 Chapter 1 Tune-up and routine maintenance

Engine compartment underside components

1. Engine drivebelt (Section 12)
2. Lower radiator hose (Section 14)
3. Automatic transaxle drain plug (Section 30)
4. Lower balljoint (Section 20)
5. Driveaxle boot (Section 21)
6. Exhaust system (Section 29)
7. Engine oil pan drain plug (Section 8)
8. Disc brake caliper (Section 16)

1 Introduction

This Chapter is designed to help the home mechanic maintain the Nissan Sentra and 200SX for peak performance, economy, safety and long life.

On the following pages is a master maintenance schedule, followed by sections dealing specifically with each item on the schedule. Visual checks, adjustments, component replacement and other helpful items are included. Refer to the **accompanying illustrations** of the engine compartment and the underside of the vehicle for the location of various components.

Servicing your Sentra or 200SX in accordance with the mileage/time maintenance schedule and the following Sections will provide it with a planned maintenance program that should result in a long and reliable service life. This is a comprehensive plan, so maintaining some items but not others at the specified service intervals will not produce the same results.

As you service your Sentra, you will discover that many of the procedures can - and should - be grouped together because of the nature of the particular procedure you're performing or because of the close proximity of two otherwise unrelated components to one another.

For example, if the vehicle is raised for any reason, you should inspect the exhaust, suspension, steering and fuel systems while you're under the vehicle. When you're rotating the tires, it makes good sense to check the brakes and wheel bearings since the wheels are already removed.

Finally, let's suppose you have to borrow or rent a torque wrench. Even if you only need to tighten the spark plugs, you might as well check the torque of as many critical fasteners as time allows.

The first step of this maintenance program is to prepare yourself before the actual work begins. Read through all sections pertinent to the procedures you're planning to do, then make a list of and gather together all the parts and tools you will need to do the job. If it looks as if you might run into problems during a particular segment of some procedure, seek advice from your local parts man or dealer service department.

Chapter 1 Tune-up and routine maintenance

1-5

Typical rear underside components

1 Exhaust system (Section 29)
2 Fuel tank (Section 18)
3 Suspension strut and spring (Section 20)
4 Fuel filler hose (Section 18)
5 Rear axle (Chapter 10)
6 Rear axle lateral link (Chapter 10)
7 Rear drum brake (Section 16)
8 Rear axle trailing arm (Chapter 10)

2 Nissan Sentra and 200SX Maintenance schedule

The maintenance intervals in this manual are provided with the assumption that you, not the dealer, will be doing the work. These are the minimum maintenance intervals recommended by the factory for Sentras that are driven daily. If you wish to keep your vehicle in peak condition at all times, you may wish to perform some of these procedures even more often. Because frequent maintenance enhances the efficiency, performance and resale value of your car, we encourage you to do so. If you drive in dusty areas, tow a trailer, idle or drive at low speeds for extended periods or drive for short distances (less than four miles) in below freezing temperatures, shorter intervals are also recommended.

When your vehicle is new, it should be serviced by a factory authorized dealer service department to protect the factory warranty. In many cases, the initial maintenance check is done at no cost to the owner.

Every 250 miles or weekly, whichever comes first

Check the engine oil level (Section 4)
Check the engine coolant level (Section 4)
Check the windshield washer fluid level (Section 4)
Check the brake fluid level (Section 4)
Check the tires and tire pressures (Section 5)

Every 3000 miles or 3 months, whichever comes first

All items listed above plus:
Check the power steering fluid level (Section 6)
Check the automatic transaxle fluid level (Section 7)
Change the engine oil and oil filter (Section 8)

Every 7500 miles or 6 months, whichever comes first

Inspect and replace if necessary the windshield wiper blades (Section 9)
Check the clutch pedal and release lever for proper freeplay (Section 10)
Check and service the battery (Section 11)
Check and adjust if necessary the engine drivebelts (Section 12)
Inspect and replace if necessary all underhood hoses (Section 13)
Check the cooling system (Section 14)
Rotate the tires (Section 15)

Every 15,000 miles or 12 months, whichever comes first

All items listed above plus:
Inspect the brake system (Section 16)*
Replace the air filter (Section 17)
Inspect the fuel system (Section 18)
Check the manual transaxle lubricant level (Section 19)
Inspect the suspension and steering components (Section 20)
Check the driveaxle boots (Section 21)

Every 30,000 miles or 24 months, whichever comes first

All items listed above plus:
Replace the fuel filter (Section 22)
Check and replace if necessary the spark plugs (conventional spark plugs) (Section 23)
Inspect and replace if necessary the spark plug wires, distributor cap and rotor (Section 24)
Service the cooling system (drain, flush and refill) (Section 25)
Check and adjust if necessary, the engine ignition timing (Section 26)
Check and adjust if necessary, the engine idle speed (Section 27)
Inspect the evaporative emissions control system (Section 28)
Check the exhaust system (Section 29)
Change the automatic transaxle fluid (Section 30)**
Change the manual transaxle lubricant (Section 31)**
Check and replace if necessary the PCV valve (Section 32)

Every 60,000 miles or 48 months, whichever comes first

Check and adjust if necessary, the valve clearances (GA16DE engine - only if the valves are making excessive noise) (Section 33)
Check and replace if necessary the spark plugs (platinum spark plugs) (Section 23)

* This item is affected by "severe" operating conditions as described below. If your vehicle is operated under "severe" conditions, perform all maintenance indicated with an asterisk (*) at 3000 mile/3 month intervals. Severe conditions are indicated if you mainly operate your vehicle under one or more of the following conditions:

Operating in dusty areas
Towing a trailer
Idling for extended periods and/or low speed operation
Operating when outside temperatures remain below freezing and when most trips are less than 4 miles

** If operated under one or more of the following conditions, change the manual or automatic transaxle fluid lubricant every 15,000 miles:

In heavy city traffic where the outside temperature regularly reaches 90-degrees F (32-degrees C) or higher
In hilly or mountainous terrain
Frequent trailer pulling

Chapter 1 Tune-up and routine maintenance

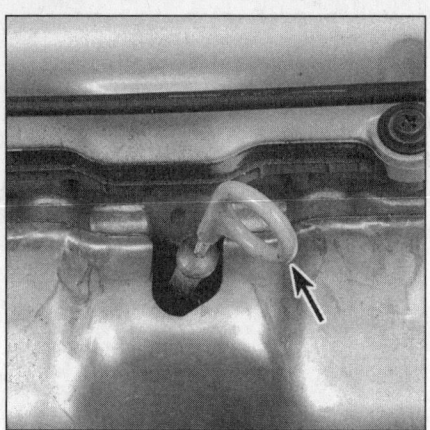

4.2 The engine oil dipstick (arrow) is located on the front side of the engine

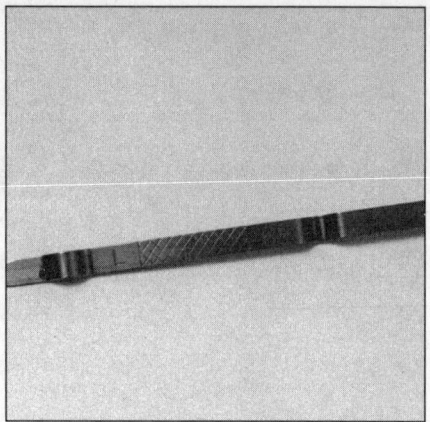

4.4 The oil level should be at or near the H mark - if it isn't, add enough oil to bring the level to near the H mark (it takes one full quart to raise the level from the L to the H mark)

4.6 The threaded oil filler cap is located on the valve cover - always make sure the area around the opening is clean before unscrewing the cap to prevent dirt from contaminating the engine

3 Tune-up general information

The term tune-up is used in this manual to represent a combination of individual operations rather than one specific procedure.

If, from the time the vehicle is new, the routine maintenance schedule is followed closely and frequent checks are made of fluid levels and high wear items, as suggested throughout this manual, the engine will be kept in relatively good running condition and the need for additional work will be minimized.

More likely than not, however, there will be times when the engine is running poorly due to lack of regular maintenance. This is even more likely if a used vehicle, which has not received regular and frequent maintenance checks, is purchased. In such cases, an engine tune-up will be needed outside of the regular routine maintenance intervals.

The first step in any tune-up or engine diagnosis to help correct a poor running engine would be a cylinder compression check. A check of the engine compression (Chapter 2 Part B) will give valuable information regarding the overall performance of many internal components and should be used as a basis for tune-up and repair procedures. If, for instance, a compression check indicates serious internal engine wear, a conventional tune-up will not help the running condition of the engine and would be a waste of time and money.

The following series of operations are those most often needed to bring a generally poor running engine back into a proper state of tune.

Minor tune-up

Check all engine related fluids (Section 4)
Clean, inspect and test the battery (Section 11)
Check and adjust the drivebelts (Section 12)
Check all underhood hoses (Section 13)
Check the cooling system (Section 14)
Check the air filter (Section 17)
Replace the spark plugs (Section 23)

Inspect the distributor cap and rotor (Section 24)
Inspect the spark plug and coil wires (Section 24)
Check and adjust the idle speed (Section 27)

Major tune-up

All items listed under Minor tune-up, plus . . .
Replace the air filter (Section 17)
Check the fuel system (Section 18)
Check the charging system (Chapter 5)
Replace the spark plug wires (Section 24)
Replace the distributor cap and rotor (Section 24)
Check the ignition system (Section 26)

4 Fluid level checks (every 250 miles or weekly)

1 Fluids are an essential part of the lubrication, cooling, brake and other systems. Because these fluids gradually become depleted and/or contaminated during normal operation of the vehicle, they must be periodically replenished. See *Recommended lubricants* and *fluids* and *capacities* at the beginning of this Chapter before adding fluid to any of the following components. **Note:** *The vehicle must be on level ground before fluid levels can be checked.*

Engine oil

Refer to illustrations 4.2, 4.4 and 4.6

2 The engine oil level is checked with a dipstick located at the front side of the engine **(see illustration)**. The dipstick extends through a metal tube from which it protrudes down into the engine oil pan.

3 The oil level should be checked before the vehicle has been driven, or about 15 minutes after the engine has been shut off. If the oil is checked immediately after driving the vehicle, some of the oil will remain in the upper engine components, producing an inaccurate reading on the dipstick.

4 Pull the dipstick from the tube and wipe all the oil from the end with a clean rag or paper towel. Insert the clean dipstick all the way back into its metal tube and pull it out again. Observe the oil at the end of the dipstick. At its highest point, the level should be between the L and H marks **(see illustration)**.

5 It takes one quart of oil to raise the level from the L mark to the H mark on the dipstick. Do not allow the level to drop below the L mark or oil starvation may cause engine damage. Conversely, overfilling the engine (adding oil above the H mark) may cause oil fouled spark plugs, oil leaks or oil seal failures.

6 Remove the threaded cap from the valve cover to add oil **(see illustration)**. Use a funnel to prevent spills. After adding the oil, install the filler cap hand tight. Start the engine and look carefully for any small leaks around the oil filter or drain plug. Stop the engine and check the oil level again after it has had sufficient time to drain from the upper block and cylinder head galleys.

7 Checking the oil level is an important preventive maintenance step. A continually dropping oil level indicates oil leakage through damaged seals, from loose connections, or past worn rings or valve guides. If the oil looks milky in color or has water droplets in it, a cylinder head gasket may be blown. The engine should be checked immediately. The condition of the oil should also be checked. Each time you check the oil level, slide your thumb and index finger up the dipstick before wiping off the oil. If you see small dirt or metal particles clinging to the dipstick, the oil should be changed (see Section 8).

Engine coolant

Refer to illustration 4.8

Warning: *Do not allow antifreeze to come in contact with your skin or painted surfaces of the vehicle. Flush contaminated areas immediately with plenty of water. Don't store new*

1-8   Chapter 1 Tune-up and routine maintenance

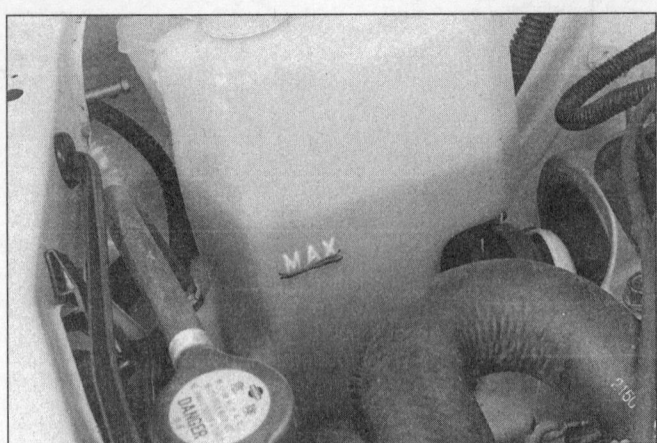

4.8 The coolant reservoir is located in the right or left front corner of the engine compartment - make sure the level is near the MAX mark on the reservoir

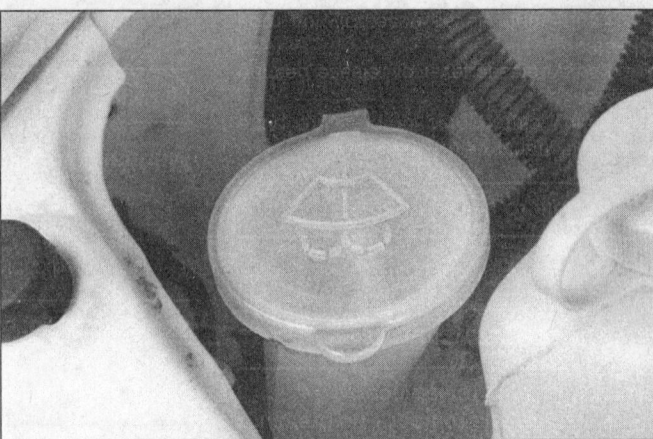

4.14 The windshield washer fluid reservoir is located on the side of the engine compartment

coolant or leave old coolant lying around where it's accessible to children or pets – they're attracted by its sweet smell. Ingestion of even a small amount of coolant can be fatal! Wipe up garage floor and drip pan spills immediately. Keep antifreeze containers covered and repair cooling system leaks as soon as they're noticed.

8   All vehicles covered by this manual are equipped with a pressurized coolant recovery system. A white coolant reservoir located in the right or left (depending on model year) front corner of the engine compartment is connected by a hose to the base of the radiator cap **(see illustration)**. If the coolant gets too hot during engine operation, coolant can escape through the pressurized radiator cap, then through a connecting hose into the reservoir. As the engine cools, the coolant is automatically drawn back into the cooling system to maintain the correct level.

9   The coolant level should be checked regularly. It must be between the Max and Min lines on the tank. The level will vary with the temperature of the engine. When the engine is cold, the coolant level should be at or slightly above the Min mark on the tank. Once the engine has warmed up, the level should be at or near the Max mark. If it isn't, allow the fluid in the tank to cool, then remove the cap from the reservoir and add coolant to bring the level up to the Max line. Use only ethylene/glycol type coolant and water in the mixture ratio recommended by your owner's manual. Do not use supplemental inhibitor additives. If only a small amount of coolant is required to bring the system up to the proper level, water can be used. However, repeated additions of water will dilute the recommended antifreeze and water solution. In order to maintain the proper ratio of antifreeze and water, it is advisable to top up the coolant level with the correct mixture. Refer to your owner's manual for the recommended ratio.

10   If the coolant level drops within a short time after replenishment, there may be a leak in the system. Inspect the radiator, hoses, radiator cap, drain plugs, air bleeder plugs and water pump. If no leak is evident, have the radiator cap pressure tested by your dealer. **Warning:** *Never remove the radiator cap or the coolant recovery reservoir cap when the engine is running or has just been shut down, because the cooling system is hot. Escaping steam and scalding liquid could cause serious injury.*

11   If it is necessary to open the radiator cap, wait until the system has cooled completely, then wrap a thick cloth around the cap and turn it to the first stop. If any steam escapes, wait until the system has cooled further, then remove the cap.

12   When checking the coolant level, always note its condition. It should be relatively clear. If it is brown or rust colored, the system should be drained, flushed and refilled. Even if the coolant appears to be normal, the corrosion inhibitors wear out with use, so it must be replaced at the specified intervals.

13   Do not allow antifreeze to come in contact with your skin or painted surfaces of the vehicle. Flush contacted areas immediately with plenty of water.

Windshield washer fluid

Refer to illustration 4.14

14   Fluid for the windshield washer system is stored in a plastic reservoir which is located on the right (passenger) side of the engine compartment **(see illustration)**. In milder climates, plain water can be used to top up the reservoir, but the reservoir should be kept no more than two-thirds full to allow for expansion should the water freeze. In colder climates, the use of a specially designed windshield washer fluid, available at your dealer and any auto parts store, will help lower the freezing point of the fluid. Mix the solution with water in accordance with the manufacturer's directions on the container. Do not use regular antifreeze. It will damage the vehicle's paint.

Battery electrolyte

15   On models not equipped with a sealed battery, check the electrolyte level of all six battery cells. It must be between the upper and lower levels. If the level is low, remove the filler/vent cap and add distilled water. Install and securely re-tighten the cap. **Caution:** *Overfilling the cells may cause electrolyte to spill over during periods of heavy charging, causing corrosion or damage.*

Brake fluid

Refer to illustration 4.17

16   The brake master cylinder is mounted on the front of the power booster unit in the engine compartment.

17   To check the fluid level of the brake master cylinder, simply look at the MAX and MIN marks on the reservoir **(see illustration)**. The level should be within the specified distance from the maximum fill line.

18   If the level is low, wipe the top of the reservoir with a clean rag to prevent contamination of the brake system before removing the cap.

19   Add only the specified brake fluid to the brake reservoir (refer to *Recommended lubricants and fluids* at the front of this Chapter or to your owner's manual). Mixing different types of brake fluid can damage the system. Fill the brake master cylinder reservoir only to the MAX line. **Warning:** *Use caution when filling either reservoir - brake fluid can harm your eyes and damage painted surfaces. Do not use brake fluid that has been opened for more than one year or has been left open. Brake fluid absorbs moisture from the air. Excess moisture can cause a dangerous loss of braking.*

20   While the reservoir cap is removed, inspect the master cylinder reservoir for contamination. If deposits, dirt particles or water droplets are present, the system should be drained and refilled.

21   After filling the reservoir to the proper level, make sure the lid is properly seated to prevent fluid leakage and/or system pressure loss.

22   The brake fluid in the master cylinder will drop slightly as the brake pads at each wheel wear down during normal operation. If

Chapter 1 Tune-up and routine maintenance

4.17 The brake fluid level should be kept between the MIN and MAX marks on the translucent plastic reservoir

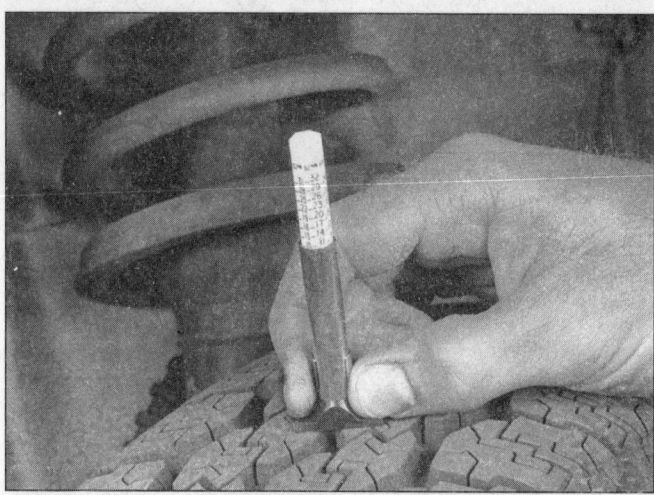

5.2 A tire tread depth indicator should be used to monitor tire wear - they are available at auto parts stores and service stations and cost very little

the master cylinder requires repeated replenishing to keep it at the proper level, this is an indication of leakage in the brake system, which should be corrected immediately. Check all brake lines and connections, along with the wheel cylinders and booster (see Section 16 for more information).

23 If, upon checking the master cylinder fluid level, you discover one or both reservoirs empty or nearly empty, the brake system should be bled (see Chapter 9).

5 Tire and tire pressure checks (every 250 miles or weekly)

Refer to illustrations 5.2, 5.3, 5.4a, 5.4b and 5.8

1 Periodic inspection of the tires may spare you from the inconvenience of being stranded with a flat tire. It can also provide you with vital information regarding possible problems in the steering and suspension systems before major damage occurs.

2 Normal tread wear can be monitored with a simple, inexpensive device known as a tread depth indicator **(see illustration)**. When the tread depth reaches the specified minimum, replace the tire(s).

3 Note any abnormal tread wear **(see illustration)**. Tread pattern irregularities such as cupping, flat spots and more wear on one side than the other are indications of front end alignment and/or balance problems. If

UNDERINFLATION

INCORRECT TOE-IN OR EXTREME CAMBER

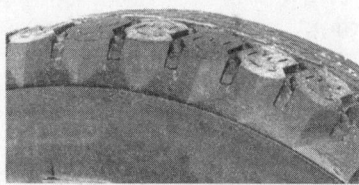

CUPPING

Cupping may be caused by:
- Underinflation and/or mechanical irregularities such as out-of-balance condition of wheel and/or tyre, and bent or damaged wheel.
- Loose or worn steering tie-rod or steering idler arm.
- Loose, damaged or worn front suspension parts.

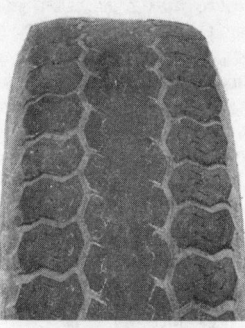

OVERINFLATION

FEATHERING DUE TO MISALIGNMENT

5.3 This chart will help you determine the condition of your tires, the probable cause(s) of abnormal wear and the corrective action necessary

1-10 Chapter 1 Tune-up and routine maintenance

5.4a If a tire loses air on a steady basis, check the valve core first to make sure it's snug (special inexpensive wrenches are commonly available at auto parts stores)

5.4b If the valve core is tight, raise the corner of the vehicle with the low tire and spray a soapy water solution onto the tread as the tire is turned slowly - slow leaks will cause small bubbles to appear

any of these conditions are noted, take the vehicle to a tire shop or service station to correct the problem.

4 Look closely for cuts, punctures and embedded nails or tacks. Sometimes a tire will hold its air pressure for a short time or leak down very slowly even after a nail has embedded itself into the tread. If a slow leak persists, check the valve stem core to make sure it is tight **(see illustration)**. Examine the tread for an object that may have embedded itself into the tire or for a "plug" that may have begun to leak (radial tire punctures are repaired with a plug that is installed in a puncture). If a puncture is suspected, it can be easily verified by spraying a solution of soapy water onto the puncture area **(see illustration)**. The soapy solution will bubble if there is a leak. Unless the puncture is inordinately large, a tire shop or gas station can usually repair the punctured tire.

5 Carefully inspect the inner sidewall of each tire for evidence of brake fluid leakage. If you see any, inspect the brakes immediately.

6 Correct tire air pressure adds miles to the lifespan of the tires, improves mileage and enhances overall ride quality. Tire pressure cannot be accurately estimated by looking at a tire, particularly if it is a radial. A tire pressure gauge is therefore essential. Keep an accurate gauge in the glove box. The pressure gauges fitted to the nozzles of air hoses at gas stations are often inaccurate.

7 Always check tire pressure when the tires are cold. "Cold," in this case, means the vehicle has not been driven over a mile in the three hours preceding a tire pressure check. A pressure rise of four to eight pounds is not uncommon once the tires are warm.

8 Unscrew the valve cap protruding from the wheel or hubcap and push the gauge firmly onto the valve **(see illustration)**. Note the reading on the gauge and compare this figure to the recommended tire pressure shown on the tire placard on the left door. Be sure to reinstall the valve cap to keep dirt and moisture out of the valve stem mechanism. Check all four tires and, if necessary, add enough air to bring them up to the recommended pressure levels.

9 Don't forget to keep the spare tire inflated to the specified pressure (consult your owner's manual). Note that the air pressure specified for the compact spare is significantly higher than the pressure of the regular tires.

6 Power steering fluid level check (every 3000 miles or 3 months)

Refer to illustration 6.4

1 Unlike manual steering, the power steering system relies on fluid which may, over a period of time, require replenishing.

2 The fluid reservoir for the power steering pump is located on the inner fender panel near the front of the engine.

3 For the check, the front wheels should be pointed straight ahead and the engine should be off.

4 The fluid level in the white plastic reservoir can be checked without removing the cap. The level should be near the upper mark on the reservoir which is marked so the fluid can be checked either cold or hot **(see**

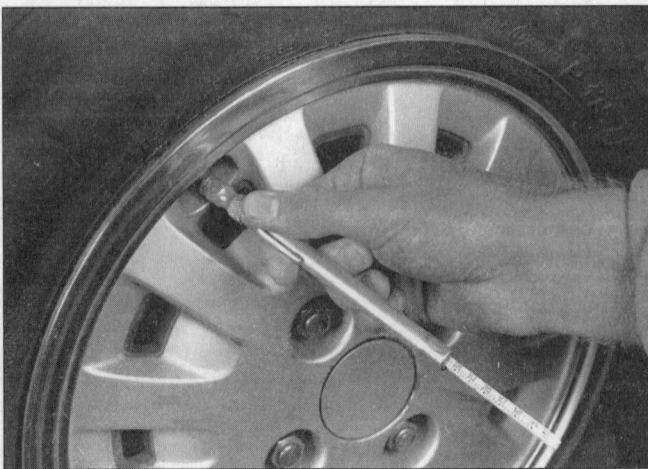

5.8 To extend the life of your tires, check the air pressure at least once a week with an accurate gauge (don't forget the spare!)

6.4 The translucent plastic reservoir allows check of the power steering fluid without removing the cap - the fluid level varies with temperature, so the fluid can be checked hot or cold

Chapter 1 Tune-up and routine maintenance

1-11

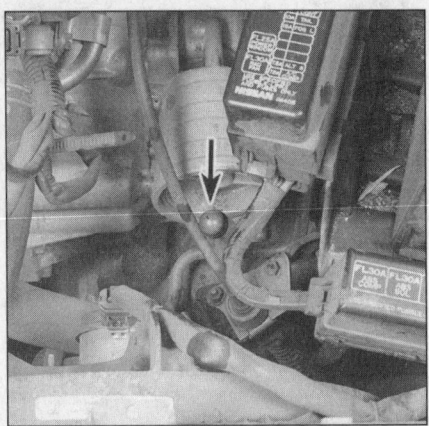

7.4a The automatic transaxle dipstick (arrow) is located in a tube which extends forward from the transaxle

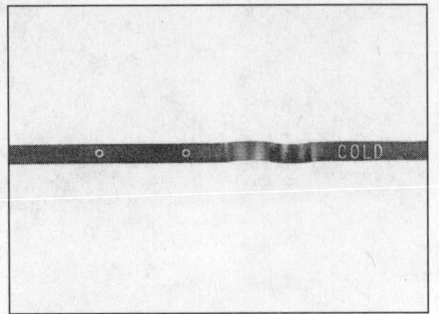

7.4b If the automatic transaxle fluid is cold, the level should be between the two lower marks

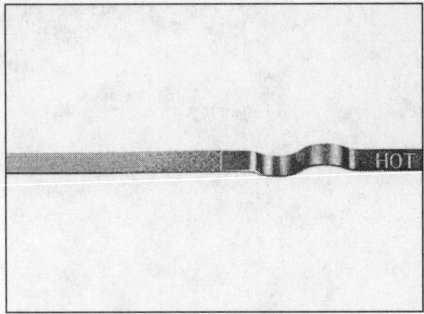

7.4c If the automatic transaxle fluid is hot, the level should be near the top of the shaded area on the dipstick

illustration). At no time should the fluid level drop below the upper mark for each heat range.
5 If additional fluid is required, pour the specified type directly into the reservoir, using a funnel to prevent spills. Use a clean rag to wipe off the reservoir cap and the area around the cap. This will help prevent any foreign matter from entering the reservoir during the check.
6 If the reservoir requires frequent fluid additions, all power steering hoses, hose connections, the power steering pump and the rack-and-pinion assembly should be carefully checked for leaks.

7 Automatic transaxle fluid level check (every 3000 miles or 3 months)

Refer to illustrations 7.4a, 7.4b and 7.4c
1 The level of the automatic transaxle fluid should be carefully maintained. Low fluid level can lead to slipping or loss of drive, while overfilling can cause foaming, loss of fluid and transaxle damage.
2 The transaxle fluid level should only be checked when the transaxle is hot (at its normal operating temperature). If the vehicle has just been driven over 10 miles (15 miles in a frigid climate), and the fluid temperature is 160 to 175-degrees F, the transaxle is hot. **Caution:** *If the vehicle has just been driven for a long time at high speed or in city traffic in hot weather, or if it has been pulling a trailer, an accurate fluid level reading cannot be obtained. Allow the fluid to cool down for about 30 minutes.*
3 If the vehicle has not just been driven, park the vehicle on level ground, set the parking brake and start the engine. While the engine is idling, depress the brake pedal and move the selector lever through all the gear ranges, beginning and ending in Park.
4 With the engine still idling, remove the dipstick from its tube **(see illustration).** Check the level of the fluid on the dipstick

(see illustrations) and note its condition.
5 Wipe the fluid from the dipstick with a clean rag and reinsert it back into the filler tube until the cap seats.
6 Pull the dipstick out again and note the fluid level. If the transaxle is cold, the level should be in the COLD or COLD range on the dipstick. If it is hot, the fluid level should be in the HOT range. If the level is at the low side of either range, add the specified automatic transmission fluid through the dipstick tube with a funnel.
7 Add just enough of the recommended fluid to fill the transaxle to the proper level. It takes about one pint to raise the level from the low mark to the high mark when the fluid is hot, so add the fluid a little at a time and keep checking the level until it is correct.
8 The condition of the fluid should also be checked along with the level. If the fluid at the end of the dipstick is black or a dark reddish brown color, or if it emits a burned smell, the fluid should be changed (see Section 30). If you are in doubt about the condition of the fluid, purchase some new fluid and compare the two for color and smell.

8 Engine oil and oil filter change (every 3000 miles or 3 months)

Refer to illustrations 8.2, 8.7, 8.13a, 8.13b and 8.15
1 Frequent oil changes are the best preventive maintenance the home mechanic can give the engine, because aging oil becomes diluted and contaminated, which leads to premature engine wear.
2 Make sure that you have all the necessary tools before you begin this procedure **(see illustration).** You should also have plenty of rags or newspapers handy for mopping up any spills.
3 Access to the underside of the vehicle is greatly improved if the vehicle can be lifted on a hoist, driven onto ramps or supported by jackstands. **Warning:** *Do not work under a vehicle which is supported only by a bumper, hydraulic or scissors-type jack.*
4 If this is your first oil change, get under the vehicle and familiarize yourself with the location of the oil drain plug. The engine and exhaust components will be warm during the

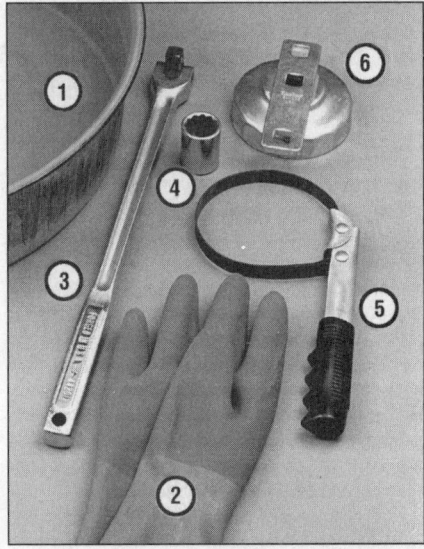

8.2 These tools are required when changing the engine oil and filter

1 **Drain pan** - It should be fairly shallow in depth, but wide in order to prevent spills
2 **Rubber gloves** - When removing the drain plug and filter, it is inevitable that you will get oil on your hands (the gloves will prevent burns)
3 **Breaker bar** - Sometimes the oil drain plug is pretty tight and a long breaker bar is needed to loosen it
4 **Socket** – To be used with the breaker bar or a ratchet (must be the correct size to fit the drain plug)
5 **Filter wrench** - This is a metal band-type wrench, which requires clearance around the filter to be effective
6 **Filter wrench** - This type fits on the bottom of the filter and can be turned with a ratchet or breaker bar (different size wrenches are available for different types of filters)

actual work, so try to anticipate any potential problems before the engine and accessories are hot.
5 Park the vehicle on a level spot. Start the engine and allow it to reach its normal operating temperature (the needle on the temperature gauge should be at least above

Chapter 1 Tune-up and routine maintenance

8.7 Use a proper size box-end wrench or socket to remove the oil drain plug and avoid rounding it off

8.13a Since the oil filter is usually on very tight, you'll need a special wrench for removal - DO NOT use the wrench to tighten the new filter (GA16DE engine)

the bottom mark). Warm oil and sludge will flow out more easily. Turn off the engine when it's warmed up. Remove the filler cap in the valve cover.

6 Raise the vehicle and support it on jackstands. **Warning:** *To avoid personal injury, never get beneath the vehicle when it is supported by only by a jack. The jack provided with your vehicle is designed solely for raising the vehicle to remove and replace the wheels. Always use jackstands to support the vehicle when it becomes necessary to place your body underneath the vehicle.*

7 Being careful not to touch the hot exhaust components, place the drain pan under the drain plug in the bottom of the pan and remove the plug **(see illustration)**. You may want to wear gloves while unscrewing the plug the final few turns if the engine is really hot.

8 Allow the old oil to drain into the pan. It may be necessary to move the pan farther under the engine as the oil flow slows to a trickle. Inspect the old oil for the presence of metal shavings and chips.

9 After all the oil has drained, wipe off the drain plug with a clean rag. Even minute metal particles clinging to the plug would immediately contaminate the new oil.

10 Clean the area around the drain plug opening, reinstall the plug and tighten it securely, but do not strip the threads.

11 Move the drain pan into position under the oil filter.

12 Remove all tools, rags, etc. from under the vehicle, being careful not to spill the oil in the drain pan, then lower the vehicle.

13 Loosen the oil filter **(see illustrations)** by turning it counterclockwise with the filter wrench. Any standard filter wrench should work. Once the filter is loose, use your hands to unscrew it from the block. Just as the filter is detached from the block, immediately tilt the open end up to prevent the oil inside the filter from spilling out. **Warning:** *The engine exhaust manifold may still be hot, so be careful.*

14 With a clean rag, wipe off the mounting surface on the block. If a residue of old oil is allowed to remain, it will smoke when the block is heated up. It will also prevent the new filter from seating properly. Also make sure that the none of the old gasket remains stuck to the mounting surface. It can be removed with a scraper if necessary.

15 Compare the old filter with the new one to make sure they are the same type. Smear some engine oil on the rubber gasket of the new filter and screw it into place **(see illustration)**. Because over-tightening the filter will damage the gasket, do not use a filter wrench to tighten the filter. Tighten it by hand until the gasket contacts the seating surface. Then seat the filter by giving it an additional 3/4-turn.

16 Add new oil to the engine through the oil filler cap in the valve cover. Use a spout or funnel to prevent oil from spilling onto the top of the engine. Pour three quarts of fresh oil into the engine. Wait a few minutes to allow the oil to drain into the pan, then check the level on the oil dipstick (see Section 4 if necessary). If the oil level is at or near the H mark, install the filler cap hand tight, start the engine and allow the new oil to circulate.

17 Allow the engine to run for about a minute. While the engine is running, look under the vehicle and check for leaks at the oil pan drain plug and around the oil filter. If either is leaking, stop the engine and tighten the plug or filter slightly.

18 Wait a few minutes to allow the oil to trickle down into the pan, then recheck the level on the dipstick and, if necessary, add enough oil to bring the level to the H mark.

19 During the first few trips after an oil change, make it a point to check frequently for leaks and proper oil level.

20 The old oil drained from the engine cannot be reused in its present state and should be discarded. Check with your local refuse disposal company, disposal facility or environmental agency to see if they will accept the oil for recycling. Don't pour used oil into drains or onto the ground. After the oil has cooled, it can be drained into a suitable container (capped plastic jugs, topped bottles, milk cartons, etc.) for transport to one of these disposal sites.

8.13b Oil filter location - SR20DE engine

8.15 Lubricate the oil filter gasket with clean engine oil before installing the filter on the engine

Chapter 1 Tune-up and routine maintenance   1-13

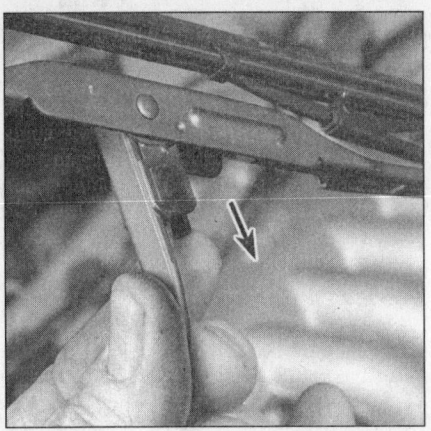

9.5 Press on the release tab and push the blade assembly down out of the hook in the arm

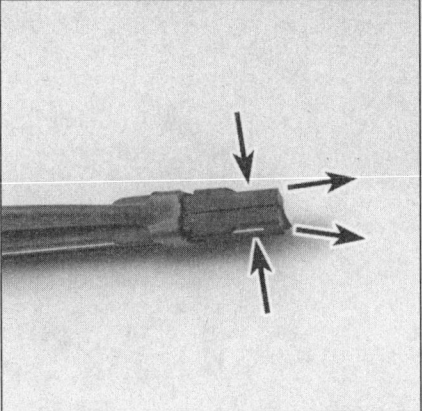

9.6 On most models, use needle-nose pliers to pull out the two metal rods, then slide the old element out - slide the new element in and lock in place with the metal rods - some models use a squeeze-type locking device

9 Windshield wiper blade inspection and replacement (every 7500 miles or 6 months)

Refer to illustrations 9.5 and 9.6

1   The windshield wiper and blade assembly should be inspected periodically for damage, loose components and cracked or worn blade elements.
2   Road film can build up on the wiper blades and affect their efficiency, so they should be washed regularly with a mild detergent solution.
3   The action of the wiping mechanism can loosen bolts, nuts and fasteners, so they should be checked and tightened, as necessary, at the same time the wiper blades are checked.
4   If the wiper blade elements are cracked, worn or warped, or no longer clean adequately, they should be replaced with new ones.
5   Lift the arm assembly away from the glass for clearance, press on the release lever, then slide the wiper blade assembly out of the hook in the end of the arm (see illustration).
6   Use needle-nose pliers to pull out the two metal rods from the blade element, then slide the element out of the frame and discard it (see illustration).
7   Installation is the reverse of removal.

10 Clutch pedal height, freeplay and release lever freeplay check and adjustment (every 7500 miles or 6 months)

Pedal height

Refer to illustration 10.1

1   With the clutch pedal fully released, measure the distance from the top of the pad to the floor (see illustration).
2   If the height is not as listed in the Specifications at the beginning of this Chapter it must be adjusted.
3   Loosen the locknut on the pedal adjustment bolt or cruise control switch (see illustration 10.1).
4   Turn the pedal adjustment bolt or cruise control switch until the pedal height is correct.
5   Tighten the locknut.
6   After adjusting the pedal height, check the release lever freeplay.

Release lever freeplay

Refer to illustration 10.7

7   In the engine compartment, move the end of the release lever back-and-forth and measure the freeplay (see illustration). Compare the freeplay measurement to the one in the Specifications Section at the beginning of this Chapter.
8   If freeplay is incorrect, loosen the locknut and turn the adjusting nut until the height is correct. Tighten the locknut.
9   After checking the release lever freeplay, check the clutch pedal freeplay.

Pedal freeplay

10   Press down lightly on the clutch pedal and measure the distance that it moves freely before the clutch resistance is felt (see illustration 10.1). The freeplay should be within the specified limits. If it isn't, it must be adjusted.
11   Loosen the locknut on the pedal adjustment bolt or cruise control switch (see illustration 10.1).

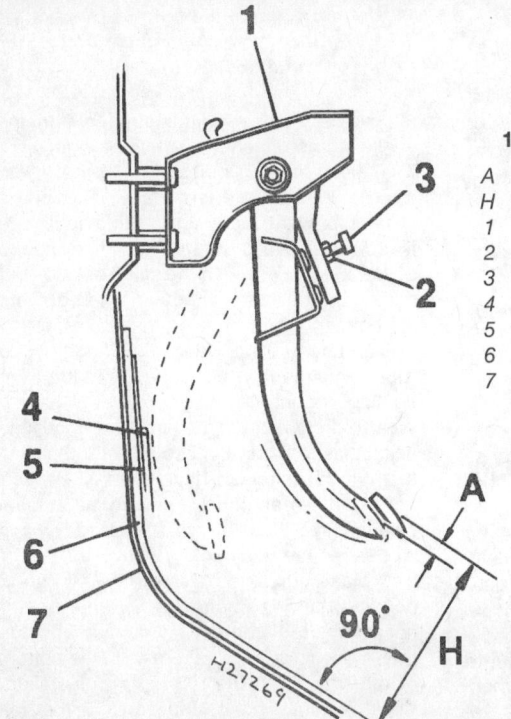

10.1 Clutch pedal checking details

A	Pedal freeplay measurement
H	Pedal height measurement
1	Pedal mounting bracket
2	Locknut
3	Pedal height adjustment bolt
4	Carpet
5	Insulator sheet
6	Insulator sheet
7	Floor panel

10.7 Move the release lever back-and-forth to check the freeplay - adjustments are made by loosening the locknut and turning the threaded adjusting nut (arrow)

12 Turn the pedal adjustment bolt or cruise control switch until the pedal freeplay is correct.
13 Tighten the locknut.

Clutch interlock switch freeplay

14 The interlock switch, mounted at the top of the clutch pedal arm under the dash, is designed to keep the engine from being started with the transmission in gear. The switch freeplay must be must be checked once all other adjustments are made.
15 Measure the clearance (freeplay) between the pedal stopper rubber and the tip of the interlock switch.
16 If the freeplay isn't as listed in the Specifications Section at the beginning of this Chapter, adjust it by loosening the locknut on the switch. Rotate the switch to achieve the specified freeplay and tighten the locknut.

11 Battery check, maintenance and charging (every 7500 miles or 6 months)

Refer to illustrations 11.1, 11.6a, 11.6b, 11.7a and 11.7b
Warning: *Certain precautions must be followed when checking and servicing the battery. Hydrogen gas, which is highly flammable, is always present in the battery cells, so keep lighted tobacco and all other open flames and sparks away from the battery. The electrolyte inside the battery is actually dilute sulfuric acid, which will cause injury if splashed on your skin or in your eyes. It will also ruin clothes and painted surfaces. When removing the battery cables, always detach the negative cable first and hook it up last!*

1 A routine preventive maintenance program for the battery in your vehicle is the only way to ensure quick and reliable starts. But before performing any battery maintenance, make sure that you have the proper equipment necessary to work safely around the battery **(see illustration)**.
2 There are also several precautions that should be taken whenever battery maintenance is performed. Before servicing the battery, always turn the engine and all accessories off and disconnect the cable from the negative terminal of the battery.
3 The battery produces hydrogen gas, which is both flammable and explosive. Never create a spark, smoke or light a match around the battery. Always charge the battery in a ventilated area.
4 Electrolyte contains poisonous and corrosive sulfuric acid. Do not allow it to get in your eyes, on your skin or on your clothes. Never ingest it. Wear protective safety glasses when working near the battery. Keep children away from the battery.
5 Note the external condition of the battery. If the positive terminal and cable clamp on your vehicle's battery is equipped with a rubber protector, make sure it isn't torn or

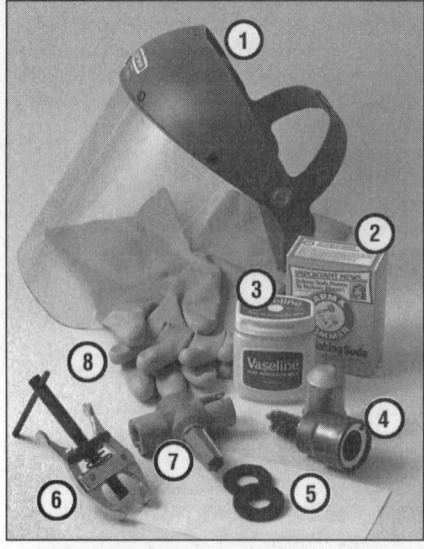

11.1 Tools and materials required for battery maintenance

1 **Face shield/safety goggles** - *When removing corrosion with a brush, the acidic particles can easily fly up into your eyes*
2 **Baking soda** - *A solution of baking soda and water can be used to neutralize corrosion*
3 **Petroleum jelly** - *A layer of this on the battery posts will help prevent corrosion*
4 **Battery post/cable cleaner** - *This wire brush cleaning tool will remove all traces of corrosion from the battery posts and cable clamps*
5 **Treated felt washers** - *Placing one of these on each post, directly under the cable clamps, will help prevent corrosion*
6 **Puller** - *Sometimes the cable clamps are very difficult to pull off the posts, even after the nut/bolt has been completely loosened. This tool pulls the clamp straight up and off the post without damage*
7 **Battery post/cable cleaner** - *Here is another cleaning tool which is a slightly different version of number 4 above, but it does the same thing*
8 **Rubber gloves** - *Another safety item to consider when servicing the battery; remember that's acid inside the battery*

damaged. It should completely cover the terminal. Look for any corroded or loose connections, cracks in the case or cover or loose hold-down clamps. Also check the entire length of each cable for cracks and frayed conductors.
6 If corrosion, which looks like white, fluffy deposits **(see illustration)** is evident, particularly around the terminals, the battery should be removed for cleaning. Loosen the cable clamp bolts with a wrench, being careful to remove the ground cable first, and slide them off the terminals **(see illustration)**. Then disconnect the hold-down clamp bolt and nut,

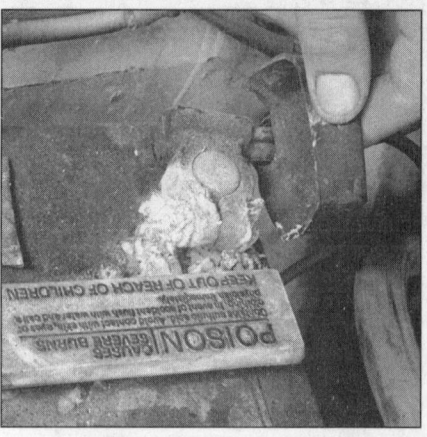

11.6a Battery terminal corrosion usually appears as light, fluffy powder

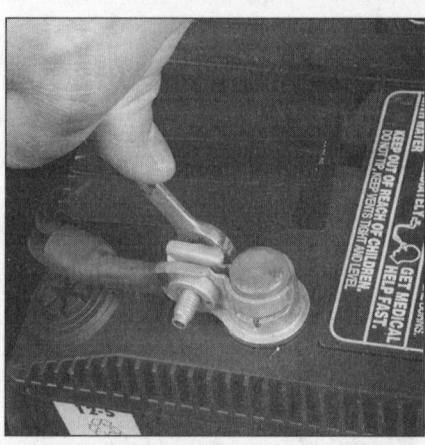

11.6b Removing a cable from the battery post with a wrench - sometimes special battery pliers are required for this procedure if corrosion has caused deterioration of the nut hex (always remove the ground cable first and hook it up last!)

remove the clamp and lift the battery from the engine compartment.
7 Clean the cable clamps thoroughly with a battery brush or a terminal cleaner and a solution of warm water and baking soda **(see illustration)**. Wash the terminals and the top of the battery case with the same solution but make sure that the solution doesn't get into the battery. When cleaning the cables, terminals and battery top, wear safety goggles and rubber gloves to prevent any solution from coming in contact with your eyes or hands. Wear old clothes too - even diluted, sulfuric acid splashed onto clothes will burn holes in them. If the terminals have been extensively corroded, clean them up with a terminal cleaner **(see illustration)**. Thoroughly wash all cleaned areas with plain water.
8 Make sure the battery tray is in good condition and the hold-down clamp bolt or nut is tight. If the battery is removed from the tray, make sure no parts remain in the bottom of the tray when the battery is reinstalled. When reinstalling the hold-down clamp bolt or nut, do not over-tighten it.

Chapter 1 Tune-up and routine maintenance   1-15

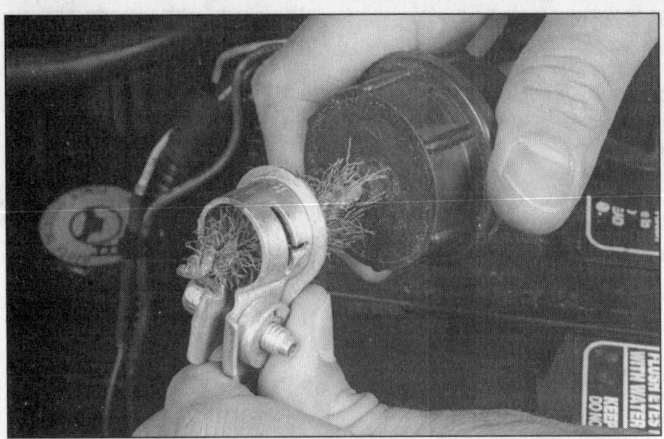

11.7a When cleaning the cable clamps, all corrosion must be removed

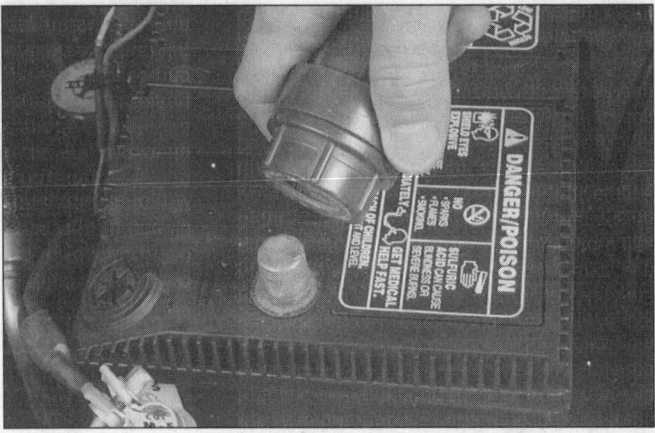

11.7b Regardless of the type of tool used to clean the battery posts, a clean, shiny surface should be the result

9 Information on removing and installing the battery can be found in Chapter 5. Information on jump starting can be found at the front of this manual. For more detailed battery checking procedures, refer to the *Haynes Automotive Electrical Manual*.

Cleaning

10 Corrosion on the hold-down components, battery case and surrounding areas can be removed with a solution of water and baking soda. Thoroughly rinse all cleaned areas with plain water.
11 Any metal parts of the vehicle damaged by corrosion should be covered with a zinc-based primer, then painted.

Charging

Warning: *When batteries are being charged, hydrogen gas, which is very explosive and flammable, is produced. Do not smoke or allow open flames near a charging or a recently charged battery. Wear eye protection when near the battery during charging. Also, make sure the charger is unplugged before connecting or disconnecting the battery from the charger.*
12 Slow-rate charging is the best way to restore a battery that's discharged to the point where it will not start the engine. It's also a good way to maintain the battery charge in a vehicle that's only driven a few miles between starts. Maintaining the battery charge is particularly important in the winter when the battery must work harder to start the engine and electrical accessories that drain the battery are in greater use.
13 It's best to use a one or two-amp battery charger (sometimes called a "trickle" charger). They are the safest and put the least strain on the battery. They are also the least expensive. For a faster charge, you can use a higher amperage charger, but don't use one rated more than 1/10th the amp/hour rating of the battery. Rapid boost charges that claim to restore the power of the battery in one to two hours are hardest on the battery and can damage batteries not in good condition. This type of charging should only be used in emergency situations.
14 The average time necessary to charge a battery should be listed in the instructions that come with the charger. As a general rule, a trickle charger will charge a battery in 12 to 16 hours.

12 Drivebelt check, adjustment and replacement (every 7500 miles or 6 months)

Refer to illustrations 12.3a, 12.3b, 12.4, 12.5, 12.8a and 12.8b

Check

1 The alternator, power steering pump and air conditioning compressor drivebelts, also referred to as simply "fan" belts, are located at the right end of the engine. The good condition and proper adjustment of the alternator belt is critical to the operation of the engine. Because of their composition and the high stresses to which they are subjected, drivebelts stretch and deteriorate as they get older. They must therefore be periodically inspected.
2 The number of belts used on a particular vehicle depends on the accessories installed. One belt transmits power from the crankshaft to the alternator and air conditioning. If the vehicle is equipped with power steering, the pump is driven by it's own belt.
3 With the engine off, open the hood and locate the drivebelts. With a flashlight, check each belt for separation of the adhesive rubber on both sides of the core, core separation from the belt side, a severed core, separation of the ribs from the adhesive rubber, cracking or separation of the ribs, and torn or worn ribs or cracks in the inner ridges of the ribs **(see illustration)**. Also check for fraying and glazing, which gives the belt a shiny appearance. Both sides of the belt should be inspected, which means you will have to twist the belt to check the underside. Use your fingers to feel the belt where you can't see it. If any of the above conditions are evident, replace the belt (go to Step 6).
4 Check the belt tension by pushing firmly on the belt with your thumb at a distance

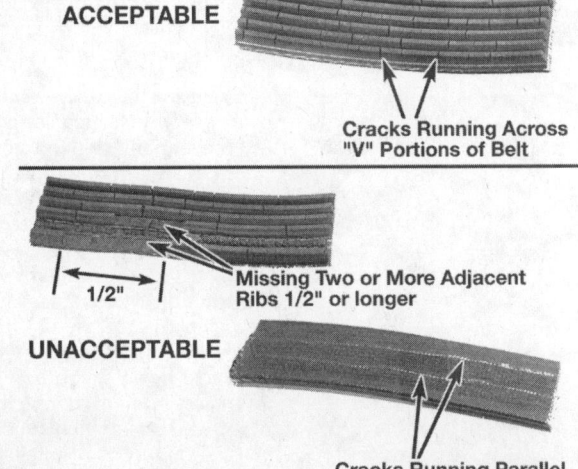

12.3 Small cracks in the underside of a V-ribbed belt are acceptable - lengthwise cracks, or missing pieces that cause the belt to make noise, are cause for replacement

1-16 Chapter 1 Tune-up and routine maintenance

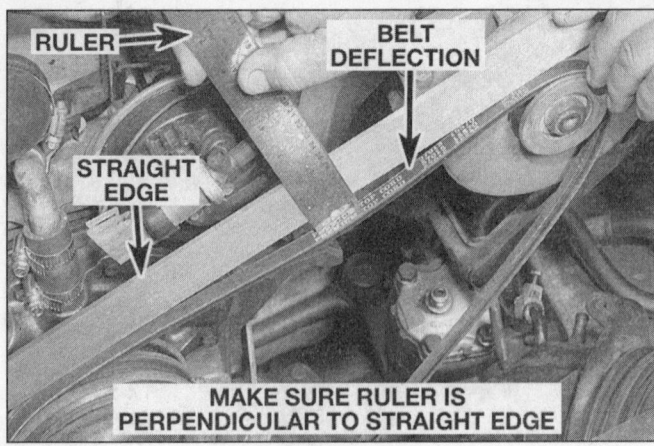

12.4 Measuring drivebelt deflection with a straightedge and ruler

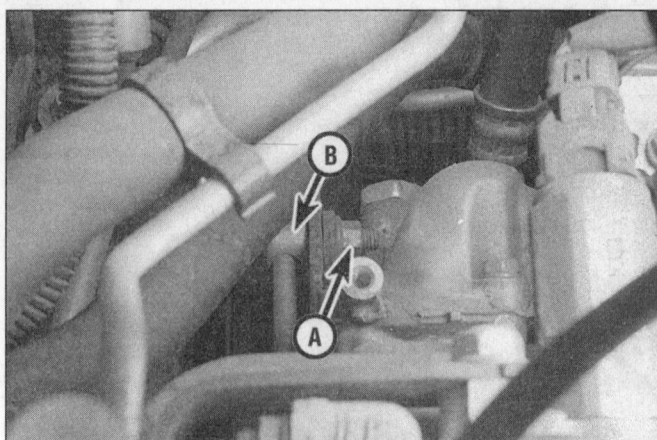

12.5 Loosen the locknut (A) and use a wrench to turn the adjusting bolt (B) (head under the hose) and adjust the tension

halfway between the pulleys and note how far the belt can be pushed (deflected). Measure this deflection with a ruler **(see illustration)**. As a rule of thumb, if the distance from pulley center-to-pulley center is between 7 and 11 inches, the belt should deflect 1/4-inch. If the belt travels between pulleys spaced 12 to 16 inches apart, the belt should deflect 1/2-inch for a V-belt or 1/4-inch for a V-ribbed belt.

Adjustment

5 Components on these models have a built-in adjusting bolts or idlers **(see illustration)**. Loosen the locknut and turn the adjusting bolt. Measure the belt tension in accordance with one of the above methods. Repeat this step until the drivebelt is adjusted.

Replacement

6 To replace a belt, follow the above procedures for drivebelt adjustment but slip the belt off the crankshaft pulley and remove it. If you are replacing the power steering pump belt, you may have to remove the air conditioning compressor belt first because of the way they are arranged on the crankshaft pulley. Because of this and because belts tend to wear out more or less together, it is a good idea to replace both belts at the same time. Mark each belt and its appropriate pulley groove so the replacement belts can be installed in their proper positions.

7 Take the old belts to the parts store in order to make a direct comparison for length, width and design.
8 After replacing ribbed drivebelts, make sure that it fits properly in the ribbed grooves in the pulleys **(see illustrations)**. It is essential that the belt be properly centered. If the belt is narrower than the pulley, be sure to route it to the inner edge of the pulley.
9 Adjust the belt(s) in accordance with the procedure outlined above.

13 Underhood hose check and replacement (every 7500 miles or 6 months)

Caution: *Replacement of air conditioning hoses must be left to a dealer service department or air conditioning shop that has the equipment to depressurize the system safely. Never remove air conditioning components or hoses until the system has been depressurized.*

General

1 High temperatures in the engine compartment can cause the deterioration of the rubber and plastic hoses used for engine, accessory and emission systems operation. Periodic inspection should be made for cracks, loose clamps, material hardening and leaks.
2 Information specific to the cooling sys-

tem hoses can be found in Section 14.
3 Some, but not all, hoses are secured to the fittings with clamps. Where clamps are used, check to be sure they haven't lost their tension, allowing the hose to leak. If clamps aren't used, make sure the hose has not expanded and/or hardened where it slips over the fitting, allowing it to leak.

Vacuum hoses

4 It's quite common for vacuum hoses, especially those in the emissions system, to be color coded or identified by colored stripes molded into them. Various systems require hoses with different wall thickness, collapse resistance and temperature resistance. When replacing hoses, be sure the new ones are made of the same material.
5 Often the only effective way to check a hose is to remove it completely from the vehicle. If more than one hose is removed, be sure to label the hoses and fittings to ensure correct installation.
6 When checking vacuum hoses, be sure to include any plastic T-fittings in the check. Inspect the fittings for cracks and the hose where it fits over the fitting for distortion, which could cause leakage.
7 A small piece of vacuum hose (1/4-inch inside diameter) can be used as a stethoscope to detect vacuum leaks. Hold one end of the hose to your ear and probe around vacuum

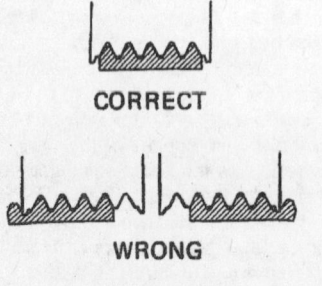

12.8a When installing a ribbed V-belt, make sure it is centered - it must not overlap either edge of the pulley

12.8b On SR20DE engines without air conditioning, make sure the ribbed drivebelt (arrow) is seated in the same corresponding grooves on both the alternator and crankshaft pulleys

Chapter 1 Tune-up and routine maintenance

hoses and fittings, listening for the "hissing" sound characteristic of a vacuum leak. **Warning:** *When probing with the vacuum hose stethoscope, be very careful not to come into contact with moving engine components such as the drivebelts, cooling fan, etc.*

Fuel hose

Warning: *There are certain precautions which must be taken when inspecting or servicing fuel system components. Work in a well ventilated area and do not allow open flames (cigarettes, appliance pilot lights, etc.) or bare light bulbs near the work area. Mop up any spills immediately and do not store fuel soaked rags where they could ignite.*

8   Check all rubber fuel lines for deterioration and chafing. Check especially for cracks in areas where the hose bends and just before fittings, such as where a hose attaches to the fuel filter.
9   High quality fuel line, designed specifically for high pressure fuel injection systems, must be used for fuel line replacement. Never, under any circumstances, use unreinforced vacuum line, clear plastic tubing or water hose for fuel lines.
10 Spring-type clamps are commonly used on fuel lines. These clamps often lose their tension over a period of time, and can be "sprung" during removal. Replace all spring-type clamps with screw clamps whenever a hose is replaced.

Metal lines

11 Sections of metal line are often used for fuel line between the fuel pump and fuel injection unit. Check carefully to be sure the line has not been bent or crimped and that cracks have not started in the line.
12 If a section of metal fuel line must be replaced, only seamless steel tubing should be used, since copper and aluminum tubing don't have the strength necessary to withstand normal engine vibration.
13 Check the metal brake lines where they enter the master cylinder and brake proportioning unit (if used) for cracks in the lines or loose fittings. Any sign of brake fluid leakage calls for an immediate thorough inspection of the brake system.

14   Cooling system check (every 7500 miles or 6 months)

Refer to illustration 14.4

1   Many major engine failures can be attributed to a faulty cooling system. If the vehicle is equipped with an automatic transaxle, the cooling system also cools the transaxle fluid and thus plays an important role in prolonging transaxle life.
2   The cooling system should be checked with the engine cold. Do this before the vehicle is driven for the day or after the engine has been shut off for at least three hours.
3   Remove the radiator cap by turning it to the left until it reaches a stop. If you hear a hissing sound (indicating there is still pressure in the system), wait until it stops. Now press down on the cap with the palm of your hand and continue turning to the left until the cap can be removed. Thoroughly clean the cap, inside and out, with clean water. Also clean the filler neck on the radiator. All traces of corrosion should be removed. The coolant inside the radiator should be relatively transparent. If it's rust colored, the system should be drained and refilled (see Section 25). If the coolant level isn't up to the top, add additional antifreeze/coolant mixture (see Section 4).
4   Carefully check the large upper and lower radiator hoses along with the smaller diameter heater hoses which run from the engine to the firewall. Inspect each hose along its entire length, replacing any hose which is cracked, swollen or shows signs of deterioration. Cracks may become more apparent if the hose is squeezed **(see illustration)**. Regardless of condition, it's a good idea to replace hoses with new ones every two years.
5   Make sure that all hose connections are tight. A leak in the cooling system will usually show up as white or rust colored deposits on the areas adjoining the leak. If wire-type clamps are used at the ends of the hoses, it may be a good idea to replace them with more secure screw-type clamps.
6   Use compressed air or a soft brush to remove bugs, leaves, etc. from the front of the radiator or air conditioning condenser. Be careful not to damage the delicate cooling fins or cut yourself on them.
7   Every other inspection, or at the first indication of cooling system problems, have the cap and system pressure tested. If you don't have a pressure tester, most gas stations and repair shops will do this for a minimal charge.

15   Tire rotation (every 7500 miles or 6 months)

Refer to illustration 15.2

1   The tires should be rotated at the specified intervals and whenever uneven wear is noticed. Since the vehicle will be raised and the tires removed anyway, check the brakes (see Section 16) at this time.
2   Radial tires must be rotated in a specific pattern **(see illustration)**.
3   Refer to the information in *Jacking and towing* at the front of this manual for the proper procedures to follow when raising the vehicle and changing a tire. If the brakes are to be checked, do not apply the parking brake as stated. Make sure the tires are blocked to prevent the vehicle from rolling.
4   Preferably, the entire vehicle should be raised at the same time. This can be done on a hoist or by jacking up each corner and then lowering the vehicle onto jackstands placed under the frame rails. Always use four jackstands and make sure the vehicle is firmly supported.
5   After rotation, check and adjust the tire

Check for a chafed area that could fail prematurely.

Check for a soft area indicating the hose has deteriorated inside.

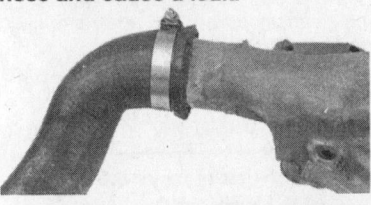
Overtightening the clamp on a hardened hose will damage the hose and cause a leak.

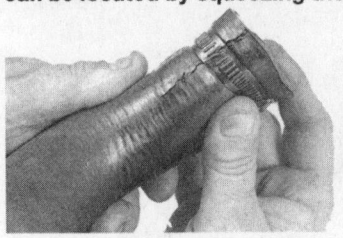
Check each hose for swelling and oil-soaked ends. Cracks and breaks can be located by squeezing the hose.

14.4 Hoses, like drivebelts, have a habit of failing at the worst possible time - to prevent the inconvenience of a blown radiator or heater hose, inspect them carefully as shown here

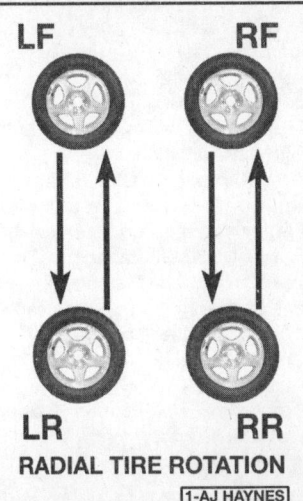

15.2 The recommended tire rotation pattern for these vehicles

16.6 You will find an inspection hole like this in each caliper - placing a ruler across the hole should enable you to determine the thickness of remaining pad material for both inner and outer pads

16.12 A quick check of the remaining drum brake shoe lining material can be made by removing the rubber plug in the backing plate and looking through the inspection hole

pressures as necessary and be sure to check the lug nut tightness.
6   For further information on the wheels and tires, refer to Chapter 10.

16   Brake check (every 15,000 miles or 12 months)

Warning: *The dust created by the brake system may contain asbestos, which is harmful to your health. Never blow it out with compressed air and don't inhale any of it. An approved filtering mask should be worn when working on the brakes. Do not, under any circumstances, use petroleum-based solvents to clean brake parts. Use brake system cleaner only! Try to use non-asbestos replacement parts whenever possible.*
Note: *For detailed photographs of the brake system, refer to Chapter 9.*

1   In addition to the specified intervals, the brakes should be inspected every time the wheels are removed or whenever a defect is suspected. Any of the following symptoms could indicate a potential brake system defect: The vehicle pulls to one side when the brake pedal is depressed; the brakes make squealing or dragging noises when applied; brake pedal travel is excessive; the pedal pulsates; brake fluid leaks, usually onto the inside of the tire or wheel.
2   The disc brake pads have built-in wear indicators which should make a high pitched squealing or scraping noise when they are worn to the replacement point. When you hear this noise, replace the pads immediately or expensive damage to the discs can result.
3   Loosen the wheel lug nuts.
4   Raise the vehicle and place it securely on jackstands.
5   Remove the wheels (see *Jacking and towing* at the front of this book, or your owner's manual, if necessary).

Disc brakes

Refer to illustration 16.6
6   There are two pads - an outer and an inner - in each caliper. The pads are visible through inspection holes in each caliper **(see illustration)**.
7   Check the pad thickness by looking at each end of the caliper and through the inspection hole in the caliper body. If the lining material is less than the thickness listed in this Chapter's Specifications, replace the pads. **Note:** *Keep in mind that the lining material is riveted or bonded to a metal backing plate and the metal portion is not included in this measurement.*
8   If it is difficult to determine the exact thickness of the remaining pad material by the above method, or if you are at all concerned about the condition of the pads, remove the caliper(s), then remove the pads from the calipers for further inspection (refer to Chapter 9).
9   Once the pads are removed from the calipers, clean them with brake cleaner and re-measure them with a ruler or a vernier caliper.
10   Measure the disc thickness with a micrometer to make sure that it still has service life remaining. If any disc is thinner than the specified minimum thickness, replace it (refer to Chapter 9). Even if the disc has service life remaining, check its condition. Look for scoring, gouging and burned spots. If these conditions exist, remove the disc and have it resurfaced (see Chapter 9).
11   Before installing the wheels, check all brake lines and hoses for damage, wear, deformation, cracks, corrosion, leakage, bends and twists, particularly in the vicinity of the rubber hoses at the calipers. Check the clamps for tightness and the connections for leakage. Make sure that all hoses and lines are clear of sharp edges, moving parts and the exhaust system. If any of the above conditions are noted, repair, reroute or replace the lines and/or fittings as necessary (see Chapter 9).

Rear drum brakes

Refer to illustrations 16.12, 16.14 and 16.16
12   On most models, it is possible to check the brake shoe lining thickness without removing the brake drums by removing the rubber plug from the backing plate and use a flashlight to inspect the linings **(see illustration)**. For a more thorough brake inspection, follow the procedure below.
13   Refer to Chapter 9 and remove the rear brake drums.
14   Note the thickness of the lining material on the rear brake shoes **(see illustration)** and look for signs of contamination by brake fluid and grease. If the lining material is within 1/16-inch of the recessed rivets or metal shoes, replace the brake shoes with new ones. The shoes should also be replaced if they are cracked, glazed (shiny lining surfaces) or contaminated with brake fluid or grease. See Chapter 9 for the replacement procedure.
15   Check the shoe return and hold-down springs and the adjusting mechanism to make sure they're installed correctly and in good condition. Deteriorated or distorted springs, if not replaced, could allow the linings to drag and wear prematurely.
16   Check the wheel cylinders for leakage by carefully peeling back the rubber boots **(see illustration)**. If brake fluid is noted behind the boots, the wheel cylinders must be replaced (see Chapter 9).
17   Check the drums for cracks, score marks, deep scratches and hard spots, which will appear as small discolored areas. If imperfections cannot be removed with emery cloth, the drums must be resurfaced by an automotive machine shop (see Chapter 9 for more detailed information).
18   Refer to Chapter 9 and install the brake drums.
19   Install the wheels and snug the wheel lug nuts finger tight.
20   Remove the jackstands and lower the vehicle.
21   Tighten the wheel lug nuts to the torque listed in this Chapter's Specifications.

Brake booster check

22   Sit in the driver's seat and perform the following sequence of tests.
23   With the brake fully depressed, start the engine - the pedal should move down a little when the engine starts.

Chapter 1 Tune-up and routine maintenance

1-19

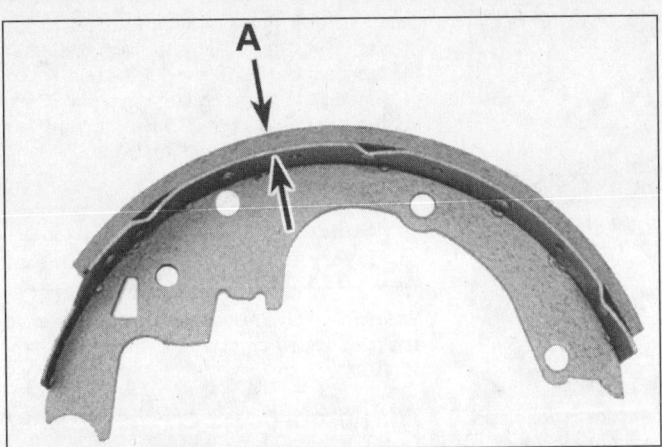

16.14 If the lining is bonded to the brake shoe, measure the lining thickness from the outer surface to the metal shoe, as shown here; if the lining is riveted to the shoe, measure from the lining outer surface to the rivet head

16.16 Carefully peel back the wheel cylinder boots and check for leaking fluid indicating that the cylinder must be replaced or rebuilt

24 With the engine running, depress the brake pedal several times - the travel distance should not change.
25 Depress the brake, stop the engine and hold the pedal in for about 30 seconds - the pedal should neither sink nor rise.
26 Restart the engine, run it for about a minute and turn it off. Then firmly depress the brake several times - the pedal travel should decrease with each application.
27 If your brakes do not operate as described above when the preceding tests are performed, the brake booster is either in need of repair or has failed. Refer to Chapter 9 for the removal procedure.

Parking brake

28 Slowly pull up on the parking brake and count the number of clicks you hear until the handle is up as far as it will go. The adjustment is correct if you hear the specified number of clicks. If you hear more or fewer clicks, it's time to adjust the parking brake (refer to Chapter 9).
29 An alternative method of checking the parking brake is to park the vehicle on a steep hill with the parking brake set and the transaxle in Neutral (be sure to stay in the vehicle during this check!). If the parking brake cannot prevent the vehicle from rolling, it is in need of adjustment (see Chapter 9).

Brake pedal height and freeplay check and adjustment

Brake pedal released height

Refer to illustration 16.30

30 With the brake pedal fully released, measure the distance from the top of the pad to the floor **(see illustration)**.
31 If the height is not as listed in the Specifications Section at the beginning of this Chapter it must be adjusted.
32 Loosen the locknut on the brake booster input rod **(see illustration 16.30)**.

33 Turn the input rod until the pedal height is correct.
34 Tighten the locknut.
35 After adjusting the pedal height, check the freeplay.

Brake pedal freeplay

36 Press down lightly on the brake pedal and, with a ruler, measure the distance that it moves freely before resistance is felt **(see illustration 16.30)**. The freeplay should be within the specified limits. If it isn't, it must be adjusted.
37 Loosen the locknut on the brake input rod **(see illustration 16.30)**.
38 Turn the input rod until the pedal freeplay is correct.
39 Tighten the locknut.

Brake pedal depressed height (all models)

40 After checking and, if necessary, adjusting the pedal released height and freeplay, the pedal depressed height must be checked.
41 With the engine running, press the brake pedal fully and measure the pedal pad-to-floor distance **(see illustration 16.30)**.
42 If the minimum depressed height is below that listed in the Specifications Section listed at the beginning of this Chapter, check the brake system for leaks or other damage.

17 Air filter replacement (every 15,000 miles or 12 months)

Refer to illustrations 17.1a and 17.1b

1 The air filter is located inside a housing at the left (driver's) side of the engine compartment. To remove the air filter, release the four spring clips that keep the two halves of the air cleaner housing together, then lift the cover up and remove the air filter element **(see illustrations)**.
2 Inspect the outer surface of the filter element. If it is dirty, replace it. If it is only moderately dusty, it can be reused by blowing it

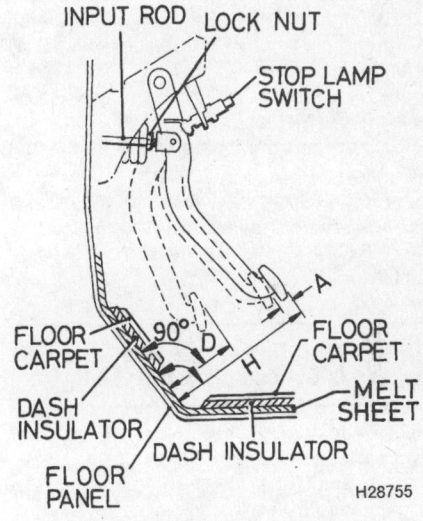

16.30 Brake pedal checking details

- A Pedal freeplay measurement
- D Depressed height measurement
- H Free height

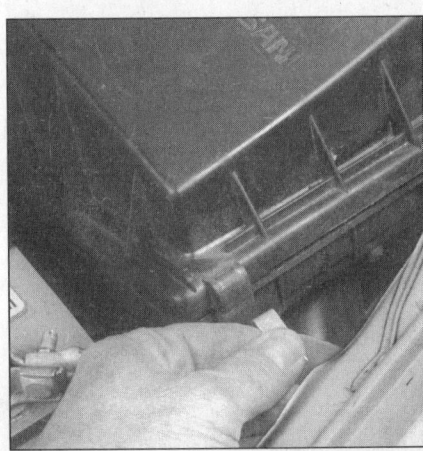

17.1a Detach the clips and separate the cover from the air cleaner housing

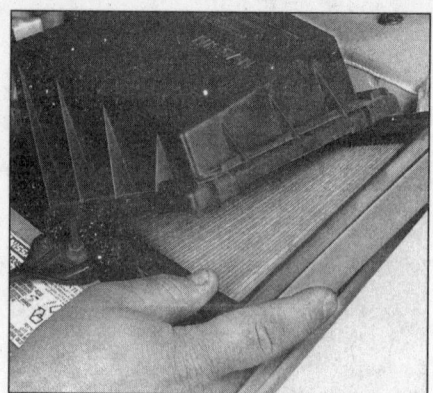

17.1b Lift the cover up and slide the element out of the housing

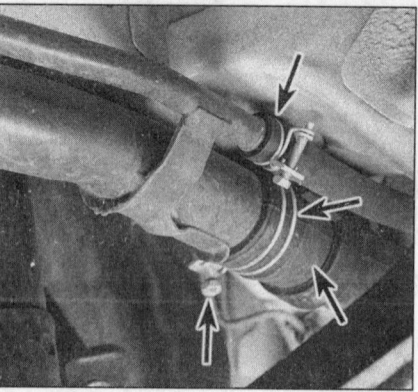

18.5 Inspect fuel filler hoses for cracks and make sure the clamps are tight (arrows)

clean from the back to the front surface with compressed air. Because it is a pleated paper type filter, it cannot be washed or oiled. If it cannot be cleaned satisfactorily with compressed air, discard and replace it. While the cover is off, be careful not to drop anything down into the housing or carburetor. **Caution:** *Never drive the vehicle with the air cleaner removed. Excessive engine wear could result and backfiring could even cause a fire under the hood.*

3 Wipe out the inside of the air cleaner housing.
4 Place the new filter into the air cleaner housing, making sure it seats properly.
5 Installation of the cover is the reverse of removal.

18 Fuel system check (every 15,000 miles or 12 months)

Refer to illustration 18.5
Warning: *Gasoline is extremely flammable, so take extra precautions when you work on any part of the fuel system. Don't smoke or allow open flames or bare light bulbs near the work area, and don't work in a garage where a natural gas-type appliance (such as a water heater or clothes dryer) with a pilot light is present. Since gasoline is carcinogenic, wear latex gloves when there's a possibility of being exposed to fuel, and, if you spill any fuel on your skin, rinse it off immediately with soap and water. Mop up any spills immediately and do not store fuel-soaked rags where they could ignite. The fuel system is under constant pressure, so, if any fuel lines are to be disconnected, the fuel pressure in the system must be relieved first (see Chapter 4 for more information). When you perform any kind of work on the fuel system, wear safety glasses and have a Class B type fire extinguisher on hand.*

1 If you smell gasoline while driving or after the vehicle has been sitting in the sun, inspect the fuel system immediately.
2 Remove the gas filler cap and inspect it for damage and corrosion. The gasket should have an unbroken sealing imprint. If the gasket is damaged or corroded, remove it and install a new one.

3 Inspect the fuel feed and return lines for cracks. Make sure the threaded flare nut type connectors which secure the metal fuel lines to the fuel injection system and the clamps which secure the hoses to the in-line fuel filter are tight.
4 Since some components of the fuel system - the fuel tank and part of the fuel feed and return lines, for example - are underneath the vehicle, they can be inspected more easily with the vehicle raised on a hoist. If that's not possible, raise the vehicle and support it securely on jackstands.
5 With the vehicle raised and safely supported, inspect the gas tank and filler neck for punctures, cracks and other damage. The connection between the filler neck and the tank is particularly critical. Sometimes a rubber filler neck will leak because of loose clamps or deteriorated rubber **(see illustration)**. These are problems a home mechanic can usually rectify. **Warning:** *Do not, under any circumstances, try to repair a fuel tank (except rubber components). A welding torch or any open flame can easily cause fuel vapors inside the tank to explode.*
6 Carefully check all rubber hoses and metal lines leading away from the fuel tank. Check for loose connections, deteriorated hoses, crimped lines and other damage. Carefully inspect the lines from the tank to the fuel injection system. Repair or replace damaged sections as necessary (see Chapter 4).

19 Manual transaxle lubricant level check (every 15,000 miles or 12 months)

1 The manual transaxle does not have a dipstick. To check the fluid level, raise the vehicle and support it securely on jackstands.
2 Remove the speedometer drive pinion from the top of the transaxle housing (see Chapter 7, Part A). Measure from the top of the opening to make sure that the oil level is at the level specified at the beginning of this Chapter.
3 If the transaxle needs more lubricant (if the level is not within the specified distance from the opening, use a funnel or a gear oil pump to add more and bring it up to the proper level. Stop filling the transaxle when the lubricant begins to run out the hole.
4 Install the speedometer drive pinion and tighten the bolt securely. Drive the vehicle a short distance, then check for leaks.

20 Steering and suspension check (every 15,000 miles or 12 months)

Note: *For detailed illustrations of the steering and suspension components, refer to Chapter 10.*

With the wheels on the ground

1 With the vehicle stopped and the front wheels pointed straight ahead, rock the steering wheel gently back and forth. If freeplay is excessive, a front wheel bearing, main shaft yoke, intermediate shaft yoke, lower arm balljoint or steering system joint is worn or the steering gear is out of adjustment or broken. Refer to Chapter 10 for the appropriate repair procedure.
2 Other symptoms, such as excessive vehicle body movement over rough roads, swaying (leaning) around corners and binding as the steering wheel is turned, may indicate faulty steering and/or suspension components.
3 Check the shock absorbers by pushing down and releasing the vehicle several times at each corner. If the vehicle does not come back to a level position within one or two bounces, the shocks/struts are worn and must be replaced. When bouncing the vehicle up and down, listen for squeaks and noises from the suspension components. Additional information on suspension components can be found in Chapter 10.

Under the vehicle

4 Raise the vehicle with a floor jack and support it securely on jackstands. See *Jacking and towing* at the front of this book for the proper jacking points.
5 Check the tires for irregular wear patterns and proper inflation. See Section 5 in this Chapter for information regarding tire wear and Chapter 10 for the wheel bearing replacement procedures.
6 Inspect the universal joint between the steering shaft and the steering gear housing. Check the steering gear housing for grease leakage or oozing. Make sure that the dust seals and boots are not damaged and that the boot clamps are not loose. Check the steering linkage for looseness or damage. Check the tie-rod ends for excessive play. Look for loose bolts, broken or disconnected parts and deteriorated rubber bushings on all suspension and steering components. While an assistant turns the steering wheel from side to side, check the steering components for free movement, chafing and binding. If the steering components do not seem to be reacting with the movement of the steering wheel, try to determine where the slack is located.

Chapter 1 Tune-up and routine maintenance

21.2 Flex the driveaxle boots by hand to check for cracks and/or leaking grease

22.3 The fuel filter is mounted in a clip below the brake master cylinder

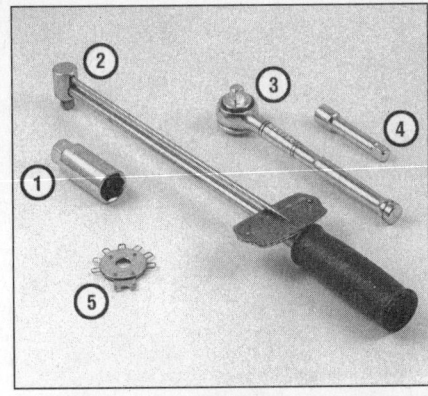

23.1 Tools required for changing spark plugs

1. *Spark plug socket* - This will have special padding inside to protect the spark plug porcelain insulator
2. *Torque wrench* - Although not mandatory, use of this tool is the best way to ensure that the plugs are tightened properly
3. *Ratchet* - Standard hand tool to fit the plug socket
4. *Extension* - Depending on model and accessories, you may need special extensions and universal joints to reach one or more of the plugs
5. *Spark plug gap gauge* - This gauge for checking the gap comes in a variety of styles. Make sure the gap for your engine is included

7 Check the balljoints moving each lower arm up and down with a pry bar to ensure that its balljoint has no play. If any balljoint does have play, replace it. See Chapter 10 for the front balljoint replacement procedure.

8 Inspect the balljoint boots for damage and leaking grease. Replace the balljoints with new ones if they are damaged (see Chapter 10).

21 Driveaxle boot check (every 15,000 miles or 12 months)

Refer to illustration 21.2

1 The driveaxle boots are very important because they prevent dirt, water and foreign material from entering and damaging the constant velocity (CV) joints. Oil and grease can cause the boot material to deteriorate prematurely, so it's a good idea to wash the boots with soap and water.

2 Inspect the boots for tears and cracks as well as loose clamps **(see illustration)**. If there is any evidence of cracks or leaking lubricant, they must be replaced as described in Chapter 8.

22 Fuel filter replacement (every 30,000 miles or 24 months)

Refer to illustration 22.3

1 The canister filter is mounted in a clip on the firewall near the brake master cylinder.

2 Depressurize the fuel system (see Chapter 4), then disconnect the cable from the negative terminal of the battery.

3 Detach the filter from the bracket, loosen the screw clamps, then detach the hoses from the top and bottom of the fuel filter and remove it **(see illustration)**.

4 Note that the inlet and outlet pipes are clearly labeled on their respective ends of the filter. Make sure the new filter is installed so that it's facing the proper direction as noted above. When correctly installed, the filter should be installed so the outlet pipe faces up and the inlet pipe faces down.

5 Install the inlet and outlet fittings and tighten the screw clamps securely. Reconnect the battery cable, start the engine and check for leaks.

23 Spark plug check and replacement (see maintenance schedule for service intervals)

Refer to illustrations 23.1, 23.4a, 23.4b, 23.6, 23.8, 23.10a and 23.10b

1 Spark plug replacement requires a spark plug socket which fits onto a ratchet handle. This socket is lined with a rubber grommet to protect the porcelain insulator of the spark plug and to hold the plug while you insert it into the spark plug hole. You will also need a wire-type feeler gauge to check and adjust the spark plug gap and a torque wrench to tighten the new plugs to the specified torque **(see illustration)**.

2 If you are replacing the plugs, purchase the new plugs, adjust them to the proper gap and then replace each plug one at a time.
Note: *When buying new spark plugs, it's essential that you obtain the correct plugs for your specific vehicle. This information can be found in the Specifications Section at the beginning of this Chapter, on the Vehicle Emissions Control Information (VECI) label located on the underside of the hood or in the owner's manual. If these sources specify different plugs, purchase the spark plug type specified on the VECI label because that information is provided specifically for your engine.*

3 Inspect each of the new plugs for defects. If there are any signs of cracks in the porcelain insulator of a plug, don't use it.

4 Check the electrode gaps of the new plugs. Check the gap by inserting the wire gauge of the proper thickness between the electrodes at the tip of the plug **(see illustration)**. The gap between the electrodes should

23.4a Spark plug manufacturers recommend using a wire-type gauge when checking the gap - if the wire does not slide between the electrodes with a slight drag, adjustment is required

23.4b To change the gap, bend the side electrode only, as indicated by the arrows, and be very careful not to crack or chip the porcelain insulator surrounding the center electrode

1-22 Chapter 1 Tune-up and routine maintenance

23.6 When removing the spark plug wires, pull only on the boot and use a twisting/pulling motion

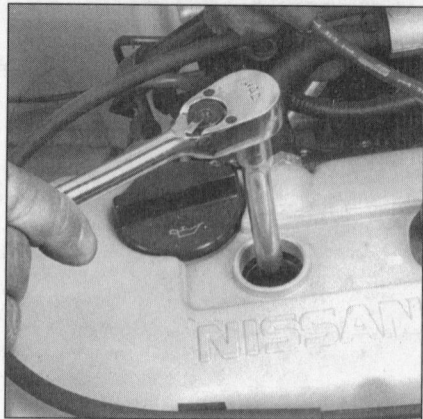

23.8 Use a socket wrench with a long extension to unscrew the spark plugs

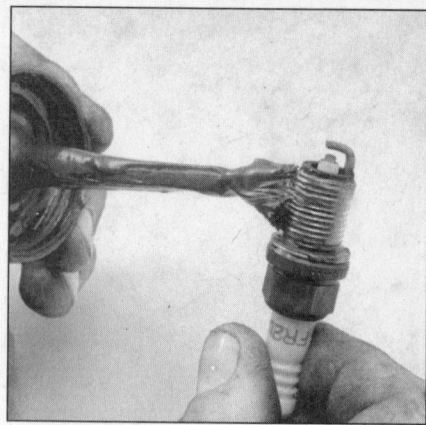

23.10a Apply a coat of anti-seize compound to the spark plug threads

be identical to that listed in this Chapter's Specifications. If the gap is incorrect, use the notched adjuster on the feeler gauge body to bend the curved side electrode slightly **(see illustration)**.

5   If the side electrode is not exactly over the center electrode, use the notched adjuster to align them. **Caution:** *If the gap of a new plug must be adjusted, bend only the base of the ground electrode - do not touch the tip.*

Removal

6   To prevent the possibility of mixing up spark plug wires, work on one spark plug at a time. Remove the wire and boot from one spark plug. Grasp the boot - not the cable - as shown, give it a half twisting motion and pull straight up **(see illustration)**.

7   If compressed air is available, blow any dirt or foreign material away from the spark plug area before proceeding (a common bicycle pump will also work).

8   Remove the spark plug **(see illustration)**.

9   Whether you are replacing the plugs at this time or intend to reuse the old plugs, compare each old spark plug with those shown in the accompanying photos to determine the overall running condition of the engine.

Installation

10 Prior to installation, apply a coat of anti-seize compound to the plug threads. It's often difficult to insert spark plugs into their holes without cross-threading them. To avoid this possibility, fit a short piece of snug-fitting rubber hose over the end of the spark plug **(see illustrations)**. The flexible hose acts as a universal joint to help align the plug with the plug hole. Should the plug begin to cross-thread, the hose will slip on the spark plug, preventing thread damage. Tighten the plug to the torque listed in this Chapter's Specifications.

11 Attach the plug wire to the new spark plug, again using a twisting motion on the boot until it is firmly seated on the end of the spark plug.

12 Follow the above procedure for the remaining spark plugs, replacing them one at a time to prevent mixing up the spark plug wires.

24 Spark plug wire, distributor cap and rotor check and replacement (every 30,000 miles or 24 months)

Refer to illustrations 24.11a, 24.11b, 24.12a and 24.12b

1   The spark plug wires should be checked whenever new spark plugs are installed.

2   Begin this procedure by making a visual check of the spark plug wires while the engine is running. In a darkened garage (make sure there is ventilation) start the engine and observe each plug wire. Be careful not to come into contact with any moving engine parts. If there is a break in the wire, you will see arcing or a small spark at the damaged area. If arcing is noticed, make a note to obtain new wires, then allow the engine to cool and check the distributor cap and rotor.

3   The spark plug wires should be inspected one at a time to prevent mixing up the order, which is essential for proper engine operation. Each original plug wire should be numbered to help identify its location. If the number is illegible, a piece of tape can be marked with the correct number and wrapped around the plug wire.

4   Disconnect the plug wire from the spark plug. A removal tool can be used for this purpose or you can grasp the rubber boot, twist the boot half a turn and pull the boot free. Do not pull on the wire itself.

5   Check inside the boot for corrosion, which will look like a white crusty powder.

6   Push the wire and boot back onto the end of the spark plug. It should fit tightly onto the end of the plug. If it doesn't, remove the wire and use pliers to carefully crimp the metal connector inside the wire boot until the

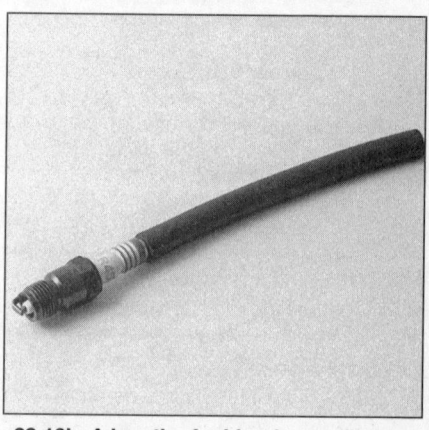

23.10b A length of rubber hose will save time and prevent damaged threads when installing the spark plugs

fit is snug.

7   Using a clean rag, wipe the entire length of the wire to remove built-up dirt and grease. Once the wire is clean, check for burns, cracks and other damage. Do not bend the wire sharply, because the conductor might become damaged.

8   Disconnect the wire from the distributor. Again, pull only on the rubber boot. Check for corrosion and a tight fit. Replace the wire in the distributor.

9   Inspect the remaining spark plug wires, making sure that each one is securely fastened at the distributor and spark plug when the check is complete.

10 If new spark plug wires are required, purchase a set for your specific engine model. Pre-cut wire sets with the boots already installed are available. Remove and replace the wires one at a time to avoid mix-ups in the firing order.

11 Detach the distributor cap by loosening the three cap retaining screws **(see illustration)**. Look inside it for cracks, carbon tracks and worn, burned or loose contacts **(see illustration)**.

12 Loosen the retaining screw and pull the rotor off the distributor shaft and examine it for cracks and carbon tracks **(see illustrations)**. Replace the cap and rotor if any dam-

Chapter 1 Tune-up and routine maintenance

1-23

24.11a Remove the three screws (arrows) and detach the cap and wires from the distributor housing

24.11b Check the distributor cap terminals and center electrode (arrows) for wear and damage as well as the cap itself for carbon tracks (if in doubt about its condition, install a new one)

24.12a Remove the screw and pull the ignition rotor off the shaft

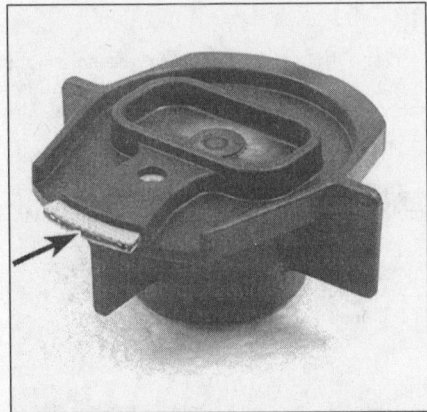

24.12b The ignition rotor should be checked for wear and corrosion of the tip (arrow) and any other cracks or damage (if in doubt about its condition, buy a new one)

age or defects are noted.
13   It is common practice to install a new cap and rotor whenever new spark plug wires are installed, but if you wish to continue using the old cap, check the resistance between the spark plug wires and the cap first. If the indicated resistance is more than the maximum value listed in this Chapter's Specifications, replace the cap and/or wires.
14   When installing a new cap, remove the wires from the old cap one at a time and attach them to the new cap in the exact same location - do not simultaneously remove all the wires from the old cap or firing order mix-ups may occur.

25 Cooling system servicing (draining, flushing and refilling) (every 30,000 miles or 24 months)

Warning: *Do not allow engine coolant (antifreeze) to come in contact with your skin or painted surfaces of the vehicle. Rinse off spills immediately with plenty of water. Antifreeze is highly toxic if ingested. Never leave antifreeze laying around in an open container or in puddles on the floor; children and pets are attracted by it's sweet smell and may drink it. Check with local authorities about disposing of used antifreeze. Many communities have collection centers which will see that antifreeze is disposed of safely.*

1   Periodically, the cooling system should be drained, flushed and refilled to replenish the antifreeze mixture and prevent formation of rust and corrosion, which can impair the performance of the cooling system and cause engine damage. When the cooling system is serviced, all hoses and the radiator cap should be checked and replaced if necessary.

Draining

Refer to illustrations 25.3a, 25.3b, 25.3c, 25.4, 25.5a and 25.5b

2   Apply the parking brake and block the wheels. If the vehicle has just been driven, wait several hours to allow the engine to cool down before beginning this procedure.
3   Move the temperature lever on the heater control panel to HOT. Remove the radiator cap and remove the air bleed plug(s) **(see illustrations)**.
4   Move a large container under the radiator drain to catch the coolant, then open the

25.3a The GA16DE engine has an air bleed plug at the end of the cylinder head, near the distributor (the distributor is removed here for clarity) . . .

25.3b . . . and one at the rear of the engine (arrow)

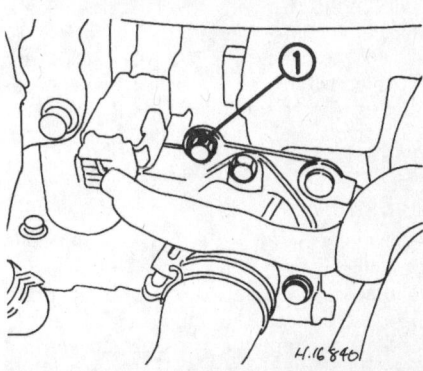

25.3c SR20DE models have a bleeder plug (1) on the thermostat housing - on some models it's on the opposite side of the thermostat cover

25.4 On most models you will have to remove a cover for access to the radiator drain fitting (arrow) located at the bottom of the radiator

25.5a After draining the radiator on the GA16DE engine, be sure to fully drain the cooling system by removing the block drain plug located on the side of the engine block (arrow)

25.5b On SR20DE engines, the block coolant drain (arrow) is accessible from below

drain fitting (a large screwdriver is needed) **(see illustration)**.
5 After the coolant stops flowing out of the radiator, move the container under the engine block drain plug **(see illustrations)**. Loosen the plug and allow the coolant in the block to drain.
6 While the coolant is draining, check the condition of the radiator hoses, heater hoses and clamps (refer to Section 14 if necessary).
7 Replace any damaged clamps or hoses (see Chapter 3).

Flushing

8 Once the system is completely drained, flush the radiator with fresh water from a garden hose until water runs clear at the drain. The flushing action of the water will remove sediments from the radiator but will not remove rust and scale from the engine and cooling tube surfaces.
9 These deposits can be removed by the chemical action of a cleaner. Follow the procedure outlined in the manufacturer's instructions. If the radiator is severely corroded, dam-

aged or leaking, it should be removed (see Chapter 3) and taken to a radiator repair shop.
10 Remove the overflow hose from the coolant recovery reservoir. Drain the reservoir and flush it with clean water, then reconnect the hose.

Refilling

11 Close and tighten the radiator drain. Install and tighten the block drain plug.
12 Make sure the heater temperature control is in the maximum heat position.
13 Slowly add new coolant (a 50/50 mixture of water and antifreeze) to the radiator until it's full. Fill the system until coolant runs out the air bleed hole(s), then install the air bleed bolt and/or cap. Add coolant to the reservoir up to the lower mark.
14 Leave the radiator cap off and run the engine in a well-ventilated area until the thermostat opens (coolant will begin flowing through the radiator and the upper radiator hose will become hot). Race the engine two or three times under no load.
15 Turn the engine off and let it cool. Add more coolant mixture to bring the level back up to the lip on the radiator filler neck.
16 Squeeze the upper radiator hose to expel air, then add more coolant mixture if necessary. Replace the radiator cap.
17 Start the engine, allow it to reach normal operating temperature and check for leaks.

26 Ignition timing check and adjustment (every 30,000 miles or 24 months)

Refer to illustrations 26.6 and 26.9
Note: *After checking the ignition timing, be sure to check the idle speed (see Section 27), since the two procedures are closely related).*
1 All vehicles are equipped with an Emissions Control Information label inside the engine compartment. The label contains important ignition timing specifications and the proper timing procedure for your specific

vehicle. If the information on the emissions label is different from the information included in this Section, follow the procedure on the label.
2 At the specified intervals, or when the distributor has been removed, the ignition timing must be checked and adjusted if necessary. Tools required for this procedure include an inductive pick-up timing light and a tachometer.
3 Make sure the engine is at normal operating temperature.
4 Clear any codes before checking the idle speed (see Chapter 6).
5 With the engine off, connect a timing light in accordance with the manufacturer's instructions. Usually, the light must be connected to the battery and the number one spark plug in some fashion. The number one spark plug wire or terminal should be marked at the distributor; trace it back to the spark plug and attach the timing light lead near the plug.
6 Locate the nail-like timing pointer on the front cover of the engine **(see illustration)**.
7 Locate the notches in the crankshaft pulley or the flywheel. It may be necessary to have an assistant temporarily turn the ignition on and off in short bursts without starting the engine in order to bring the grooves into a position where they can easily be cleaned and marked. **Warning:** *Stay clear of all moving engine components when the engine is turned over in this manner.*
8 Use white soapstone, chalk or paint to mark the proper groove in the pulley.
9 With the engine at normal operating temperature, run it at 1000 rpm for two minutes, then rev it up two or three times, allow it idle for one minute and shut it off. Unplug the throttle position sensor **(see illustration)**.
10 Start the engine and increase it's speed to 2000 to 3000 rpm two or three times, then let it drop to idle speed. Aim the timing light at the pointer, again being careful not to come into contact with moving parts. The mark on the pulley should appear stationary.

Chapter 1 Tune-up and routine maintenance

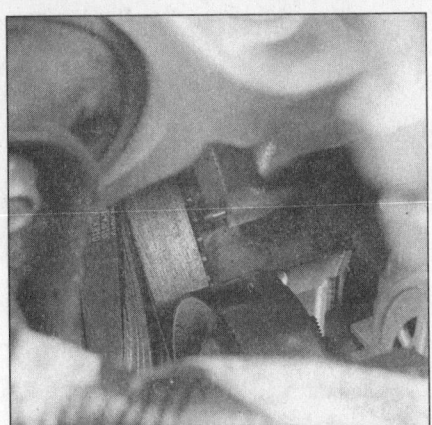

26.6 The timing marks on the crankshaft pulley are graduated in 5-degree increments, starting with -5 degrees (the farthest mark on the left as you face the pulley)

26.9 Disconnect the fuel injection throttle position sensor

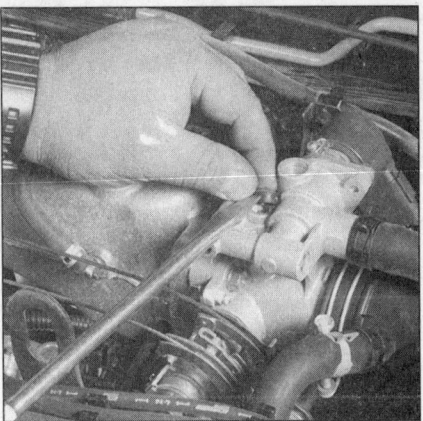

27.9a On GA16DE engines, remove the rubber plug from the adjusting screw opening

If the mark is in alignment with the pointer, the timing is correct. If the mark is not aligned, turn off the engine.
11   Loosen the hold-down bolt or nut at the base of the distributor. Loosen the bolt/nut only slightly, just enough to turn the distributor (see Chapter 5).
12   Now restart the engine and turn the distributor very slowly until the timing marks are aligned.
13   Shut off the engine and tighten the distributor bolt/nut, being careful not to move the distributor.
14   Start the engine and recheck the timing to make sure the marks are still in alignment.
15   Disconnect the timing light and tachometer and reconnect any components which were disconnected for this procedure. Reconnect the throttle position sensor electrical connector.
16   On some models, disconnecting the throttle position sensor may have turned on the CHECK ENGINE light. If this happened, clear the trouble code as discussed in Chapter 6.

27   Engine idle speed check and adjustment (every 30,000 miles or 24 months)

Refer to illustrations 27.9a, 27.9b and 27.9c
Note: For best results, check the ignition timing (Section 26) before beginning this procedure.
1   Engine idle speed is the speed at which the engine operates when no accelerator pedal pressure is applied, as when stopped at a traffic light. This speed is critical to the performance of the engine itself, as well as many engine subsystems.
2   Set the parking brake firmly and block the wheels to prevent the vehicle from rolling. Put the transaxle in Neutral. Unplug the engine fan electrical connector(s). If the fan(s) should come on during the idle adjustment procedure, idle speed is affected.
3   Connect a hand-held tachometer. **Caution:** *Don't allow the tachometer to touch ground or damage to the igniter and/or the ignition coil may occur.* **Note:** *Some tachometers may not be compatible with this ignition system. It is recommended that you consult the manufacturer of the tool.*
4   Start the engine and allow it to reach normal operating temperature.
5   Check, and adjust if necessary, the ignition timing (see Section 26).
6   Clear any codes before checking the idle speed (see Chapter 6).
7   With the engine at normal operating temperature, run it at 1000 rpm for two minutes, then rev it up two or three times, allow it idle for one minute and shut it off. Unplug the throttle position sensor **(see illustration 26.9)**.
8   Start the engine and increase the speed to 2000 to 3000 rpm two or three times and let it drop to idle speed. Check the engine idle speed on the tachometer and compare it to those listed in the Specifications Section at the beginning of this Chapter or the Vehicle Emission Control Information label in the engine compartment. If there is a difference between the Specifications Section and the VECI label, always assume that the information on the label is correct.
9   If the idle speed is too low or too high, use a screwdriver to turn the idle speed adjusting screw until the specified curb idle speed is obtained **(see illustrations)**.
10   Turn off the engine, disconnect the tachometer and reconnect any components which were disconnected.

27.9b Use a screwdriver to turn the adjusting screw - note that the idle speed is quite sensitive to even the slightest turn of this screw (turning it counterclockwise increases the engine speed, turning it clockwise decreases it) (GA16DE engine shown)

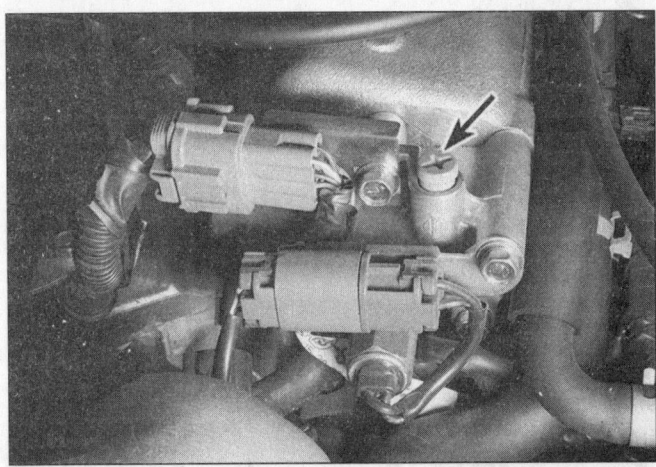

27.9c Idle speed screw location on the SR20DE engine

28 Evaporative emissions control (EVAP) system check (every 30,000 miles or 24 months)

Refer to illustration 28.2

1 The function of the evaporative emissions control system is to draw fuel vapors from the gas tank and fuel system, store them in a charcoal canister and then burn them during normal engine operation.
2 The most common symptom of a fault in the evaporative emissions system is a strong fuel odor in the engine compartment (1997 and earlier models) or at the rear near the fuel tank (1998 and later models). If a fuel odor is detected, inspect the charcoal canister, located at the front of the engine compartment (1997 and earlier models) or at the rear near the fuel tank (1998 and later models). Check the canister and all hoses for damage and deterioration **(see illustration)**.
3 On 1998 and later models, problems with the fuel tank cap will cause a trouble code to set (see Chapter 6) and the CHECK ENGINE light to illuminate.
4 The evaporative emissions control system is explained in more detail in Chapter 6.

29 Exhaust system check (every 30,000 miles or 24 months)

Refer to illustration 29.4

1 With the engine cold (at least three hours after the vehicle has been driven), check the complete exhaust system from its starting point at the engine to the end of the tailpipe. This should be done on a hoist where unrestricted access is available.
2 Check the pipes and connections for evidence of leaks, severe corrosion or damage. Make sure that all brackets and hangers are in good condition and tight.
3 At the same time, inspect the underside of the body for holes, corrosion, open seams, etc. which may allow exhaust gases to enter the passenger compartment. Seal all body openings with silicone or body putty.
4 Rattles and other noises can often be traced to the exhaust system, especially the

28.2 Check the charcoal canister for damage and the hose connections (arrows) for cracks and damage

mounts and hangers **(see illustration)**. Try to move the pipes, muffler and catalytic converter. If the components can come in contact with the body or suspension parts, secure the exhaust system with new mounts.
5 Check the running condition of the engine by inspecting inside the end of the tailpipe. The exhaust deposits here are an indication of engine state-of-tune. If the pipe is black and sooty or coated with white deposits, the engine is in need of a tune-up, including a thorough fuel system inspection.

30 Automatic transaxle fluid change (every 30,000 miles or 24 months)

Refer to illustration 30.7

1 At the specified time intervals, the automatic transaxle fluid should be drained and replaced.
2 Before beginning work, purchase the specified transmission fluid (see *Recommended fluids and lubricants* at the front of this Chapter).
3 Other tools necessary for this job include jackstands to support the vehicle in a raised position, a wrench, a drain pan capable of holding at least four quarts, newspapers and clean rags.
4 The fluid should be drained immediately

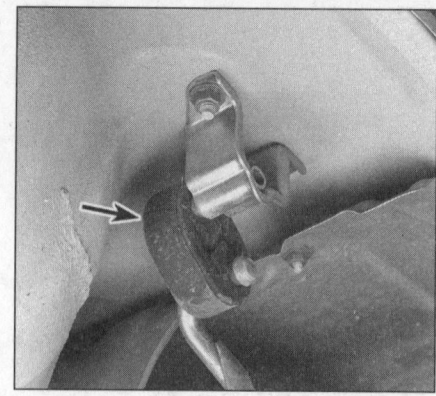

29.4 Check the exhaust system hangers (arrow) for damage and cracks

after the vehicle has been driven. Hot fluid is more effective than cold fluid at removing built up sediment. **Warning:** *Fluid temperature can exceed 350-degrees F in a hot transaxle. Wear protective gloves.*
5 After the vehicle has been driven to warm up the fluid, raise it and place it on jackstands for access to the transaxle drain plugs.
6 Move the necessary equipment under the vehicle, being careful not to touch any of the hot exhaust components.
7 Place the drain pan under the drain plug in the transaxle housing or fluid pan and remove the drain plug **(see illustration)**. Be sure the drain pan is in position, as fluid will come out with some force. Once the fluid is drained, reinstall the drain plug securely.
8 Lower the vehicle.
9 With the engine off, add new fluid to the transaxle through the dipstick tube (see *Recommended fluids and lubricants* for the recommended fluid type and capacity). Use a funnel to prevent spills. It is best to add a little fluid at a time, continually checking the level with the dipstick (see Section 7). Allow the fluid time to drain into the pan.
10 Start the engine and shift the selector into all positions from P through L, then shift into P and apply the parking brake.
11 With the engine idling, check the fluid level. Add fluid up to the Cool level on the dipstick.

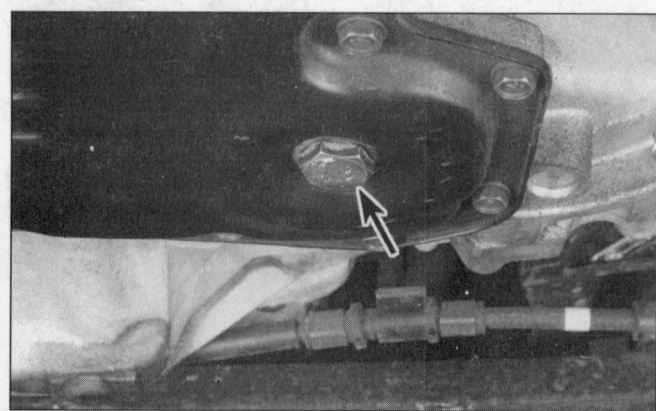

30.7 The automatic transaxle drain plug (arrow) is located on the bottom of the fluid pan

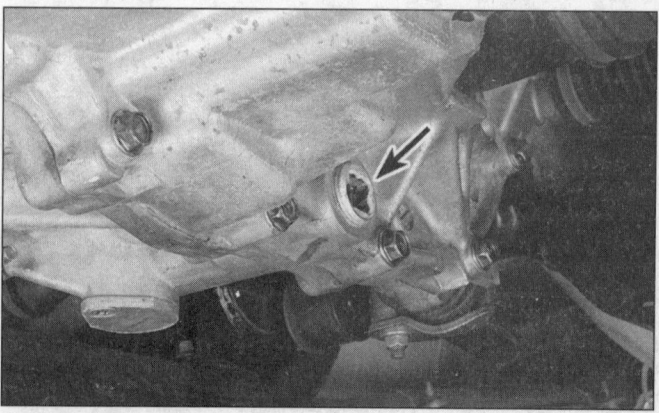

31.1 Use a ratchet or breaker bar to remove the manual transaxle drain plug

Chapter 1 Tune-up and routine maintenance   1-27

32.2 Pull the PVC valve out of the valve cover and verify that vacuum is felt at the valve with the engine running

33.6a When the no. 1 piston is at TDC on the compression stroke, the valve clearances for the no. 1 and no. 3 cylinder exhaust valves and the no. 1 and no. 2 cylinder intake valves can be measured

31 Manual transaxle lubricant change (every 30,000 miles or 24 months)

Refer to illustration 31.1

1   Remove the drain plug and drain the fluid **(see illustration)**.
2   Reinstall the drain plug securely.
3   Remove the speedometer drive pinion and add new fluid until it reaches the proper level (see Section 19). See *Recommended lubricants and fluids* for the specified lubricant type.
4   Install the speedometer drive pinion and tighten the bolt securely.

32 Positive Crankcase Ventilation (PCV) valve and hose check and replacement (every 30,000 miles or 24 months)

Refer to illustration 32.2

1   The PCV valve is located in the valve cover.

2   Grasp the PCV valve and pull it out of the valve cover **(see illustration)**.
3   With the engine idling at normal operating temperature, place your finger over the end of the valve. If there's no vacuum at the valve, check for a plugged hose or valve. Replace any plugged or deteriorated hoses.
4   When purchasing a replacement PCV valve, make sure it's for your particular vehicle and engine size. Compare the old valve with the new one to make sure they're the same.

33 Valve clearance check and adjustment (GA16DE engine only) (every 60,000 miles or 48 months)

Refer to illustrations 33.6a, 33.6b, 33.7, 33.9a, 33.9b, 33.9c and 33.10

Note: *The manufacturer recommends adjusting the valve clearance at the specified interval only if the valve train is making excessive noise. The following procedure requires the use of special valve lifter tools. The tools are available from specialty tool manufacturers*

and auto parts stores. It is impossible to perform this task without them.

1   Disconnect the cable from the negative terminal of the battery.
2   Remove the valve cover (see Chapter 2, Part A).
3   On manual transaxle vehicles set the parking brake and place the transaxle in the neutral position.
4   Remove the spark plugs (see Section 23).
5   Position the number 1 piston at TDC on the compression stroke and align the timing marks. Refer to the Top Dead Center Section in Chapter 2, Part A.
6   Measure the clearance of the indicated valves with a feeler gauge **(see illustrations)**. Record each measurement and compare your measurements with the desired valve clearance found in this Chapter's Specifications. Note which are out of specification, this data will be used later to determine the required replacement shims.
7   Turn the crankshaft one complete revolution and realign the timing marks. Measure and record the clearances of the remaining valves **(see illustration)**.

33.6b Measure the clearance for each valve with a feeler gauge of the specified thickness - if the clearance is correct, you should feel a slight drag as you pull it out

33.7 When the no. 4 piston is at TDC on the compression stroke, the valve clearances for the no. 2 and no. 4 exhaust valves and the no. 3 and no. 4 intake valves can be measured

33.9a Install the valve lifter tool as shown and squeeze the handles together to lower the valve lifter so the shim can be removed

33.9b With the small tool wedged between the lifter and the camshaft pry the shim up with a small screwdriver at the notch

33.9c Remove the shim with a small screwdriver, a pair of tweezers or a magnet

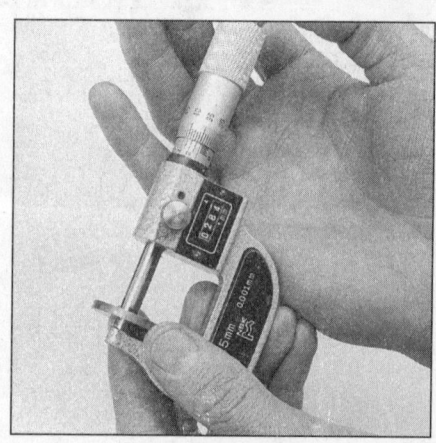

32.10 Measure the shim thickness with a micrometer

8   After the clearance of all the valves have been measured, rotate the crankshaft pulley until the camshaft lobe above the first valve which you intend to adjust is pointing up, away from the lifter.

9   Align the notch in the valve lifter with the notch in the cylinder head casting. Place the special valve lifter tool in position as shown, with the upper jaw over the camshaft, next to the lobe and the lower jaw on top of the shim **(see illustrations)**. Depress the valve lifter by squeezing the handles of the valve lifter tool together and rotating the tool away from the camshaft. Insert the small tool between the edge of the lifter and the camshaft and release the lifter. Remove the adjusting shim with a small screwdriver and a magnet or a pair of tweezers **(see illustration)**.

10   Measure the thickness of the shim with a micrometer **(see illustration)**. To calculate the correct thickness of a replacement shim that will place the valve clearance within the specified value, use the following formula:

Intake side: $N = R + (M - 0.0146\text{-inch}$
$[0.37 \text{ mm}])$

Exhaust side: $N = R + (M - 0.0157\text{-inch}$
$[0.40 \text{ mm}])$

R = thickness of the old shim
M = valve clearance measured
N = thickness of the new shim

11   Select a shim with a thickness as close as possible to the valve clearance calculated. Shims, which are available in 50 sizes in increments of 0.0008-inch (0.02 mm). Available shims range in size from 0.0787 inch (2.00 mm) to 0.1055 inch (2.98 mm) (see accompanying chart below). **Note:** *Through careful analysis of the shim sizes needed to bring all the out-of-specification valve clearances within specification, it is often possible to simply move a shim that has to come out anyway to another valve lifter requiring a shim of that particular size, thereby reducing the number of new shims that must be purchased.*

Valve adjusting shim thickness chart

New valve shims are identified by a number stamped on the face of the shim which corresponds with the millimeter size. Example: the number 198 = 1.98 mm (0.0780 inch).

2.00 mm (0.0787 inch)
2.02 mm (0.0795 inch)
2.04 mm (0.0803 inch)
2.06 mm (0.0811 inch)
2.08 mm (0.0819 inch)
2.10 mm (0.0827 inch)
2.12 mm (0.0835 inch)
2.14 mm (0.0843 inch)
2.16 mm (0.0850 inch)
2.18 mm (0.0858 inch)
2.20 mm (0.0866 inch)
2.22 mm (0.0874 inch)
2.24 mm (0.0882 inch)
2.26 mm (0.0890 inch)
2.28 mm (0.0898 inch)
2.30 mm (0.0906 inch)
2.32 mm (0.0913 inch)
2.34 mm (0.0921 inch)
2.36 mm (0.0929 inch)
2.38 mm (0.0937 inch)
2.40 mm (0.0945 inch)
2.42 mm (0.0953 inch)
2.44 mm (0.0961 inch)
2.46 mm (0.0969 inch)
2.48 mm (0.0976 inch)
2.50 mm (0.0984 inch)
2.52 mm (0.0992 inch)
2.54 mm (0.1000 inch)
2.56 mm (0.1008 inch)
2.58 mm (0.1016 inch)
2.60 mm (0.1024 inch)
2.62 mm (0.1031 inch)
2.64 mm (0.1039 inch)
2.66 mm (0.1047 inch)
2.68 mm (0.1055 inch)
2.70 mm (0.1063 inch)
2.72 mm (0.1071 inch)
2.74 mm (0.1079 inch)
2.76 mm (0.1087 inch)
2.78 mm (0.1094 inch)
2.80 mm (0.1102 inch)
2.82 mm (0.1110 inch)
2.84 mm (0.1118 inch)
2.86 mm (0.1126 inch)
2.88 mm (0.1134 inch)
2.90 mm (0.1142 inch)
2.92 mm (0.1150 inch)
2.94 mm (0.1157 inch)
2.96 mm (0.1165 inch)
2.98 mm (0.1173 inch)

Chapter 2 Part A Engine

Contents

	Section
Camshafts - removal, inspection and installation	9
Crankshaft front oil seal - replacement	8
Cylinder compression check	See Chapter 2B
Cylinder head - removal and installation	11
Drivebelt check, adjustment and replacement	See Chapter 1
Engine mounts - check and replacement	16
Engine oil and filter change	See Chapter 1
Engine overhaul - general information	See Chapter 2B
Engine - removal and installation	See Chapter 2B
Exhaust manifold - removal and installation	6
Flywheel/driveplate - removal and installation	14
General information	1

	Section
Intake manifold - removal and installation	5
Oil pan - removal and installation	12
Oil pump - removal, inspection and installation	13
Rear main oil seal - replacement	15
Repair operations possible with the engine in the vehicle	2
Spark plug replacement	See Chapter 1
Timing chain - removal, inspection and installation	7
Top Dead Center (TDC) for number one piston - locating	3
Valves - servicing	See Chapter 2B
Valve cover - removal and installation	4
Valve springs, retainers and seals - replacement	10
Water pump - removal and installation	See Chapter 3

Specifications

General

Displacement
- GA16DE .... 1597cc (97.45 cubic inches)
- SR20DE .... 1998cc (121.92 cubic inches)

Cylinder numbers (drivebelt end-to-transaxle end) 1-2-3-4
Firing order 1-3-4-2

Warpage limits

Cylinder head-to-block surface .... 0.004 inch

Camshaft
- Thrust clearance (endplay)
  - Standard
    - SR20DE .... 0.055 to 0.139 inches
    - GA16DE .... 0.115 to 0.188 inches
  - Limit .... 0.20 inch
- Camshaft journal diameter
  - SR20DE .... 1.0998 to 1.1006 inches
  - GA16DE
    - No. 1 .... 1.0998 to 1.1006 inches
    - No. 2 through 5 .... 0.9423 to 0.9431 inch
- Camshaft bearing inside diameter
  - SR20DE .... 1.1024 to 1.1032 inches
  - GA16DE
    - No. 1 .... 1.1024 to 1.1032 inches
    - No. 2 through 5 .... 0.9449 to 0.9457 inch
- Bearing oil clearance
  - Standard .... 0.0018 to 0.0034 inch
  - Service limit .... 0.0059 inch
- Runout limit .... 0.004 inch maximum
- Intake lobe height
  - SR20DE .... 1.4783 to 1.4858 inches
  - GA16DE .... 1.5988 to 1.6063 inches

Cylinder location and distributor rotation diagram

Warpage limits (continued)

Exhaust lobe height
- SR20DE .... 1.4929 to 1.5004 inches
- GA16DE .... 1.5713 to 1.5787 inches
- Cam lobe wear limit .... 0.0079 inch

Oil pump
- SR20DE
  - Body-to-outer gear clearance .... 0.0045 to 0.0079 inch
  - Inner gear-to-outer gear tip clearance .... less than 0.0071 inch
  - Body-to-inner gear clearance .... 0.0020 to 0.0035 inch
  - Body-to-outer gear clearance .... 0.0020 to 0.0043 inch
  - Inner gear to brazed part of housing clearance ... 0.0018 to 0.0036 inch
- GA16DE
  - Body-to-outer gear clearance .... 0.0043 to 0.0079 inch
  - Inner gear-to-crescent clearance .... 0.0085 to 0.0129 inch
  - Outer gear-to-crescent clearance .... 0.0083 to 0.0126 inch
  - Body-to-inner gear clearance .... 0.0020 to 0.0035 inch
  - Body-to-outer gear clearance .... 0.0020 to 0.0043 inch
  - Inner gear to brazed part of housing clearance ... 0.0018 to 0.0036 inch

Torque specifications   Ft-lbs (unless otherwise indicated)

Intake manifold bolts/nuts
- SR20DE .... 13 to 15
- GA16DE .... 12 to 15

Exhaust manifold-to-cylinder head bolts/nuts
- SR20DE .... 27 to 35
- GA16DE .... 19 to 21

Crankshaft pulley-to-crankshaft bolt
- GA16DE .... 98 to 112
- SR20DE .... 105 to 112

Flywheel bolts .... 61 to 69
Driveplate bolts .... 69 to 76

Cylinder head bolts
- GA16DE
  - Step 1 (bolts 1 through 10) 22
  - Step 2 (bolts 1 through 10) 43
  - Step 3 (bolts 1 through 10) .... Loosen all bolts completely
  - Step 4 (bolts 1 through 10) 22
  - Step 5 (bolts 1 through 10) .... 40 to 47
  - Step 6 (bolts 11 through 15) .... 56 to 74 in-lbs
- SR20DE
  - Step 1 29
  - Step 2 58
  - Step 3 .... Loosen all bolts completely
  - Step 4 .... 25 to 33
  - Step 5 .... Tighten an additional 90 to 95-degrees
  - Step 6 .... Tighten an additional 90 to 95-degrees

Camshaft bearing cap bolts
- SR20DE
  - Step 1 .... 52 in-lbs
  - Step 2 .... 86 to 104 in-lbs
  - Step 3 (bolts 8 and 9 only) .... 13 to 19
- GA16DE
  - Step 1 (all bolts) .... 52 in-lbs
  - Step 2 (bolts 1 through 14) .... 84 to 108 in-lbs
  - Step 3 (bolt 15) .... 56 to 74 in-lbs

Camshaft sprocket bolt
- GA16DE .... 80 to 87
- SR20DE .... 101 to 116

Idler sprocket bolt (GA16DE) .... 32 to 58
Engine mount through-bolts .... 33 to 40

Oil pump-to-front cover bolts
- GA16DE .... 55 to 73 in-lbs
- SR20DE .... 56 to 66 in-lbs

Oil pump-to-front cover screws .... 32 to 44 in-lbs
Oil pick-up (strainer) nuts/bolts .... 55 to 73 in-lbs

Oil pan-to-block bolts
- GA16DE .... 55 to 73 in-lbs

Chapter 2 Part A Engine

Torque specifications (continued) **Ft-lbs** (unless otherwise indicated)

SR20DE
    Aluminum section-to-block
        Large bolts .... 12 to 14
        Two rear small nuts .... 56 to 66 in-lbs
    Steel pan-to-aluminum section .... 56 to 66 in-lbs
Valve cover bolts/nuts
    GA16DE .... 17 to 34 in-lbs
    SR20DE .... 69 to 87 in-lbs

1 General information

This Part of Chapter 2 is devoted to in-vehicle repair procedures. All information concerning engine removal and installation and engine block and cylinder head overhaul can be found in Part B of this Chapter.

The following repair procedures are based on the assumption that the engine is installed in the vehicle. If the engine has been removed from the vehicle and mounted on a stand, many of the steps outlined in this Part of Chapter 2 will not apply.

The Specifications included in this Part of Chapter 2 apply only to the procedures contained in this Part. Part B of Chapter 2 contains the Specifications necessary for cylinder head and engine block rebuilding.

The standard engine type in the Sentra is the GA16DE (1597cc). The SR20DE (1998cc) is also available. Both have dual overhead camshafts (DOHC) and four-valves-per-cylinder. The main difference between them is that the GA16DE has two timing chains, and the SR20DE has only one.

2 Repair operations possible with the engine in the vehicle

Warning: *These models are equipped with airbags. The airbag is armed and can deploy (inflate) any time the battery is connected. To prevent accidental deployment (and possible injury), disconnect the negative and then the positive battery cables whenever working near airbag components. After the battery is disconnected, wait at least ten minutes before beginning work (the system has a back-up capacitor that must fully discharge). This procedure must be done whenever working in the vicinity of the impact sensors, steering column or the instrument panel. The airbag wiring harnesses are identified with yellow insulation. Do not use test equipment on any of these wires or tamper with them in any way. See Chapter 12 for more information.*

Many major repair operations can be accomplished without removing the engine from the vehicle.

Clean the engine compartment and the exterior of the engine with some type of degreaser before any work is done. It will make the job easier and help keep dirt out of the internal areas of the engine.

Depending on the components involved, it may be helpful to remove the hood to improve access to the engine as repairs are performed (see Chapter 11 if necessary). Cover the fenders to prevent damage to the paint. Special pads are available, but an old bedspread or blanket will also work.

If vacuum, exhaust, oil or coolant leaks develop, indicating a need for gasket or seal replacement, the repairs can generally be made with the engine in the vehicle. The intake and exhaust manifold gaskets, oil pan gasket, crankshaft oil seals and cylinder head gasket are all accessible with the engine in place.

Exterior engine components, such as the intake and exhaust manifolds, the oil pan, the oil pump, the water pump, the starter motor, the alternator, the distributor and the fuel system components can be removed for repair with the engine in place.

Since the cylinder head can be removed without pulling the engine, camshaft and valve component servicing can also be accomplished with the engine in the vehicle. Replacement of the timing chain and sprockets is also possible with the engine in the vehicle.

In extreme cases caused by a lack of necessary equipment, repair or replacement of piston rings, pistons, connecting rods and rod bearings is possible with the engine in the vehicle. However, this practice is not recommended because of the cleaning and preparation work that must be done to the components involved.

3 Top Dead Center (TDC) for number one piston - locating

Refer to illustrations 3.4 and 3.8
Note: *The following procedure is based on the assumption that the distributor is correctly installed. If you are trying to locate TDC to install the distributor correctly, piston position must be determined by feeling for compression at the number one spark plug hole, then aligning the ignition timing marks as described in Step 8.*

3.4 The crankshaft can be turned with a ratchet and socket on the crankshaft pulley bolt. It is necessary on some models to remove the plastic splash shield for access

1 Top Dead Center (TDC) is the highest point in the cylinder that each piston reaches as it travels up-and-down when the crankshaft turns. Each piston reaches TDC on the compression stroke and again on the exhaust stroke, but TDC generally refers to piston position on the compression stroke.
2 Positioning the number one piston at TDC is an essential part of many procedures, such as camshaft and timing chain removal and distributor removal.
3 Before beginning this procedure, be sure to place the transmission in Neutral and apply the parking brake or block the rear wheels. Also, disable the ignition system by disconnecting the primary (low voltage) electrical connectors at the distributor (see Chapter 5). Remove the spark plugs (see Chapter 1).
4 In order to bring any piston to TDC, the crankshaft must be turned using one of the methods outlined below. When looking at the front of the engine (timing chain end), normal crankshaft rotation is clockwise.
    a) *The preferred method is to turn the crankshaft with a socket and ratchet attached to the bolt threaded into the front of the crankshaft* **(see illustration).**

Chapter 2 Part A Engine

3.8 Timing mark location (typical)

4.2 Remove the PCV hoses, remove the throttle cable from its brackets and remove the retaining screws (arrows) (GA16DE engine)

4.7a On SR20DE engines, apply RTV sealant to the cut-out areas (arrows)

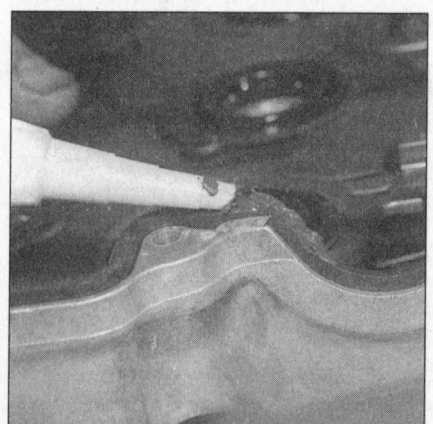

4.7b On GA16DE engines, apply RTV sealant to the exhaust camshaft bearing cut-out

4.7c Install new spark plug seals (GA16DE engine shown)

b) *A remote starter switch, which may save some time, can also be used. Follow the instructions included with the switch. Once the piston is close to TDC, use a socket and ratchet as described in the previous paragraph.*

c) *If an assistant is available to turn the ignition switch to the Start position in short bursts, you can get the piston close to TDC without a remote starter switch. Make sure your assistant is out of the vehicle, away from the ignition switch, then use a socket and ratchet as described in Paragraph a) to complete the procedure.*

5 Note the position of the terminal for the number one spark plug wire on the distributor cap. If the terminal isn't marked, follow the plug wire from the number one cylinder spark plug to the cap.
6 Use a felt-tip pen or chalk to make a mark on the distributor body directly under the number one terminal.
7 Detach the cap from the distributor and set it aside (see Chapter 1 if necessary).
8 Turn the crankshaft (see Step 4) until the notch in the crankshaft pulley is aligned with the 0 on the timing plate (located at the front of the engine) **(see illustration)**.
9 Look at the distributor rotor - it should be pointing directly at the mark you made on the distributor body.
10 If the rotor is 180-degrees off, the number one piston is at TDC on the exhaust stroke.
11 To get the piston to TDC on the compression stroke, turn the crankshaft one complete revolution (360-degrees) clockwise. The rotor should now be pointing at the mark on the distributor. When the rotor is pointing at the number one spark plug wire terminal in the distributor cap and the ignition timing marks are aligned, the number one piston is at TDC on the compression stroke. **Note:** *If it's impossible to align the ignition timing marks when the rotor is pointing at the mark on the distributor body, the timing chain may have jumped the teeth on the sprockets or may have been installed incorrectly.*
12 After the number one piston has been positioned at TDC on the compression stroke, TDC for any of the remaining pistons can be located by turning the crankshaft and following the firing order (180-degrees at a time). Mark the remaining spark plug wire terminals on the distributor body just like you did for the number one terminal, then number the marks to correspond with the cylinder numbers. Also make a mark on the crankshaft pulley 180-degrees from the TDC mark. As you turn the crankshaft, the rotor will also turn. When it's pointing directly at one of the marks on the distributor and one of the marks on the crankshaft pulley, the piston for that particular cylinder is at TDC on the compression stroke.

4 **Valve cover - removal and installation**

Removal

Refer to illustrations 4.2, 4.7a, 4.7b, 4.7c, 4.9a and 4.9b

1 Disconnect the battery cable from the negative battery terminal.
2 Remove the spark plug wires from the spark plugs, remove the PCV hoses, and remove the throttle cable from its brackets and position it out of the way **(see illustration)**.

Chapter 2 Part A Engine

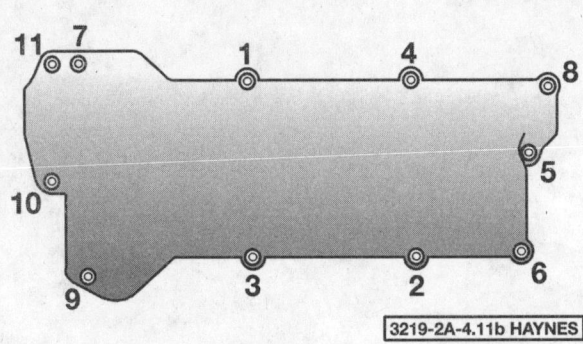

4.9a Valve cover fastener tightening sequence for GA16DE engines

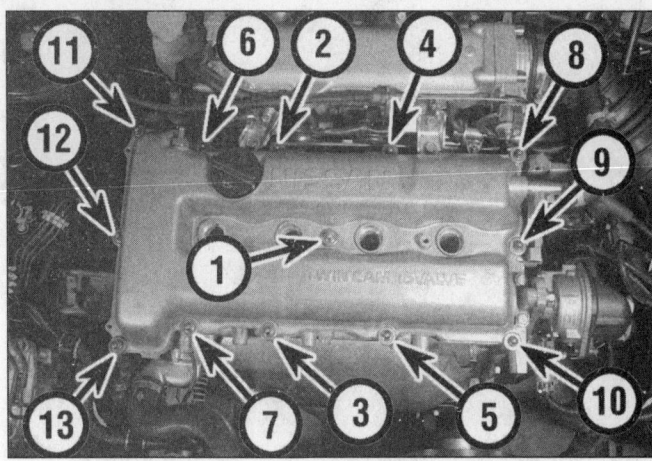

4.9b Valve cover fastener tightening sequence for SR20DE engines

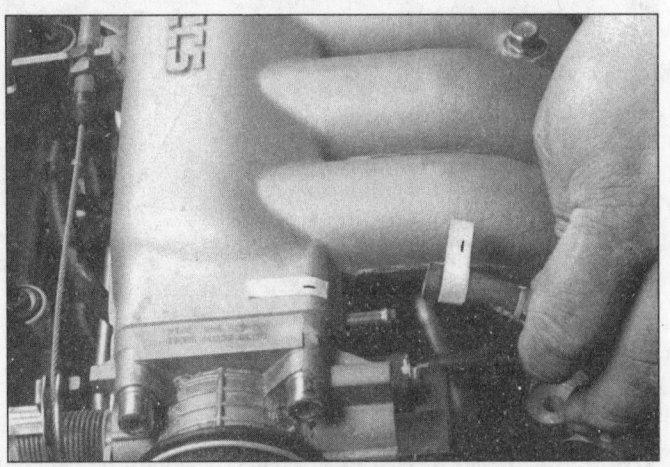

5.3 The various hoses should be marked to ensure correct reinstallation

5.6 Remove the air intake plenum, giving better access to wires and hoses underneath

3 Detach the spark plug wires from the clips on the valve cover and any other interfering components.

4 Remove the fasteners around the cover's perimeter. If the cover is stuck to the head, bump the end with a wood block and a hammer to jar it loose. If that doesn't work, try to slip a flexible putty knife between the head and cover to break the seal. **Caution:** *Don't pry at the cover or housing-to-head joint or damage to the sealing surfaces may occur, leading to oil leaks after the cover is reinstalled.*

5 Remove the gasket from the valve cover.

Installation

6 The mating surfaces of the valve cover and cylinder head must be clean when the cover is installed. Use a gasket scraper to remove all traces of sealant and old gasket material, then clean the mating surfaces with lacquer thinner or acetone. If there's residue or oil on the mating surfaces when the cover is installed, oil leaks may develop.

7 On SR20DE engines, apply silicone sealant to the new valve cover gasket and the cylinder head around the cutout areas **(see illustrations)**. Install new spark plug seals, if used **(see illustration)**.

8 Install the valve cover with its new gasket.

9 Tighten the fasteners to the torque listed in this Chapter's Specifications, following the recommended tightening sequence **(see illustrations)**.

5 Intake manifold - removal and installation

Warning: *The engine must be completely cool before beginning this procedure.*

Removal

Refer to illustrations 5.3 and 5.6

1 Relieve the fuel system pressure (see Chapter 4). Disconnect the cable from the negative battery terminal.

2 Drain the cooling system (see Chapter 1).

3 Label and detach all wire harnesses, control cables, coolant and vacuum hoses connected to the intake manifold **(see illustration)**.

4 Remove the fuel rail and injectors (see Chapter 4).

5 Unbolt the intake manifold from the brace under it (at the timing chain end of the manifold). There are three supports on the SR20DE engine.

6 Remove the mounting nuts/bolts, then detach the manifold from the engine. **Note:** *The intake manifold is a two-piece design. It is easier to remove the manifold in sections, rather than as one piece. Remove the plenum first, giving greater access to hose and pipes connected to the lower section* **(see illustration)**.

Installation

7 Use a scraper to remove all traces of old gasket material and sealant from the manifold and cylinder head, then clean the mating surfaces with lacquer thinner or acetone.

8 Install a new gasket, then position the manifold on the head and install the nuts/bolts.

9 Tighten the nuts/bolts in steps to the

Chapter 2 Part A Engine

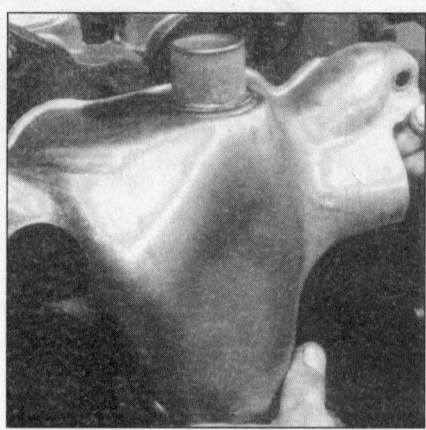

6.3 Remove the exhaust manifold heat shield

6.8 Remove the nuts (arrows) retaining the exhaust pipe to the exhaust manifold

6.10 Remove the nuts and/or bolts attaching the exhaust manifold to the cylinder head and remove the manifold

7.18 Remove the cam sprocket cover

7.21 Remove the lower chain tensioner (arrow)

torque listed in this Chapter's Specifications. Work from the center out towards the ends to avoid warping the manifold.
10 Install the remaining parts in the reverse order of removal. Refill the cooling system (see Chapter 1).
11 Before starting the engine, check the throttle linkage for smooth operation.
12 Run the engine and check for coolant and vacuum leaks.
13 Road test the vehicle and check for proper operation of all accessories.

6 Exhaust manifold - removal and installation

Warning: *The engine must be completely cool before beginning this procedure.*

Removal

Refer to illustrations 6.3, 6.8 and 6.10

1 Disconnect the cable from the negative battery terminal.
2 Remove the oxygen sensor (see Chapter 6).
3 Remove the heat shield from the manifold **(see illustration)**.
4 Block the rear wheels and set the parking brake.
5 Support the front of the vehicle securely on jackstands (see Chapter 1).
6 Remove the engine splash shields.
7 Apply penetrating oil to the exhaust manifold-to-head and manifold-to-exhaust pipe mounting nuts/bolts.
8 From beneath the vehicle, disconnect the exhaust pipe from the exhaust manifold **(see illustration)**.
9 Unbolt the exhaust pipe brace at the engine mount, if equipped.
10 Remove the manifold-to-head nuts/bolts and detach the manifold and gaskets **(see illustration)**.

Installation

11 Use a scraper to remove all traces of old gasket material and carbon deposits from the manifold and cylinder head mating surfaces.
12 Position the new exhaust manifold gaskets over the cylinder head studs.
13 Install the manifold and thread the mounting nuts/bolts into place.
14 Working from the center out, tighten the nuts/bolts to the torque listed in this Chapter's Specifications in three or four equal steps.
15 Reinstall the remaining parts in the reverse order of removal.
16 Run the engine and check for exhaust leaks.

7 Timing chain - removal, inspection and installation

Caution: *The engine must be completely cool before beginning this procedure.*

GA16DE engine

Refer to illustrations 7.18, 7.21, 7.22a, 7.22b, 7.24a, 7.24b, 7.25, 7.28, 7.30a, 7.30b, 7.33, 7.38, 7.39, 7.41, 7.42 and 7.43

Removal

1 Disconnect the cable from the negative battery terminal.
2 Block the rear wheels and set the parking brake.
3 Loosen the lug nuts on the right front wheel and raise the vehicle. Support the front of the vehicle securely on jackstands.

Chapter 2 Part A Engine   2A-7

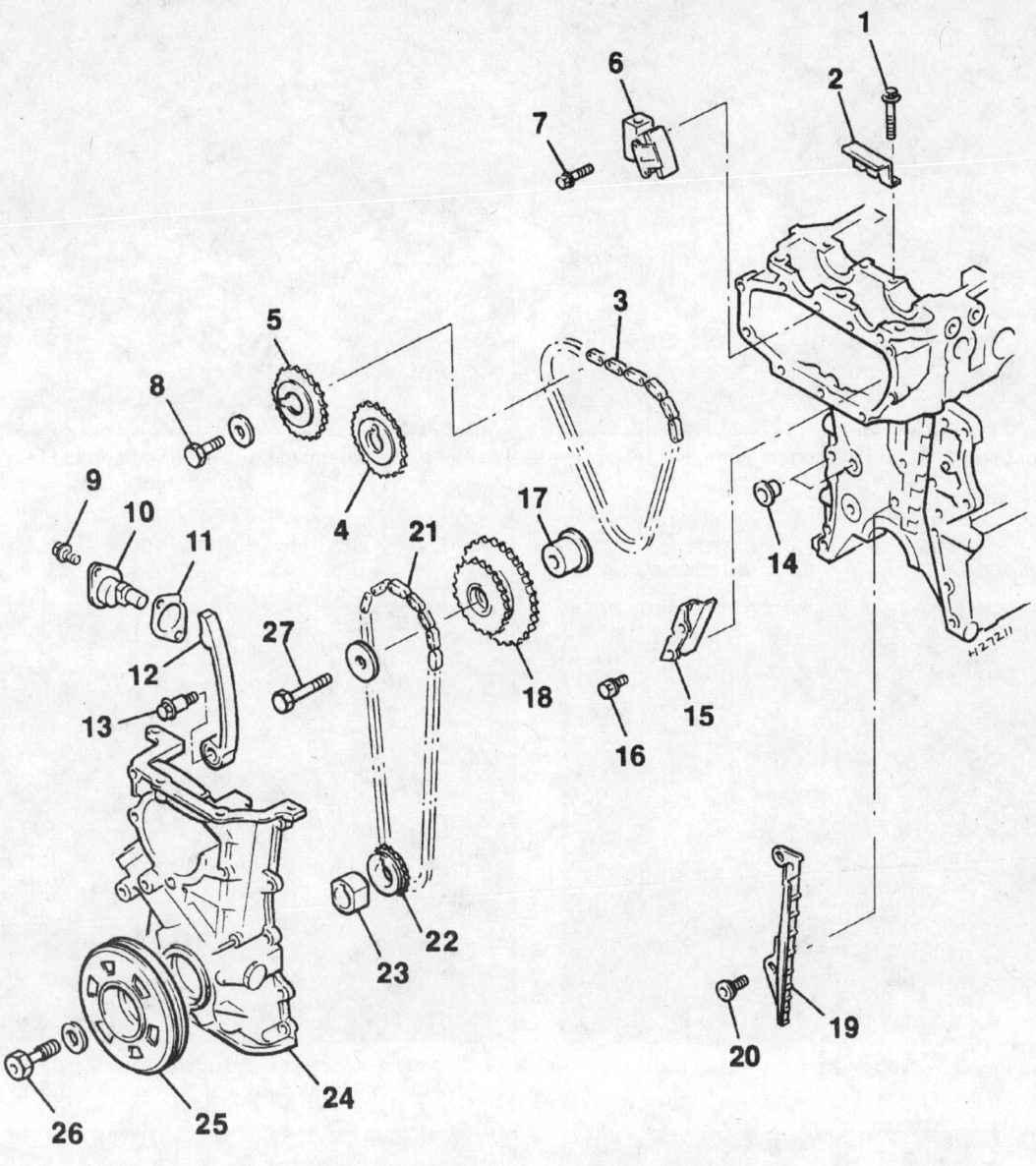

7.22a Timing chain components (GA16DE engine)

1	Bolt	7	Bolt	15	Upper chain front	21	Lower timing chain
2	Guide	8	Bolt		guide (early models)	22	Crankshaft sprocket
3	Upper timing chain	9	Bolt	16	Bolt	23	Oil pump drive spacer
4	Exhaust camshaft	10	Lower chain tensioner	17	Idler sprocket shaft	24	Cover
	sprocket	11	Gasket	18	Idler sprocket	25	Crankshaft pulley
5	Intake camshaft	12	Guide	19	Lower chain front	26	Bolt
	sprocket	13	Pivot bolt		guide	27	Bolt
6	Upper tensioner	14	Dowel and O-ring	20	Bolt		

Remove the right-front wheel.
4   Remove the right (passenger's) side engine splash shield.
5   Remove the lower splash cover.
6   Remove the valve cover (see Section 4).
7   Remove the drivebelts (see Chapter 1).
8   Relieve the system fuel pressure (see Chapter 4).
9   Remove the power steering pump and bracket (see Chapter 10).
10 Drain the cooling system (see Chapter 1).

11 Remove the front exhaust pipe (see Section 6).
12 Remove the support bracket for the front of the cylinder head.
13 Remove the air duct connected to the intake manifold. (see Chapter 3).
14 Remove the spark plugs (see Chapter 1).
15 Remove the support for the lower intake manifold.
16 Position the engine at TDC for number one cylinder (see Section 3).

17 Remove the distributor (see Chapter 5).
18 Remove the cam sprocket cover from the cylinder head **(see illustration)**.
19 Remove the water pump pulley.
20 Remove the thermostat housing (see Chapter 3).
21 Remove the lower chain tensioner from the front cover **(see illustration)**.
22 Remove the upper chain tensioner and the timing chain guide **(see illustrations)**.
23 Loosen the idler sprocket bolt.

Chapter 2 Part A Engine

7.22b Upper timing chain components to be removed

1. Valve Timing Control (VTC) cam sprocket
2. Upper chain guide
3. Idler chain guide
4. Upper chain tensioner
5. Cover gusset stud

7.24a Hold the camshaft with a wrench on the flats to prevent it from turning

7.24b Remove the VTC sprocket from the intake camshaft

7.25 Note the numbers and positions of the camshaft bearing caps, unbolt the caps and then remove the camshafts

7.28 Lift the head off with the manifolds in place

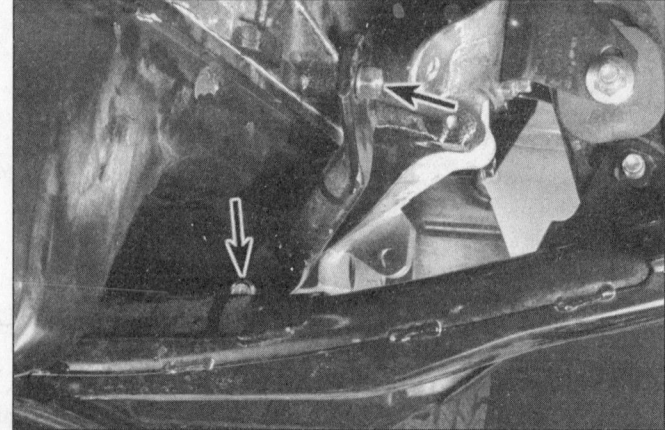

7.30a Remove the bolts (arrows) from the front engine-to-transaxle brace and remove the brace

24 Apply dabs of paint to the timing chain and the sprockets to mark them as an aid to installation. There should be marks stamped into the sprockets and colored chain links which line up with them. Remove the camshaft sprockets **(see illustrations)**.

25 Mark the camshafts left and right. Loosen the camshaft bearing caps in two or three steps, in numerical sequence (see Section 9). The camshaft bearing caps have arrows indicating the front of the engine, are

Chapter 2 Part A Engine

7.30b Remove the rear engine-to-transaxle brace in a similar manner

7.33 With the engine supported by a jack, remove the front engine mount and the brace

7.38 Examine the chain guides for signs of wear such as that shown here

7.39 Put the chain on the crankshaft sprocket so that the colored link meets with the mark on the sprocket

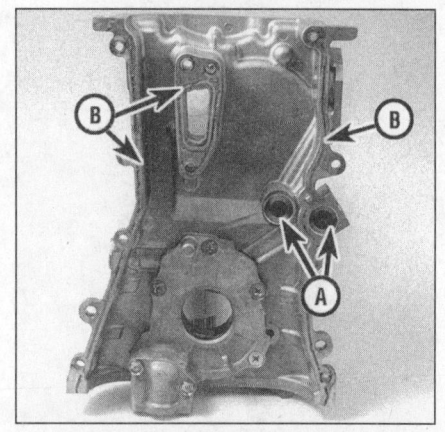

7.41 Install new O-rings (A) and apply sealant to the edges of the timing cover and to the water sealing areas (B)

7.42 Alignment details of the upper chain and sprockets (GA16DE engine)

numbered from 1 to 5, and are stamped with an "I" or an "E", to indicate intake or exhaust. **Caution:** *Keep the caps in order. They must go back in the same location they were removed from.* Number these caps if there are no identification marks. Remove the bearing caps and the camshafts **(see illustration)**.

26 Remove the bolt from the idler sprocket.
27 Following the reverse order of the tightening sequence, loosen the head bolts in several steps and remove them (see Section 14).
28 Lift off the cylinder head with the manifolds attached **(see illustration)**.
29 Remove the idler sprocket shaft and the upper timing chain.
30 Remove the two engine-to-transaxle braces **(see illustrations)**.
31 Drain the engine oil and remove the oil pan and oil pump pick-up (see Section 12).
32 Lower the vehicle and support the engine with a jack and wood block placed under the front main bearing saddle.
33 Remove the engine mount at the front (passenger side) of the engine **(see illustration)**.
34 Remove the crankshaft pulley.
35 Remove the engine front cover.

36 Apply paint to the chain to mark its location on the idler sprocket. The sprocket has a stamped mark on it. Remove the idler sprocket and the lower chain.
37 Remove the chain guides, the oil pump drive spacer and the crankshaft sprocket.

Inspection

38 Inspect camshaft, idler and crankshaft sprockets for wear of the teeth and keyways. Inspect the chains for cracks or excessive wear of the rollers. Refer to the Specifications in this Chapter. Inspect the facing of the chain guides for excessive wear **(see illustration)**.

Installation

39 Install the lower chain with either of its two mating (silver-colored) links aligned with the mark on the crank sprocket **(see illustration)**.
40 Position the idler sprocket in place with its mating mark aligned with the silver-colored link on the lower chain.
41 Apply a bead of RTV sealant to the timing cover sealing surface and install the cover. **Note:** *Be sure to apply sealant to the water passage sealing area, and make sure new O-rings are in place* **(see illustration)**.
42 Position the upper chain in place on the idler sprocket with its gold-colored link aligned with the mating mark on the idler's smaller front sprocket **(see illustration)**.
43 Install the cylinder head with a new gasket (see Section 11) **(see illustration)**. Install the camshafts (see Section 9).
44 The remainder of the installation is the

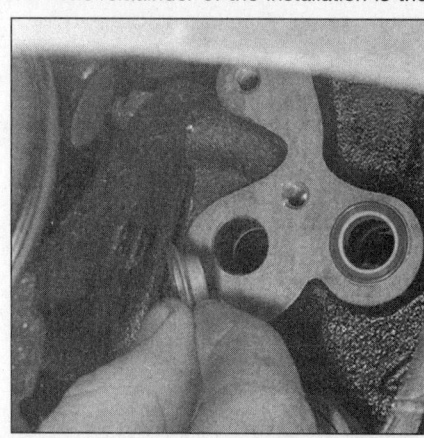

7.43 There are special oil sealing collars which must be installed to the block on the GA16DE engine during reassembly

Chapter 2 Part A Engine

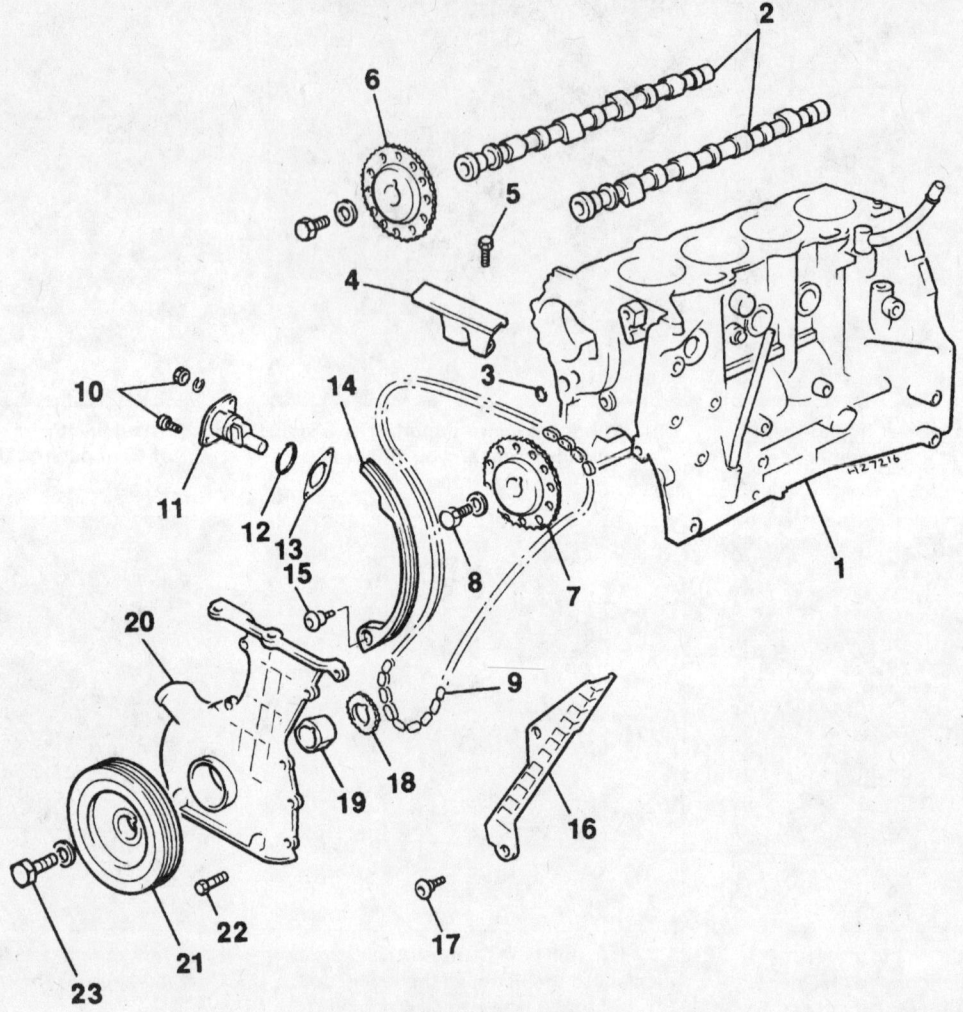

7.45 Exploded view of the timing chain assembly on the SR20DE engine

1	Block	9	Timing chain	17	Bolt
2	Camshafts	10	Nut and bolt	18	Crankshaft sprocket
3	O-ring	11	Tensioner	19	Oil pump drive spacer
4	Upper chain guide	12	O-ring	20	Cover
5	Bolt	13	Gasket	21	Pulley
6	Intake camshaft sprocket	14	Rear guide	22	Bolt
7	Exhaust camshaft sprocket	15	Bolt	23	Bolt
8	Bolt	16	Front guide		

reverse of the disassembly sequence. Install the cam sprockets and upper chain with the camshafts at TDC and the upper chain installed with its silver links aligned with the sprocket mating marks. Do not install the lower chain tensioner until after the cam and idler sprockets and chains are properly aligned. Tighten all the sprocket bolts to the torque listed in this Chapter's Specifications.

SR20DE engine

Refer to illustrations 7.45, 7.46, 7.50a, 7.50b, 7.57a, 7.57b, 7.57c, 7.59, 7.60, 7.62a, 7.62b, 7.62c, 7.63a, 7.63b and 7.63c

Removal

45 Perform Steps 1 through 11. **Note:** *This engine has only one timing chain* **(see illustration)**.
46 Rotate the engine until the left and right camshaft sprocket marks are aligned **(see illustration)**.
47 Remove the crankshaft pulley.
48 Remove the chain tensioner.
49 Remove the distributor.
50 Remove the upper chain guide and, using a suitable tool to hold the camshafts, remove the camshaft sprockets **(see illustrations)**. Remove the front and rear chain guides.
51 Remove the starter (see Chapter 5).
52 Loosen the camshaft bearing caps in two or three steps, in numerical sequence (see Section 9). Carefully lift out the camshafts, oil tubes and baffle plate.
53 Remove the knock sensor harness connector.
54 Loosen the cylinder head bolts in sequence and remove them (see Section 11).
55 Lift off the cylinder head with intake manifold lower section and exhaust manifold connected.
56 Drain the engine oil and remove the oil pan, the aluminum section, the oil pump and pick-up the baffle (see Section 12).
57 Remove the front cover. Remove the timing chain and the two chain guides **(see illustrations)**.

Inspection

58 Inspect camshaft, and crankshaft sprockets for wear of the teeth and keyways. Refer to the Specifications in this Chapter.

Chapter 2 Part A Engine

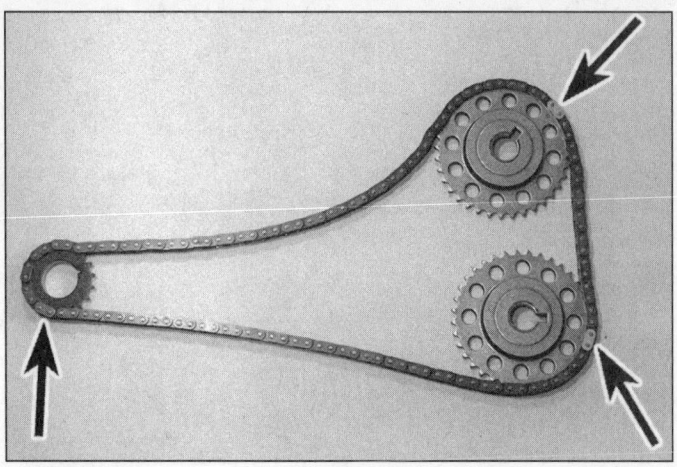

7.46 Timing chain sprocket marks and the colored chain links which mate with them

7.50a Remove the two bolts . . .

7.50b . . . and remove the upper chain guide

7.57a Remove the rear guide

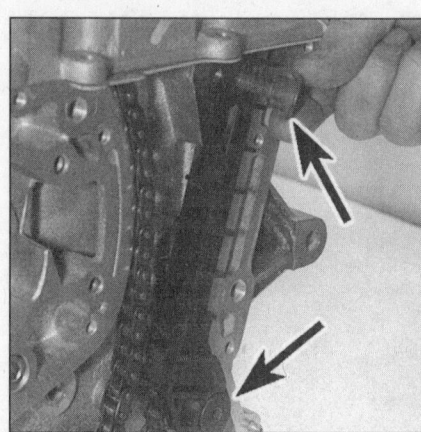

7.57b Remove the two bolts (arrows) . . .

7.57c . . . and remove the timing chain front guide

7.59 Engage the colored link on the chain with the mark on the crankshaft sprocket

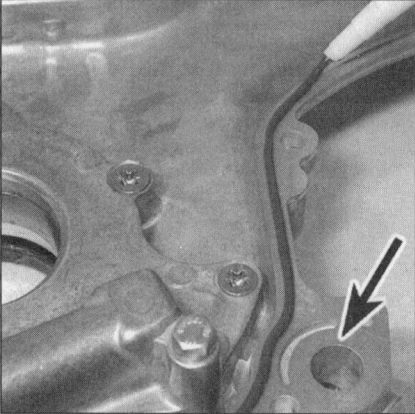

7.60 Apply a bead of RTV sealant to the cover. Do not put sealant around the oil gallery and groove (arrow)

Inspect the chain for cracks or excessive wear of the rollers. Inspect the facing of the chain guides for excessive wear.

Installation

59 Install the timing chain with the bottom mating mark aligned with the mark on the crankshaft sprocket **(see illustration)**.

60 Install the chain guides and the front cover, keeping the excess upper chain laying over the right-hand guide (as you face the front of the engine). Apply a bead of RTV sealant to the outer perimeter of the cover, but not to the groove indicated **(see illustration)**.

61 Install the cylinder head (see Section 11). Install the camshafts (see Section 9).

62 Install the camshaft sprockets, aligning

2A-12　　　Chapter 2 Part A　Engine

7.62a Engage the exhaust camshaft sprocket with the chain, aligning the timing mark with the colored link . . .

7.62b . . . then perform the same operation with the intake camshaft sprocket

7.62c Tighten both sprocket bolts to the proper torque - hold the camshaft with a wrench on one of the flats

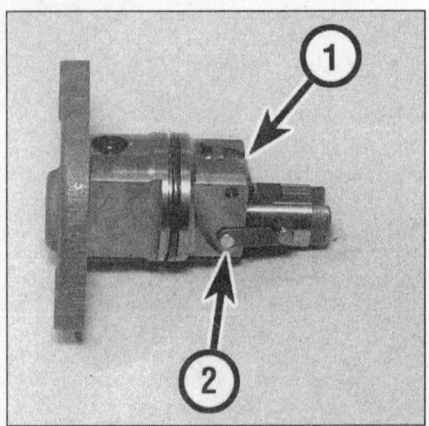

7.63a Press the detent lever (1), then push the plunger back into its housing and hold it in position with the hook (2)

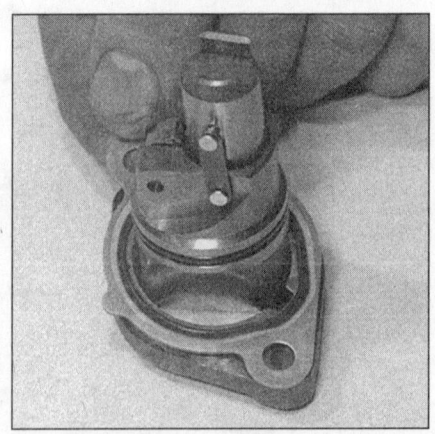

7.63b Install a new O-ring and gasket to the tensioner

7.63c Install the tensioner to the head, making sure that its arrow is pointing away from the head

8.3 Remove the pulley bolt and its washer. The pulley can now be removed

8.10 Align the pulley groove (arrow) with the key and slide the pulley onto the crankshaft

the silver-colored chain links with the mating marks on the cam sprockets **(see illustrations)**. Tighten the camshaft sprocket bolts to the torque listed in this Chapter's Specifications.
63　Retract the chain tensioner by holding the detent lever back while depressing the plunger until the hook can be latched over

the pin **(see illustrations)**. Install the chain tensioner with the (cast-in) arrow pointing to the front of the engine. The hook will be released automatically as the tensioner is bolted in.
64　The remainder of the installation is the reverse of the removal procedure.

8　Crankshaft front oil seal - replacement

Refer to illustrations 8.3 and 8.10
1　Remove the drivebelts from the engine (see Chapter 1).
2　If necessary for clearance, remove the radiator and other components.
3　Remove the bolt from the center of the crankshaft pulley **(see illustration)**. This is usually very tight. If possible, use an impact wrench to remove it. If an impact wrench is not available, it will be necessary to prevent the crankshaft from turning as torque is applied (see the next Step).
4　If the vehicle has a manual transaxle, have an assistant sit in the car with the transmission in high gear with the brakes applied. If it has an automatic transaxle, the driveplate will have to be locked using a prybar inserted through the inspection cover or the starter hole.
5　Raise the vehicle and support it securely on jackstands.
6　Remove the crankshaft pulley by hand or using a special puller, if necessary. Inspect the portion of the pulley which contacts the

Chapter 2 Part A Engine

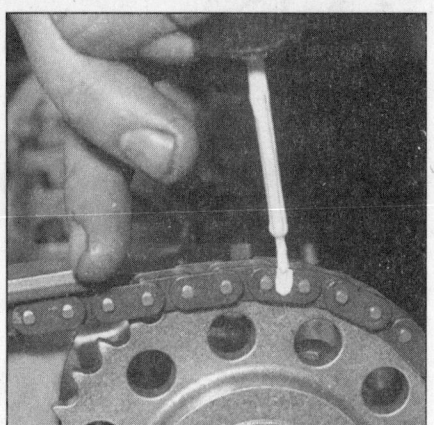

9.6a If the camshafts are to be removed without first removing the chain, make alignment marks on the sprockets and chain

9.6b The sprockets can be tied to the chain with wire ties (arrow)

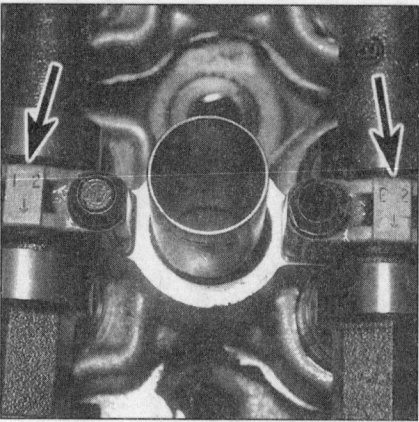

9.7a Inlet and exhaust camshaft bearing cap markings on the GA16DE engine (arrows)

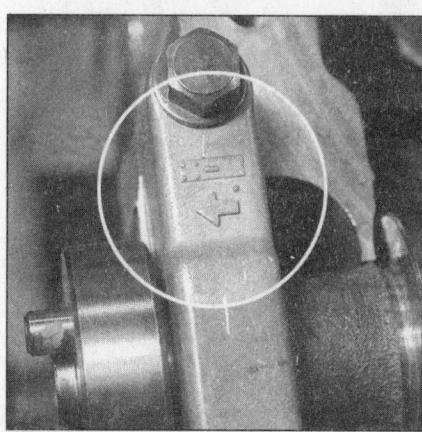

9.7b Camshaft bearing cap markings on the SR20DE engine

seal. If it has a groove worn in it, replace it.
7 Wrap the tip of a small screwdriver with tape, and use it to pry out the seal, being careful not to damage the seal bores.
8 Clean the bore in the housing and coat the outer edge of the new seal with engine oil or multi-purpose grease. Also lubricate the seal lips.
9 Using a socket with an outside diameter slightly smaller than the outside diameter of the seal, carefully drive the new seal into place with a hammer. Make sure it's installed squarely and driven in to the same depth as the original. If a socket isn't available, a short section of large diameter pipe will also work. Check the seal after installation to make sure the internal garter spring didn't pop out of place.
10 Installation is the reverse of removal **(see illustration)**. Run the engine and check for oil leaks.

9 Camshafts - removal, inspection and installation

Removal

Refer to illustrations 9.6a, 9.6b, 9.7a, 9.7b and 9.9

1 Detach the cable from the negative terminal of the battery.
2 Remove the valve cover (see Section 4).
3 Remove the cam sprocket cover and cover gusset (stud).
4 Use a dial indicator to check camshaft endplay. Mount the dial indicator so the gauge tip can be placed at the end of the camshaft. Move the camshaft all the way to the rear and zero the dial indicator. Next, use a screwdriver to pry it all the way forward. If the endplay (the total amount of movement), exceeds the limit listed in this Chapter's Specifications, replace the camshaft and/or cylinder head.
5 Position the engine at TDC for number one cylinder (see Section 3) and remove the distributor (see Chapter 5).
6 Mark the timing chain next to the marks on the sprockets if the colored links on the chain are not visible **(see illustrations)**. This link must line up with the marks on the sprockets. Remove the camshaft sprockets and position the timing chain out of the way.
7 Loosen the camshaft bearing caps in two or three steps, in the *opposite* order of the tightening sequence **(see illustrations 9.19a and 9.19b)**. The camshaft bearing caps have arrows indicating the front of the engine **(see illustrations)**. They are numbered from 1 to 5, and are stamped with an "I" or an "E", to indicate intake or exhaust. **Caution:** *Keep the caps in order. They must go back in the same location they were removed from.* Number these caps if there are no identification marks.
8 Remove the bearing caps and camshafts.
9 On GA16DE engines, remove the Valve Timing Control (VTC) solenoid, located on the right rear of the cylinder head, for inspection **(see illustration)**. If you are going to disassemble the head any further, see Chapter 2B.

Inspection

Refer to illustrations 9.11a and 9.11b

10 Visually examine the camshaft lobes, journals, bearing caps, pivot points and

9.9 Location of the VTC solenoid on the front/intake side of the GA16DE cylinder head

metal-to-metal contact areas. Check for score marks, pitting and evidence of overheating (blue, discolored areas). If wear is excessive or damage is evident, the component will have to be replaced.
11 Using a micrometer, measure camshaft journal diameters and lobe heights (see illustrations), and compare your measurements to this Chapter's Specifications. If the lobe height is less than the minimum allowable, the camshaft is worn and must be replaced.
12 Check the oil clearance for each camshaft journal as follows:

a) *Clean the bearing caps and the camshaft journals with lacquer thinner or acetone.*
b) *Carefully lay the camshafts in place in the head. DON'T use any lubrication.*
c) *Lay a strip of Plastigage on each journal.*
d) *Install the bearing caps with the arrows pointing toward the front (timing chain end) of the engine.*
e) *Tighten the bolts in sequence* **(see illustrations 9.19a and 9.19b)** *to the torque listed in this Chapter's Specifications in 1/4-turn increments.*
Caution: *Don't turn the camshaft while the Plastigage is in place.*

2A-14　　Chapter 2 Part A Engine

9.11a Measuring the camshaft bearing journal diameter with a micrometer

9.11b Use the same micrometer to check the camshaft lobe heights

9.17a On the SR20DE engine, lay the camshafts in the head with the knock pins in the positions shown (arrows)

9.17b Install the bearing caps, making sure that the marks are correct

9.17c On the SR20DE engine, the exhaust camshaft cap must have RTV sealant applied to the area shaded

9.18a Install the baffle plate on SR20DE engines to the number 2 exhaust camshaft bearing cap

9.18b Install the oil pipes and cap bolts (SR20DE engine shown)

f) Remove the bolts, in the proper sequence, and detach the bearing caps
g) Compare the width of the crushed Plastigage (at it's widest point) to the scale on the Plastigage envelope.
h) If the clearance is greater than specified, replace the camshaft and/or cylinder head.

13 Scrape off the Plastigage with your fingernail or the edge of a credit card - don't scratch or nick the journals or bearing caps.
14 The VTC solenoid controls oil flow to the VTC unit incorporated into the intake cam sprocket. With the VTC solenoid removed from the head, it can be tested by applying battery voltage to the terminals of the solenoid. With battery voltage applied, the plunger shaft should protrude from the solenoid body.

Installation

Refer to illustrations 9.17a, 9.17b, 9.17c, 9.18a, 9.18b, 9.19a and 9.19b
15 On GA16DE engines, install the VTC solenoid.
16 Apply moly-based engine assembly lubricant to the camshaft lobes and journals. Also apply the same lubricant to the rocker arm contact surfaces on SR20DE engines.
17 Install the camshafts in their original positions (see illustration). Note: *The slotted end of the exhaust camshaft is the rear (transaxle end). The distributor is driven by the camshaft when inserted in this slot.* On SR20DE engines, position the knock pin of the intake camshaft at approximately twelve o'clock and the knock pin on the exhaust camshaft also at approximately twelve o'clock. On GA16DE engines, the pin on the intake camshaft must be 10 to 11-degrees before 9 o'clock and the pin on the exhaust camshaft must be at 12 o'clock. The colored links on the chain must line up with the marks on the sprockets. Install the bearing caps (see illustrations).
18 On SR20DE engines, install the baffle plate (see illustration) and the oil pipes (see illustration). Install the bearing cap bolts and tighten them hand tight, tighten the front cap bolts (near the sprockets) first, then tighten the remainder of the bolts in sequence (see illustrations 9.19a and 9.19b).
19 Tighten the bolts to the torque listed in this Chapter's Specifications, using the proper tightening sequence (see illustrations).
20 Install the camshaft sprockets and timing chain (see Section 7). Hold the camshafts with a suitable wrench as you tighten the sprocket bolts to the specified torque.

Chapter 2 Part A Engine

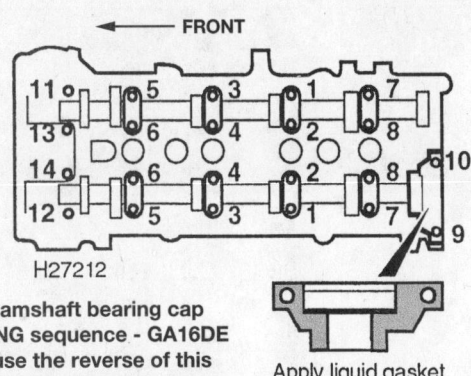

9.19a Camshaft bearing cap TIGHTENING sequence - GA16DE engine (use the reverse of this sequence for disassembly). Be sure to apply RTV sealant to the area shown on installation

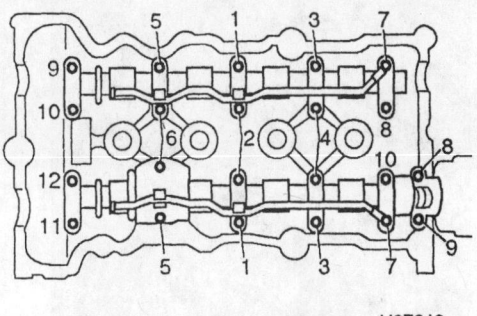

9.19b SR20DE engine camshaft cap TIGHTENING sequence. Use the reverse of this sequence for disassembly

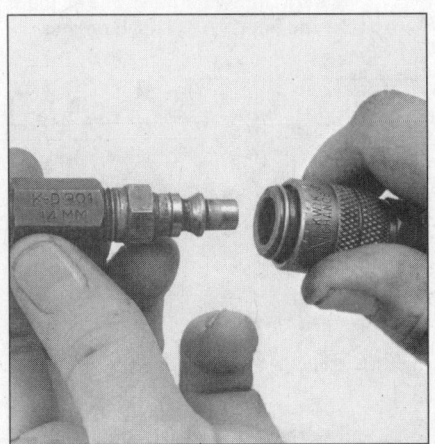

10.4 This is what the air hose adapter that threads into the spark plug hole looks like - they are commonly available at auto parts stores

10.9 Compress the valve spring and remove the two keepers (arrow) with needle-nose pliers or a magnet

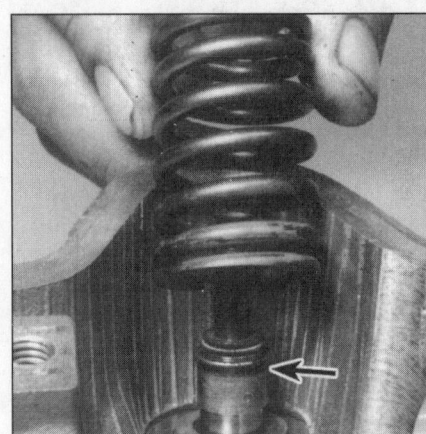

10.10 Remove the retainer and valve spring, then use pliers to remove the stem seal (arrow)

21 The remainder of installation is the reverse of removal. If any part of the valve train was replaced, check and adjust, if necessary, the valve clearance (see Chapter 1).

10 Valve springs, retainers and seals - replacement

Refer to illustrations 10.4, 10.9, 10.10, 10.15 and 10.17

Note: *Broken valve springs and leaking/worn out valve stem seals can be replaced without removing the cylinder head. Two special tools and a compressed air source are normally required to perform this operation, so read through this Section carefully and rent or buy the tools before beginning the job. If compressed air isn't available, a length of nylon rope can be used to keep the valves from falling into the cylinder during this procedure.*

1 Refer to Section 9 and remove the camshafts.
2 Remove the spark plug from the cylinder which has the defective component. If all of the valve stem seals are being replaced, all of the spark plugs should be removed.
3 Rotate the crankshaft until the piston in the affected cylinder is at Top Dead Center (TDC) on the compression stroke (see Section 3). If you're replacing all of the valve stem seals, begin with cylinder number one and work on the valves for one cylinder at a time. Move from cylinder-to-cylinder following the firing order sequence (see the Specifications).
4 Thread an adapter into the spark plug hole **(see illustration)** and connect an air hose from a compressed air source to it. Most auto parts stores can supply the air hose adapter. **Note:** *Many cylinder compression gauges utilize a screw-in fitting that may work with your air hose quick-disconnect fitting.*
5 Apply compressed air to the cylinder. **Warning:** *The piston may be forced down by compressed air, causing the crankshaft to turn suddenly. If the wrench used when positioning the number one piston at TDC is still attached to the bolt in the crankshaft nose, it could cause damage or injury when the crankshaft moves.*
6 The valves should be held in place by the air pressure.
7 If you don't have access to compressed air, an alternative method can be used. Position the piston at a point approximately 45-degrees before TDC on the compression stroke, then feed a long piece of nylon rope through the spark plug hole until it fills the combustion chamber. Be sure to leave the end of the rope hanging out of the engine so it can be removed easily.
8 Use a large ratchet and socket to rotate the crankshaft in the normal direction of rotation (clockwise, viewed from the front) until slight resistance is felt.
9 Stuff shop rags into the cylinder head areas, above and below the valves, to prevent parts and tools from falling into the engine, then use a valve spring compressor to compress the spring **(see illustration)**. Remove the keepers with small needle-nose pliers or a magnet.
10 Remove the spring retainer, valve spring and shim(s) then remove the stem oil seal with pliers **(see illustration)**. **Note:** *If air pressure fails to hold the valve in the closed position during this operation, the valve face and/or seat is probably damaged. If so, the cylinder head will have to be removed for additional repair operations.*
11 Wrap a rubber band or tape around the top of the valve stem so the valve won't fall into the combustion chamber, then release the air pressure. **Note:** *If a rope was used instead of air pressure, turn the crankshaft slightly in the direction opposite normal rotation.*
12 Inspect the valve stem for damage. Rotate the valve in the guide and check the

2A-16 Chapter 2 Part A Engine

10.15 Place each new seal squarely on the valve guide and tap it into place with a hammer and socket. Do not continue to hammer once the seal is seated

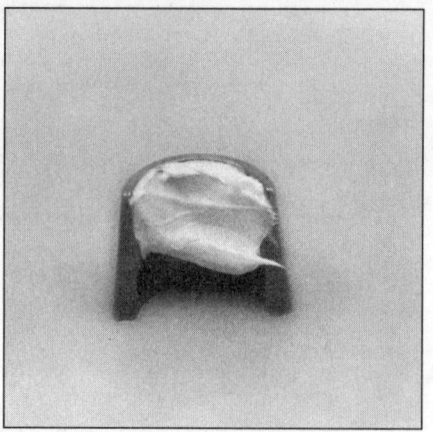

10.17 Apply a small dab of grease to each keeper to hold it in place unit the spring is released

11.11 There are two small bolts at the front of the head on SR20DE engines

11.15 Carefully remove all traces of old gasket material from the sealing surfaces

end for eccentric movement, which would indicate that the valve is bent.

13 Move the valve up-and-down in the guide and make sure it doesn't bind. If the valve stem binds, either the valve is bent or the guide is damaged. In either case, the head will have to be removed for repair.

14 Reapply air pressure to the cylinder to retain the valve in the closed position, then remove the tape or rubber band from the valve stem. If a rope was used instead of air pressure, rotate the crankshaft in the normal direction of rotation until slight resistance is felt.

15 Lubricate the valve stem with engine oil and install the new oil seal, using an installation tool or deep socket **(see illustration)**.

16 Install the spring in position over the valve. The tapered (progressively-wound) springs should be installed with the smaller (tighter-wound) ends down, or with the paint marks toward the head. **Caution:** *Make sure the valve spring seat is in place before the spring is installed.*

17 Install the valve spring retainer. Compress the valve spring and carefully position the keepers in the groove. Apply a small dab of grease to the inside of each keeper to hold it in place if necessary **(see illustration)**.

18 Remove the pressure from the spring tool and make sure the keepers are seated.

19 Disconnect the air hose and remove the adapter from the spark plug hole. If a rope was used in place of air pressure, pull it out of the cylinder.

20 The procedure is the same for all valve stem seals, both intake and exhaust.

21 Refer to the appropriate Section in this Chapter and install the camshafts.

22 The rest of the installation is the reverse of the removal procedure.

23 Start and run the engine, then check for oil leaks and unusual sounds coming from the valve cover area.

11 Cylinder head - removal and installation

Caution: *The engine must be completely cool before beginning this procedure.*

Removal

Refer to illustration 11.11

1 Relieve the fuel system pressure (see Chapter 4), then disconnect the cable from the negative battery terminal.

2 Drain the coolant from the engine block and radiator (see Chapter 1).

3 Drain the engine oil and remove the oil filter (see Chapter 1).

4 Remove the fuel rail.

5 Remove the intake manifold (see Section 5).

6 Remove the exhaust manifold (see Section 6).

7 Remove the timing chain(s) (see Section 7). **Note:** *If you're very careful, you can unbolt the camshaft sprockets from the camshafts and support them with pieces of wire, making it unnecessary to completely remove the timing chain(s). Tension must be maintained on the chain, however, to prevent it from becoming disengaged at the lower end.*

8 Remove the alternator and distributor (see Chapter 5).

9 Remove the power steering pump (see Chapter 10) and set it aside without disconnecting the hoses.

10 Label and remove any remaining items attached to the cylinder head, such as coolant fittings, tubes, cables, hoses or wires.

11 Using a breaker bar and the appropriate sized socket, loosen the cylinder head bolts in 1/4-turn increments until they can be removed by hand. Loosen the bolts in a pattern *opposite* that of the tightening sequence **(see illustrations 11.24a and 11.24b)** to avoid warping or cracking the head. **Note:** *On GA series engines, you will need a 10mm Allen wrench at least one and a half inches long to remove the head bolts under the camshafts. There are 6 mm bolts at the front of the head on SR20DE engines* **(see illustration)**.

12 Lift the cylinder head off the engine block. If it's stuck, very carefully pry up at the transaxle end, beyond the gasket surface, at a casting protrusion.

13 Remove all external components from the head to allow for thorough cleaning and inspection. **Note:** *See Chapter 2, Part B, for cylinder head inspection and servicing procedures.*

Installation

Refer to illustrations 11.15, 11.18, 11.21, 11.23, 11.24a, 11.24b, 11.24c and 11.24d

14 The mating surfaces of the cylinder head and block must be perfectly clean when the head is installed.

15 Use a gasket scraper to remove all traces of carbon and old gasket material **(see illustration)**, then clean the mating surfaces with lacquer thinner or acetone. If there's oil on the mating surfaces when the head is installed, the gasket may not seal correctly and leaks could develop. When working on the block, stuff the cylinders with clean shop rags to keep out debris. Use a vacuum cleaner to remove material that falls into the cylinders.

16 Check the block and head mating surfaces for nicks, deep scratches and other

Chapter 2 Part A Engine

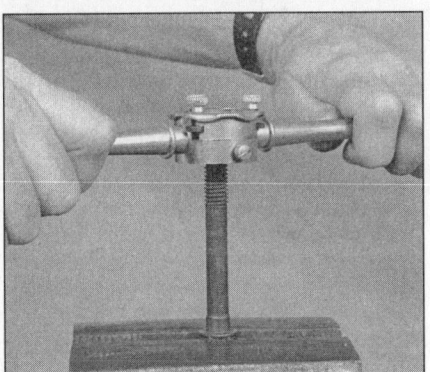

11.18 A die should be used to remove sealant and corrosion from the head bolts prior to installation

11.21 Make sure that the head gasket is placed on the block so that all holes line up correctly

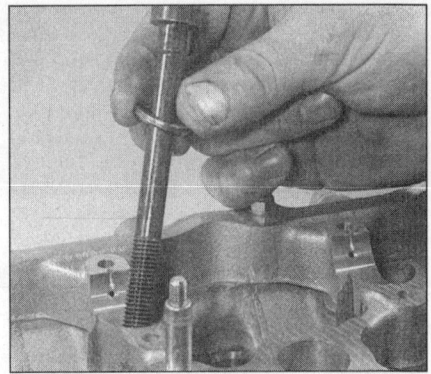

11.23 Oil the bolt threads and the washers, then install them. The sharp outer corners of the washers face towards the head

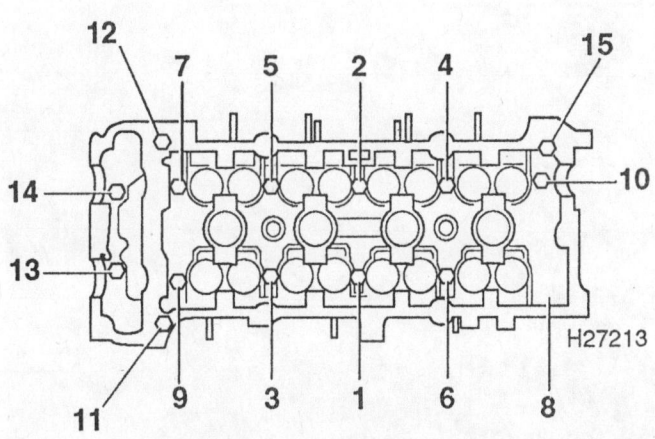

11.24a GA16DE head bolt TIGHTENING sequence (use the reverse order for loosening the bolts)

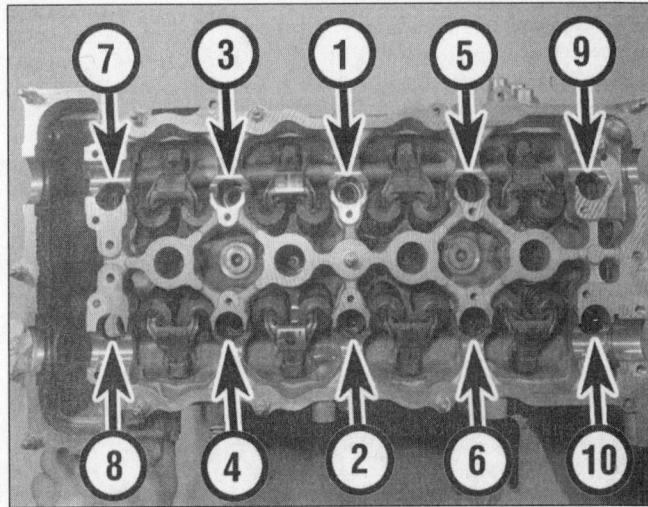

11.24b SR20DE head bolt TIGHTENING sequence (use the reverse order for loosening the bolts)

damage. If damage is slight, it can be removed with a file; if it's excessive, machining may be the only alternative.

17 Use a tap of the correct size to chase the threads in the head bolt holes, then clean the holes with compressed air - make sure that nothing remains in the holes. **Warning:** *Wear eye protection when using compressed air!*

18 Mount each bolt in a vise and run a die down the threads to remove corrosion and restore the threads **(see illustration)**. Dirt, corrosion, sealant and damaged threads will affect torque readings.

19 Check the cylinder head for warpage (see Chapter 2B). Check the head gasket, intake and exhaust manifold surfaces.

20 Install the components that were removed from the head.

21 Position the new cylinder head gasket over the dowel pins in the block **(see illustration)**.

22 Carefully set the head on the block without disturbing the gasket.

23 Before installing the head bolts, apply a small amount of clean engine oil to the threads and hardened washers (if equipped). The chamfered side of the washers must face the

11.24c Tighten the bolts in stages to the correct torque values . . .

bolt heads **(see illustration)**.

24 Install the bolts in their original locations and tighten them finger tight. Then tighten all the bolts, following the proper sequence **(see illustrations)**, to the torque listed in this Chapter's Specifications. **Caution:** *On GA16DE engines, the small hex-head bolts (no. 11*

11.24d . . . then use an angle torque meter and tighten the bolts the additional degree listed in this Chapter's Specifications

through 15) should only be tightened to 56 to 74 inch pounds.

25 The remaining installation steps are the reverse of removal.

26 Check and adjust the valves as necessary (see Chapter 1).

Chapter 2 Part A Engine

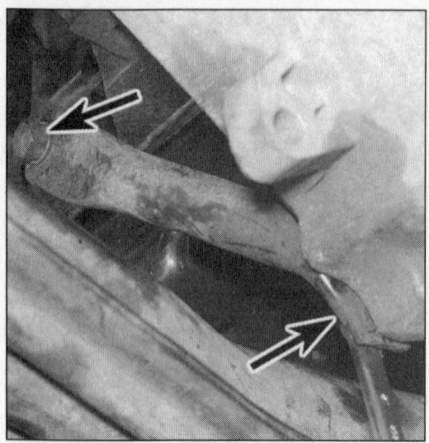

12.8 Engine/transaxle front support bolts (arrows)

12.9 Removing the oil pan from the GA16DE engine

12.10 On the SR20DE engine, the small stamped steel pan must be removed first

12.11a Remove the bolts (arrows) . . .

12.11b . . . then remove the baffle from the base of the aluminum casting (SR20DE engine)

27 Refill the cooling system, install a new oil filter and add oil to the engine (see Chapter 1).
28 Start the engine and set the ignition timing (see Chapter 5). Road test the vehicle and check for leaks.

12 Oil pan - removal and installation

Removal

GA16DE engine

Refer to illustrations 12.8 and 12.9

1 Disconnect the cable from the negative battery terminal.
2 Set the parking brake and block the rear wheels.
3 Raise the front of the vehicle and support it securely on jackstands.
4 Remove the splash shields under the engine and the right side cover.
5 Drain the engine oil (see Chapter 1).
6 Disconnect the front exhaust pipe from the exhaust manifold.
7 Remove the center crossmember after supporting the engine with a hoist.
8 Remove the transaxle support braces

and, on models with an automatic transaxle, the driveplate cover **(see illustration)**.
9 Remove the nuts and bolts and separate the oil pan from the engine block **(see illustration)**. Use a soft-faced hammer to break the oil pan-to-block seal. Do not pry with a screwdriver between the pan and the block or the pan flange could be distorted and cause an oil leak. In extreme cases, a knife blade will have to be used to cut the seal between the block and the oil pan.

SR20DE engine

Refer to illustrations 12.10, 12.11a, 12.11b, 12.13a, 12.13b, 12.13c, 12.15a and 12.15b

10 The oil pan is a two piece design. A steel pan is attached to an aluminum section which is bolted to the engine block **(see illustration)**. Perform steps 1 through 5 and remove the steel pan from the aluminum section. **Caution:** *Do not pry between the steel pan and the aluminum flange or damage to the sealing surface may result.*
11 Remove the internal baffle **(see illustrations)**.
12 Disconnect the exhaust pipe from the exhaust manifold. Support the engine from

12.13a Remove the two bolts (arrows)

above with an engine hoist and support the transaxle from below with a floor jack. Remove the center crossmember.
13 Disconnect the automatic transaxle cable. Remove the small rear cover **(see illustrations)**.
14 Remove the air conditioning compressor bracket-to-oil pan braces and the rear cover

Chapter 2 Part A Engine

12.13b ... then remove the plate from the pan flange on SR20DE engines ...

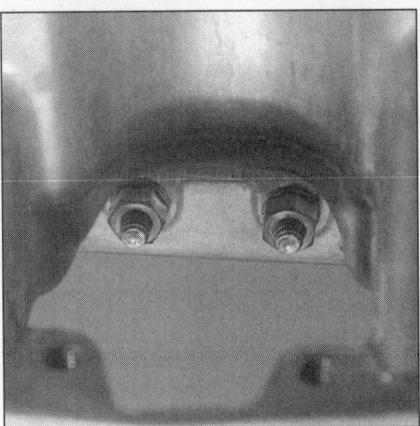

12.13c ... to access these two nuts

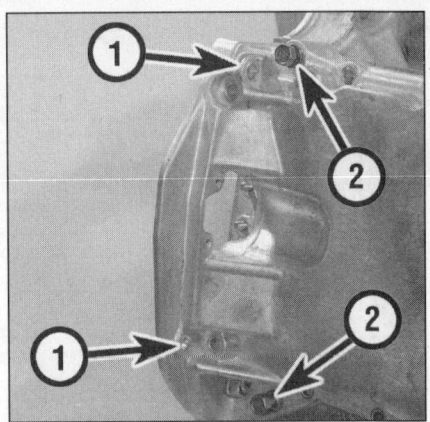

12.15a Remove the bolts from the pan flange (1) and tighten them into the holes (2) ...

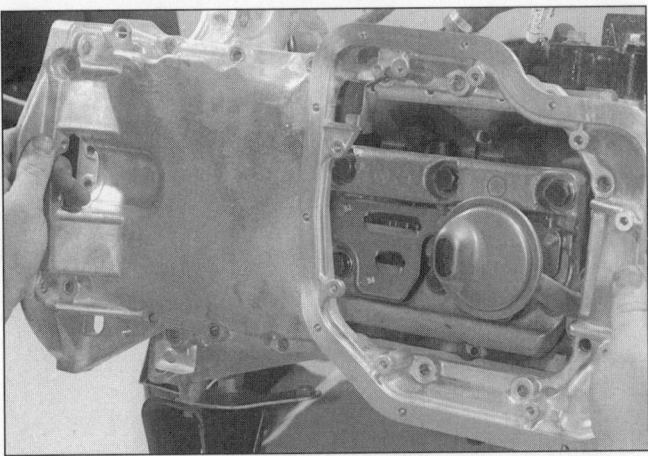

12.15b ... to separate the aluminum casting from the block

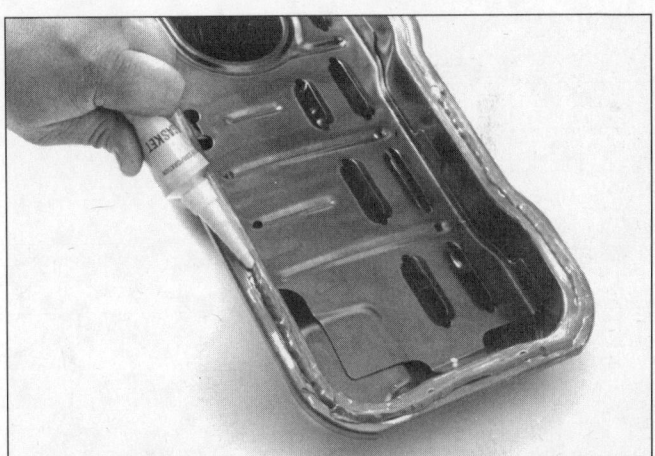

12.19a Apply an even bead of RTV sealant to the oil pan flange prior to installation (GA16DE engine shown)

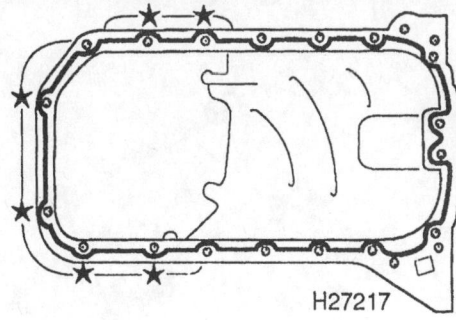

12.19b On the SR20DE engine, the sealant bead must be around the outside of all bolt holes marked with a star and around the inside of all others

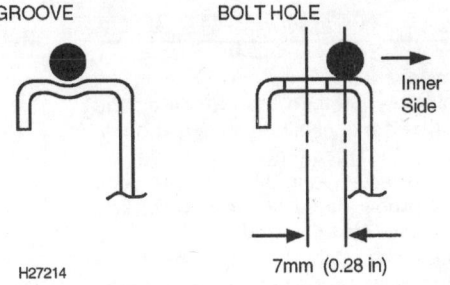

12.19c The sealant must be applied around the inside of all bolt holes on the GA16DE engine and in the groove between the holes

plate. Remove the bolts attaching the aluminum section to the block, working from the ends toward the center to prevent warpage.

15   To loosen the aluminum section, remove two transaxle bolts and insert them into two empty, threaded holes in the oil pan (see illustrations). Tightening these bolts into the indicated holes will separate the pan from the block. Remove the two transaxle bolts and reinstall them in their original location.

Installation

Refer to illustrations 12.19a, 12.19b, 12.19c, 12.20a, 12.20b and 12.20c

16   Use a scraper to remove all traces of old gasket material and sealant from the block and oil pan. Clean the mating surfaces with lacquer thinner or acetone. **Caution:** *Be careful not to scratch or gouge the gasket surface of the block or oil pan. A leak could develop after the repairs have been completed.*

17   Make sure the threaded bolt holes in the block are clean.
18   Check the oil pan flange for distortion, particularly around the bolt holes. If necessary, place the pan on a wood block and use a hammer to flatten and restore the gasket surface.
19   Apply a 3/16-inch wide bead of RTV sealant around the oil pan flange (see illustrations). **Note:** *The oil pan must be installed within 15 minutes once the sea-*

Chapter 2 Part A Engine

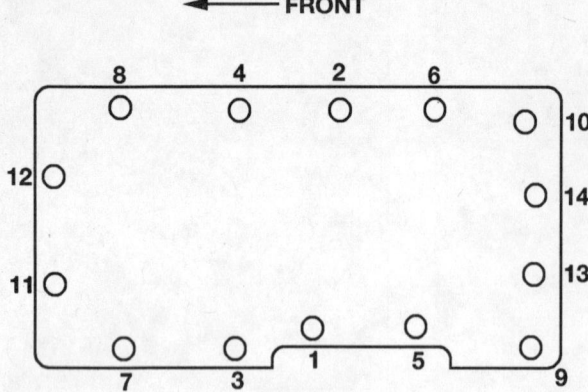

12.20a GA16DE engine oil pan bolt tightening sequence. Always use the reverse order when removing the oil pan on any engine

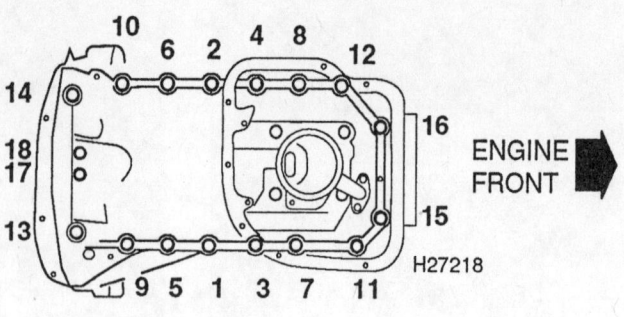

12.20b This is the correct tightening sequence for the bolts which secure the aluminum oil pan casting on the SR20DE engine

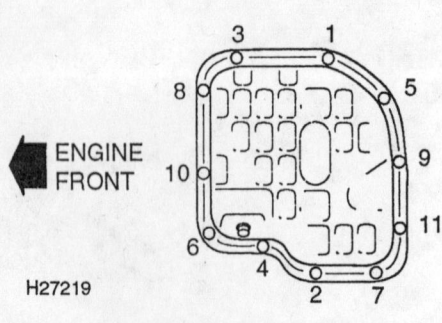

12.20c The stamped steel oil pan on the SR20DE engine is installed using this sequence for tightening the bolts. Always use the reverse order when removing the oil pan on any engine

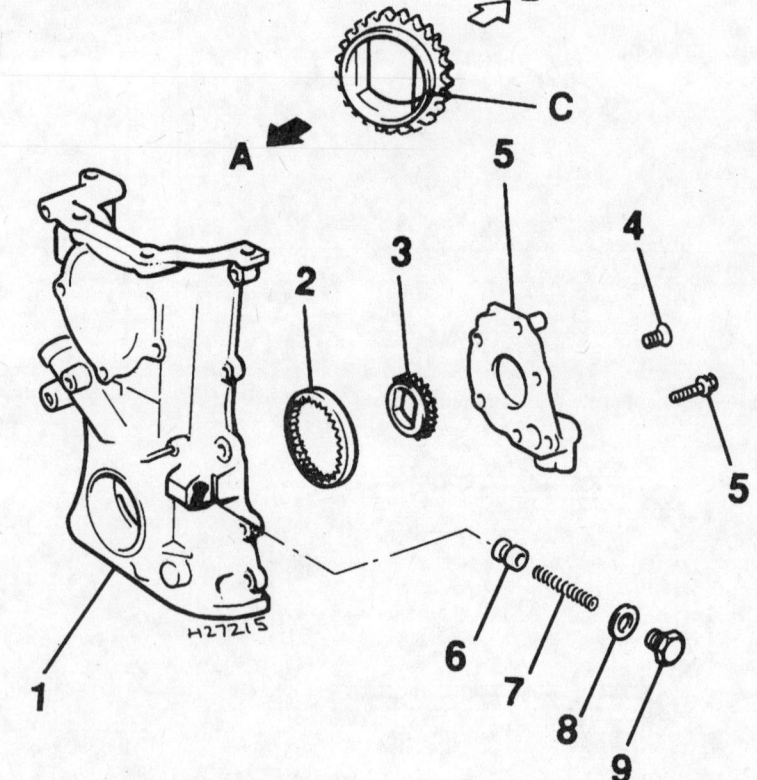

13.7a Exploded view of the oil pump (GA16DE engine)

1	Cover	6	Pressure regulator piston
2	Outer gear	7	Spring
3	Inner gear	8	Washer
4	Screw	9	Pressure regulator valve bolt
5	Bolt		

has been applied.

20 Carefully position the oil pan on the engine block and install the nuts and bolts. Following the recommended sequence, tighten the fasteners in three or four steps to the torque listed in this Chapter's Specifications **(see illustrations)**.

21 The remainder of installation is the reverse of removal. Be sure to install a new oil filter (see Chapter 1) and wait at least thirty minutes before adding oil.

13 Oil pump - removal, inspection and installation

Removal

Refer to illustration 13.7a, 13.7b and 13.7c

1 The oil pump is located inside the lower part of the engine front cover, and is driven from the crankshaft.
2 Drain the engine oil.
3 Remove the drivebelts (see Chapter 1).
4 Remove the cylinder head (refer to Section 11).
5 Remove the entire oil pan assembly and pick-up.
6 Remove the front cover.
7 Remove the bolts retaining the pump cover to the pump body **(see illustrations)**, but do not disassemble the inner rotor and drive gear. **Caution:** *Be very careful with these components. Any nicks or other damage will require replacement of the complete pump assembly.*

Chapter 2 Part A Engine

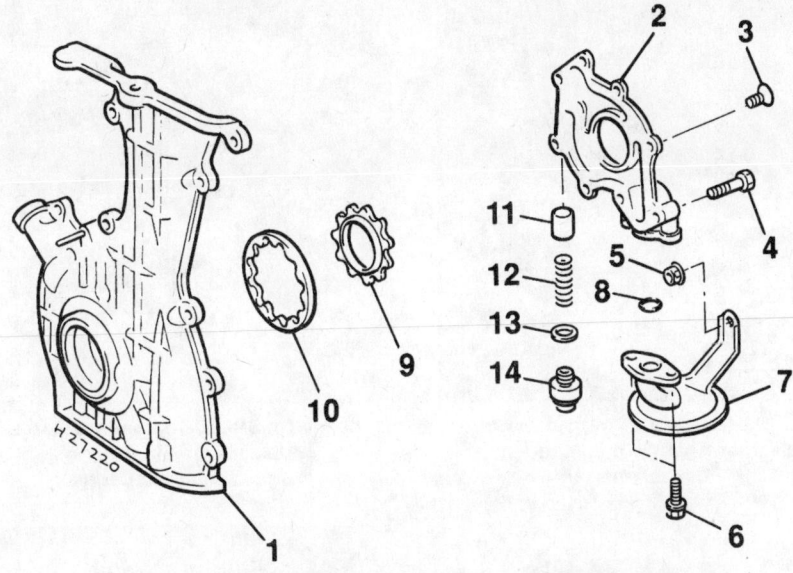

13.7b Exploded view of the oil pump (SR20DE engine)

1	Cover	9	Inner rotor
2	Oil pump cover	10	outer rotor
3	Screw	11	Pressure regulator valve
4	Bolt		piston
5	Nut	12	Spring
6	Bolt	13	Washer
7	Strainer/pick-up	14	Pressure regulator valve cap
8	O-ring		

13.7c Remove the screws from the pump cover (GA16DE engine shown)

13.9 Remove the pressure regulator assembly for inspection and cleaning

Inspection

Refer to illustrations 13.9, 13.10a, 13.10b, 13.10c, 13.10d, 13.10e and 13.10f

8 Clean all components, including the pump and block gasket surfaces, with solvent, then inspect all surfaces for excessive wear and/or damage.

9 Disassemble the relief valve by removing the cap, washer, spring and regulator valve **(see illustration)**. The regulator valve is in the pump cover on SR20DE engines, and in the engine front cover on all GA series engines. Check the oil pressure regulator valve sliding surface and valve spring. The regulator, when clean and oiled, should slide easily in the valve hole. If either the spring or the valve is damaged, they must be replaced as a set. If no damage is found reassemble the relief valve parts, coating the parts with oil, and reinstall it in the front cover.

10 Check the oil pump component clearances with a feeler gauge **(see illustrations)** and compare the results to the specifications at the beginning of this Chapter.

Installation

Refer to illustration 13.14

11 Assemble the cover to the pump body and tighten the screws to the torque listed in this Chapter's Specifications.

12 Install the front cover and any other components removed for disassembly, to the engine. **Note:** *The flats inside the oil pump must be turned to align with the flats on the crankshaft when installing the front cover.*

13 Install a new oil filter and add engine oil to the crankcase, if necessary (see Chapter 1).

13.10a Measuring outer gear-to-cover clearance on the GA16DE engine

13.10b The outer gear-to-crescent clearance is measured like this

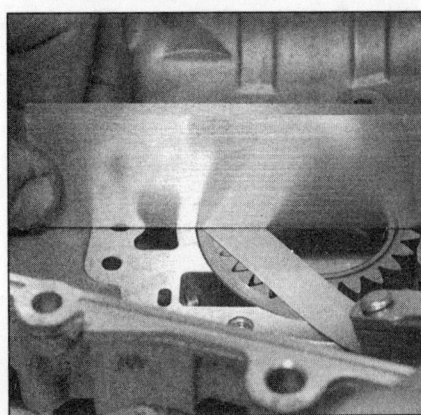

13.10c Check the gear endplay on the GA16DE engine using a straightedge

Chapter 2 Part A Engine

13.10d Check the outer rotor-to-body clearance on the SR20DE oil pump like this

13.10e The inner rotor tip-to-outer rotor clearance can be measured in this manner

13.10f On SR20DE engines, the rotor endplay is checked using a precision straightedge

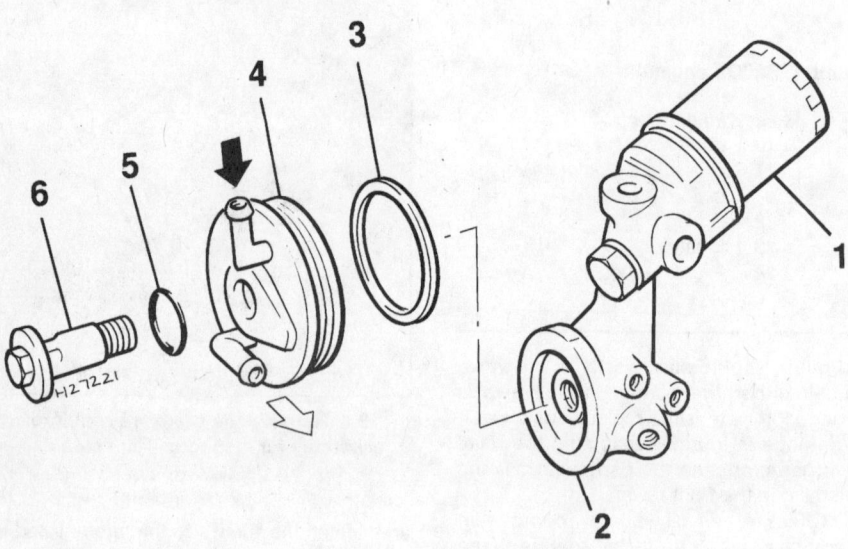

13.14 Oil cooler components for the SR20DE engine

1. Oil filter
2. Filter housing
3. Sealing ring
4. Oil cooler
5. O-ring
6. Bolt

14.3 Mark the flywheel for realignment (arrow) and use a prybar against two clutch cover bolts to keep the flywheel from turning when removing the bolts

14 Start the engine and check for oil pressure and leaks. If any oil leaks occur, check the oil cooler assembly on SR20DE engines **(see illustration)**.
15 Recheck the engine oil level.

14 Flywheel/driveplate - removal and installation

Refer to illustration 14.3

Removal

1 Raise the vehicle and support it securely on jackstands (see Chapter 1), then remove the transaxle (see Chapter 7). If it's leaking, now would be a very good time to replace the front pump seal/O-ring (automatic transaxle only).

2 Remove the pressure plate and clutch disc (see Chapter 8) (manual transaxle models). Now is a good time to check/replace the clutch components and bearing.
3 Make alignment marks on the flywheel/driveplate and crankshaft to ensure correct alignment during reinstallation **(see illustration)**.
4 Remove the bolts that secure the flywheel/driveplate to the crankshaft. If the crankshaft turns, reinstall two clutch cover bolts and use a large prybar across them to hold the flywheel. **(see illustration 14.3)**. On automatic transaxle models, a large screwdriver can be used to secure the driveplate.
5 Remove the flywheel/driveplate from the crankshaft. **Warning:** *Since the flywheel is heavy, be sure to support it while removing the last bolt. Flywheel teeth can be sharp, so wear gloves or use rags to hold the flywheel.*

Installation

6 Clean the flywheel to remove grease and oil. Inspect the surface for cracks, rivet grooves, burned areas and score marks. Light scoring can be removed with emery cloth. Deep grooves can only be corrected by machining. Check for cracked and broken ring gear teeth. If the ring gear is damaged, a specialty automotive shop can install a new one. Lay the flywheel on a flat surface and use a straightedge to check for warpage.
7 Clean and inspect the mating surfaces of the flywheel/driveplate and the crankshaft. If the crankshaft rear seal is leaking, replace it before reinstalling the flywheel/driveplate.
8 Position the flywheel/driveplate against the crankshaft. Be sure to align the marks made during removal. Note that some engines have an alignment dowel or staggered bolt holes to ensure correct installation. Before installing the bolts, apply thread locking compound to the threads.
9 Wedge a screwdriver in the ring gear teeth to keep the flywheel/driveplate from turning as you tighten the bolts to the torque listed in this Chapter's Specifications. Follow

Chapter 2 Part A Engine

15.3a Remove the rear oil seal retainer

15.3b Removing the crankshaft oil seal with the retainer in place (SR20DE engine shown)

15.5 After removing the retainer from the block, support it between two wood blocks and drive out the old seal with a punch and hammer

a criss-cross pattern and work up to the final torque in three or four steps.
10   The remainder of installation is the reverse of the removal procedure.

15 Rear main oil seal - replacement

Refer to illustrations 15.3a, 15.3b, 15.5 and 15.6

1   Remove the transaxle (see Chapter 7).
2   Remove the flywheel/driveplate (see Section 14).
3   Remove the rear oil seal retainer from the block **(see illustration)**. **Note:** *If the seal retainer is removed, several oil pan bolts will have to be removed if the pan is still in place, and it may disturb the pan gasket.* It is possible to remove the seal from the retainer with a screwdriver if great care is taken **(see illustration)**.
4   Scrape any sealant or gasket material from the retainer, oil pan and the block.
5   Position the seal and retainer assembly between two wood blocks, to evenly support the aluminum housing, and drive the old seal out with a hammer and punch **(see illustration)**. A screwdriver can also be used to pry the seal out, being careful not to gouge or nick the housing during seal removal.
6   Place the new seal squarely on the retainer and drive it into the retainer with a wood block **(see illustration)** or a section of pipe slightly smaller in diameter than the outside diameter of the seal.
7   Lubricate the crankshaft seal journal and the lip of the new seal with multi-purpose grease.
8   Apply a thin, uniform layer of RTV sealant to the seal retainer. Also apply sealant to the bottom of the retainer (oil pan mating surface).
9   Slowly and carefully push the seal onto the crankshaft. The seal lip is stiff, so work it onto the crankshaft with a smooth object such as the end of a socket extension as you push the retainer against the block.

10   Install and tighten the retainer bolts securely. The bottom sealing flange of the retainer must not extend below the bottom sealing flange (oil pan rail) of the block.
11   The remaining steps are the reverse of removal.

16 Engine mounts - check and replacement

Note: *See Chapter 7 for transaxle mount replacement.*

1   Engine mounts seldom require attention, but broken or deteriorated mounts should be replaced immediately or the added strain placed on the driveline components may cause damage or wear.

Check

2   During the check, the engine must be raised slightly to remove the weight from the mounts.
3   Raise the vehicle and support it securely on jackstands and remove the splash shields.
4   Position a jack under the engine oil pan. Place a large wood block between the jack head and the oil pan, then carefully raise the engine just enough to take the weight off the mounts. Do not place the wood block under the oil pan drain plug. **Warning:** *DO NOT place any part of your body under the engine when it's supported only by a jack!*
5   Check the mounts to see if the rubber is cracked, hardened or separated from the metal plates. Sometimes the rubber will split right down the center.
6   Check for relative movement between the mount plates and the engine or frame (use a large screwdriver or prybar to attempt to move the mounts). If movement is noted, lower the engine and tighten the mount fasteners.
7   Rubber preservative should be applied to the mounts to slow deterioration.

15.6 Drive the new seal into the retainer with a wood block or a section of pipe, if you have one large enough - make sure that you don't cock the seal in the retainer bore

Replacement

8   Disconnect the negative battery cable from the battery, then raise the vehicle and support it securely on jackstands (if not already done). Support the engine as described in Step 3.
9   To remove the right engine mount (at the timing chain end of the engine), remove the nut and withdraw the through-bolt from the frame bracket.
10   Remove the mount-to-bracket nuts and detach the mount.
11   To remove the front engine mount, remove the engine mount through bolt.
12   Remove the engine block-to-mount bolts and remove the mount.
13   Installation is the reverse of removal. Use thread locking compound on the mount bolts/nuts and be sure to tighten them securely.

Notes

Chapter 2 Part B
General engine overhaul procedures

Contents

	Section
Compression check	4
Crankshaft - inspection	19
Crankshaft - installation and main bearing oil clearance check	23
Crankshaft - removal	14
Cylinder head - cleaning and inspection	10
Cylinder head - disassembly	9
Cylinder head - reassembly	12
Cylinder honing	17
Engine - removal and installation	6
Engine block - cleaning	15
Engine block - inspection	16
Engine overhaul - disassembly sequence	8
Engine overhaul - general information	2
Engine overhaul - reassembly sequence	21

	Section
Engine rebuilding alternatives	7
Engine removal - methods and precautions	5
General information	1
Initial start-up and break-in after overhaul	26
Main and connecting rod bearings - inspection and selection	20
Piston rings - installation	22
Pistons/connecting rods - inspection	18
Pistons/connecting rods - installation and rod bearing oil clearance check	25
Pistons/connecting rods - removal	13
Rear main oil seal - installation	24
Vacuum gauge diagnostic checks	3
Valves - servicing	11

Specifications

General

Cylinder compression pressure
- GA16DE .... 199 psi (171 psi minimum)
- SR20DE .... 178 psi (149 psi minimum)

Oil pump pressure (at normal operating temperature)

GA16DE
- Idle .... 7 to 27 psi
- 3000 rpm .... 50 to 64 psi

SR20DE
- Idle .... 11 psi
- 3200 rpm .... 46 to 57 psi

Engine block

Deck surface warpage .... 0.004 inch

Bore diameter
- GA16DE
  - Grade 1 .... 2.9921 to 2.9925 inches
  - Grade 2 .... 2.9925 to 2.9929 inches
  - Grade 3 .... 2.9929 to 2.9933 inches
- SR20DE
  - Grade 1 .... 3.3858 to 3.3862 inches
  - Grade 2 .... 3.3862 to 3.3866 inches
  - Grade 3 .... 3.3866 to 3.3870 inches

Out-of-round
- Service limit .... 0.0006 inch
- Taper .... 0.0004 inch

Crankshaft and connecting rods

Crankshaft runout	0.0020 inch maximum
Endplay	
GA16DE	0.012 inch maximum
SR20DE	0.0118 inch maximum
Main bearing journal	
Diameter	
GA16DE	
Grade 0	1.9668 to 1.9671 inches
Grade 1	1.9665 to 1.9668 inches
Grade 2	1.9661 to 1.9665 inches
SR20DE	
Grade 0	2.1643 to 2.1646 inches
Grade 1	2.1641 to 2.1643 inches
Grade 2	2.1639 to 2.1641 inches
Grade 3	2.1636 to 2.1639 inches
Taper	0.0002 inch maximum
Out-of-round	0.0002 inch maximum
Main bearing oil clearance	
GA16DE	
Standard	0.0007 to 0.0017 inch
Limit	0.004
SR20DE	
Standard	0.0002 to 0.0009 inch
Limit	0.002 inch
Connecting rod journal	
Diameter	
GA16DE	
Grade 0	1.5735 to 1.5738 inches
Grade 1	1.5733 to 1.5735 inches
Grade 2	1.5731 to 1.5733 inches
SR20DE	
Grade 0	1.8885 to 1.8887 inches
Grade 1	1.8883 to 1.8885 inches
Grade 2	1.8880 to 1.8883 inches
Connecting rod oil clearance	
GA16DE	
Standard	0.0004 to 0.0014 inch
Limit	0.004 inch
SR20DE	
Standard	0.0008 to 0.0018 inch
Limit	0.00256 inch
Connecting rod side clearance (endplay)	
GA16DE	
Standard	0.0079 to 0.0185 inch
Limit	0.0197 inch
SR20DE	
Standard	0.0079 to 0.0138 inch
Limit	0.020 inch

Piston and rings

Piston diameter	
GA16DE	
Grade 1	2.9911 to 2.9915 inches
Grade 2	2.9915 to 2.9919 inches
Grade 3	2.9919 to 2.9923 inches
SR20DE	
Grade 1	3.3850 to 3.3854 inches
Grade 2	3.3854 to 3.3858 inches
Grade 3	3.3858 to 3.3862 inches
0.20 oversize	3.3929 to 3.3941 inches
Piston-to-bore clearance	
GA16DE	0.0006 to 0.0014 inch
SR20DE	0.0004 to 0.0012 inch
Piston ring end gap	
GA16DE	
Top	0.0079 to 0.0157 inch
Oil ring	0.0079 to 0.0236 inch

SR20DE
- No. 1 (top ring) .... 0.0079 to 0.0118 inch
- No. 2 (middle ring) .... 0.0138 to 0.0197 inch
- Oil ring .... 0.0079 to 0.0236 inch

Piston ring side clearance
- GA16DE (top) .... 0.0016 to 0.0031 inch

SR20DE
- No. 1 (top ring) .... 0.0012 to 0.0026 inch
- No. 2 (middle ring) .... 0.0012 to 0.0026 inch
- Service limit (all models) .... 0.008 inch

Valves and related components

Valve face angle 45-degrees

Valve spring free length
- GA16DE .... 1.6217 inches
- SR20DE .... 1.9433 inches

Valve spring out of square limit
- GA16DE .... 0.0709 inch
- SR20DE .... 0.087 inch

Valve margin width
- GA16DE .... 0.035 to 0.043 inch
- SR20DE
  - Intake .... 0.043 inch
  - Exhaust .... 0.051 inch

Valve stem diameter
- GA16DE
  - Intake .... 0.2152 to 0.2157 inch
  - Exhaust .... 0.2144 to 0.2150 inch
- SR20DE
  - Intake .... 0.2348 to 0.2354 inch
  - Exhaust .... 0.2341 to 0.2346 inch

Valve stem-to-guide clearance
- GA16DE
  - Intake .... 0.0008 to 0.0020 inch
  - Exhaust .... 0.0016 to 0.0028 inch
- SR20DE
  - Intake .... 0.0008 to 0.0021 inch
  - Exhaust .... 0.0016 to 0.0029 inch

Torque specifications*

Ft-lbs (unless otherwise indicated)

Main bearing bolts
- GA16DE .... 34 to 38
- SR20DE
  - Step 1 .... 20 to 24
  - Step 2 .... Tighten an additional 75 to 80-degrees
  - Step 3 .... Loosen completely
  - Step 4 .... 24 to 28
  - Step 5 .... Tighten an additional 45 to 50-degrees

Connecting rod nuts
- GA16DE
  - Step 1 .... 120 to 144 in-lbs
  - Preferred step 2 .... Tighten an additional 35 to 40-degrees
  - Alternate step 2 .... 17 to 21
- SR20DE
  - Step 1 .... 120 to 144 in-lbs
  - Preferred step 2 .... Tighten an additional 60 to 65-degrees
  - Alternate step 2 .... 28 to 33

*Note: *Refer to Part A for additional torque specifications*

Chapter 2 Part B General engine overhaul procedures

1 General information

Included in this portion of Chapter 2 are the general overhaul procedures for the cylinder head and internal engine components.

The information ranges from advice concerning preparation for an overhaul and the purchase of replacement parts to detailed, step-by-step procedures covering removal and installation of internal engine components and the inspection of parts.

The following Sections have been written based on the assumption that the engine has been removed from the vehicle. For information concerning in-vehicle engine repair, as well as removal and installation of the external components necessary for the overhaul, see Section 7 and Chapter 2A.

The Specifications included in this Part are only those necessary for the inspection and overhaul procedures which follow. Refer to Part A for additional Specifications.

2 Engine overhaul - general information

It's not always easy to determine when, or if, an engine should be completely overhauled, as a number of factors must be considered.

High mileage is not necessarily an indication that an overhaul is needed, while low mileage doesn't preclude the need for an overhaul. Frequency of servicing is probably the most important consideration. An engine that's had regular and frequent oil and filter changes, as well as other required maintenance, will most likely give many thousands of miles of reliable service. Conversely, a neglected engine may require an overhaul very early in its life.

Excessive oil consumption is an indication that piston rings, valve seals and/or valve guides are in need of attention. Make sure that oil leaks aren't responsible before deciding that the rings and/or guides are bad. Perform a cylinder compression check to determine the extent of the work required (see Section 4).

Check the oil pressure with a gauge installed in place of the oil pressure sending unit and compare it to the Specifications. If it's extremely low, the bearings and/or oil pump are probably worn out.

Loss of power, rough running, knocking or metallic engine noises, excessive valve train noise and high fuel consumption rates may also point to the need for an overhaul, especially if they're all present at the same time. If a complete tune-up doesn't remedy the situation, major mechanical work is the only solution.

An engine overhaul involves restoring the internal parts to the specifications of a new engine. During an overhaul, the piston rings are replaced and the cylinder walls are reconditioned (rebored and/or honed). If a rebore is done by an automotive machine shop, new oversize pistons will also be installed. The main bearings, connecting rod bearings and camshaft bearings are generally replaced with new ones and, if necessary, the crankshaft may be reground to restore the journals. Generally, the valves are serviced as well, since they're usually in less-than-perfect condition at this point. The end result should be a like-new engine that will give many trouble free miles. **Note:** *Critical cooling system components such as the hoses, drivebelts, thermostat and water pump MUST be replaced with new parts when an engine is overhauled. The radiator should be checked carefully to ensure that it isn't clogged or leaking* (see Chapter 3). *If you purchase a rebuilt engine or shortblock, some rebuilders will not warranty their engines unless the radiator has been professionally flushed.*

Before beginning the engine overhaul, read through the entire procedure to familiarize yourself with the scope and requirements of the job. Overhauling an engine isn't difficult, but it is time consuming. Plan on the vehicle being tied up for a minimum of two weeks, especially if parts must be taken to an automotive machine shop for repair or reconditioning. Check on availability of parts and make sure that any necessary special tools and equipment are obtained in advance. Most work can be done with typical hand tools, although a number of precision measuring tools are required for inspecting parts to determine if they must be replaced. Often an automotive machine shop will handle the inspection of parts and offer advice concerning reconditioning and replacement. **Note:** *Always wait until the engine has been completely disassembled and all components, especially the engine block, have been inspected before deciding what service and repair operations must be performed by an automotive machine shop. Since the block's condition will be the major factor to consider when determining whether to overhaul the original engine or buy a rebuilt one, never purchase parts or have machine work done on other components until the block has been thoroughly inspected.* As a general rule, time is the primary cost of an overhaul, so it doesn't pay to install worn or substandard parts.

As a final note, to ensure maximum life and minimum trouble from a rebuilt engine, everything must be assembled with care in a spotlessly clean environment.

3 Vacuum gauge diagnostic checks

A vacuum gauge provides valuable information about what is going on in the engine at a low-cost. You can check for worn rings or cylinder walls, leaking head or intake manifold gaskets, incorrect carburetor adjustments, restricted exhaust, stuck or burned valves, weak valve springs, improper ignition or valve timing and ignition problems.

Unfortunately, vacuum gauge readings are easy to misinterpret, so they should be used in conjunction with other tests to confirm the diagnosis.

Both the absolute readings and the rate of needle movement are important for accurate interpretation. Most gauges measure vacuum in inches of mercury (in-Hg). The following references to vacuum assume the diagnosis is being performed at sea level. As vacuum increases (or atmospheric pressure decreases), the reading will decrease. For every 1,000 foot increase in elevation above approximately 2000 feet, the gauge readings will decrease about one inch of mercury.

Connect the vacuum gauge directly to intake manifold vacuum, not to ported (carburetor) vacuum. Be sure no hoses are left disconnected during the test or false readings will result.

Before you begin the test, allow the engine to warm up completely. Block the wheels and set the parking brake. With the transmission in neutral (or Park, on automatics), start the engine and allow it to run at normal idle speed. **Warning:** *Carefully inspect the fan blades for cracks or damage before starting the engine. Keep your hands and the vacuum tester clear of the fan and do not stand in front of the vehicle or in line with the fan when the engine is running.*

Read the vacuum gauge; an average, healthy engine should normally produce between 17 and 22 inches of vacuum with a fairly steady needle. Refer to the following vacuum gauge readings and what they indicate about the engine's condition:

1 A low steady reading usually indicates a leaking gasket between the intake manifold and carburetor or throttle body, a leaky vacuum hose, late ignition timing or incorrect camshaft timing. Check ignition timing with a timing light and eliminate all other possible causes, utilizing the tests provided in this Chapter before you remove the timing chain cover to check the timing marks.

2 If the reading is three to eight inches below normal and it fluctuates at that low reading, suspect an intake manifold gasket leak at an intake port or a faulty injector (on port-injected models only).

3 If the needle has regular drops of about two to four inches at a steady rate the valves are probably leaking. Perform a compression or leak-down test to confirm this.

4 An irregular drop or down-flick of the needle can be caused by a sticking valve or an ignition misfire. Perform a compression or leak-down test and read the spark plugs.

5 A rapid vibration of about four in.-Hg vibration at idle combined with exhaust smoke indicates worn valve guides. Perform a leak-down test to confirm this. If the rapid vibration occurs with an increase in engine speed, check for a leaking intake manifold gasket or head gasket, weak valve springs, burned valves or ignition misfire.

6 A slight fluctuation, say one inch up and down, may mean ignition problems. Check all the usual tune-up items and, if necessary, run

Chapter 2 Part B General engine overhaul procedures

4.6 A compression gauge with a threaded fitting for the spark plug hole is preferred over the type that requires hand pressure to maintain the seal - be sure to open the throttle valve as far as possible during the compression check

the engine on an ignition analyzer.

7 If there is a large fluctuation, perform a compression or leak-down test to look for a weak or dead cylinder or a blown head gasket.

8 If the needle moves slowly through a wide range, check for a clogged PCV system, incorrect idle fuel mixture, carburetor/throttle body or intake manifold gasket leaks.

9 Check for a slow return after revving the engine by quickly snapping the throttle open until the engine reaches about 2,500 rpm and let it shut. Normally the reading should drop to near zero, rise above normal idle reading (about 5 in.-Hg over) and then return to the previous idle reading. If the vacuum returns slowly and doesn't peak when the throttle is snapped shut, the rings may be worn. If there is a long delay, look for a restricted exhaust system (often the muffler or catalytic converter). An easy way to check this is to temporarily disconnect the exhaust ahead of the suspected part and redo the test.

4 Compression check

Refer to illustration 4.6

1 A compression check will tell you what mechanical condition the upper end (pistons, rings, valves, head gasket) of your engine is in. Specifically, it can tell you if the compression is down due to leakage caused by worn piston rings, defective valves and seats or a blown head gasket. **Note:** *The engine must be at normal operating temperature and the battery must be fully charged for this check.*

2 Begin by cleaning the area around the spark plugs before you remove them (compressed air should be used, if available, otherwise a small brush or even a bicycle tire pump will work). The idea is to prevent dirt from getting into the cylinders as the compression check is being done.

3 Remove all of the spark plugs from the engine (see Chapter 1).
4 Block the throttle wide open.
5 Detach the primary (low voltage) wiring from the distributor. The fuel pump fuse should also be removed.
6 Install the compression gauge in the spark plug hole **(see illustration)**.
7 Crank the engine over at least five compression strokes and watch the gauge. The compression should build up quickly in a healthy engine. Low compression on the first stroke, followed by gradually increasing pressure on successive strokes, indicates worn piston rings. A low compression reading on the first stroke, which doesn't build up during successive strokes, indicates leaking valves or a blown head gasket (a cracked head could also be the cause). Deposits on the undersides of the valve heads can also cause low compression. Record the highest gauge reading obtained.
8 Repeat the procedure for the remaining cylinders and compare the results to the Specifications.
9 Add some engine oil (about three squirts from a plunger-type oil can) to each cylinder, through the spark plug holes, and repeat the test.
10 If the compression increases after the oil is added, the piston rings are definitely worn. If the compression doesn't increase significantly, the leakage is occurring at the valves or head gasket. Leakage past the valves may be caused by burned valve seats and/or faces or warped, cracked or bent valves.
11 If two adjacent cylinders have equally low compression, there's a strong possibility that the head gasket between them is blown. The appearance of coolant in the combustion chambers or the crankcase would verify this condition.
12 If one cylinder is 20-percent lower than the others, and the engine has a slightly rough idle, a worn exhaust lobe on the camshaft could be the cause.
13 If the compression is unusually high, the combustion chambers are probably coated with carbon deposits. If that's the case, the cylinder head(s) should be removed and decarbonized.
14 If compression is way down or varies greatly between cylinders, it would be a good idea to have a leak-down test performed by an automotive repair shop. This test will pinpoint exactly where the leakage is occurring and how severe it is.
15 After the compression test has been performed, restore the ignition and fuel pump functions and be sure to unblock the throttle.

5 Engine removal - methods and precautions

If you've decided that an engine must be removed for overhaul or major repair work, several preliminary steps should be taken.

Locating a suitable place to work is extremely important. Adequate work space, along with storage space for the vehicle, will be needed. If a shop or garage isn't available, at the very least a flat, level, clean work surface made of concrete or asphalt is required.

Cleaning the engine compartment and engine before beginning the removal procedure will help keep tools clean and organized.

An engine hoist or A-frame will also be necessary. Make sure the equipment is rated in excess of the combined weight of the engine and transaxle. Safety is of primary importance, considering the potential hazards involved in lifting the engine out of the vehicle.

If the engine is being removed by a novice, a helper should be available. Advice and aid from someone more experienced would also be helpful. There are many instances when one person cannot simultaneously perform all of the operations required when lifting the engine out of the vehicle.

Plan the operation ahead of time. Arrange for or obtain all of the tools and equipment you'll need prior to beginning the job. Some of the equipment necessary to perform engine removal and installation safely and with relative ease are (in addition to an engine hoist) a heavy duty floor jack, complete sets of wrenches and sockets as described in the front of this manual, wooden blocks and plenty of rags and cleaning solvent for mopping up spilled oil, coolant and gasoline. If the hoist must be rented, make sure that you arrange for it in advance and perform all of the operations possible without it beforehand. This will save you money and time.

Plan for the vehicle to be out of use for quite a while. A machine shop will be required to perform some of the work which the do-it-yourselfer can't accomplish without special equipment. These shops often have a busy schedule, so it would be a good idea to consult them before removing the engine in order to accurately estimate the amount of time required to rebuild or repair components that may need work.

Always be extremely careful when removing and installing the engine. Serious injury can result from careless actions. Plan ahead, take your time and a job of this nature, although major, can be accomplished successfully.

6 Engine - removal and installation

Refer to illustrations 6.7, 6.18, 6.22 and 6.23
Warning: *Gasoline is extremely flammable, so take extra precautions when disconnecting any part of the fuel system. Don't smoke or allow open flames or bare light bulbs in or near the work area and don't work in a garage where a natural gas appliance (such as a clothes dryer or water heater) is installed. If you spill gasoline on your skin, rinse it off immediately. Have a fire extinguisher rated for*

6.7 Be sure to disconnect all of the ground cables from the cylinder head and engine block

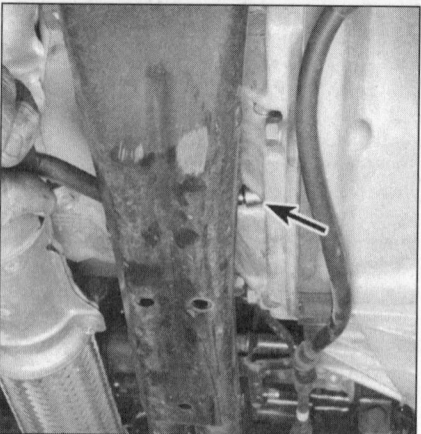

6.18 On automatic transaxle models, the converter nuts must be removed before the engine can be disconnected from the transaxle. Access to the bolts is through a dust cover (arrow) or the starter hole

6.22 Mark the position of the front engine mount bolt on its bracket before removing it

6.23 The transaxle can be left attached to the vehicle (and supported by a jack) and the engine pulled out separately, as on this model. Guide the engine/transaxle assembly as it comes up and out to clear any components in the engine compartment

gasoline fires handy and know how to use it.
Note: *Read through the entire Section before beginning this procedure. The engine and transaxle may be removed as a unit, to be separated afterward on the garage floor, or either may be supported with jacks and/or chain hoists while the other is removed separately. For more on transaxle removal, see Chapter 7.*

Removal

1 Relieve the fuel system pressure (see Chapter 4). Disconnect the cable from the negative battery terminal.
2 Place protective covers on the fenders and cowl and remove the hood (see Chapter 11).
3 Remove the battery and battery tray (see Chapter 5).
4 Remove the air cleaner assembly (see Chapter 4).
5 Remove the cruise control actuator cable, if equipped.
6 Raise the vehicle and support it securely on jackstands. Drain the cooling system and engine oil and remove the drivebelts (see Chapter 1). **Note:** *Don't raise the vehicle any higher than necessary.*
7 Clearly label and disconnect all vacuum lines, coolant and emissions hoses, electrical connectors, ground straps and fuel lines (for fuel line removal see Chapter 4). Masking tape and/or a touch up paint applicator work well for marking items. Take instant photos or sketch the locations of components and brackets, if necessary **(see illustration)**.
8 Remove the cooling fan(s), shroud and radiator (see Chapter 3).
9 Release the residual fuel pressure in the tank by removing the gas cap, then disconnect the fuel lines between the engine and chassis (see Chapter 4). Plug or cap all open fittings.
10 Disconnect the throttle linkage and transaxle shift control cable (if equipped) from the throttle body (see Chapter 4).
11 Remove the alternator and starter (see Chapter 5).

12 On power steering equipped vehicles, unbolt the power steering pump (see Chapter 10). Tie the pump aside without disconnecting the hoses.
13 On air-conditioned models, unbolt the compressor and set it aside (see Chapter 3). **Warning:** *Do not disconnect the refrigerant hoses.*
14 Unbolt the exhaust pipe from the exhaust manifold (see Chapter 2A).
15 Unbolt and remove the engine-to-transmission reinforcement brackets **(see Chapter 2A, illustrations 9.6a and 9.6b)**.
16 Remove the right front wheel and the right driveaxle (see Chapter 8).
17 On automatic transaxle equipped models, detach the torque converter cover from the lower bellhousing. On some automatic transaxle models, the rear engine plate is one-piece and access to the converter bolts is through the starter hole in the plate.
18 Remove the torque converter-to-driveplate fasteners **(see illustration)** and push the converter back slightly into the bellhousing.
19 Attach a lifting sling or chain to the brackets on the engine. Position a hoist and connect the sling or chain to it. Take up the slack until there is slight tension on the hoist. **Warning:** *Do not place any part of your body under the engine/transaxle when it's supported only by a hoist or other lifting device.*
20 Recheck to be sure nothing except the mounts are still connecting the engine/transaxle to the vehicle. Disconnect anything still remaining.
21 Place a floor jack under the transmission bellhousing area, to support the transaxle when the engine is unbolted and removed. Be sure to place a wood block on the jack head to serve as a cushion.
22 Remove the remaining engine mounts **(see illustration)**, but leave the transaxle mounts connected.
23 Slowly raise the engine out of the vehicle **(see illustration)**. It may be necessary to tilt the engine up at the front, using the jack under the transaxle to tilt it up as well. Have an assistant twist the engine sling to angle the engine as needed as it is raised with the hoist.
24 Move the engine away from the vehicle and carefully lower the hoist until the engine is on the floor, supported by wood blocks, or remove the flywheel or driveplate and mount the engine on an engine stand.

Installation

25 Check the engine/transaxle mounts. If they're worn or damaged, replace them.
26 On manual transaxle equipped models, inspect the clutch components (see Chapter 8) and, on automatic transaxle models, inspect the converter seal and bushing.
27 On automatic transaxle equipped models, apply a dab of grease to the nose of the torque converter and to the seal lips.
28 Attach the hoist to the engine and carefully guide the engine into place, following the

procedure outlined in Chapter 7A or 7B. **Caution:** *Do not use the bolts to force the engine and transaxle into alignment. It may crack or damage major components.* Install the engine-to-transaxle bolts and tighten them to the torque listed in Chapter 7A or 7B Specifications.

29 If you're working on a model equipped with an automatic transaxle, install the torque converter-to-driveplate fasteners and tighten them to the torque listed in the Chapter 7B Specifications.

30 Install the mount bolts and tighten them securely.

31 Reinstall the remaining components and fasteners in the reverse order of removal.

32 Add coolant, oil, power steering and transmission fluids as needed (see Chapter 1).

33 Run the engine and check for proper operation and leaks. Shut off the engine and recheck the fluid levels.

7 Engine rebuilding alternatives

The do-it-yourselfer is faced with a number of options when performing an engine overhaul. The decision to replace the engine block, piston/connecting rod assemblies and crankshaft depends on a number of factors, with the number one consideration being the condition of the block. Other considerations are cost, access to machine shop facilities, parts availability, time required to complete the project and the extent of prior mechanical experience on the part of the do-it-yourselfer.

Some of the rebuilding alternatives include:

Individual parts - If the inspection procedures reveal that the engine block and most engine components are in reusable condition, purchasing individual parts may be the most economical alternative. The block, crankshaft and piston/connecting rod assemblies should all be inspected carefully. Even if the block shows little wear, the cylinder bores should be surface honed.

Short block - A short block consists of an engine block with a crankshaft and piston/connecting rod assemblies already installed. All new bearings are incorporated and all clearances will be correct. The existing camshaft, valve train components, cylinder head(s) and external parts can be bolted to the short block with little or no machine shop work necessary.

Long block - A long block consists of a short block plus an oil pump, oil pan, cylinder head, camshaft and valve train components, timing pulleys and belt. All components are installed with new bearings, seals and gaskets incorporated throughout. The installation of manifolds and external parts is all that's necessary.

Give careful thought to which alternative is best for you and discuss the situation with local automotive machine shops, auto parts dealers and experienced rebuilders before ordering or purchasing replacement parts.

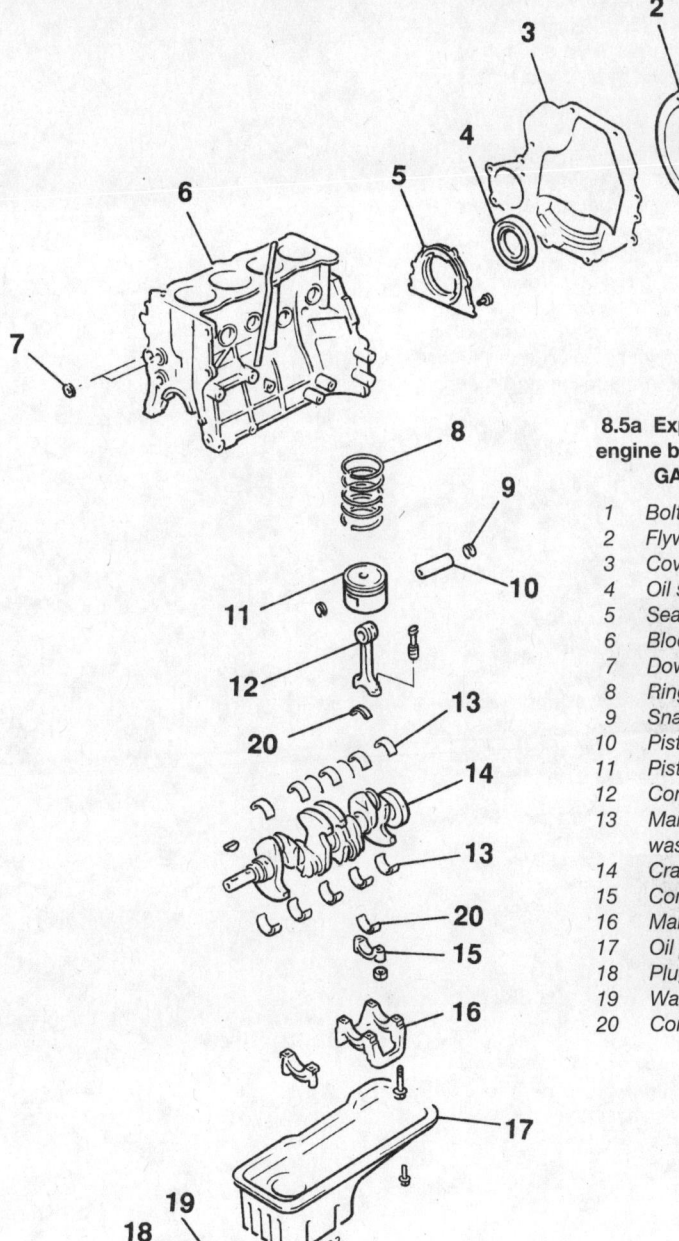

8.5a Exploded view of the engine block components - GA16DE engine

1 Bolt
2 Flywheel
3 Cover
4 Oil seal
5 Seal housing
6 Block
7 Dowels and O-rings
8 Rings
9 Snap-rings
10 Piston pin
11 Piston
12 Connecting rod
13 Main bearings and thrust washers
14 Crankshaft
15 Connecting rod cap
16 Main bearing caps
17 Oil pan
18 Plug
19 Washer
20 Connecting rod bearings

8 Engine overhaul - disassembly sequence

Refer to illustrations 8.5a and 8.5b

1 It's much easier to disassemble and work on the engine if it's mounted on a portable engine stand. A stand can often be rented quite cheaply from an equipment rental yard. Before the engine is mounted on a stand, the flywheel/driveplate and rear oil seal retainer should be removed from the engine.

2 If a stand isn't available, it's possible to disassemble the engine with it blocked up on the floor. Be extra careful not to tip or drop the engine when working without a stand.

3 If you're going to obtain a rebuilt engine, all external components must come off first, to be transferred to the replacement engine, just as they will if you're doing a complete engine overhaul yourself. These include:

Emissions control components
Distributor, spark plug wires and spark plugs
Thermostat and housing
Water pump
EFI components
Intake/exhaust manifolds
Oil filter (always use a new filter)
Engine mounts
Clutch and flywheel/driveplate
Engine rear plate

Note: *When removing the external compo-*

2B-8 Chapter 2 Part B General engine overhaul procedures

nents from the engine, pay close attention to details that may be helpful or important during installation. Note the installed position of gaskets, seals, spacers, pins, brackets, washers, bolts and other small items.

4 If you're obtaining a short block, which consists of the engine block, crankshaft, pistons and connecting rods all assembled, then the cylinder head, oil pan, front timing cover and oil pump will have to be removed as well, to be reused on the rebuilt short block. See Section 7 for additional information regarding the different possibilities to be considered.

5 If you're planning a complete overhaul, the engine must be disassembled and the components removed in the following order (see illustrations).

Valve cover
Oil pan and pick-up tube
Timing chain cover
Oil pump
Timing chain and sprockets
Camshafts
Cylinder head
Piston/connecting rod assemblies
Crankshaft rear oil seal retainer
Crankshaft and main bearings

6 Before beginning the disassembly and overhaul procedures, make sure the following items are available. Also, refer to Section 21 for a list of tools and materials needed for engine reassembly.

Common hand tools
Small cardboard boxes or plastic bags for storing parts
Gasket scraper
Ridge reamer
Crankshaft pulley puller
Micrometers
Telescoping gauges
Dial indicator set
Valve spring compressor
Cylinder surfacing hone
Piston ring groove cleaning tool
Electric drill motor
Tap and die set
Wire brushes
Oil gallery brushes
Cleaning solvent

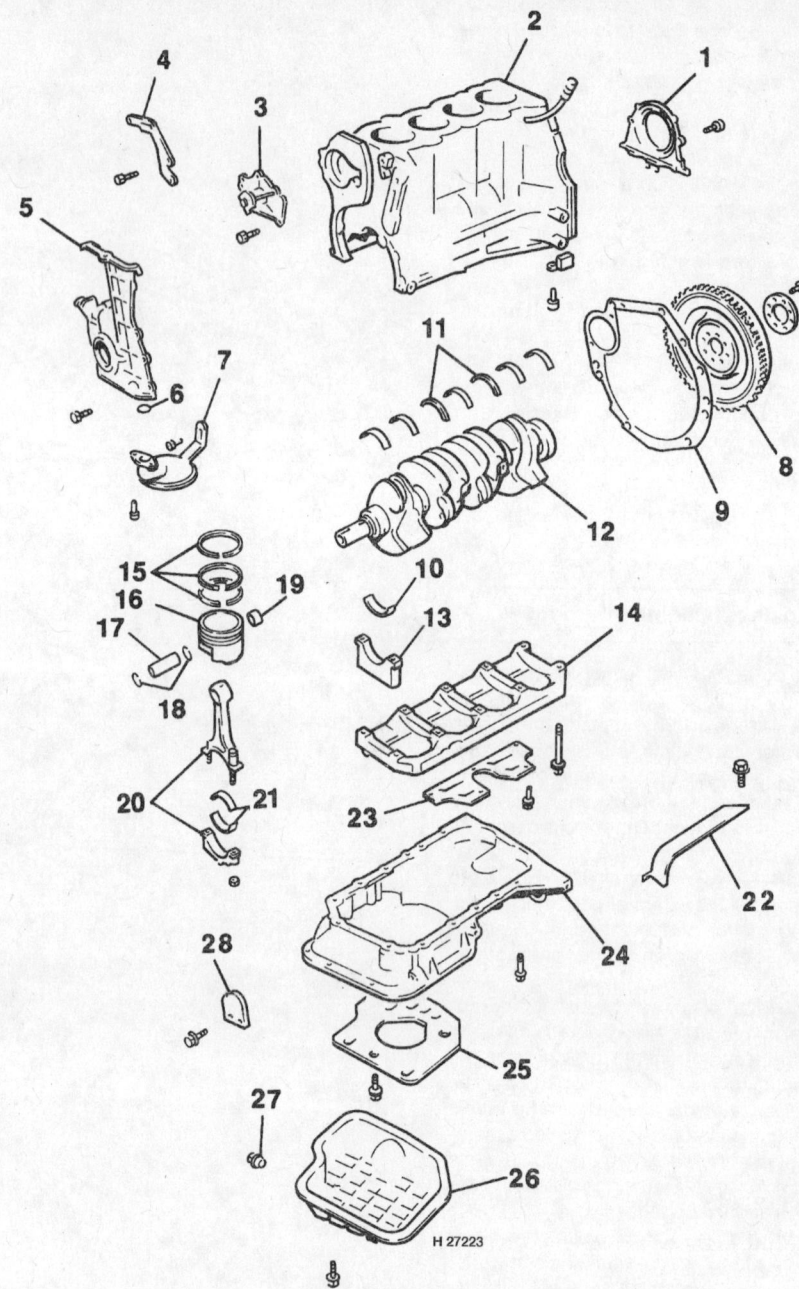

8.5b Exploded view of the engine block components - SR20DE engine

1	Oil seal housing	11	Thrustwashers	21	Connecting rod bearings
2	Block	12	Crankshaft	22	Baffle
3	Water pump	13	Main bearing caps	23	Baffle
4	Alternator bracket	14	Main bearing beam	24	Oil pan, aluminum
5	Cover	15	Rings	25	Baffle
6	O-ring	16	Piston	26	Oil pan, steel
7	Strainer and pick-up	17	Piston pin	27	Plug
8	Flywheel	18	Snap-rings	28	Cover
9	Cover	19	Bushing		
10	Main bearings	20	Connecting rod cap		

9 Cylinder head - disassembly

Refer to illustrations 9.2, 9.3, 9.4a and 9.4b
Note: *New and rebuilt cylinder heads are commonly available for most engines at dealerships and auto parts stores. Due to the fact that some specialized tools are necessary for the disassembly and inspection procedures, and replacement parts may not be readily available, it may be more practical and economical for the home mechanic to purchase a replacement head rather than taking the time to disassemble, inspect and recondition the original.*

1 Cylinder head disassembly involves removal of the intake and exhaust valves and related components. It's assumed that the camshafts have already been removed (see Part A as needed). On SR20DE engines, remove the lash adjusters and store them in order vertically with the smaller end up or else immersed in a container of engine oil. This will prevent the entry of air. On GA16DE engines, make certain to keep the valve shims in order.

2 Before the valves are removed, arrange to label and store them, along with their related components, so they can be kept

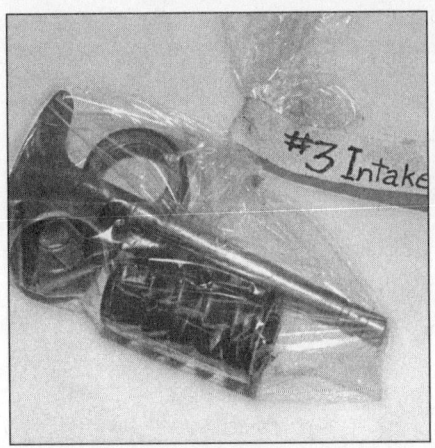

9.2 A small plastic bag, with an appropriate label, can be used to store the valve train components so they can be kept together and reinstalled in the original position

9.3 Use a valve spring compressor to compress the spring, then remove the keepers (arrow) from the valve stem

9.4a If the valve won't pull through the guide, deburr the edge of the stem end and the area around the top of the keeper groove with a file or whetstone

9.4b Use pliers to remove the seal from the guide

separate and reinstalled in the same valve guides they are removed from **(see illustration)**.

3 Compress the springs on the first valve with a spring compressor and remove the keepers **(see illustration)**. Carefully release the valve spring compressor and remove the retainer, the spring and the spring seat.

4 Pull the valve out of the head and remove the oil seal from the guide (in either order). **Note:** *It is difficult to remove the seals on the GA16 engine. A special tool is available, but you might be able to remove the seals with needle-nose pliers.* If the valve binds in the guide (won't pull through), push it back into the head and deburr the area around the keeper groove with a fine file or whetstone **(see illustrations)**.

5 Repeat the procedure for the remaining valves. Remember to keep all the parts for each valve together so they can be reinstalled in the same locations.

6 Once the valves and related components have been removed and stored in an organized manner, the head should be thoroughly cleaned and inspected. If a complete engine overhaul is being done, finish the engine disassembly procedures before beginning the cylinder head cleaning and inspection process.

10 Cylinder head - cleaning and inspection

1 Thorough cleaning of the cylinder head and related valve train components, followed by a detailed inspection, will enable you to decide how much valve service work must be done during the engine overhaul. **Note:** *If the engine was severely overheated, the cylinder head is probably warped, if not cracked (see Step 12).*

Cleaning

2 Scrape all traces of old gasket material and sealing compound off the head gasket, intake manifold and exhaust manifold sealing surfaces. Be very careful not to gouge the cylinder head. Special gasket removal solvents that soften gaskets and make removal much easier are available at auto parts stores.

3 Remove all built-up scale from the coolant passages.

4 Run a stiff wire brush through the various holes to remove deposits that may have formed in them.

5 Run an appropriate size tap into each of the threaded holes to remove corrosion and thread sealant that may be present. If compressed air is available, use it to clear the holes of debris produced by this operation. **Warning:** *Wear eye protection when using compressed air!*

6 Clean the exhaust and intake manifold stud threads with a wire brush.

7 Clean the cylinder head with solvent and dry it thoroughly. Compressed air will speed the drying process and ensure that all holes and recessed areas are clean. **Note:** *Decarbonizing chemicals are available and may prove very useful when cleaning cylinder heads and valve train components. They are very caustic and should be used with caution. Be sure to follow the instructions on the container.*

8 Clean the lifters (if used) with solvent and dry them thoroughly (don't mix them up during the cleaning process). Compressed air will speed the drying process and can be used to clean out the oil passages. **Caution:** *There are hydraulic lash adjusters on SR20DE engines. These must be kept in a vertical (as installed) plane or oil may bleed out and air may enter. They may be stored on their side submerged in a pan of engine oil. They are very difficult to bleed down in engine operation if air is allowed in during disassembly procedures.*

9 Clean all the valve springs, spring seats, keepers and retainers with solvent and dry them thoroughly. Do the components from one valve at a time to avoid mixing up the parts. Number the valve lifters and their shims and keep them together.

10 Scrape off any heavy deposits that may have formed on the valves, then use a motorized wire brush to remove deposits from the valve heads and stems. Again, make sure the valves don't get mixed up.

Inspection

Note: *Be sure to perform all of the following inspection procedures before concluding that machine shop work is required. Make a list of the items that need attention. The inspection procedures for the lifters and rocker arms, as well as the camshaft(s), can be found in Part A.*

Cylinder head

Refer to illustrations 10.12, 10.14a and 10.14b

11 Inspect the head very carefully for cracks, evidence of coolant leakage and other damage. If cracks are found, check with an automotive machine shop concerning repair. If repair isn't possible, a new cylinder head should be obtained. If the engine has been overheated, have an automotive

2B-10   Chapter 2 Part B   General engine overhaul procedures

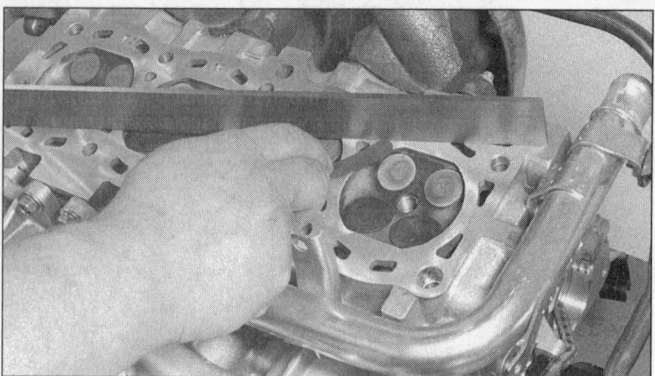

10.12 Check the cylinder head gasket surface for warpage by trying to slip a feeler gauge under the straightedge (see this Chapter's Specifications for the maximum warpage allowed and use a feeler gauge of that thickness)

10.14a Measuring valve stem diameter

10.14b A dial indicator can be used to determine the valve stem-to-guide clearance (move the valve stem as indicated by the arrows)

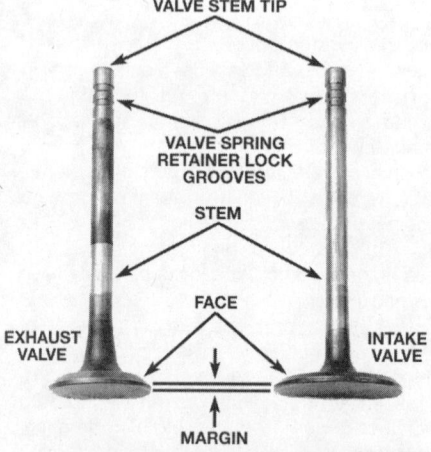

10.15 Check for valve wear at the points shown here

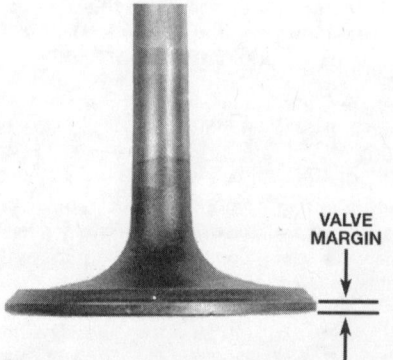

10.16 The margin width on each valve must be as specified (if no margin exists, the valve cannot be reused)

10.17 Measure the free length of each valve spring with a dial or vernier caliper

machine shop inspect the head for cracks which are invisible to the naked eye.

12 Using a straightedge and feeler gauge, check the head gasket mating surface for warpage **(see illustration)**. If the warpage exceeds the specified limit, it can be resurfaced at an automotive machine shop.

13 Examine the valve seats in each of the combustion chambers. If they're pitted, cracked or burned, the head will require valve service that's beyond the scope of the home mechanic.

14 Measure the valve guide inside diameter with a small hole gauge and micrometer then measure the valve stem diameter with a micrometer and subtract it from the valve guide inside diameter to obtain the stem-to-guide clearance **(see illustration)**. Also, check the valve stem deflection crosswise (parallel to the rocker arm) with a dial indicator attached securely to the head **(see illustration)**. The valve must be in the guide and approximately 1/16-inch off the seat. The total valve stem movement indicated by the gauge needle must be divided by two to determine the actual clearance. If it exceeds the specified stem-to-guide clearance limit, the valve guides should be replaced. After this is done, if there's still some doubt regarding the condition of the valve guides they should be checked by an automotive machine shop (the cost should be minimal).

Valves

Refer to illustrations 10.15 and 10.16

15 Carefully inspect each valve face for uneven wear, deformation, cracks, pits and burned areas. Check the valve stem for scuffing and galling and the neck for cracks. Rotate the valve and check for any obvious indication that it's bent. Look for pits and excessive wear on the end of the stem **(see illustration)**. The presence of any of these conditions indicates the need for valve service by an automotive machine shop.

16 Measure the margin width on each valve **(see illustration)**. Any valve with a margin narrower than specified will have to be replaced with a new one.

Valve components

Refer to illustrations 10.17 and 10.18

17 Check each valve spring for wear (on the ends) and pits. Measure the free height and compare it to the Specifications **(see illustration)**. Any springs that are shorter than specified have sagged and should not be reused. The tension of all springs should be checked with a special fixture before deciding that they're suitable for use in a rebuilt engine (take the springs to an automotive machine shop for this check).

18 Stand each spring on a flat surface and check it for squareness **(see illustration)**. If any of the springs are distorted or sagged, replace all of them with new parts.

19 Check the spring retainers and keepers for obvious wear and cracks. Any questionable parts should be replaced with new ones, as extensive damage will occur if they fail

Chapter 2 Part B General engine overhaul procedures

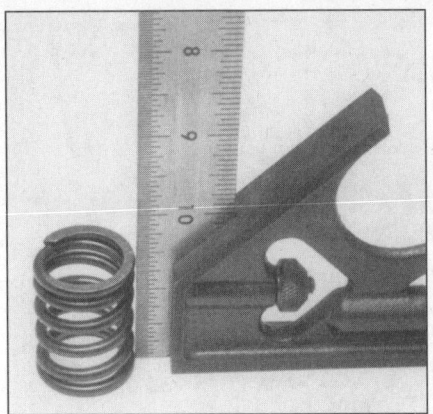

10.18 Check each valve spring for squareness, if it is bent it should be replaced

12.5a Install the spring seat . . .

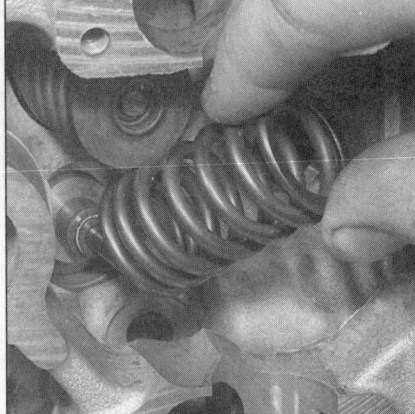

12.5b . . . then place the spring on - the tighter-wound coils must go at the bottom

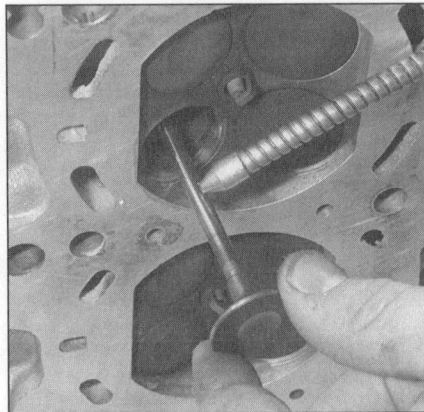

12.5c Oil the valve stem before sliding it through the guide

12.5d Install the retainer and compress the spring

12.6 Use a little grease to hold the keepers in place while you install them

during engine operation.
20 Any damaged or excessively worn parts must be replaced with new ones.
21 If the inspection process indicates that the valve components are in generally poor condition and worn beyond the limits specified, which is usually the case in an engine that's being overhauled, reassemble the valves in the cylinder head and refer to valve servicing recommendations (see Section 11).

11 Valves - servicing

1 Because of the complex nature of the job and the special tools and equipment needed, servicing of the valves, the valve seats and the valve guides, commonly known as a valve job, should be done by a professional.
2 The home mechanic can remove and disassemble the head, do the initial cleaning and inspection, then reassemble and deliver it to a dealer service department or an automotive machine shop for the actual service work. Doing the inspection will enable you to see what condition the head and valvetrain components are in and will ensure that you know what work and new parts are required when dealing with an automotive machine shop.
3 The dealer service department, or automotive machine shop, will remove the valves and springs, recondition or replace the valves and valve seats, check and replace the valve guides (as necessary), check and replace the valve springs, spring retainers and keepers (as necessary), replace the valve seals with new ones, reassemble the valve components and make sure the installed spring height is correct. The cylinder head gasket surface will also be resurfaced (within specification limits) if it's warped.
4 After the valve job has been performed by a professional, the head will be in like-new condition. When the head is returned, be sure to clean it again before installation on the engine to remove any metal particles and abrasive grit that may still be present from the valve service or head resurfacing operations. Use compressed air, if available, to blow out all the oil holes and passages.

12 Cylinder head - reassembly

Refer to illustrations 12.5a, 12.5b, 12.5c, 12.5d and 12.6

1 Regardless of whether or not the head was sent to an automotive repair shop for valve servicing, make sure it's clean before beginning reassembly.
2 If the head was sent out for valve servicing, the valves and related components will already be in place.
3 If the machine shop did not install them, install new seals on each of the valve guides. Gently tap each intake valve seal into place until it's seated on the guide (see Chapter 2A). **Caution:** *Don't hammer on the valve seals once they're seated or you may damage them. Don't twist or cock the seals during installation or they won't seat properly on the valve stems.*
4 Beginning at one end of the head, lubricate and install the first valve. Apply moly-base grease or clean engine oil to the valve stem.
5 Drop the spring seat or shim(s) (if equipped) over the valve guide and set the valve spring and retainer in place **(see illustrations)**.
6 Compress the springs with a valve spring compressor and carefully install the keepers in the upper groove, then slowly release the compressor and make sure the keepers seat properly. Apply a small dab of grease to each keeper to hold it in place if necessary.

2B-12   Chapter 2 Part B General engine overhaul procedures

13.1 A ridge reamer is required to remove the ridge from the top of each cylinder - do this before removing the pistons!

13.3 Check the connecting rod side clearance (endplay) with a feeler gauge as shown here

13.4 If not already marked, mark the rod bearing caps in order from the front of the engine to the rear (one mark for the front cap, two for the second one and so on) - the marks should always be on the same side of the rod and cap

13.5 Removing a connecting rod cap and bearing

13.6 To prevent damage to the crankshaft journals and cylinder walls, slip sections of hose over the rod bolts before removing the piston/rod assembly

7   Repeat the procedure for the remaining valves. Be sure to return the components to their original locations - don't mix them up!

13 Pistons/connecting rods - removal

Refer to illustrations 13.1, 13.3, 13.4, 13.5 and 13.6

Note: *Prior to removing the piston/connecting rod assemblies, remove the cylinder head, the oil pan and the oil pump pick-up tube by referring to the appropriate Sections in Part A.*

1   Use your fingernail to feel if a ridge has formed at the upper limit of ring travel (about 1/4-inch down from the top of each cylinder). If carbon deposits or cylinder wear have produced ridges, they must be completely removed with a special tool **(see illustration)**. Follow the manufacturer's instructions provided with the tool. Failure to remove the ridges before attempting to remove the piston/connecting rod assemblies may result in piston breakage. **Caution:** *When removing a ridge in the cylinder bore, DO NOT damage the top of the cylinder bore or the block deck surface with the ridge removal tool.*

2   After the cylinder ridges have been removed, turn the engine upside-down so the crankshaft is facing up.

3   Before the connecting rods are removed, check the endplay (side clearance) with feeler gauges. Slide them between each connecting rod and the crankshaft throw until the play is removed **(see illustration)**. The endplay is equal to the thickness of the feeler gauge(s). If the endplay exceeds the service limit, new connecting rods will be required. If new rods (or a new crankshaft) are installed, the endplay may fall under the specified minimum (if it does, the rods will have to be machined to restore it - consult an automotive machine shop for advice if necessary). Repeat the procedure for the remaining connecting rods.

4   Check the connecting rods and caps for identification marks. If they aren't plainly marked, use a small center-punch to make the appropriate number of indentations on each rod and cap (1, 2, 3, etc., depending on the engine type and cylinder they're associated with) **(see illustration)**.

5   Loosen each of the connecting rod cap nuts 1/2-turn at a time until they can be removed by hand **(see illustration)**. Remove the number one connecting rod cap and bearing insert. Don't drop the bearing insert out of the cap.

6   Slip a short length of plastic or rubber hose over each connecting rod cap bolt to protect the crankshaft journal and cylinder wall as the piston is removed **(see illustration)**.

7   Remove the bearing insert and push the connecting rod/piston assembly out through the top of the engine. Use a wooden hammer handle to push on the upper bearing surface in the connecting rod. If resistance is felt, double-check to make sure that all of the ridge was removed from the cylinder.

8   Repeat the procedure for the remaining cylinders.

9   After removal, reassemble the connecting rod caps and bearing inserts in their respective connecting rods and install the

Chapter 2 Part B General engine overhaul procedures

2B-13

14.1 Checking crankshaft endplay with a dial indicator

14.3 Checking crankshaft endplay with a feeler gauge

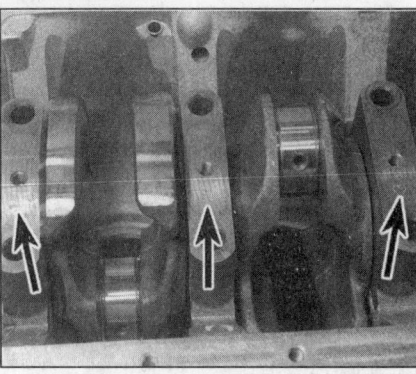

14.4a On SR20DE engines, the main bearing caps should be numbered from 1 to 5 (arrows)

14.4b Removing the main bearing cap support beam from a SR20DE engine

14.6 Move cautiously while lifting the crankshaft from the block

cap nuts finger tight. Leaving the old bearing inserts in place until reassembly will help prevent the connecting rod bearing surfaces from being accidentally nicked or gouged.

10   Don't separate the pistons from the connecting rods (see Section 18).

14 Crankshaft - removal

Refer to illustrations 14.1, 14.3, 14.4a, 14.4b and 14.6

Note: *The crankshaft can be removed only after the engine has been removed from the vehicle. It's assumed that the flywheel or driveplate, crankshaft pulley, timing chain, oil pan, oil pick-up tube, oil pump and piston/connecting rod assemblies have already been removed. The rear main oil seal retainer must be unbolted and separated from the block before proceeding with crankshaft removal.*

1   Before the crankshaft is removed, check the endplay. Mount a dial indicator with the stem in line with the crankshaft and touching the crankshaft end **(see illustration)**.

2   Pry the crankshaft all the way to the rear and zero the dial indicator. Next, pry the crankshaft to the front as far as possible and check the reading on the dial indicator. The distance that it moves is the endplay. If it's greater than specified, check the crankshaft thrust surfaces for wear. If no wear is evident, new thrust bearings should correct the endplay.

3   If a dial indicator isn't available, feeler gauges can be used. Gently pry or push the crankshaft all the way to the front of the engine. Slip feeler gauges between the crankshaft and the front face of the thrust main bearing to determine the clearance **(see illustration)**. The thrust bearing is the number three (center) bearing.

4   Check the main bearing caps to see if they're marked to indicate their locations. They should be numbered consecutively from the front of the engine to the rear **(see illustration)**. If they aren't, mark them with number stamping dies or a center-punch.

Main bearing caps generally have a cast-in arrow, which points to the front of the engine. Loosen the main bearing cap bolts 1/4-turn at a time each, using the sequence shown **(see illustrations 23.12a and 23.12b)** until they can be removed by hand. If any studs are used, make sure they're returned to their original locations when the crankshaft is reinstalled. **Note:** *SR20DE engines have separate main caps for each main bearing, but the GA series engines have two bearing saddles, which incorporate two bearing caps in each, plus a separate center bearing cap. The separate main bearing caps on the SR20DE engine are supported by a full length beam* **(see illustration)**.

5   Gently tap the caps with a soft-face hammer, then separate them from the engine block. If necessary, use the bolts as levers to remove the caps. Try not to drop the bearing inserts if they come out with the caps.

6   Carefully lift the crankshaft out of the engine **(see illustration)**. It may be a good idea to have an assistant available, since the crankshaft is quite heavy. With the bearing inserts in place in the engine block and main bearing caps or cap assembly, return the caps to their respective locations on the engine block and tighten the bolts finger tight.

15 Engine block- cleaning

Refer to illustrations 15.1a, 15.1b, 15.8 and 15. 10

Caution: *The core plugs (also known as freeze or soft plugs) may be difficult or impossible to retrieve if they're driven into the block coolant passages.*

1   Using the wide end of a punch **(see illustration)** tap in on one side of the outer edge of the core plug to turn the plug sideways in the bore. Then, using a pair of pliers,

15.1a A hammer and a large punch can be used to knock the core plugs sideways in their bores

Chapter 2 Part B General engine overhaul procedures

15.1b Pull the core plugs from the block with pliers

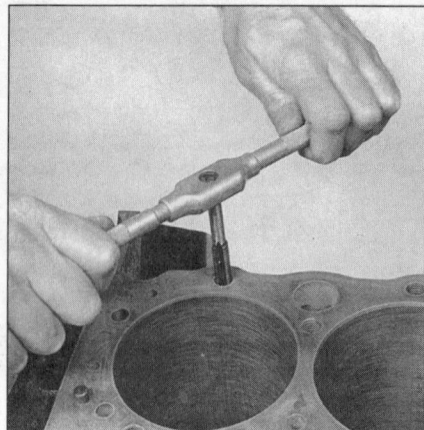

15.8 All bolt holes in the block - particularly the main bearing cap and head bolt holes - should be cleaned and restored with a tap (be sure to remove debris from the holes after this is done)

15.10 A large socket on an extension can be used to drive the new core plugs into the bores

pull the core plug from the engine block **(see illustration)**. Don't worry about the condition of the old core plugs as they are being removed because they will be replaced on reassembly with new plugs.

2 Using a gasket scraper, remove all traces of gasket material from the engine block. **Caution:** *Be EXTREMELY CAREFUL not to nick or gouge the gasket sealing surfaces.*

3 Remove the main bearing caps or cap assembly and separate the bearing inserts from the caps and the engine block. Tag the bearings, indicating which cylinder they were removed from and whether they were in the cap or the block, then set them aside.

4 Remove all of the threaded oil gallery plugs from the block. The plugs are usually very tight - they may have to be drilled out and the holes retapped. Use new plugs when the engine is reassembled.

5 If the engine is extremely dirty it should be taken to an automotive machine shop for cleaning.

6 After the block is returned, clean all oil holes and oil galleries one more time. Brushes specifically designed for this purpose are available at most auto parts stores. Flush the passages with warm water until the water runs clear, dry the block thoroughly and wipe all machined surfaces with a light, rust preventive oil. If you have access to compressed air, use it to speed the drying process and to blow out all the oil holes and galleries. **Warning:** *Wear eye protection when using compressed air!*

7 If the block isn't extremely dirty or sludged up, you can do an adequate cleaning job with hot soapy water and a stiff brush. Take plenty of time and do a thorough job. Regardless of the cleaning method used, be sure to clean all oil holes and galleries very thoroughly, dry the block completely and coat all machined surfaces with light oil.

8 The threaded holes in the block must be clean to ensure accurate torque readings during reassembly. Run the proper size tap into each of the holes to remove rust, corrosion, thread sealant or sludge and restore damaged threads **(see illustration)**. If possible, use compressed air to clear the holes of debris produced by this operation. Now is a good time to clean the threads on the head bolts and the main bearing cap bolts as well.

9 Reinstall the main bearing caps and tighten the bolts finger tight.

10 After coating the sealing surfaces of the new core plugs with a non-hardening sealant, such as Permatex no. 2 sealant or equivalent, install them in the engine block **(see illustration)**. Make sure they're driven in straight or leakage could result. Special tools are available for this purpose, but a large socket, with an outside diameter that will just slip into the core plug, a 1/2-inch drive extension and a hammer will work just as well.

11 Apply non-hardening sealant (such as Permatex no. 2 or Teflon pipe sealant) to the new oil gallery plugs and thread them into the holes in the block. Make sure they're tightened securely.

12 If the engine isn't going to be reassembled right away, cover it with a large plastic trash bag to keep it clean.

16 Engine block - inspection

Refer to illustrations 16.4a, 16.4b and 16.12

1 Before the block is inspected, it should be cleaned (see Section 15).

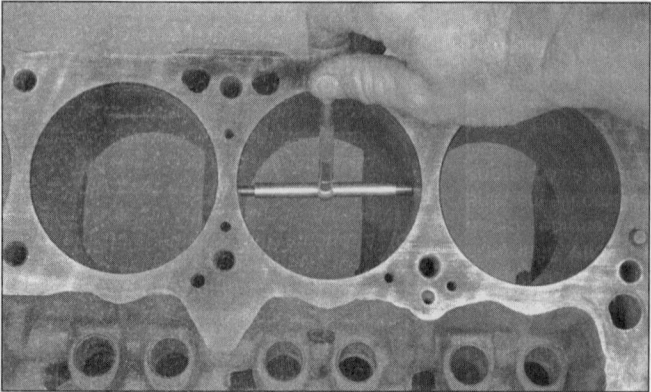

16.4a Measure the diameter of each cylinder - the ability to feel when the telescoping gauge is at the correct point will be developed over time, so work slowly and repeat the check until you're satisfied the bore measurement is accurate

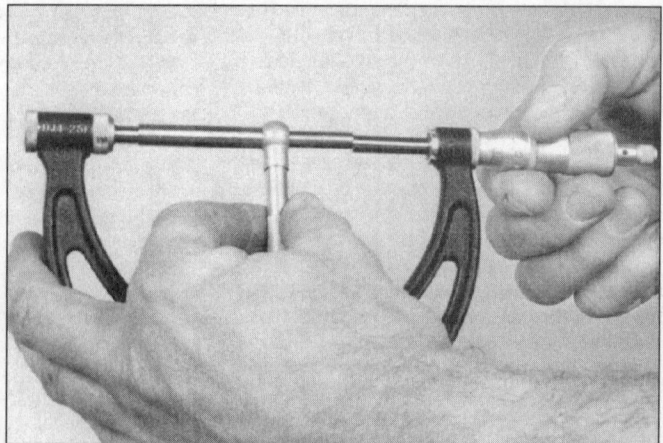

16.4b The gauge is then measured with a micrometer to determine the bore size

Chapter 2 Part B General engine overhaul procedures

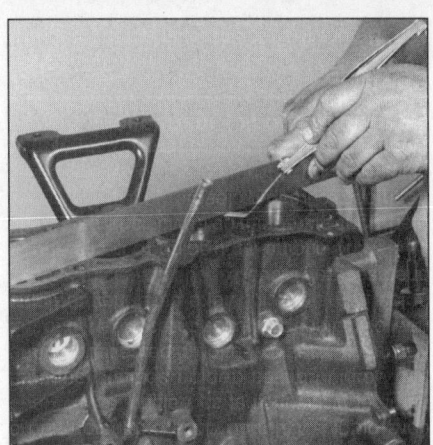

16.12 Check the block deck for distortion with a precision straightedge and a feeler gauge - lay the straightedge across the block, diagonally and from end-to-end when making the check

17.3a A "bottle brush" hone will produce better results if you've never honed cylinders before

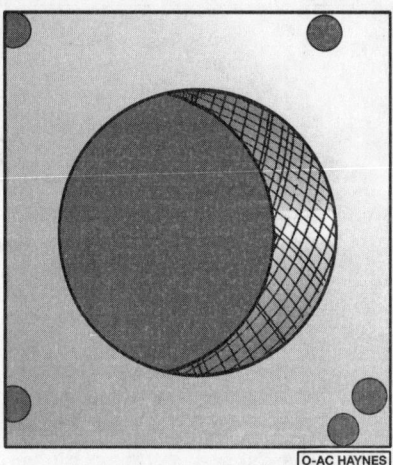

17.3b The cylinder hone should leave a smooth, crosshatch pattern with the lines intersecting at approximately a 60-degree angle

2   Visually check the block for cracks, rust and corrosion. Look for stripped threads in the threaded holes. It's also a good idea to have the block checked for hidden cracks by an automotive machine shop that has the special equipment to do this type of work. If defects are found, have the block repaired, if possible, or replaced. The blocks are marked with a "grade" indication, which refers to the sizes of the bores and other specifications (see Section 20). This information can be important when ordering bearings and other parts.
3   Check the cylinder bores for scuffing and scoring.
4   Measure the diameter of each cylinder at the top (just under the ridge area), center and bottom of the cylinder bore, parallel to the crankshaft axis **(see illustrations)**.
5   Next, measure each cylinder's diameter at the same three locations across the crankshaft axis. Compare the results to the Specifications.
6   If the required precision measuring tools aren't available, the piston-to-cylinder clearances can be obtained, though not quite as accurately, using feeler gauge stock. Feeler gauge stock comes in 12-inch lengths and various thickness and is generally available at auto parts stores.
7   To check the clearance, select a feeler gauge the same thickness as the recommended piston-to-bore clearance (see this chapter's specifications) and slip it into the cylinder along with the matching piston (with piston rings removed). The piston must be positioned exactly as it normally would be. The feeler gauge must be between the piston and cylinder on one of the thrust faces (90-degrees to the piston pin bore).
8   The piston should slip through the cylinder (with the feeler gauge in place) with moderate force.
9   If it falls through or slides through easily, the clearance is excessive and a new piston will be required. If the piston binds at the lower end of the cylinder and is loose toward the top, the cylinder is tapered. If tight spots are encountered as the piston/feeler gauge is rotated in the cylinder, the cylinder is out-of-round.
10   Repeat the procedure for the remaining pistons and cylinders.
11   If the cylinder walls are badly scuffed or scored, or if they're out-of-round or tapered beyond the limits given in this Chapter's Specifications, have the engine block rebored and honed at an automotive machine shop. If a rebore is done, oversize pistons and rings will be required.
12   Using a precision straightedge and a feeler gauge, check the block deck (the surface that mates with the cylinder head) for distortion **(see illustration)**. If it's distorted beyond the specified limit, it can be resurfaced by an automotive machine shop.
13   If the cylinders are in reasonably good condition and not worn to the outside of the limits, and if the piston-to-cylinder clearances can be maintained properly, then they don't have to be rebored. Honing is all that's necessary (see Section 17).

17 Cylinder honing

Refer to illustrations 17.3a and 17.3b

1   Prior to engine reassembly, the cylinder bores must be honed so the new piston rings will seat correctly and provide the best possible combustion chamber seal. **Note:** *If you don't have the tools or don't want to tackle the honing operation, most automotive machine shops will do it for a reasonable fee.*
2   Before honing the cylinders, install the main bearing caps (without bearing inserts) and tighten the bolts to the torque listed in this Chapter's Specifications.
3   Two types of cylinder hones are commonly available - the flex hone or "bottle brush" type and the more traditional surfacing hone with spring-loaded stones. Both will do the job, but for the less experienced mechanic the "bottle brush" hone will probably be easier to use. You'll also need some kerosene or honing oil, rags and an electric drill motor. Proceed as follows:

a) Mount the hone in the drill motor, compress the stones and slip it into the first cylinder **(see illustration)**. *Be sure to wear safety goggles or a face shield!*
b) Lubricate the cylinder with plenty of honing oil, turn on the drill and move the hone up-and-down in the cylinder at a pace that will produce a fine crosshatch pattern on the cylinder walls. Ideally, the crosshatch lines should intersect at approximately a 60-degrees angle **(see illustration)**. *Be sure to use plenty of lubricant and don't take off any more material than is absolutely necessary to produce the desired finish.* **Note:** *Piston ring manufacturers may specify a smaller crosshatch angle than the traditional 60-degrees - read and follow any instructions included with the new rings.*
c) Don't withdraw the hone from the cylinder while it's running. Instead, shut off the drill and continue moving the hone up-and-down in the cylinder until it comes to a complete stop, then compress the stones and withdraw the hone. If you're using a "bottle brush" type hone, stop the drill motor, then turn the chuck in the normal direction of rotation while withdrawing the hone from the cylinder.
d) Wipe the oil out of the cylinder and repeat the procedure for the remaining cylinders.

4   After the honing job is complete, chamfer the top edges of the cylinder bores with a small file so the rings won't catch when the pistons are installed. Be very careful not to nick the cylinder walls with the end of the file.
5   The entire engine block must be washed again very thoroughly with warm, soapy water to remove all traces of the abrasive grit produced during the honing operation. **Note:** *The bores can be considered clean when a*

Chapter 2 Part B General engine overhaul procedures

18.4a The piston ring grooves can be cleaned with a special tool, as shown here . . .

18.4b . . . or a section of a broken ring

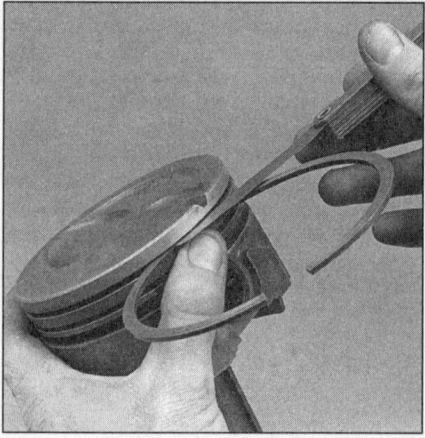

18.10 Check the ring side clearance with a feeler gauge at several points around the groove

lint-free white cloth - dampened with clean engine oil - used to wipe them out doesn't pick up any more honing residue, which will show up as gray areas on the cloth. Be sure to run a brush through all oil holes and galleries and flush them with running water.

6 After rinsing, dry the block and apply a coat of light rust preventive oil to all machined surfaces. Wrap the block in a plastic trash bag to keep it clean and set it aside until reassembly.

18 Pistons/connecting rods - inspection

Refer to illustrations 18.4a, 18.4b, 18.10 and 18.11

1 Before the inspection process can be carried out, the piston/connecting rod assemblies must be cleaned and the original piston rings removed from the pistons. **Note:** *Always use new piston rings when the engine is reassembled.*

2 Using a piston ring installation tool, carefully remove the rings from the pistons. Be careful not to nick or gouge the pistons in the process.

3 Scrape all traces of carbon from the top of the piston. A hand-held wire brush or a piece of fine emery cloth can be used once the majority of the deposits have been scraped away. Do not, under any circumstances, use a wire brush mounted in a drill motor to remove deposits from the pistons. The piston material is soft and may be eroded away by the wire brush.

4 Use a piston ring groove cleaning tool to remove carbon deposits from the ring grooves **(see illustration)**. If a tool isn't available, a piece broken off the old ring will do the job. Be very careful to remove only the carbon deposits - don't remove any metal and do not nick or scratch the sides of the ring grooves **(see illustration)**.

5 Once the deposits have been removed, clean the piston/rod assemblies with solvent and dry them with compressed air (if available). **Warning:** *Wear eye protection when using compressed air!* Make sure the oil return holes in the back sides of the ring grooves and the oil hole in the lower end of each rod are clear.

6 If the pistons and cylinder walls aren't damaged or worn excessively, and if the engine block is not rebored, new pistons won't be necessary. Normal piston wear appears as even vertical wear on the piston thrust surfaces and slight looseness of the top ring in its groove. New piston rings, however, should always be used when an engine is rebuilt.

7 Carefully inspect each piston for cracks around the skirt, at the pin bosses and at the ring lands.

8 Look for scoring and scuffing on the thrust faces of the skirt, holes in the piston crown and burned areas at the edge of the crown. If the skirt is scored or scuffed, the engine may have been suffering from overheating and/or abnormal combustion, which caused excessively high operating temperatures. The cooling and lubrication systems should be checked thoroughly. A hole in the piston crown is an indication that abnormal combustion (preignition) was occurring. Burned areas at the edge of the piston crown are usually evidence of spark knock (detonation). If any of the above problems exist, the causes must be corrected or the damage will occur again. The causes may include intake air leaks, incorrect fuel/air mixture, incorrect ignition timing and EGR system malfunctions.

9 Corrosion of the piston, in the form of small pits, indicates that coolant is leaking into the combustion chamber and/or the crankcase. Again, the cause must be corrected or the problem may persist in the rebuilt engine.

10 Measure the piston ring side clearance by laying a new piston ring in each ring groove and slipping a feeler gauge in beside it **(see illustration)**. Check the clearance at three or four locations around each groove. Be sure to use the correct ring for each groove - they are different. If the side clearance is greater than specified, new pistons

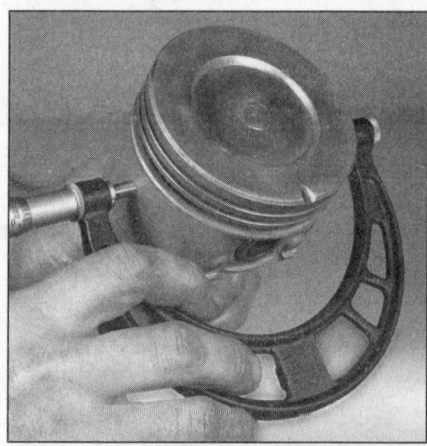

18.11 Measure the piston diameter at a 90-degree angle to the piston pin below the bottom of the piston pin bore

will have to be used.

11 Check the piston-to-bore clearance by measuring the bore (see Section 16) and the piston diameter. Make sure the pistons and bores are correctly matched. Measure the piston across the skirt, at a 90-degree angle to the piston pin and at approximately the bottom of the pin bore **(see illustration)**. Subtract the piston diameter from the bore diameter to obtain the clearance. If it's greater than specified, the block will have to be rebored and new pistons and rings installed.

12 Check the piston-to-rod clearance by twisting the piston and rod in opposite directions. Any noticeable play indicates excessive wear, which must be corrected. The piston/connecting rod assemblies should be taken to an automotive machine shop to have the pistons and rods re-sized and new pins installed.

13 If the pistons must be removed from the connecting rods for any reason, they should be taken to an automotive machine shop. While they are there, have the connecting rods checked for bend and twist, since automotive machine shops have special equip-

Chapter 2 Part B General engine overhaul procedures

19.1 The oil holes should be chamfered so sharp edges don't gouge or scratch the new bearings

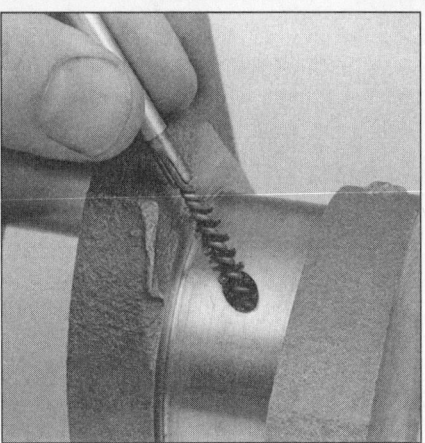

19.2 Use a wire or stiff plastic bristle brush to clean the oil passages in the crankshaft

19.4 Rubbing a penny lengthwise on each journal will reveal its condition - if copper rubs off and is embedded in the crankshaft, the journals should be reground

ment for this purpose. **Note:** *Unless new pistons and/or connecting rods must be installed, do not disassemble the pistons and connecting rods.*

14 Check the connecting rods for cracks and other damage. A machine shop can inspect the bores to determine if they require resizing. Temporarily remove the rod caps, lift out the old bearing inserts, wipe the rod and cap bearing surfaces clean and inspect them for nicks, gouges and scratches. After checking the rods, replace the old bearings, slip the caps into place and tighten the nuts finger tight. **Note:** *If the engine is being rebuilt because of a connecting rod knock, be sure to install new rods or have the old ones resized.*

19 Crankshaft - inspection

Refer to illustrations 19.1, 19.2, 19.4, 19.6 and 19.8

1 Remove all burrs from the crankshaft oil holes with a stone, file or scraper **(see illustration)**.
2 Clean the crankshaft with solvent and dry it with compressed air (if available). Be sure to clean the oil holes with a stiff brush and flush them with solvent **(see illustration)**.
3 Check the main and connecting rod bearing journals for uneven wear, scoring, pits and cracks.
4 Rub a penny across each journal several times **(see illustration)**. If a journal picks up copper from the penny, it's too rough and must be reground.
5 Check the rest of the crankshaft for cracks and other damage. It should be magnafluxed to reveal hidden cracks - an automotive machine shop will handle the procedure.
6 Using a micrometer, measure the diameter of the main and connecting rod journals and compare the results to this chapters specifications **(see illustration)**. By measuring the diameter at a number of points around each journal's circumference, you'll be able to determine whether or not the journal is out-of-round. Take the measurement at each end of the journal, near the crank throws, to determine if the journal is tapered. Crankshaft runout should be checked also, but large V-blocks and a dial indicator are needed to do it correctly. If you don't have the equipment, have a machine shop check the runout.
7 If the crankshaft journals are damaged, tapered, out-of-round or worn beyond the limits given in the Specifications, have the crankshaft reground by an automotive machine shop. Be sure to use the correct size bearing inserts if the crankshaft is reconditioned.
8 Check the oil seal journals at each end of the crankshaft for wear **(see illustration)**. If the seal has worn a groove in the journal, or if it's nicked or scratched, the new seal may leak when the engine is reassembled. In some cases, an automotive machine shop may be able to repair the journal by pressing on a thin sleeve. If repair isn't feasible, a new or different crankshaft should be installed.
9 Examine the main and rod bearing inserts (see Section 20).
10 If the crankshaft requires replacement make certain that you match the original crankshaft with any replacement purchased.

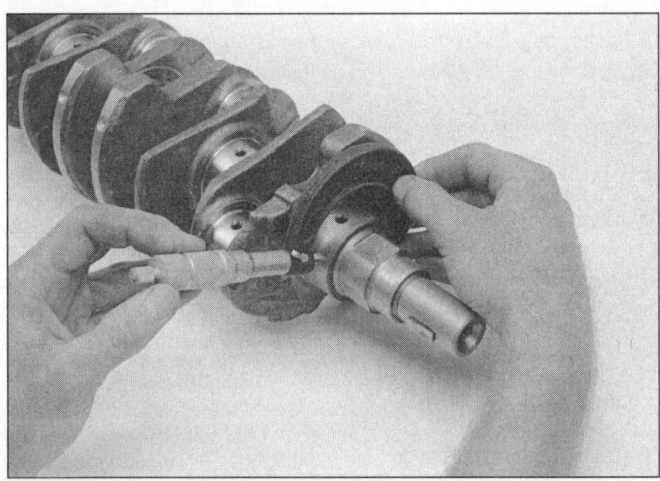

19.6 Measure the diameter of each crankshaft main journal at several points to detect taper and out-of-round conditions

19.8 The surface which contacts the rear oil seal must be perfectly smooth

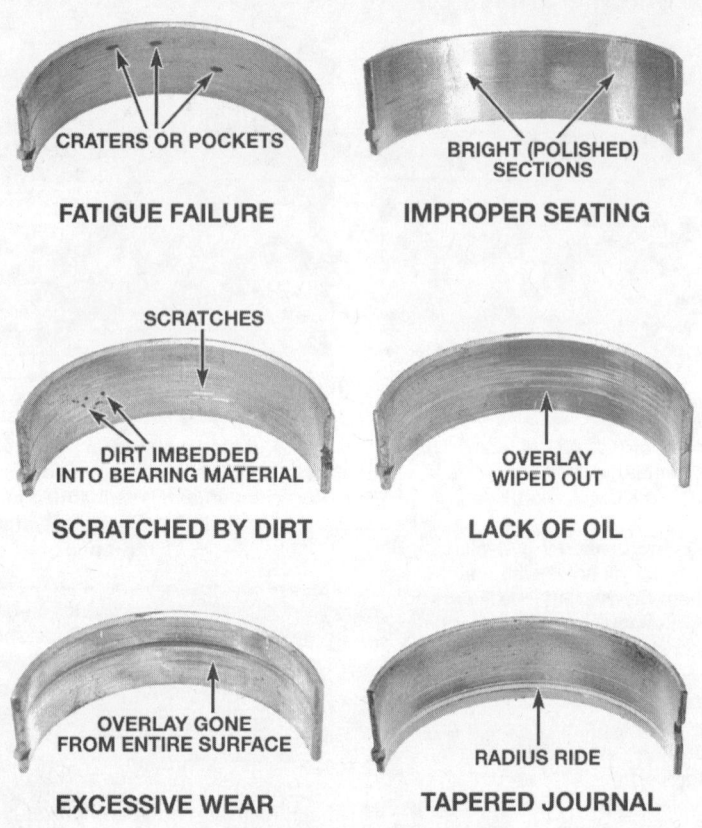

20.1 Typical bearing failures

20.10a The engine block is stamped at the rear of the oil pan surface with numbers indicating the "grade" or size of the cylinder bores and main journals (GA series engine shown, SA series has only one row indicating main journal grade)

1. Cylinder bore grade number (front to rear, cylinders 1 to 4)
2. Main journal grade number (front to rear, journals 1 to 5)

20.10b The main and rod journal grades are stamped on the first crankshaft throw - on GA series engines the upper row (A) indicates the rod journal grades and the lower row (B) indicates the main journal grades (read from right to left on GA series) - on SR series Type I crankshafts the upper row indicates the main journal grade and the lower row indicates the connecting rod journal grade, on Type II crankshafts the main journal grades are stamped on the first crankshaft throw and the connecting rod journal grades are stamped on the last crankshaft throw (read from left to right on SR series)

20 Main and connecting rod bearings - inspection and selection

Inspection

Refer to illustration 20.1

1 Even though the main and connecting rod bearings should be replaced with new ones during the engine overhaul, the old bearings should be retained for close examination, as they may reveal valuable information about the condition of the engine **(see illustration)**.

2 Bearing failure occurs because of lack of lubrication, the presence of dirt or other foreign particles, overloading the engine and corrosion. Regardless of the cause of bearing failure, it must be corrected before the engine is reassembled to prevent it from happening again.

3 When examining the bearings, remove them from the engine block, the main bearing caps, the connecting rods and the rod caps and lay them out on a clean surface in the same general position as their location in the engine. This will enable you to match any bearing problems with the corresponding crankshaft journal.

4 Dirt and other foreign particles get into the engine in a variety of ways. It may be left in the engine during assembly, or it may pass through filters or the PCV system. It may get into the oil, and from there into the bearings. Metal chips from machining operations and normal engine wear are often present. Abrasives are sometimes left in engine components after reconditioning, especially when parts are not thoroughly cleaned using the proper cleaning methods. Whatever the source, these foreign objects often end up embedded in the soft bearing material and are easily recognized. Large particles will not embed in the bearing and will score or gouge the bearing and journal. The best prevention for this cause of bearing failure is to clean all parts thoroughly and keep everything spotlessly clean during engine assembly. Frequent and regular engine oil and filter changes are also recommended.

5 Lack of lubrication (or lubrication breakdown) has a number of interrelated causes. Excessive heat (which thins the oil), overloading (which squeezes the oil from the bearing face) and oil leakage or throw off (from excessive bearing clearances, worn oil pump or high engine speeds) all contribute to lubrication breakdown. Blocked oil passages, which usually are the result of misaligned oil holes in a bearing shell, will also oil starve a bearing and destroy it. When lack of lubrication is the cause of bearing failure, the bearing material is wiped or extruded from the steel backing of the bearing. Temperatures may increase to the point where the steel backing turns blue from overheating.

6 Driving habits can have a definite effect on bearing life. Low speed operation in too high a gear (lugging the engine) puts very high loads on bearings, which tends to squeeze out the oil film. These loads cause the bearings to flex, which produces fine cracks in the bearing face (fatigue failure). Eventually the bearing material will loosen in

Chapter 2 Part B General engine overhaul procedures

SR series bearing selection charts

CRANKSHAFT MAIN JOURNAL GRADE NUMBER	CYLINDER BLOCK MAIN JOURNAL GRADE NUMBER			
	0	1	2	3
0	0 BLACK	1 BROWN	2 GREEN	3 YELLOW
1	1 BROWN	2 GREEN	3 YELLOW	4 BLUE
2	1 GREEN	3 YELLOW	4 BLUE	5 PINK
3	3 YELLOW	4 BLUE	5 PINK	6 NONE

Connecting rod journal

Connecting rod journal grade number	Connecting rod bearing grade number
0	0 (No Color)
1	1 (Black)
2	2 (Brown)

20.10c SR series bearing selection charts

GA series bearing selection charts

Main journal

Crankshaft main journal grade number	Cylinder block main journal grade number		
	0	1	2
0	Black	Brown	Green
1	Brown	Green	Yellow
2	Green	Yellow	Blue

Connecting rod journal

Connecting rod journal grade number	Connecting rod bearing grade color
0	—
1	Brown
2	Green

20.10d GA series bearing selection charts

---

pieces and tear away from the steel backing. Short trip driving leads to corrosion of bearings because insufficient engine heat is produced to drive off the condensed water and corrosive gases. These products collect in the engine oil, forming acid and sludge. As the oil is carried to the engine bearings, the acid attacks and corrodes the bearing material.

7 Incorrect bearing installation during engine assembly will lead to bearing failure as well. Tight fitting bearings leave insufficient bearing oil clearance and will result in oil starvation. Dirt or foreign particles trapped behind a bearing insert result in high spots on the bearing which lead to failure.

Selection

Refer to illustrations 20.10a, 20.10b, 20.10c and 20.10d

8 If the original bearings are worn or damaged, or if the oil clearances are incorrect (see Section 20 or 23) and the crankshaft is undamaged or worn excessively, new bearings should be purchased for engine reassembly. However, if the crankshaft has been reground, new undersize bearings must be installed. The automotive machine shop that reconditions the crankshaft will provide or help you select the correct size bearings. Regardless of how the bearing sizes are determined, measure the oil clearance with Plastigage to ensure the bearings are the right size.

9 If you need to use STANDARD size bearings, install bearings that have the same grade number as the original bearings.

10 The engine block and crankshaft are stamped with grade numbers indicating different size crankshaft journals **(see illustrations)**. This allows for very precise fitting of bearing inserts. If the cylinder block or crankshaft require replacement, determine the correct grade bearing insert from the accompanying charts **(see illustrations)**. The bearing inserts are color coded.

11 Remember, the oil clearance is the final judge when selecting new bearing sizes. If you have any questions or are unsure which bearings to use, get help from a dealer parts or service department or other parts supplier.

21 Engine overhaul - reassembly sequence

1 Before beginning engine reassembly, make sure you have all the necessary new parts, gaskets and seals as well as the following items on hand:
Common hand tools
A torque wrench
Piston ring installation tool
Piston ring compressor
Short lengths of rubber or plastic hose to fit over
connecting rod bolts
Plastigage
Feeler gauges
A fine-tooth file
New engine oil
Engine assembly lube or moly-base grease
Gasket sealant
Thread locking compound

2 In order to save time and avoid problems, engine reassembly must be done in the following general order:
Piston rings
Crankshaft and main bearings
Piston/connecting rod assemblies
Rear main oil seal
Timing chain, sprockets and cover (Part A)
Cylinder head and camshaft(s) (Part A)
Oil pump (Part A)
Oil pick-up tube
Oil pan (Part A)
Valve cover (Part A)
Intake and exhaust manifolds (Part A)
Flywheel/driveplate

22 Piston rings - installation

Refer to illustrations 22.3, 22.4, 22.5, 22.7a, 22.7b and 22.10

1 Before installing the new piston rings, the ring end gaps must be checked. It's assumed that the piston ring side clearance has been checked and verified correct (see Section 18).

2 Lay out the piston/connecting rod assemblies and the new ring sets so the ring sets will be matched with the same piston and cylinder during the end gap measurement and engine assembly.

3 Insert the top (number one) ring into the first cylinder and square it up with the cylinder walls by pushing it in with the top of the piston **(see illustration)**. The ring should be near the bottom of the cylinder, at the lower limit of ring travel.

4 To measure the end gap, slip feeler gauges between the ends of the ring until a

22.3 When checking piston ring end gap, the ring must be square in the cylinder bore (this is done by pushing the ring down with the top of a piston as shown)

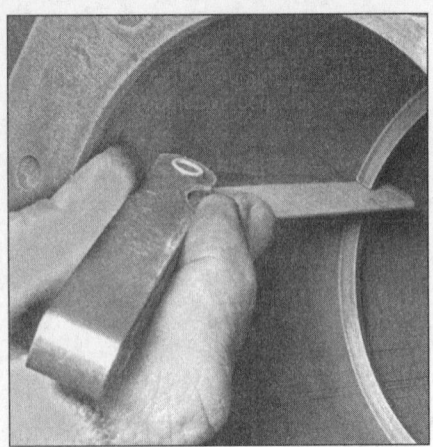

22.4 With the ring square in the cylinder, measure the end gap with a feeler gauge

22.5 File in an inward direction and proceed slowly when enlarging ring end gaps

22.7a Installing the spacer/expander in the oil control ring groove

22.7b DO NOT use a piston ring installation tool when installing the oil ring side rails

gauge equal to the gap width is found (see illustration). The feeler gauge should slide between the ring ends with a slight amount of drag. Compare the measurement to the Specifications. If the gap is larger or smaller than specified, double-check to make sure you have the correct rings before proceeding.

5   Repeat the procedure for each ring that will be installed in the first cylinder and for each ring in the remaining cylinders. Remember to keep rings, pistons and cylinders matched up. If any rings have insufficient end clearance, they may be filed carefully (see illustration).

6   Once the ring end gaps have been checked the rings can be installed on the pistons.

7   The oil control ring (lowest one on the piston) is installed first. It's composed of three separate components. Slip the spacer/expander into the groove (see illustration). If an anti-rotation tang is used, make sure it's inserted into the drilled hole in the ring groove. Next, install the lower side rail. Don't use a piston ring installation tool on the oil ring side rails, as they may be damaged. Instead, place one end of the side rail into the groove between the spacer/expander and the ring land, hold it firmly in place and slide a finger around the piston while pushing the rail into the groove (see illustration). Next, install the upper side rail in the same manner.

8   After the three oil ring components have been installed, check to make sure that both the upper and lower side rails can be turned smoothly in the ring groove.

9   The number two (middle) ring (if used) is installed next. It's usually stamped with a mark which must face up, toward the top of the piston. **Note:** *Always follow the instructions printed on the ring package or box - different manufacturers may require different approaches. Do not mix up the top and middle rings, as they have different cross sections (GA16DE engines use only one compression ring).*

10   Use a piston ring installation tool and make sure the identification mark is facing the top of the piston, then slip the ring into the middle groove on the piston. Don't expand the ring any more than necessary to slide it over the piston (see illustration).

11   Install the number one (top) ring in the same manner. Make sure the mark is facing up. Be careful not to confuse the number one and number two rings on SR20DE engines.

12   Repeat the procedure for the remaining pistons and rings.

22.10 Installing the compression rings with a ring expander - the mark (arrow) must face up

23.10 Lay the Plastigage strips (arrow) on the main bearing journals, parallel to the crankshaft centerline

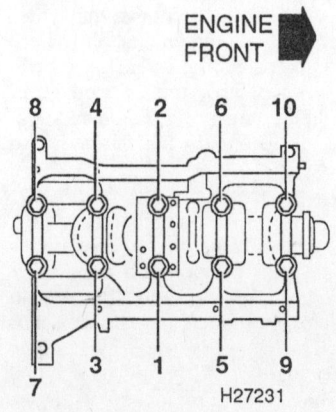

23.12a Main bearing cap tightening sequence for the GA16DE engine

Chapter 2 Part B General engine overhaul procedures

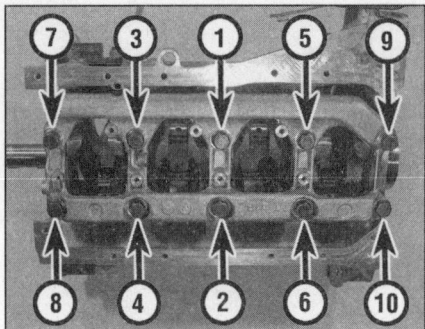

23.12b Use this sequence when tightening the main bearing caps on the SR20DE engine

23.14 Compare the width of the crushed Plastigage to the scale on the envelope to determine the main bearing oil clearance (always take the measurement at the widest point of the Plastigage); be sure to use the correct scale - standard and metric ones are included

23.19a On SR20DE engines, install the main bearing caps so that the marks are in the correct orientation

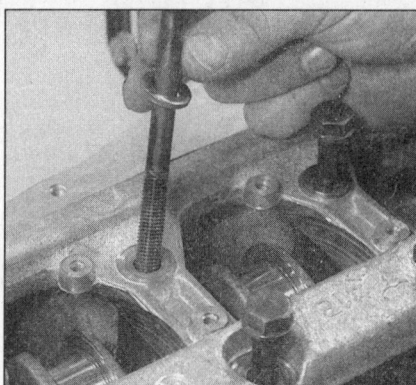

23.19b Place the main bearing beam on the engine and install the bolts

23 Crankshaft - installation and main bearing oil clearance check

Refer to illustrations 23.10, 23.12a, 23.12b, 23.14, 23.19a, 23.19b, 23.20a and 23.20b

1 Crankshaft installation is the first major step in engine reassembly. It's assumed at this point that the engine block and crankshaft have been cleaned, inspected and repaired or reconditioned.
2 Position the engine with the bottom facing up.
3 Remove the main bearing cap bolts and lift out the caps. Lay the caps out in the proper order to ensure correct installation.
4 If they're still in place, remove the old bearing inserts from the block and the main bearing caps. Wipe the main bearing surfaces of the block and caps with a clean, lint-free cloth. They must be kept spotlessly clean!

Main bearing oil clearance check

5 Clean the back sides of the new main bearing inserts and lay the bearing half with the oil groove and hole in each main bearing saddle in the block (on all engines covered by this manual, the bearings with oil holes go in the block and those without oil holes go in the caps). Lay the other bearing half from each bearing set in the corresponding main bearing cap. Make sure the tab on each bearing insert fits into the recess in the block or cap. Also, the oil holes in the block must line up with the oil holes in the bearing insert. **Caution:** *Do not hammer the bearings into place and don't nick or gouge the bearing faces. No lubrication should be used at this time.*
6 The thrust bearings (washers) must be installed in the number three main bearing saddle (on the cylinder block side). Be sure to install them with the oil grooves facing out (away from the main bearing saddle. All engines have a three-piece arrangement (bearing plus two thrust washers). Apply a thin film of moly-based grease or engine assembly lube to the back sides of the thrust bearings to hold them in place.
7 Clean the faces of the bearings in the block and the crankshaft main bearing jour-

nals with a clean, lint-free cloth. Check or clean the oil holes in the crankshaft, as any dirt here can go only one way - straight through the new bearings.
8 Once you're certain the crankshaft is clean, carefully lay it in position in the main bearings. No lubricant should be used at this time.
9 Before the crankshaft can be permanently installed, the main bearing oil clearance must be checked.
10 Trim several pieces of the appropriate size Plastigage (they must be slightly shorter than the width of the main bearings) and place one piece on each crankshaft main bearing journal, parallel with the journal axis **(see illustration)**.
11 Clean the faces of the bearings in the caps and install the caps in their respective positions (don't mix them up) with the arrows pointing toward the front of the engine. The main bearing caps are numbered, starting with number one at the timing chain end of the block. Don't disturb the Plastigage. Apply a light coat of oil to the bolt threads and the under sides of the bolt heads, then install them.
12 Following the recommended sequence **(see illustrations)**, tighten the main bearing cap bolts, in three steps, to the torque listed in this Chapter's Specifications. Don't rotate the crankshaft at any time during this operation!
13 Remove the bolts and carefully lift off the main bearing caps. Keep them in order. Don't disturb the Plastigage or rotate the crankshaft. If any of the main bearing caps are difficult to remove, tap them gently from side-to-side with a soft-face hammer to loosen them.
14 Compare the width of the crushed Plastigage on each journal to the scale printed on the Plastigage envelope to obtain the main bearing oil clearance **(see illustration)**. Check the Specifications to make sure it's correct.
15 If the clearance is not as specified, the

bearing inserts may be the wrong size which means different ones will be required (see Section 20). Before deciding that different inserts are needed, make sure that no dirt or oil was between the bearing inserts and the caps or block when the clearance was measured. If the Plastigage is noticeably wider at one end than the other, the journal may be tapered (see Section 19).
16 Carefully scrape all traces of the Plastigage material off the main bearing journals and/or the bearing faces. Don't nick or scratch the bearing faces.

Final crankshaft installation

17 Carefully lift the crankshaft out of the engine. Clean the bearing faces in the block, then apply a thin, uniform layer of clean moly-base grease or engine assembly lube to each of the bearing surfaces. Coat the thrust washers as well (thrust flanges on one-piece thrust bearings).
18 Lubricate the crankshaft surfaces that contact the oil seals with moly-base grease, engine assembly lube or clean engine oil.
19 Make sure the crankshaft journals are clean, then lay the crankshaft back in place in the block. Clean the faces of the bearings in the caps or cap assembly, then apply the same lubricant to them. Install the caps in their respective positions with the arrows pointing toward the front of the engine **(see illustrations)**.

2B-22   Chapter 2 Part B General engine overhaul procedures

23.20a Use a torque wrench to tighten the bolts to the first (torque) setting

23.20b The next stage requires an angle meter (see the text and the torque specifications in this Chapter)

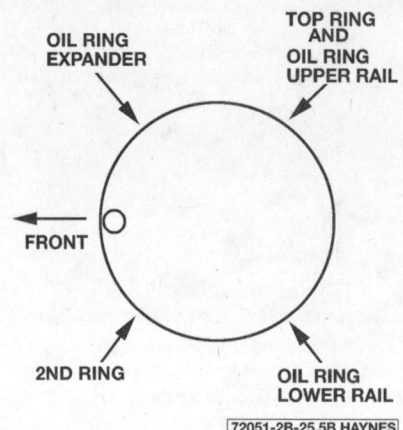

25.5 Position the rings on the piston using this pattern

20   Apply a light coat of oil to the bolt threads and the under sides of the bolt heads, then install them. Tighten the bolts to 10-to-12 ft-lbs following the recommended sequence **(see illustrations 23.12a and 23.12b)**. Tap the ends of the crankshaft forward and backward with a brass hammer to line up the thrust washer and crankshaft surfaces. Re-tighten all main bearing cap bolts to the specified torque, following the recommended sequence **(see illustrations)**.
21   Rotate the crankshaft a number of times by hand to check for any obvious binding.
22   Check the crankshaft endplay with a feeler gauge or a dial indicator (see Section 14). The endplay should be correct if the crankshaft thrust faces aren't worn or damaged and new thrust bearings/washers have been installed.
23   Install a new rear main oil seal, then bolt the retainer to the block (see Section 24).

24 Rear main oil seal - installation

The crankshaft must be installed first and the main bearing caps bolted in place, then the new seal should be installed in the retainer and the retainer bolted to the block (refer to Chapter 2A for the procedure and illustrations). **Note:** *Depending on the design of the engine stand being used, you may not be able to install the rear main seal retainer with the engine on the stand. In this case, install the rear main seal retainer as the last step before engine installation, when the engine is off the stand and on the hoist.*

25 Pistons/connecting rods - installation and rod bearing oil clearance check

Refer to illustrations 25.5, 25.9, 25.11, 25.13, 25.17, 25.21a and 25.21b
1   Before installing the piston/connecting rod assemblies, the cylinder walls must be perfectly clean, the top edge of each cylinder must be chamfered, and the crankshaft must be in place.
2   Remove the cap from the end of the number one connecting rod (refer to the marks made during removal). Remove the original bearing inserts and wipe the bearing surfaces of the connecting rod and cap with a clean, lint-free cloth. They must be kept spotlessly clean.

Connecting rod bearing oil clearance check

3   Clean the back side of the new upper bearing insert, then lay it in place in the connecting rod. Make sure the tab on the bearing fits into the recess in the rod so the oil holes line up. Don't hammer the bearing insert into place and be very careful not to nick or gouge the bearing face. Don't lubricate the bearing at this time.
4   Clean the back side of the other bearing insert and install it in the rod cap. Again, make sure the tab on the bearing fits into the recess in the cap, and don't apply any lubricant. It's critically important that the mating surfaces of the bearing and connecting rod are perfectly clean and oil free when they're assembled.
5   Position the piston ring gaps at staggered intervals around the piston **(see illustration)**.
6   Slip a section of plastic or rubber hose over each connecting rod cap bolt **(see illustration 13.6)**.
7   Lubricate the piston and rings with clean engine oil and attach a piston ring compressor to the piston. Leave the skirt protruding about 1/4-inch to guide the piston into the cylinder. The rings must be compressed until they're flush with the piston.
8   Rotate the crankshaft until the number one connecting rod journal is at BDC (bottom dead center) and apply a coat of engine oil to the cylinder walls.
9   With the mark on top of the piston facing the front (timing chain end) of the engine **(see illustration)**, gently insert the

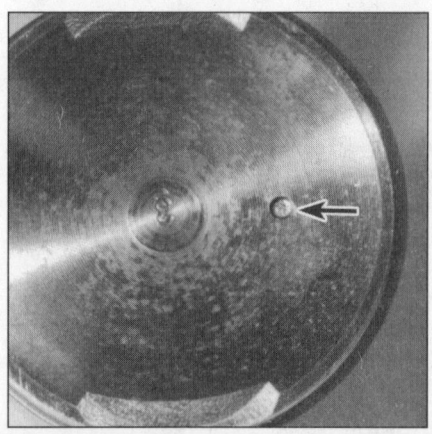

25.9 Position the piston, when installing it, with the mark/notch (arrow) in the piston facing the front of the engine

piston/connecting rod assembly into the number one cylinder bore and rest the bottom edge of the ring compressor on the engine block.
10   Tap the top edge of the ring compressor to make sure it's contacting the block around its entire circumference.
11   Gently tap on the top of the piston with the end of a wooden or plastic hammer handle while guiding the end of the connecting rod into place on the crankshaft journal **(see illustration)**. The piston rings may try to pop out of the ring compressor just before entering the cylinder bore, so keep some pressure on the ring compressor. Work slowly, and if any resistance is felt as the piston enters the cylinder, stop immediately. Find out what's hanging up and fix it before proceeding. Do not, for any reason, force the piston into the cylinder - you might break a ring and/or the piston.
12   Once the piston/connecting rod assembly is installed, the connecting rod bearing oil clearance must be checked before the rod cap is permanently bolted in place.
13   Cut a piece of the appropriate size Plastigage slightly shorter than the width of the connecting rod bearing and lay it in place on

Chapter 2 Part B General engine overhaul procedures 2B-23

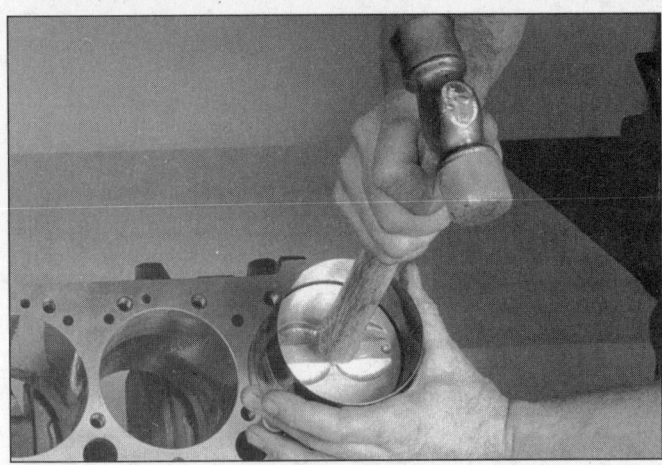

25.11 With the ring compressor firmly seated against the block, the piston can be driven gently into the cylinder bore with the end of a wooden or plastic hammer handle

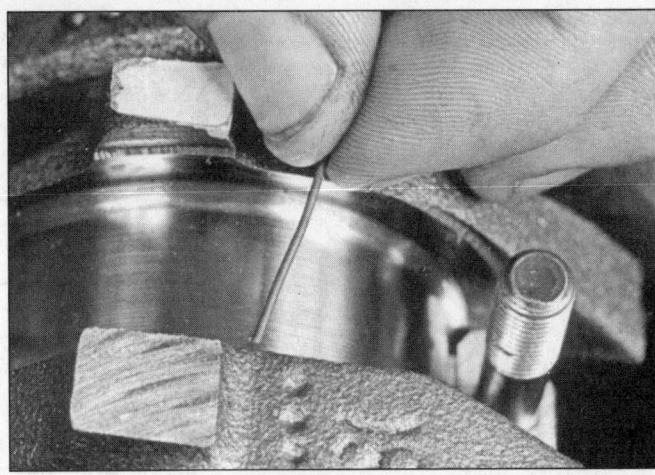

25.13 Lay a strip of Plastigage on each rod bearing journal, parallel to the crankshaft centerline

the number one connecting rod journal, parallel with the journal axis **(see illustration)**.

14 Clean the connecting rod cap bearing face, remove the protective hoses from the connecting rod bolts and install the rod cap. Make sure the mating mark on the cap is on the same side as the mark on the connecting rod. Check the cap to make sure the front mark is facing the timing belt/chain end of the engine.

15 Apply a light coat of oil to the under sides of the nuts, then install and tighten them to the torque listed in this Chapter's Specifications. Use a thin-wall socket to avoid erroneous torque readings that can result if the socket is wedged between the rod cap and nut. If the socket tends to wedge itself between the nut and the cap, lift up on it slightly until it no longer contacts the cap. Do not rotate the crankshaft at any time during this operation.

16 Remove the nuts and detach the rod cap, being very careful not to disturb the Plastigage.

17 Compare the width of the crushed Plastigage to the scale printed on the Plastigage envelope to obtain the oil clearance **(see illustration)**. Compare it to the Specifications to make sure the clearance is correct.

18 If the clearance is not as specified, the bearing inserts may be the wrong size (which means different ones will be required). Before deciding that different inserts are needed, make sure that no dirt or oil was between the bearing inserts and the connecting rod or cap when the clearance was measured. Also, recheck the journal diameter. If the Plastigage was wider at one end than the other, the journal may be tapered (see Section 19).

Final connecting rod installation

19 Carefully scrape all traces of the Plastigage material off the rod journal and/or bearing face. Be very careful not to scratch the bearing - use your fingernail or the edge of a credit card.

20 Make sure the bearing faces are perfectly clean, then apply a uniform layer of clean moly-base grease or engine assembly lube to both of them. You'll have to push the piston into the cylinder to expose the face of the bearing insert in the connecting rod - be sure to slip the protective hoses over the rod bolts first.

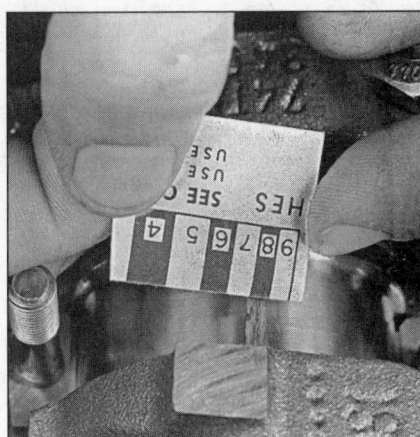

25.17 Measure the width of the crushed Plastigage to determine the rod bearing oil clearance (be sure to use the correct scale - standard and metric ones are included)

21 Slide the connecting rod back into place on the journal, remove the protective hoses from the rod cap bolts, install the rod cap and tighten the nuts to the torque listed in this Chapter's Specifications **(see illustrations)**.

25.21a Tighten the nuts to the specified torque setting . . .

25.21b . . . and on SR20DE engines, use an angle meter (see this Chapter Specifications)

22 Repeat the entire procedure for the remaining pistons/connecting rods.
23 The important points to remember are:
a) Keep the back sides of the bearing inserts and the insides of the connecting rods and caps perfectly clean when assembling them.
b) Make sure you have the correct piston/rod assembly for each cylinder.
c) The mark on the piston top must face the front (timing belt/chain end) of the engine.
d) Lubricate the cylinder walls with clean oil.
e) Lubricate the bearing faces when installing the rod caps after the oil clearance has been checked.

24 After all the piston/connecting rod assemblies have been properly installed, rotate the crankshaft a number of times by hand to check for any obvious binding.
25 As a final step, the connecting rod side clearance (endplay) must be checked (see Section 13).
26 Compare the measured side clearance to the Specifications to make sure it's correct. If it was correct before disassembly and the original crankshaft and rods were reinstalled, it should still be right. If new rods or a new crankshaft were installed, the side clearance may be inadequate. If so, the rods will have to be removed and taken to an automotive machine shop for re-sizing.

26 Initial start-up and break-in after overhaul

Warning: *Have a fire extinguisher handy when starting the engine for the first time.*

1 Once the engine has been installed in the vehicle, double-check the engine oil and coolant levels.
2 With the spark plugs out of the engine and the ignition system disabled (see Section 4), crank the engine until oil pressure registers on the gauge or the light goes out.
3 Install the spark plugs, hook up the plug wires and reconnect the ignition coil wire, if it was removed earlier for the compression test in Section 4.
4 Start the engine. It may take a few moments for the fuel system to build up pressure, but the engine should start without a great deal of effort.
5 After the engine starts, it should be allowed to warm up to normal operating temperature. While the engine is warming up, make a thorough check for fuel, oil and coolant leaks.
6 Shut the engine off and recheck the engine oil and coolant levels.
7 Drive the vehicle to an area with minimum traffic, accelerate from 30 to 50 mph, then allow the vehicle to slow to 30 mph with the throttle closed. Repeat the procedure 10 or 12 times. This will load the piston rings and cause them to seat properly against the cylinder walls. Check again for oil and coolant leaks.
8 Drive the vehicle gently for the first 500 miles (no sustained high speeds) and keep a constant check on the oil level. It is not unusual for an engine to use oil during the break-in period.
9 At approximately 500 to 600 miles, change the oil and filter.
10 For the next few hundred miles, drive the vehicle normally. Do not pamper it or abuse it.
11 After 2000 miles, change the oil and filter again and consider the engine broken in.

Chapter 3
Cooling, heating and air conditioning systems

Contents

	Section		Section
Air conditioner and heater control assembly - removal and installation	12	Cooling system check	See Chapter 1
Air conditioning compressor - removal and installation	15	Cooling system servicing (draining, flushing and refilling)	See Chapter 1
Air conditioning condenser - removal and installation	16	Drivebelt check, adjustment and replacement	See Chapter 1
Air conditioning evaporator and expansion valve - removal and installation	17	Engine cooling fan(s) and circuit(s) - check and component replacement	4
Air conditioning receiver/drier - removal and installation	14	General information	1
Air conditioning and heating system - check and maintenance	13	Heater core - removal and installation	11
Antifreeze - general information	2	Radiator - removal and installation	5
Blower motor and circuit - check, removal and component replacement	10	Thermostat - check and replacement	3
Coolant level check	See Chapter 1	Underhood hose - check and replacement	See Chapter 1
Coolant reservoir - removal and installation	6	Water pump - check	7
Coolant temperature gauge sending unit - check and replacement	9	Water pump - replacement	8

Specifications

General

Radiator cap pressure rating
    Standard.. 11 to 14 psi
    Limit... 9 to 14 psi
Cooling system test pressure... 23 psi
Coolant temperature sensor resistance
    At 140-degrees F... 70 to 90 ohms
    At 212-degrees F... 21 to 24 ohms
Thermostat rating (opening temperature)..................................... 170-degrees F
Refrigerant type.. R-134a
Refrigerant capacity... 1.32 to 1.54 lbs

Torque specifications

Thermostat housing bolts... 55 to 73 in-lbs
Water pump-to-engine bolts
    GA16DE.. 55 to 73 in-lbs
    SR20DE... 144 to 180 in-lbs

3-2   Chapter 3 Cooling, heating and air conditioning systems

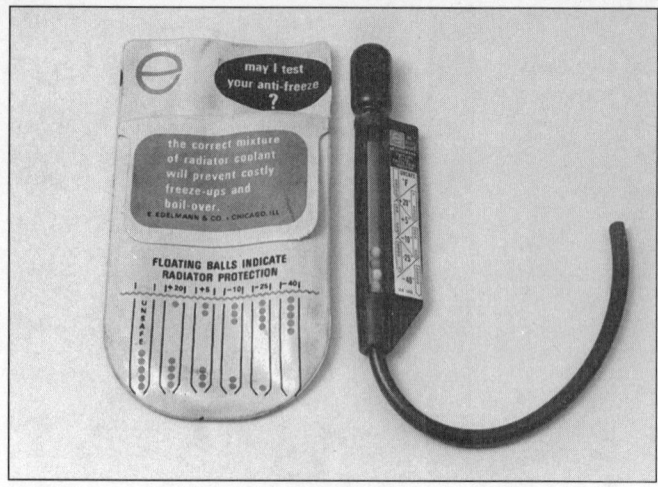

2.5 Inexpensive coolant hydrometers can be purchased at most auto parts stores

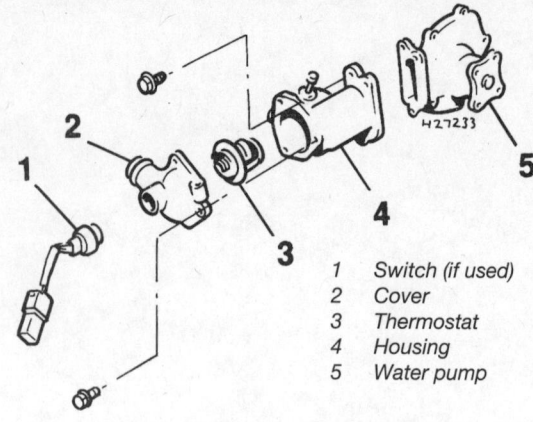

1 Switch (if used)
2 Cover
3 Thermostat
4 Housing
5 Water pump

3.7a Thermostat housing and water pump components - GA16DE engine

1 General information

Engine cooling system

All vehicles covered by this manual employ a pressurized engine cooling system with thermostatically controlled coolant circulation. An impeller type water pump mounted on the drivebelt end of the block pumps coolant through the engine. The coolant flows around each cylinder and toward the transaxle end of the engine. Cast-in coolant passages direct coolant around the intake and exhaust ports, near the spark plug areas and in close proximity to the exhaust valve guides.

A wax pellet type thermostat is located near the water pump (at the rear of the engine on the water pipes on the SR20DE engines). During warm up, the closed thermostat prevents coolant from circulating through the radiator. As the engine nears normal operating temperature, the thermostat opens and allows hot coolant to travel through the radiator, where it's cooled before returning to the engine.

The cooling system is sealed by a pressure type radiator cap, which raises the boiling point of the coolant and increases the cooling efficiency of the radiator. If the system pressure exceeds the cap pressure relief value, the excess pressure in the system forces the spring-loaded valve inside the cap off its seat and allows the coolant to escape through the overflow tube into a coolant reservoir. When the system cools, the excess coolant is automatically drawn from the reservoir back into the radiator.

The coolant reservoir serves as both the point at which fresh coolant is added to the cooling system to maintain the proper fluid level and as a holding tank for overheated coolant.

This type of cooling system is known as a closed design because coolant that escapes past the pressure cap is saved and re-used.

Heating system

The heating system consists of a blower fan and heater core located in the heater box, the hoses connecting the heater core to the engine cooling system and the heater/air conditioning control head on the dashboard. Hot engine coolant is circulated through the heater core. When the heater mode is activated, a flap opens to expose the heater box to the passenger compartment. A fan switch on the control head activates the blower motor, which forces air through the core, heating the air.

Air conditioning system

The air conditioning system consists of a condenser mounted in front of the radiator, an evaporator mounted adjacent to the heater core, a compressor mounted on the engine, a receiver-drier and the plumbing connecting all of the above components.

A blower fan forces the warmer air of the passenger compartment through the evaporator core (sort of a radiator-in-reverse), transferring the heat from the air to the refrigerant. The liquid refrigerant boils off into low pressure vapor, taking the heat with it when it leaves the evaporator.

2 Antifreeze - general information

Refer to illustration 2.5

Warning: *Do not allow antifreeze to come in contact with your skin or painted surfaces of the vehicle. Rinse off spills immediately with plenty of water. Antifreeze is highly toxic if ingested. Never leave antifreeze lying around in an open container or in puddles on the floor; children and pets are attracted by it's sweet smell and may drink it. Check with local authorities about disposing of used antifreeze. Many communities have collection centers which will see that antifreeze is disposed of safely. Never dump used anti-freeze on the ground or into drains.*

Note: *Non-toxic coolant is available at most auto parts stores. Although the coolant is non-toxic, proper disposal is still required.*

1 The cooling system should be filled with a water/ethylene glycol-based antifreeze solution, which will prevent freezing down to at least -20-degrees F, or lower if local climate requires it. It also provides protection against corrosion and increases the coolant boiling point.

2 The cooling system should be drained, flushed and refilled at the specified intervals (see Chapter 1). Old or contaminated antifreeze solutions are likely to cause damage and encourage the formation of corrosion and scale in the system. Use distilled water with the antifreeze.

3 Before adding antifreeze, check all hose connections, because antifreeze tends to leak through very minute openings. Engines don't normally consume coolant, so if the level goes down, find the cause and correct it.

4 The exact mixture of antifreeze to water which you should use depends on the relative weather conditions. The mixture should contain at least 50-percent antifreeze, but should never contain more than 70-percent antifreeze. Consult the mixture ratio chart on the antifreeze container before adding coolant.

Antifreeze/coolant testing

Warning: *Do not remove the radiator cap to take a sample until the engine has cooled completely. The system is under pressure and extremely hot during and after running the engine, and will severely scald skin it comes in contact with.*

5 Hydrometers used to test the condition of the coolant are available at most auto parts stores **(see illustration)**. They are inexpensive and are very easy to use. To test coolant, draw a small amount from the radiator, or coolant reservoir, with the hydrometer until all the balls are submerged (most of them will probably float). **Note:** *It is preferable to take the coolant sample from the radiator. The mixture in the coolant reservoir may be slightly diluted if any water has recently been added.* The strength of the mixture is shown by the number of balls

Chapter 3 Cooling, heating and air conditioning systems

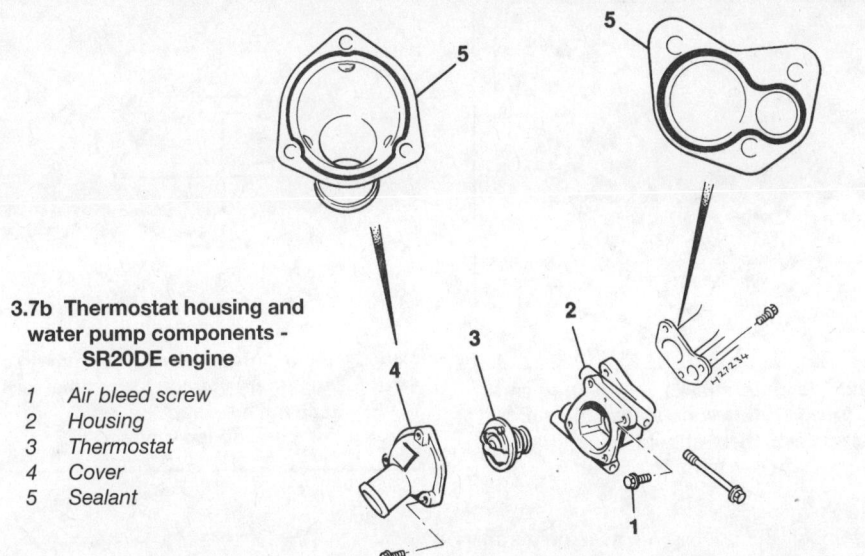

3.7b Thermostat housing and water pump components - SR20DE engine

1. Air bleed screw
2. Housing
3. Thermostat
4. Cover
5. Sealant

that are floating. Exact temperature protection indicated by the hydrometer is clearly described on the package instructions. Always use antifreeze which meets the vehicle manufacturer's specifications.

3 Thermostat - check and replacement

Warning: *Do not remove the radiator cap, drain the coolant or replace the thermostat until the engine has cooled completely. Do not allow antifreeze to come in contact with your skin or painted surfaces of the vehicle. Rinse off spills immediately with plenty of water. Antifreeze is highly toxic if ingested. Never leave antifreeze lying around in an open container or in puddles on the floor; children and pets are attracted by it's sweet smell and may drink it. Check with local authorities about disposing of used antifreeze. Many communities have collection centers which will see that antifreeze is disposed of safely. Never dump used anti-freeze on the ground or into drains.*

Note: *Non-toxic coolant is available at most auto parts stores. Although the coolant is non-toxic, proper disposal is still required.*

Check

1 Before assuming the thermostat is to blame for a cooling system problem, check the coolant level, drivebelt tension (see Chapter 1) and temperature gauge operation.
2 If the engine seems to be taking a long time to warm up (based on heater output or temperature gauge operation), the thermostat is probably stuck open. Replace the thermostat with a new one.
3 If the engine runs hot, use your hand to check the temperature of the lower radiator hose. If the hose isn't hot, but the engine is, the thermostat is probably stuck closed, preventing the coolant from circulating through the radiator. Replace the thermostat. **Caution:** *Don't drive the vehicle without a thermostat. The engine may never reach the required temperature to allow the computer to go to closed loop operation. Performance, emissions and fuel economy will probably be poor.*
4 If the radiator-to-thermostat hose is hot, it means that the coolant is flowing and the thermostat is open. Consult the *Troubleshooting* section at the front of this manual for cooling system diagnosis.

Replacement

Refer to illustrations 3.7a, 3.7b, 3.8, 3.12 and 3.14

5 Disconnect the battery cable from the negative battery terminal.
6 Drain the cooling system (see Chapter 1). If the coolant is relatively new, or tests in good condition (see Section 2), save it and re-use it.
7 Follow the lower radiator hose to the engine to locate the thermostat housing **(see illustrations)**.
8 Loosen the hose clamp and detach the hose from the fitting **(see illustration)**. If the hose is stuck, grasp it near the end with a pair of large adjustable pliers and twist it to break the seal, then pull it off. If the hose is old or deteriorated, cut it off and install a new one.
9 If the outer surface of the large fitting that mates with the hose is deteriorated (corroded, pitted, etc.) it may be damaged further by hose removal. If it is, the thermostat housing cover will have to be replaced.
10 If there are any sensors in the thermostat housing, unplug each of the electrical connections at the thermostat housing. **Note:** *Each of the connections should have matching colors on the male and female connector ends, to help in reassembly.*
11 Remove the bolts/nuts and detach the housing cover. If the cover is stuck, tap it with a soft-face hammer to jar it loose. Be prepared for some coolant to spill as the gasket seal is broken.
12 Note the position of the air bleed valve or air bleed hole **(see illustration)**.
13 Install the new thermostat in the housing. Make sure the air bleed or jiggle valve on the thermostat is at the top and the spring end is directed toward the engine.
14 Clean the housing and the cover sealing surfaces thoroughly. If using RTV sealant instead of a gasket, wipe the surfaces with acetone or lacquer thinner. Position a new gasket or apply RTV sealant to the thermostat cover **(see illustration)**.

3.8 Loosen the clamp and detach the hose from the thermostat cover

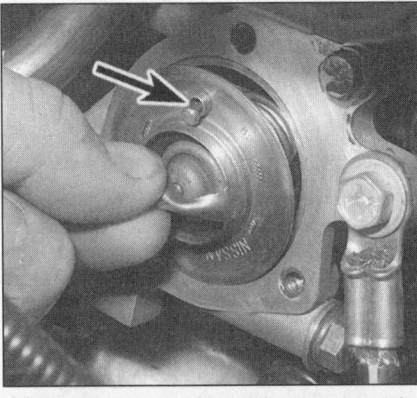

3.12 Note the position of the thermostat's air bleed valve (arrow) aligned with the casting and how the thermostat is installed (the spring end goes into the engine)

3.14 Apply a bead of RTV sealant to the thermostat cover

Chapter 3 Cooling, heating and air conditioning systems

4.2a This cover identifies the locations of the relays at the underhood relay center

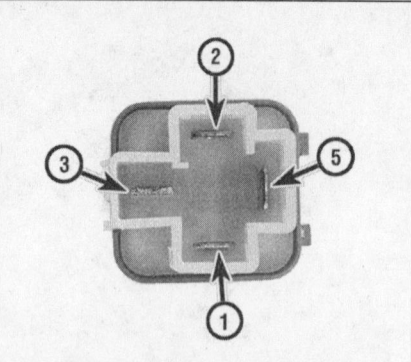

4.2b When terminal 1 is grounded and fused battery voltage is applied to terminal 2, there should be continuity between terminals 3 and 5

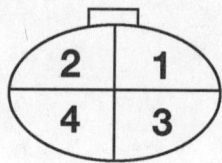

4.3a Connect a ground lead to terminal 1 and fused battery voltage to terminal 2 to check fan operation on air-conditioned models

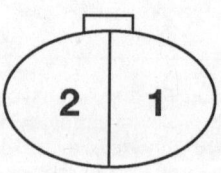

4.3b On models without air conditioning, apply fused battery voltage to terminal 1 and ground terminal 2 for the fan check

15 Install the cover and bolts. Tighten the bolts to the torque listed in this Chapter's Specifications.
16 Reattach the hose to the fitting and tighten the hose clamp securely.
17 Refill the cooling system (see Chapter 1).
18 Start the engine and allow it to reach normal operating temperature, then check for leaks and proper thermostat operation (as described in Steps 2 through 4).

4 Engine cooling fan(s) and circuit(s) - check and component replacement

1 The engine cooling fan is controlled by the ECCS control module (the ECM), using the coolant temperature sensor (part of the engine management system), the vehicle speed sensor and the air conditioner On signal as factors affecting fan operation.

Check

Refer to illustrations 4.2a, 4.2b, 4.3a and 4.3b
2 First, check the fuses (see Chapter 12). The fans are also controlled by relays which are located in the relay center under the hood **(see illustration)**. Manual transaxle models use only one relay, automatic models use three. Use an ohmmeter to check for continuity between terminals 3 and 5 of the relay when voltage is applied between terminals 1 and 2 **(see illustration)**.
3 To test a fan motor, unplug the electrical connector and use fused jumper wires to connect the fan directly to the battery **(see illustrations)**. If the fan still does not work, replace the motor.
4 If the motor tested OK in Step 4, check for trouble codes stored in the computer (see Chapter 6). If the coolant temperature sensor and vehicle speed sensor check out OK, check the wiring.

Replacement

Engine cooling fan

Refer to illustrations 4.6, 4.7 and 4.8
5 Disconnect the cable from the negative battery terminal.
6 Unbolt the cooling fan assembly from the radiator and remove it **(see illustration)**.
7 Remove the nut and detach the fan blade assembly from the motor shaft **(see illustration)**.
8 Take out the screws holding the fan motor to the mounting bracket, then detach the motor **(see illustration)**.
19 Installation is the reverse of removal.

Condenser fan (air-conditioned models only)

10 Air-conditioned models have an additional fan located next to the main engine cooling fan. The condenser fan is usually smaller than the cooling fan. It is also controlled by the ECM.
11 Disconnect the battery cable from the negative battery terminal.
12 Unplug the electrical connector.
13 Remove the mounting bolts holding the fan assembly to the radiator support and remove the fan assembly.
14 Installation is the reverse of the removal procedure.

5 Radiator - removal and installation

Refer to illustrations 5.7, 5.8 and 5.9
Warning: *Do not start this procedure until the engine is completely cool. Do not allow antifreeze to come in contact with your skin or painted surfaces of the vehicle. Rinse off spills immediately with plenty of water. Antifreeze is highly toxic if ingested. Never leave antifreeze lying around in an open container or in puddles on the floor; children and pets are attracted by it's sweet smell and may drink it. Check with local authorities about disposing of used antifreeze. Many communities have collection centers which will see that antifreeze is disposed of safely. Never dump used anti-freeze on the ground or into drains.*
Note: *Non-toxic coolant is available at most auto parts stores. Although the coolant is non-toxic, proper disposal is still required.*

Removal

1 Disconnect the cable from the negative battery terminal.
2 Raise the front of the vehicle and support it securely on jackstands. Remove the lower splash shields, if necessary to gain access to the radiator drain plug.
3 Drain the cooling system (see Chapter 1). If the coolant is relatively new, or tests in good condition, save it and reuse it (see Section 2).
4 Loosen the upper and lower hose clamps, then detach the radiator hoses from the fittings. If they're stuck, grasp each hose near the end with a pair of adjustable pliers and twist it to break the seal, then pull it off - be careful not to damage the radiator fittings! If the hoses are old or deteriorated, cut them off and install new ones.
5 Disconnect the coolant reservoir hose from the radiator.
6 Disconnect the cooling fan electrical connector(s).
7 Remove the bolts and mounting brackets **(see illustration)**.
8 If the vehicle is equipped with an automatic transaxle, disconnect the transmission oil cooler lines **(see illustration)** and plug the lines and fittings.
9 Carefully lift out the radiator **(see illustration)**. Don't spill coolant on the vehicle or scratch the paint. Remove the bolts securing the cooling fan to the radiator and pull it free.
10 Inspect the lower radiator rubber mounting pads. Replace them if they are deteriorated, split or broken.

Chapter 3 Cooling, heating and air conditioning systems   3-5

4.6 Remove the four screws holding the fan frame to the radiator and lift the fan and housing out of the engine compartment

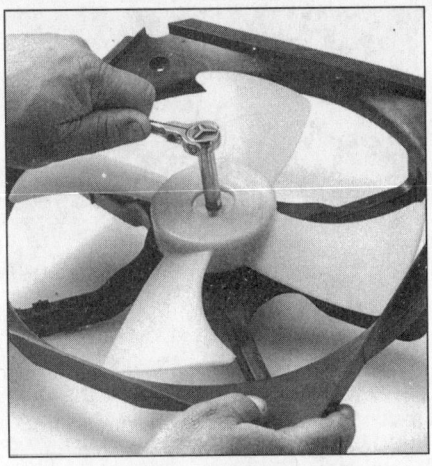

4.7 To replace the fan motor, unscrew the fan blade nut and remove the fan blades

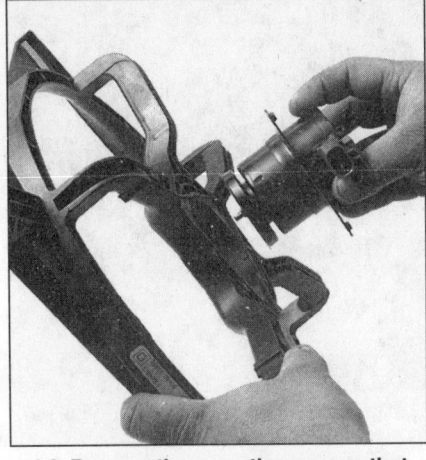

4.8 Remove the mounting screws that hold the motor to the fan housing and remove the motor

11 With the radiator removed, it can be inspected for leaks and damage. If it needs repair, have a radiator shop or dealer service department perform the work as special techniques are required.

12 Bugs and dirt can be removed from the radiator with a soft brush, followed by forcing water from a garden hose through the core from the engine side. Don't bend the cooling fins as this is done.

Installation

13 Installation is the reverse of the removal procedure. Be sure the radiator mounting pads are seated properly at the base of the radiator **(see illustration 5.9)**.

14 After installation, fill the cooling system with the proper mixture of antifreeze and water (see Section 2 and Chapter 1).

15 Start the engine and check for leaks. Allow the engine to reach normal operating temperature, indicated by the upper radiator hose becoming hot. Recheck the coolant level and add more if required.

16 If you're working on an automatic transaxle equipped vehicle, check and add transmission fluid as needed (see Chapter 1).

6 Coolant reservoir - removal and installation

Refer to illustration 6.2

Warning: *Do not start this procedure until the engine is completely cool. Do not allow antifreeze to come in contact with your skin or painted surfaces of the vehicle. Rinse off spills immediately with plenty of water. Antifreeze is highly toxic if ingested. Never leave antifreeze lying around in an open container or in puddles on the floor; children and pets are attracted by it's sweet smell and may drink it. Check with local authorities about disposing of used antifreeze. Many communities have collection centers which will see that antifreeze is disposed of safely. Never dump used anti-freeze on the ground or into drains.*

Note: *Non-toxic coolant is available at most auto parts stores. Although the coolant is non-toxic, proper disposal is still required.*

1 Disconnect the overflow hose from the reservoir, and plug it to prevent coolant from spilling if the reservoir is full.

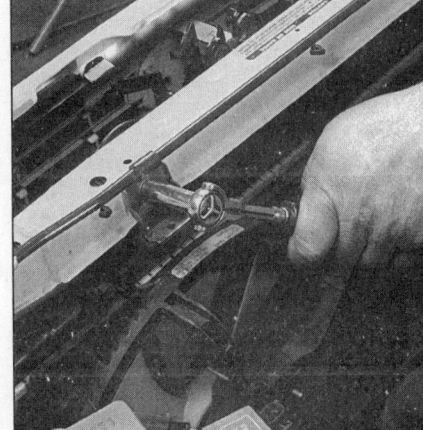

5.7 Remove the bracket hold-down bolt from the support on each side of the radiator

2 Lift the coolant reservoir and remove it from the vehicle **(see illustration)**. **Note:** *On some models, the neck of the windshield-washer tank must be pulled out to remove the coolant reservoir.*

5.8 On automatic transmission models disconnect the cooler lines - have a drain pan ready to catch spilled fluid

5.9 Carefully lift the radiator free, avoid scratching or spilling coolant on the vehicle - arrows indicate lower mounting pads

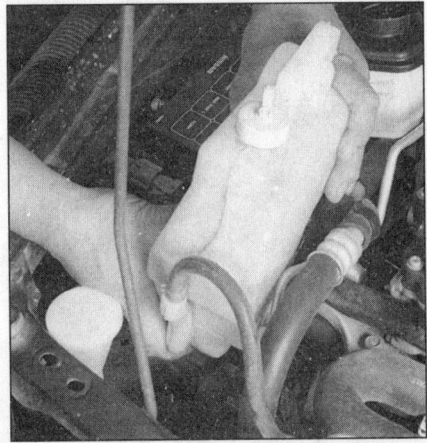

6.2 The coolant recovery bottle can be lifted out without the use of tools

3-6 Chapter 3 Cooling, heating and air conditioning systems

8.6a Removing the water pump pulley from the GA16DE engine

8.6b Unbolt the thermostat housing from the GA16DE engine's water pump

8.6c Remove the thermostat housing from the GA16DE engine before taking off the water pump

3   Pour the coolant into a container. Wash out and inspect the reservoir for cracks and chafing. Replace it if any damage is found.
4   Installation is the reverse of removal.

7 Water pump - check

1   A failure in the water pump can cause serious engine damage due to overheating.
2   There are three ways to check the operation of the water pump while it's installed on the engine. If the pump is defective, it should be replaced with a new or rebuilt unit.
3   With the engine running at normal operating temperature, squeeze the upper radiator hose. If the water pump is working properly, a pressure surge should be felt as the hose is released. **Warning:** *Keep your hands away from the fan blades! The fan can come on at any time.*
4   Water pumps are equipped with weep or vent holes. If a failure occurs in the pump seal, coolant will leak from the hole. In most cases you'll need a flashlight to find the hole on the water pump from underneath to check for leaks.
5   If the water pump shaft bearings fail there may be a howling sound at the drivebelt end of the engine while it's running. Shaft wear can be felt if the water pump pulley is rocked up and down. Don't mistake drivebelt slippage, which causes a squealing sound, for water pump bearing failure.
6   Turn the pump shaft by hand and feel for roughness in the bearing. Replace the pump if roughness is felt.

8 Water pump - replacement

Refer to illustrations 8.6a, 8.6b, 8.6c, 8.9, 8.12a, 8.12b and 8.13

Warning: *Wait until the engine is completely cool before beginning this procedure. Do not allow antifreeze to come in contact with your skin or painted surfaces of the vehicle. Rinse off spills immediately with plenty of water. Antifreeze is highly toxic if ingested. Never leave antifreeze lying around in an open container or in puddles on the floor; children and pets are attracted by it's sweet smell and may drink it. Check with local authorities about disposing of used antifreeze. Many communities have collection centers which will see that antifreeze is disposed of safely. Never dump used anti-freeze on the ground or into drains.*
Note: *Non-toxic coolant is available at most auto parts stores. Although the coolant is non-toxic, proper disposal is still required.*

1   Disconnect the battery cable from the negative battery terminal.
2   Drain the cooling system (see Chapter 1). If the coolant is relatively new, or tests in good condition (see Section 2), save it and re-use it.
3   On GA16DE engines, remove the front mounting bracket from the cylinder head.
4   Loosen (but do not remove) the bolts which secure the water pump pulley.
5   Remove any other accessory drivebelts that would interfere with removal (see Chapter 1).
6   Remove the water pump pulley **(see illustration)**. Remove the thermostat housing **(see illustrations)**.
7   On models with the SR20DE engine, remove the right front wheel, engine side cover, and front engine mount for water pump access.
8   On GA16DE engines, disconnect the coolant hoses and pull them out of the way.

8.9 Unbolting the GA16DE water pump

8.12a Apply RTV sealant to the mating surface of the water pump . . .

Chapter 3 Cooling, heating and air conditioning systems

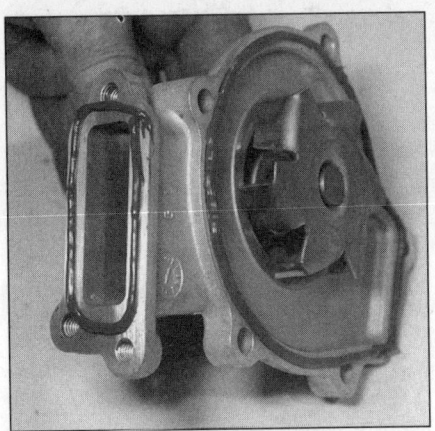

8.12b ... and to the thermostat housing of the GA16DE engine

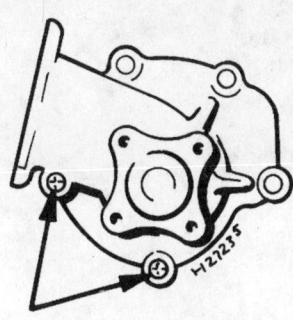

8.13 Thread locking compound must be applied to the two bolts shown on the GA16DE engine

9.3a Location of the coolant temperature sending unit (arrow) on the GA16DE engine

9.3b Location of the coolant temperature sending unit (arrow) on the SR20DE engine

9   Remove all of the water pump bolts on GA16DE engines. The water pump can then be removed along with the thermostat housing. On SR20DE engines, remove only the three lower water pump bolts and proceed with the next step **(see illustration)**.
10 On SR20DE engines, place a jack under the engine and use it to take the weight off of the mounts. Remove the front engine mount and then remove the last water pump bolt. Remove the water pump. If it is stuck, gently tap it with a soft faced hammer to break the seal.
11 Remove all traces of old gasket material from the sealing surface on the engine block. If you are using RTV sealant instead of a gasket, clean the mating surfaces with lacquer thinner or acetone.
12 Install the water pump, using a new gasket or sealant **(see illustrations)**.
13 Tighten the bolts to the torque listed in this Chapter's Specifications in 1/4-turn increments, using thread locking compound as indicated **(see illustration)**. Don't overtighten them or the pump may be damaged or leak.
14 Reinstall all parts removed for access to the pump.
15 Refill the cooling system (see Chapter 1). Run the engine and check for leaks.

9   Coolant temperature gauge sending unit - check and replacement

Refer to illustrations 9.3a and 9.3b
Note: *This single wire sensor (also known as the thermal transmitter) operates the instrument panel gauge only. Another sensor provides engine coolant temperature data to the engine management system (refer to Chapter 6).*

Check

1   If the coolant temperature gauge is inoperative, check the fuses first (see Chapter 12).
2   If the temperature indicator shows excessive temperature after running awhile, see the *Troubleshooting* section in the front of the manual.
3   If the temperature gauge indicates Hot shortly after the engine is started cold, and the engine is not up to normal operating temperature, disconnect the wire at the coolant temperature sending unit **(see illustrations)**. **Note:** *The sending unit has a single wire terminal.* If the gauge reading drops, replace the sending unit. If the reading remains high, the wire to the gauge may be shorted to ground or the gauge is faulty.
4   If the coolant temperature gauge fails to indicate after the engine has been warmed up (approximately 10 minutes) and the fuses checked out okay, shut off the engine. Disconnect the wire at the sending unit and using a jumper wire, connect it to a clean ground on the engine. Turn on the ignition without starting the engine. If the gauge now indicates Hot, replace the sending unit.
5   If the gauge still does not work, the circuit may be open or the gauge may be faulty.
6   To check the sending unit for correct calibration, connect one lead of an ohmmeter to the electrical terminal on the sender and the other lead to a good engine ground.
7   With the coolant temperature at 140-degrees F, the resistance shown on the meter should be 70 to 90 ohms. With the engine warmed completely (coolant temperature approximately 212-degrees F) the resistance should be 21 to 24 ohms.

Replacement

Warning: *The engine must be completely cool before removing the sending unit.*
8   With the engine completely cool, remove the cap from the radiator to release any pressure, then reinstall the cap. This reduces coolant loss during sending unit replacement.
9   Disconnect the electrical connector from the sending unit.
10 Prepare the new sending unit for installation by applying a light coat of sealant to the threads, or by wrapping the threads with Teflon sealing tape.
11 Unscrew the sending unit from the engine and quickly install the new one to prevent coolant loss.
12 Tighten the sending unit securely and attach the electrical connector.
13 Refill the cooling system (see Chapter 1) and run the engine. Check for leaks and proper gauge operation.

10 Blower motor and circuit - check, removal and component replacement

Blower motor

Check

Refer to illustration 10.4
1   If the blower motor speed does not correspond to the setting selected on the blower switch, or the blower motor does not operate at all, the problem could be a bad fuse, relay, switch, blower motor resistor, blower motor or blower motor circuit wiring.
2   Before checking the blower motor or circuit, always check the fuse and relay (if equipped) first (see Chapter 12).

3-8   Chapter 3   Cooling, heating and air conditioning systems

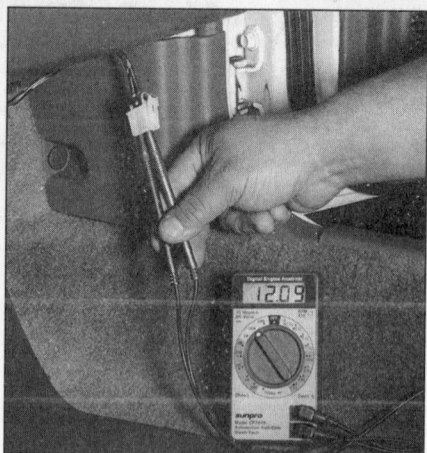

10.4 Test for voltage at the blower motor connector

10.8 Blower motor mounting details - to remove the blower motor, unplug the electrical connector (arrow), remove the screws (arrows) and pull the assembly from the housing

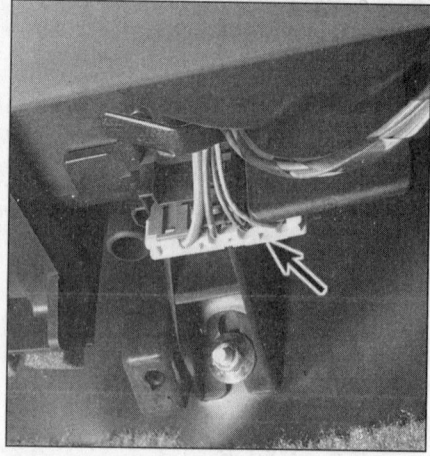

10.12 The blower motor resistor (arrow) is usually mounted on the blower case or nearby ductwork

10.14 Blower motor resistor terminals

3   Remove the glove compartment and lower dash trim (see Chapter 11) to gain access to the heater case and blower motor.
4   With the ignition key in the ON position, turn the blower switch to the faulty position(s) and, using a test light or voltmeter, check the voltage at the motor electrical connector (see illustration). If the motor is receiving voltage but not operating, either the motor ground is bad (on these models the blower switch and resistor are part of the ground circuit) or the motor itself is faulty or the fan is binding.
5   To check for a bad ground, disconnect the electrical connector from the blower motor, connect a jumper wire between the ground wire terminal on the blower motor and a good ground, then connect a fused jumper wire between the battery positive terminal and the positive terminal on the blower motor. If the motor now operates properly, the ground circuit or resistor is bad. If the motor does not operate, the fan is either binding or faulty.
6   If you suspect the blower motor fan is binding, remove the blower motor (see step 7) to check for free operation of the fan.

Replacement

Refer to illustration 10.8

7   Disconnect the battery cable from the negative battery terminal. **Warning:** *On airbag-equipped models also disconnect the positive battery cable and wait at least ten minutes after the ignition has been turned off and the battery disconnected before starting work on the vehicle. The supplemental inflatable restraint (airbag) system has electrical capacitors that must bleed down to avoid the possibility of airbag deployment.*
8   Unplug the electrical connector at the blower motor (see illustration).
9   Remove the three blower unit retaining screws and lower the unit from the housing.
10   If you are replacing the motor, detach the fan and transfer it to the new motor.
11   Installation is the reverse of removal. Run the blower and check for proper operation.

Blower motor resistor

Check

Refer to illustrations 10.12 and 10.14

12   The blower motor resistor is located near the blower motor (see illustration).
13   Verify that the resistor is getting current from the blower motor at terminal 2:
   a) *If the resistor is not getting current, check the wires and the connectors between the resistor and the motor. Check for loose or corroded connections and damaged wires.*
   b) *If the wires and connectors are good and the blower switch is getting current, verify current is flowing out of the blower speed control switch to ground. If it is not go to step 15.*
14   Check the blower resistor for correct resistance (see illustration) by connecting an ohmmeter between terminal 1 and the other three terminals in sequence. There should be 1.4 to 1.6 ohms resistance between terminals 2 and 1, 2.5 to 2.8 ohms between 2 and 3 and 0.5 to 0.6 ohms between 2 and 4. If any checks indicate infinite resistance, replace the blower resistor.

Blower speed control switch

Refer to illustrations 10.16a and 10.16b

15   Remover the interfering instrument panel components as described in Chapter 11 and pull the control unit out from the dash enough to access the rear of the fan switch. Refer to Section 12.
16   Unplug the electrical connector from the blower switch and check the continuity across the indicated switch terminals with an ohmmeter (see illustrations).
17   If continuity isn't as specified, replace the switch.

11   Heater core - removal and installation

Refer to illustrations 11.4, 11.6, 11.7, 11.9, 11.10, 11.11a and 11.11b

Warning: *Wait until the engine is completely cool before beginning this procedure.*

1   If the vehicle is equipped with air conditioning, have the air conditioning system discharged at a dealer service department or service station. **Warning:** *The air conditioning system is under high pressure. Do not loosen any fittings or remove any components until after the system has been discharged by a dealer service department or service station. Always wear eye protection when disconnecting refrigerant fittings.*
2   Disconnect the battery cable from the negative battery terminal. **Warning:** *On airbag-equipped models also disconnect the positive battery cable and wait at least ten minutes after the ignition has been turned off and the battery disconnected before starting work on the vehicle. The supplemental inflatable restraint (airbag) system has electrical capacitors that must bleed down to avoid the possibility of airbag deployment.*
3   Turn the heater control setting to HOT. Drain the cooling system (see Chapter 1). If

Chapter 3 Cooling, heating and air conditioning systems

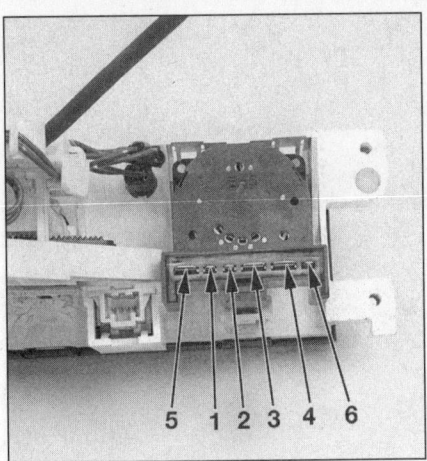

10.16a Terminal guide for the blower motor switch

Terminal	Off	1	2	3	4
1					X
2				X	
3			X		
4		X			
5		X	X	X	X
6		X	X	X	X

10.16b Blower speed control switch continuity chart. There must be continuity between the terminals marked with an X when the control is in the proper position

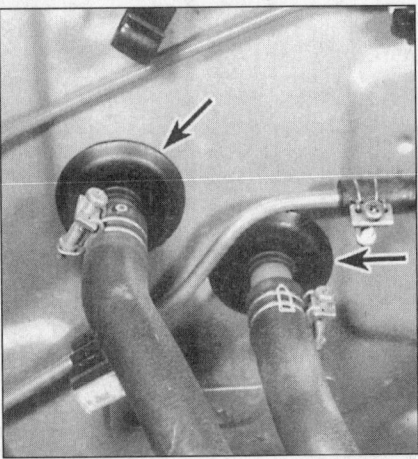

11.4 Disconnect the heater hoses at the firewall by loosening the hose clamps (arrows) and sliding them off the tubes, then remove the grommets (arrows)

the coolant is relatively new, or tests in good condition (see Section 2), save it and re-use it. **Warning:** *Do not allow antifreeze to come in contact with your skin or painted surfaces of the vehicle. Rinse off spills immediately with plenty of water. Antifreeze is highly toxic if ingested. Never leave antifreeze lying around in an open container or in puddles on the floor; children and pets are attracted by it's sweet smell and may drink it. Check with local authorities about disposing of used antifreeze. Many communities have collection centers which will see that antifreeze is disposed of safely. Never dump used anti-freeze on the ground or into drains.*

4   Working in the engine compartment, disconnect the heater hoses where they enter the firewall **(see illustration)**. **Caution:** *If the heater hoses are stuck, it is better to cut off the hoses than to twist them with pliers and risk breaking the heater core tubes.*

5   Remove the rubber grommets where the heater core tubes go through the firewall.

6   Disconnect the air conditioning refrigerant lines and rubber grommet from the evaporator core if the vehicle is so equipped **(see illustration)**. **Warning:** *Always wear eye protection when disconnecting air conditioning system fittings.*

7   Remove the entire instrument panel (refer to Chapter 11). Remove the steel horizontal instrument panel reinforcement bar **(see illustration)**. **Note:** *Heater core removal is very difficult. Take your time and don't use excessive force - there may be fasteners you haven't found yet.*

8   Remove the large cooling assembly (if equipped with air conditioning) or passenger side heater duct on models without air conditioning.

9   Remove bolts/nuts securing the heater unit to the firewall. Disconnect the control cables from the unit **(see illustration)**.

10 Pull down and back on the heater unit, making sure that all fasteners have been removed. Some twisting is required to separate the heater unit from the upper ducts **(see illustration)**.

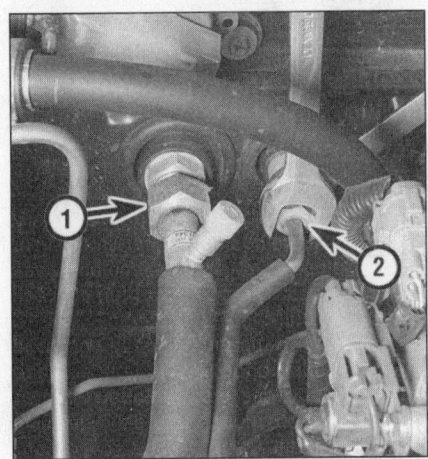

11.6 Disconnect the two air conditioning refrigerant lines (arrows) that go through the firewall to the evaporator

1   Suction tube
2   Liquid tube

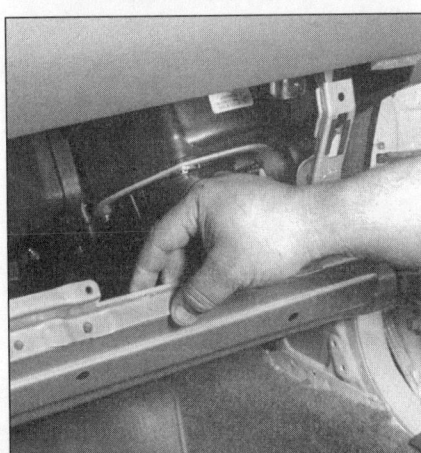

11.7 Access to the blower motor and ducts requires removing the glove box and this metal support brace

11.9 Remove the control cables (arrows) attached to the heater unit

11.10 Removing the heater unit

3-10 Chapter 3 Cooling, heating and air conditioning systems

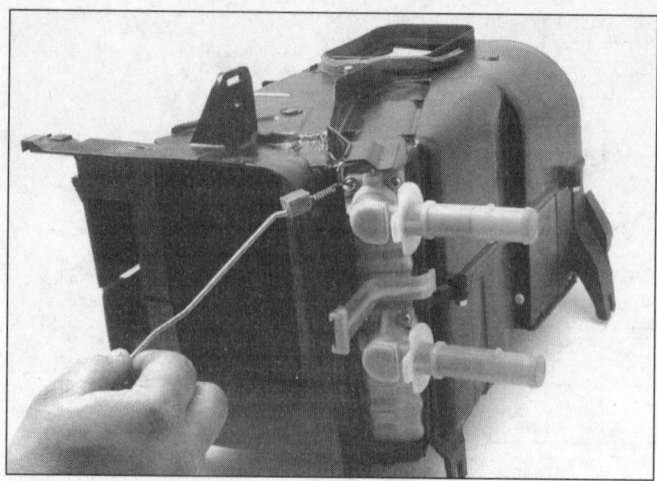

11.11a Unhook the water control rod from the top of the heating unit

11.11b Slide the heater core from the heater unit

11 Remove the heater core from the heater unit **(see illustrations)**.
12 Reassemble the heater unit and check the operation of the control flaps. If any parts bind, correct the problem before installation.
13 Reinstall the remaining parts in the reverse order of removal.
14 Refill the cooling system (see Chapter 1), reconnect the battery and run the engine. Check for leaks and proper system operation.

12 Air conditioner and heater control assembly - removal and installation

Refer to illustration 12.3

1 Disconnect the cable from the negative battery terminal. **Warning:** *On airbag-equipped models also disconnect the positive battery cable and wait at least ten minutes after the ignition has been turned off and the battery disconnected before starting work on the vehicle. The supplemental inflatable restraint (airbag) system has electrical capacitors that must bleed down to avoid the possibility of airbag deployment.* **Caution:** *If the stereo in your vehicle is equipped with an anti-theft system, make sure you have the correct activation code before disconnecting the battery.*
2 Pry off the center cluster trim panel which surrounds the radio and heater control assembly (see Chapter 11). Use a screwdriver which has been wrapped with tape for this procedure. (The rectangular cover directly above the heater control knobs will have to be pried off first).
3 Remove the control mounting screws **(see illustration)**.
4 Carefully pull and tilt the unit out of the dash. Unplug the lamps.
5 Check the outer sheath of the cables for indentations where the clamps grip them. Mark the cable sheath with paint if no indentation is visible. Remove the clamps and detach the control cables.
6 Unplug the electrical connectors and lift the control from the vehicle.
7 Installation is the reverse of removal.
8 Connect and/or adjust the heater control assembly cables as they were originally.

13 Air conditioning and heating system - check and maintenance

Refer to illustrations 13.1 and 13.10
Warning: *The air conditioning system is under high pressure. Do not loosen any fittings or remove any components until after the system has been discharged. Air conditioning refrigerant should be properly discharged into an EPA-approved container at a dealer service department or an automotive air conditioning repair facility. Always wear eye protection when disconnecting air conditioning system fittings.*

1 The following maintenance checks should be performed on a regular basis to ensure that the air conditioning system continues to operate at peak efficiency:

a) *Check the compressor drivebelt. If it's worn or deteriorated, replace it (see Chapter 1).*
b) *Check the drivebelt tension and, if necessary, adjust it (see Chapter 1).*
c) *Check the system hoses. Look for cracks, bubbles, hard spots and deterioration. Inspect the hoses and all fittings for oil bubbles and seepage. If there's any evidence of wear, damage or leaks, replace the hose(s).*
d) *Inspect the condenser fins for leaves, bugs and other debris. Use a "fin comb" or compressed air to clean the condenser.*
e) *Make sure the system has the correct refrigerant charge.*
f) *Check the evaporator housing drain tube for blockage* **(see illustration)**.

2 It's a good idea to operate the system for about 10 minutes at least once a month, particularly during the winter. Long term non-use can cause hardening, and subsequent failure, of the seals.
3 Because of the complexity of the air conditioning system and the special equip-

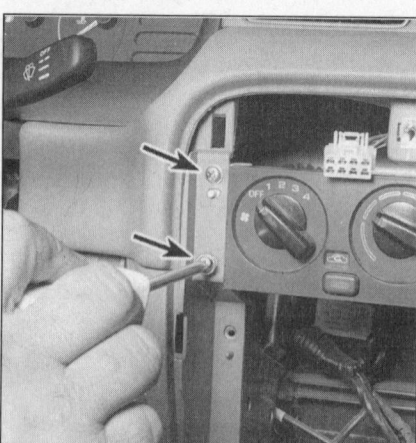

12.3 Remove the screws on each side of the control unit then pull the control unit out of the dash

13.1 The air conditioning evaporator core drain tube (arrow) is located above the right side steering gear boot - check that it is not blocked

Chapter 3 Cooling, heating and air conditioning systems   3-11

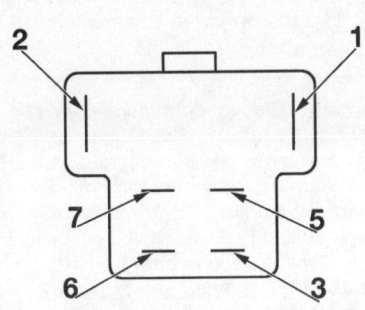

13.10 Air conditioning relay continuity check. When terminal 1 is grounded and fused battery voltage is applied to terminal 2, there should be continuity between terminals 3 and 5 and also 6 and 7

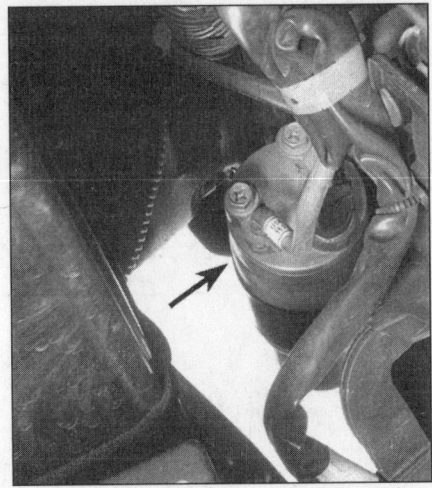

14.2 The receiver/drier (arrow) is mounted to the left of the condenser on most models

15.3 Unbolting the refrigerant lines from the compressor

ment necessary to service it, in-depth troubleshooting and repairs are not included in this manual. However, simple checks and component replacement procedures are provided in this Chapter. For more complete information on the air conditioning system, refer to the Haynes *Automotive Heating and Air Conditioning Manual*.
4   The most common cause of poor cooling is simply a low system refrigerant charge. If a noticeable drop in cool air output occurs, one of the following quick checks will help you determine if the refrigerant level is low.
5   Warm the engine up to normal operating temperature.
6   Place the air conditioning temperature selector at the coldest setting and put the blower at the highest setting. Open the doors (to make sure the air conditioning system doesn't cycle off as soon as it cools the passenger compartment).
7   With the compressor engaged - the clutch will make an audible click and the center of the clutch will rotate - inspect the sight glass. **Note:** *The sight glass is normally located in the top of the receiver/drier.* If the refrigerant looks foamy, it's low, and must be recharged.
8   If there's no sight glass, feel the inlet and outlet pipes at the compressor. One side should be much colder the other. If there's no perceptible difference between the two pipes, there's something wrong with the compressor or system. It might be a low charge - it might be something else. Take the vehicle to a dealer service department or other qualified shop.
9   If the compressor clutch doesn't engage when the air conditioning is turned on, check the air conditioning relay for continuity. The air conditioning relay is located in the main relay box next in the engine compartment.
10   Inspect the air conditioning relay for continuity using an ohmmeter **(see illustration)**. Apply battery voltage to the relay coil terminals numbered 1 and 2 - there should be

continuity between the other two pairs of terminals. If continuity doesn't exist when voltage is applied, replace the relay.

14 Air conditioning receiver/drier - removal and installation

Refer to illustration 14.2
Warning: *The air conditioning system is under high pressure. Do not loosen any fittings or remove any components until after the system has been discharged. Air conditioning refrigerant should be properly discharged into an EPA-approved container at a dealer service department or an automotive air conditioning repair facility. Always wear eye protection when disconnecting air conditioning system fittings.*
1   Have the refrigerant discharged at a dealer service department or an automotive air conditioning repair facility.
2   The receiver/drier, which acts as a reservoir and filter for the refrigerant, is located at the driver's side of the air conditioning condenser **(see illustration)**.
3   Detach the two refrigerant lines from the receiver/drier.
4   Immediately cap the open fittings to prevent the entry of dirt and moisture.
5   Unbolt the receiver/drier mounting bolt, at the clamp, and lift it from the vehicle.
6   Install new O-rings on the lines and lubricate them with clean refrigerant oil.
7   Installation is the reverse of removal.
Note: *Do not remove the sealing caps until you are ready to reconnect the lines. Do not mistake the inlet (marked IN) and the outlet connections.*
8   If a new receiver/drier is installed, add 0.2 ounces (6 cc) of refrigerant oil to the system. **Caution:** *When replacing entire components, additional refrigerant oil should be added equal to the amount that is removed*

with the component being replaced. Refrigerant oils, just like refrigerant R-12 vs. R-134a, are not compatible. Be sure to read the can before adding any oil to the system, to make sure it is compatible with the type of system being repaired.
9   Have the system evacuated, charged and leak tested by the shop that discharged it.

15 Air conditioning compressor - removal and installation

Refer to illustrations 15.3, 15.4 and 15.6
Warning: *The air conditioning system is under high pressure. Do not loosen any fittings or remove any components until after the system has been discharged. Air conditioning refrigerant should be properly discharged into an EPA-approved container at a dealer service department or an automotive air conditioning repair facility. Always wear eye protection when disconnecting air conditioning system fittings.*
Caution: *When replacing entire components, additional refrigerant oil should be added equal to the amount that is removed with the component being replaced. Refrigerant oils, just like refrigerant R-12 vs. R-134a, are not compatible. Be sure to read the can before adding any oil to the system, to make sure it is compatible with the type of system being repaired.*
1   Have the refrigerant discharged at a dealer service department or an automotive air conditioning repair facility.
2   Disconnect the cable from the negative battery terminal and remove the battery.
3   Detach the refrigerant lines from the compressor **(see illustration)** and immediately cap the open fittings to prevent the entry of dirt and moisture.
4   Disconnect the clutch wire from the

15.4 Disconnect the electrical connector (arrow) for the magnetic clutch

15.6 Once the refrigerant lines and electrical connections have been separated from the compressor, remove the mounting bolts (arrows) and the compressor from the mounting bracket

compressor **(see illustration)**.
5   Remove the accessory drivebelt (see Chapter 1).
6   Remove the compressor mounting bolts **(see illustration)** and remove the compressor from the engine compartment. **Note:** *Keep the compressor level during handling and storage. If the compressor seized or you find metal particles in the refrigerant lines, the system must be flushed out by an air conditioning technician and the receiver/drier must be replaced* (see Section 14).
7   Prior to installation, turn the center of the clutch six times to disperse any oil that has collected in the head.
8   Install the compressor in the reverse order of removal.
9   If you are installing a new compressor, the cycling clutch assembly may have to be transferred to the new compressor. **Note:** *The removal of the clutch assembly will require the use of several special tools. You may want to have the clutch assembly transferred to the new compressor by an air conditioning shop or dealer service department.* Refer to the manufacturer's instructions for adding refrigerant oil to the system. **Caution:** *When replacing entire components, additional refrigerant oil should be added equal to the amount that is removed with the component being replaced. Refrigerant oils, just like refrigerant R-12 vs. R-134a, are not compatible. Be sure to read the can before adding any oil to the system, to make sure it is compatible with the type of system being repaired.*
10   Have the system evacuated, charged and leak tested by the shop that discharged it.

16   Air conditioning condenser - removal and installation

Warning: *The air conditioning system is under high pressure. Do not loosen any fittings or remove any components until after the system has been discharged. Air condi-* *tioning refrigerant should be properly discharged into an EPA-approved container at a dealer service department or an automotive air conditioning repair facility. Always wear eye protection when disconnecting air conditioning system fittings.*

1   Have the refrigerant discharged at a dealer service department or an automotive air conditioning repair facility.
2   Remove the grille if necessary for access to the condenser mounts (refer to Chapter 11).
3   Disconnect the refrigerant lines from the condenser. Be sure to use a back-up wrench to avoid twisting the lines (if equipped with two nuts).
4   Immediately cap the open fittings to prevent the entry of dirt and moisture.
5   Unbolt the condenser mounts.
6   Unbolt the radiator mounts and pull the radiator and fan assembly toward the rear of the engine compartment until there is room to pull up the condenser and lift it out of the vehicle. **Note:** *On some models, it may be necessary to completely remove the radiator from the vehicle. Store it upright to prevent oil loss.* **Caution:** *Place a towel or section of cardboard between the radiator and condenser while removing the condenser, to prevent damage to the radiator fins.*
7   Installation is the reverse of removal.
8   If a new condenser was installed, add 2.5 ounces (75 cc) of refrigerant oil to the system. **Caution:** *When replacing entire components, additional refrigerant oil should be added equal to the amount that is removed with the component being replaced. Refrigerant oils, just like refrigerant R-12 vs. R-134a, are not compatible. Be sure to read the can before adding any oil to the system, to make sure it is compatible with the type of system being repaired.*
9   Have the system evacuated, charged and leak tested by the shop that discharged it.

17   Air conditioning evaporator and expansion valve - removal and installation

Warning: *The air conditioning system is under high pressure. Do not loosen any fittings or remove any components until after the system has been discharged. Air conditioning refrigerant should be properly discharged into an EPA-approved container at a dealer service department or an automotive air conditioning repair facility. Always wear eye protection when disconnecting air conditioning system fittings.*

1   Have the refrigerant discharged at a dealer service department or an automotive air conditioning repair facility.
2   Disconnect the cable from the negative battery terminal. **Warning:** *On airbag-equipped models also disconnect the positive battery cable and wait at least ten minutes after the ignition has been turned off and the battery disconnected before starting work on the vehicle. The supplemental inflatable restraint (airbag) system has electrical capacitors that must bleed down to avoid the possibility of airbag deployment.*
3   Working in the engine compartment, disconnect the suction and liquid lines - use a back-up wrench to avoid twisting and damaging the lines.
4   Immediately cap the open fittings to prevent the entry of dirt and moisture and remove the inlet and outlet grommets.
5   Remove the instrument panel (see Chapter 11).
6   Disconnect all electrical connectors and tubing from the assembly, remove the mounting nuts and bolts and pull the unit free.
7   To remove the evaporator, disconnect all connectors, unfasten the retaining clips holding the two halves of the housing, and separate the housing halves.
8   Check the evaporator fins for blockage; if they are dirty, clean them with compressed air - never use water for this purpose!
9   Check fittings for cracks and signs of wear; replace parts as necessary.
10   Installation is the reverse of the removal procedure. Be sure to replace all O-rings removed during disassembly with new ones.
11   If a new evaporator was installed, add 2.5 ounces (75 cc) of refrigerant oil to the system. **Caution:** *When replacing entire components, additional refrigerant oil should be added equal to the amount that is removed with the component being replaced. Refrigerant oils, just like refrigerant R-12 vs. R-134a, are not compatible. Be sure to read the can before adding any oil to the system, to make sure it is compatible with the type of system being repaired.*
12   Have the system evacuated, charged and leak tested by the shop that discharged it.

Chapter 4
Fuel and exhaust systems

Contents

	Section
Accelerator cable - removal, installation and adjustment	10
Air filter assembly - removal and installation	9
Air filter replacement	See Chapter 1
Electronic fuel injection system - check	12
Electronic fuel injection system - general information	11
Exhaust system check	See Chapter 1
Exhaust system servicing - general information	14
Fuel filter replacement	See Chapter 1
Fuel level sending unit - check and replacement	8
Fuel lines and fittings - repair and replacement	5

	Section
Fuel pressure relief procedure	2
Fuel pump/fuel pressure - check	3
Fuel pump - removal and installation	4
Fuel system check	See Chapter 1
Fuel tank cleaning and repair - general information	7
Fuel tank - removal and installation	6
General information	1
Multi Port Fuel Injection (MPFI) system components - check, removal and installation	13
Underhood hose check and replacement	See Chapter 1

Specifications

Fuel pressure	
Fuel system pressure (at idle)	
Vacuum hose attached	32 to 38 psi
Vacuum hose detached	38 to 46 psi
Fuel system hold pressure (after 5 minutes)	30 to 40 psi
Fuel pump pressure (maximum)	65 psi
Fuel pump hold pressure	50 psi
Fuel pump resistance	0.2 to 5.0 ohms
Injector resistance	10 to 14 ohms
IAC/AAC valve resistance	
GA16DE engine	
1995 through 1997	50 to 100 ohms
1998	
Cold	138 to 238 ohms
Hot	175 to 280 ohms
SR20DE engine	10 ohms (approximately)
IAC/FICD solenoid valve resistance (GA16DE engine only)	
1995 to 1997	75 to 125 ohms
1998	
Cold	138 to 238 ohms
Hot	175 to 280 ohms
IACV-air regulator resistance (SR20DE engine only)	70 to 80 ohms

Torque specifications

	Ft-lbs (unless otherwise indicated)
Air intake plenum mounting bolts	156 to 180 in-lbs
Throttle body mounting bolts	156 to 192 in-lbs
EGR valve-to-throttle body nuts	72 to 108 in-lbs
EGR base mounting assembly to air intake plenum	15 to 20
Fuel tank mounting bolts	20 to 26
Fuel tank hole cover bolts	29 to 37 in-lbs
Air regulator to intake manifold	72 to 108 in-lbs
Fuel rail mounting bolts	70 to 105 in-lbs
Exhaust pipe-to-exhaust manifold bolts	21 to 25

Chapter 4 Fuel and exhaust systems

2.1 Location of the fuel pump fuse

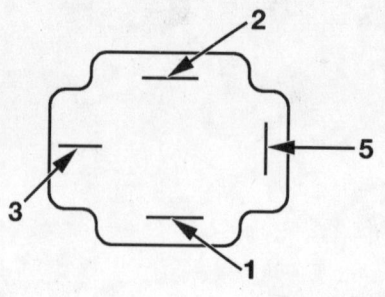

3.2 Fuel pump relay terminals. Apply battery voltage to terminal 1 and ground terminal 2. There should be a click resulting in continuity between terminals 3 and 5

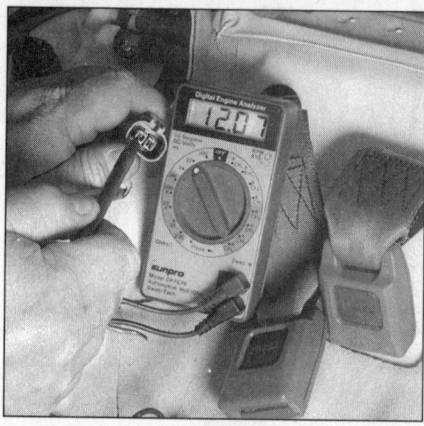

3.4 Remove the rear seat from the vehicle and check for battery voltage to the fuel pump with a voltmeter or test light. The fuel pump connector is the one with only two terminals

1 General information

The fuel system consists of a fuel tank, an electric fuel pump (located in the fuel tank), a fuel pump relay, fuel injector(s), an air filter assembly and a throttle body unit. All models are equipped with the Multi Port Fuel Injection (MPFI) system. The MPFI system uses separate injectors positioned over the intake valves, mounted in the intake manifold.

Multi Port Fuel Injection (MPFI) system

Multi Port Fuel Injection uses timed impulses to inject the fuel directly into the intake port of each cylinder. The injectors are controlled by the Engine Control Module (ECM). The ECM monitors various engine parameters and delivers the exact amount of fuel required into the intake ports. The throttle body serves only to control the amount of air passing into the system. Because each cylinder is equipped with its own injector, much better control of the fuel/air mixture ratio is possible.

Fuel pump and lines

Fuel is circulated from the fuel tank to the fuel injection system, and back to the fuel tank, through a pair of metal lines running along the underside of the vehicle. An electric fuel pump is located inside the fuel tank. A vapor return system routes all vapors back to the fuel tank through a separate return line.

The fuel pump will operate as long as the engine is cranking or running and the ECM is receiving ignition reference pulses from the electronic ignition system. If there are no reference pulses, the fuel pump will shut off after two or three seconds.

Exhaust system

The exhaust system includes an exhaust manifold fitted with exhaust oxygen sensors, a catalytic converter, an exhaust pipe, and a muffler.

The catalytic converter is an emission control device added to the exhaust system to reduce pollutants. A single-bed converter is used in combination with a three-way (reduction) catalyst. Refer to Chapter 6 for more information regarding the catalytic converter.

2 Fuel pressure relief procedure

Refer to illustration 2.1

Warning: *Gasoline is extremely flammable, so take extra precautions when you work on any part of the fuel system. Don't smoke or allow open flames or bare light bulbs near the work area, and don't work in a garage where a natural gas-type appliance (such as a water heater or a clothes dryer) with a pilot light is present. Since gasoline is carcinogenic, wear latex gloves when there's a possibility of being exposed to fuel, and, if you spill any fuel on your skin, rinse it off immediately with soap and water. Mop up any spills immediately and do not store fuel-soaked rags where they could ignite. The fuel system is under constant pressure, so, if any fuel lines are to be disconnected, the fuel pressure in the system must be relieved first. When you perform any kind of work on the fuel system, wear safety glasses and have a Class B type fire extinguisher on hand.*

Note: *After the fuel pressure has been relieved, it's a good idea to lay a shop towel over any fuel connection to be disassembled, to absorb the residual fuel that may leak out when servicing the fuel system.*

1 Remove the fuel pump fuse from the fuse panel in the lower dash panel near the hood release lever **(see illustration)**.
2 Start the engine and allow it to run until it stops. This should take only a few seconds. Disconnect the cable from the negative terminal of the battery before working on the fuel system.
3 The fuel system pressure is now relieved. When you're finished working on the fuel system, simply install the fuel pump fuse back into the fuse panel and connect the negative cable to the battery.

3 Fuel pump/fuel pressure - check

Warning: *Gasoline is extremely flammable, so take extra precautions when you work on any part of the fuel system. See the Warning in Section 2.*
Note: *To perform the fuel pressure test, you will need to obtain a fuel pressure gauge and adapter set (fuel line fittings).*

Preliminary inspection

Refer to illustrations 3.2 and 3.4
1 Should the fuel system fail to deliver the proper amount of fuel, or any fuel at all, inspect it as follows. Remove the fuel filler cap. Have an assistant turn the ignition key to the On position (engine not running) while you listen at the fuel filler opening. You should hear a whirring sound that lasts for a couple of seconds.
2 If you don't hear anything, check the fuel pump fuse (see Section 2). If the fuse is blown, replace it and see if it blows again. If it does, trace the fuel pump circuit for a short. If it isn't blown, check the fuel pump relay. The fuel pump relay is mounted in the relay box. To test it, connect a ground to terminal 1 and fused battery voltage to terminal 2 **(see illustration)**. There should be continuity between terminals 3 and 5.
3 Remove the relays and check for battery voltage to the fuel pump relay connector and the EFI relay connector. If there is no battery voltage present, trace the circuit for an open-circuit condition. If there is voltage (and the relay checked out good), reinstall the relay.
4 Remove the rear seat and check for battery voltage at the fuel pump harness connector **(see illustration)**. If there is no voltage, check the fuel pump circuit between the relay and the pump for an open-circuit condition. If there is voltage present, replace the pump (see Section 4).

Operating pressure check

Refer to illustrations 3.6 and 3.9
5 Relieve the fuel system pressure (see

3.6 Install the fuel pressure gauge between the fuel filter and the fuel rail using a T-fitting

3.9 Connect a hand-held vacuum pump to the fuel pressure regulator. As vacuum is applied, fuel pressure should decrease

Section 2). Detach the cable from the negative battery terminal.
6   Remove the fuel line from the fuel filter and attach a fuel pressure gauge with a T-fitting between the fuel filter and the fuel rail **(see illustration)**. Tighten the hose clamps securely.
7   Attach the cable to the negative battery terminal. Start the engine.
8   Note the fuel pressure and compare it with the pressure listed in this Chapter's Specifications.
9   Disconnect the vacuum hose from the fuel pressure regulator and hook up a hand-held vacuum pump to the port on the fuel pressure regulator **(see illustration)**.
10 Read the fuel pressure gauge with vacuum applied to the pressure regulator and also with no vacuum applied.
11 The fuel pressure should decrease as vacuum increases (and increase as vacuum decreases). Compare your readings with the values listed in this Chapter's Specifications.
12 Reconnect the vacuum hose to the regulator and check the fuel pressure at idle, then disconnect the vacuum hose and watch the gauge - the pressure should jump up considerably as soon as the hose is disconnected. If it doesn't, check for a vacuum signal to the fuel pressure regulator (see Step 15).
13 If the fuel pressure is low, pinch the fuel return line shut and watch the gauge. If the pressure doesn't rise, the fuel pump could be defective, there could be a fuel leak (including a leaking injector) or there is a restriction in the fuel feed line (the fuel filter could be clogged). If the pressure rises sharply, replace the pressure regulator.
14 If the fuel pressure is too high, turn the engine off. Disconnect the fuel return line and blow through it to check for a blockage. If there is no blockage, replace the fuel pressure regulator.
15 Connect a vacuum gauge to the pressure regulator vacuum hose. Start the engine and check for vacuum. If there isn't vacuum present, check for a clogged hose or vacuum port. If the amount of vacuum is adequate, but the pressure is too high, replace the fuel

pressure regulator.
16 Turn the ignition switch to Off, wait five minutes and recheck the pressure on the gauge. Compare the reading with the hold pressure listed in this Chapter's Specifications. If the hold pressure is less than specified:

a) The fuel lines may be leaking.
b) The fuel pressure regulator may be allowing the fuel pressure to bleed through to the return line.
c) A fuel injector (or injectors) may be leaking.
d) The fuel pump may be defective.

Fuel pump output pressure check

Warning: *For this test it is necessary to use a fuel pressure gauge with a bleeder valve in order to relieve the fuel pressure after the test is completed (the normal procedure for pressure relief will not work because the gauge is connected directly to the fuel pump).*
17 Relieve the system fuel pressure (see Section 2).
18 Detach the cable from the negative battery terminal.
19 Attach a fuel pressure gauge directly to the fuel feed line at the fuel tank.
20 Attach the cable to the negative battery terminal.
21 Turn the ignition key to the On position to energize the fuel pump.
22 Note the pressure reading on the gauge and compare the reading to the value listed in this Chapter's Specifications.
23 If the indicated pressure is less than specified, inspect the fuel line for leaks between the pump and gauge. If no leaks are found, replace the fuel pump.
24 Turn the ignition key to Off and wait five minutes. Note the reading on the gauge and compare it to the hold pressure listed in this Chapter's Specifications. If the hold pressure is less than specified, check the fuel line between the pump and gauge for leaks. If no leaks are found, replace the fuel pump.
25 Open the bleeder valve on the gauge

and allow the pressurized fuel to drain into an approved fuel container. Remove the gauge and reconnect the fuel line.

4   Fuel pump - removal and installation

Refer to illustrations 4.5, 4.6, 4.9, 4.10, 4.11, 4.12 and 4.13
Warning: *Gasoline is extremely flammable, so take extra precautions when you work on any part of the fuel system. See the* **Warning** *in Section 2.*
1   Unless the vehicle has been driven far enough to completely empty the tank, it's a good idea to siphon the residual fuel out before removing the fuel pump from the vehicle. **Warning:** *DO NOT start the siphoning action by mouth! Use a siphoning kit (available at most auto parts stores).*
2   Relieve the fuel pressure (see to Section 2).
3   Detach the cable from the negative terminal of the battery.
4   Remove the rear seat from the vehicle (see Chapter 11).
5   Remove the access cover for the fuel pump/fuel level sending unit assembly **(see illustration)**.

4.5 After removing the rear seat, remove the screws and pull the cover from the floor access hole

4-4 Chapter 4 Fuel and exhaust systems

6   Disconnect the fuel pump/fuel level sender electrical connectors (see illustration).
7   Disconnect the fuel hoses.
8   Remove the six screws securing the fuel pump/level sending unit to the tank. Lift the unit out with great care, maneuvering it so that it is not damaged.
9   Disconnect the hose from the fuel pump to the upper cover (see illustration).
10 Disconnect the wiring from the fuel pump (see illustration).
11 Disconnect the longer hose from the fuel pump housing (see illustration).
12 Use a small screwdriver to release the three tabs on the sides of the housing (see illustration).
13 Separate the housing and remove the fuel pump and strainer (see illustration).
14 Reassembly and installation is the reverse of removal.

5 Fuel lines and fittings - repair and replacement

Warning: *Gasoline is extremely flammable, so take extra precautions when you work on any part of the fuel system. See the **Warning** in Section 2.*

4.6 Clean the area thoroughly to prevent debris from entering the fuel tank, then disconnect the electrical connectors and fuel hoses

Inspection
Refer to illustration 5.2

1   Once in a while, you will have to raise the vehicle to service or replace some component (an exhaust pipe hanger, for example). Whenever you work under the vehicle, always inspect the fuel lines and fittings for possible damage or deterioration.

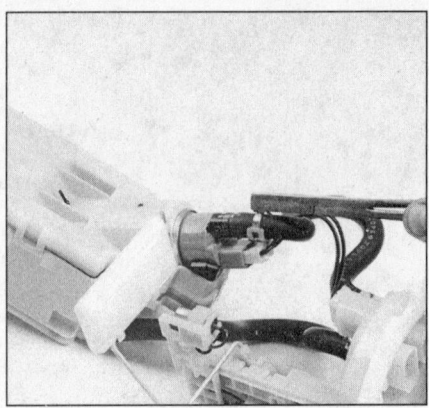

4.9 After carefully removing the assembly from the tank, disconnect the short fuel hose . . .

2   Check all hoses and pipes for cracks, kinks, deformation or obstructions (see illustration).
3   Make sure all hose and pipe clips attach their associated hoses or pipes securely to the underside of the vehicle.
4   Verify all hose clamps attaching rubber hoses to metal fuel lines or pipes are snug enough to assure a tight fit between the hoses and the metal pipes.

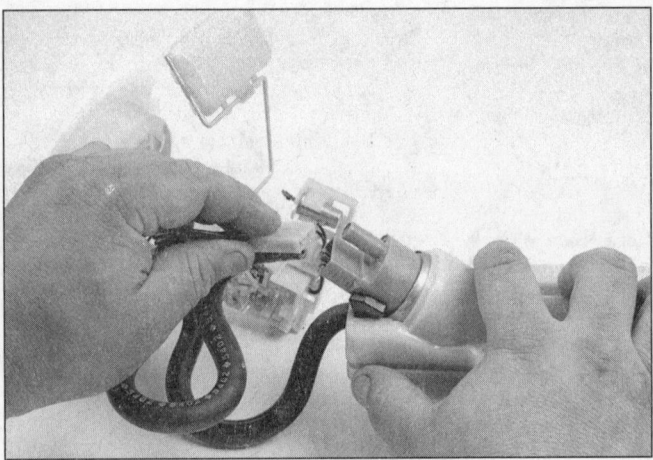

4.10 . . . unplug the electrical connector . . .

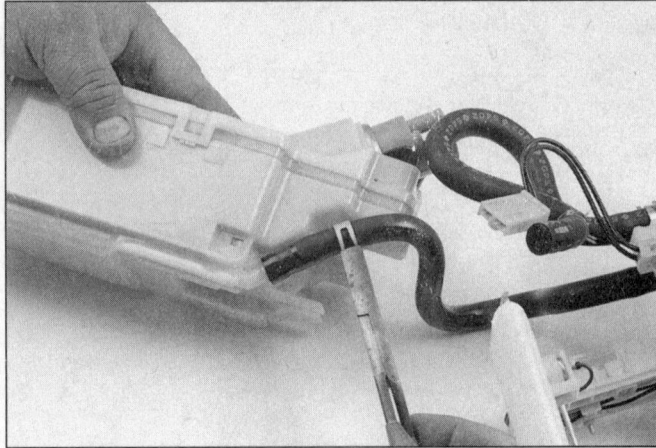

4.11 . . . and disconnect the longer hose

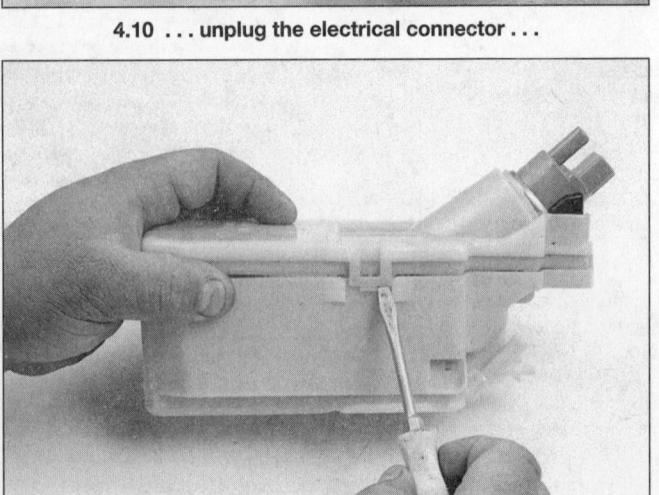

4.12 There are three tabs which secure the housing halves together. A small screwdriver can be used to unsnap them

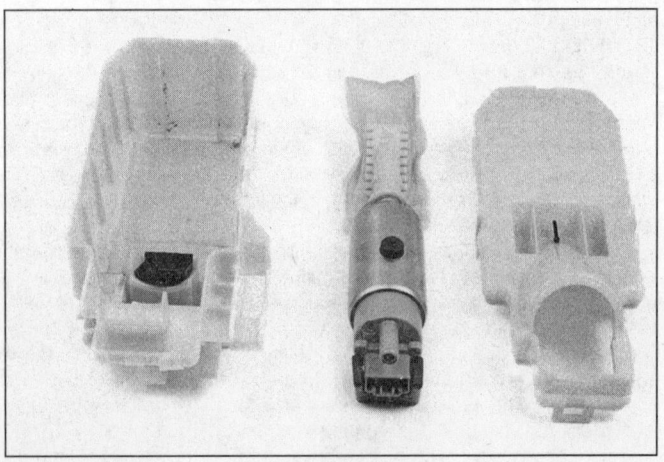

4.13 The fuel pump with its attached strainer can now be removed from the housing

Chapter 4 Fuel and exhaust systems

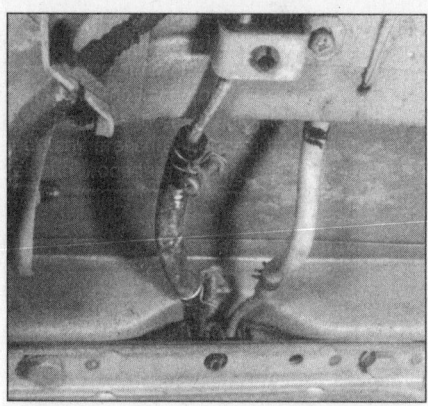

5.2 Carefully inspect the fuel lines that extend from the fuel pump to the fuel tank

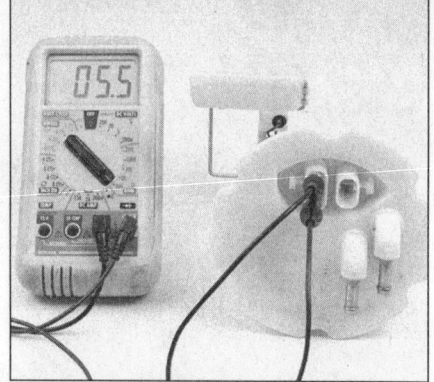

8.4a With the fuel tank float in the FULL position, the resistance of the unit should be approximately 5 ohms

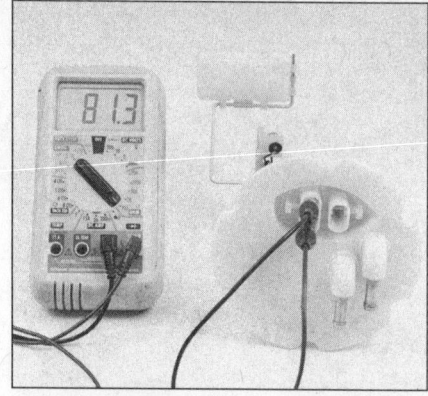

8.4b When the float is in the EMPTY position, the ohmmeter should read approximately 80 ohms

Replacement

5 If you must replace any damaged sections, use hoses approved for use in high-pressure fuel injection systems or pipes made from steel only (it's best to use an original-type pipe that's already flared and pre-bent). Do not install substitutes constructed from inferior or inappropriate material, as this could result in a fuel leak and a fire.
6 Always, before detaching or disassembling any part of the fuel line system, note the routing of all hoses and pipes and the orientation of all clamps and clips to ensure that replacement sections are installed in exactly the same manner.
7 Before detaching any part of the fuel system, be sure to relieve the fuel tank pressure by removing the fuel filler cap.
8 It's a good idea to use new hose clamps after loosening or removing them.
9 While you're under the vehicle, it's a good idea to check the following related components:
   a) Check the condition of the fuel filter - make sure that it's not damaged (see Chapter 1).
   b) Inspect the evaporative emission control (EVAP) system. Verify that all hoses are attached and in good condition (see Chapter 6).

6 Fuel tank - removal and installation

Warning: *Gasoline is extremely flammable, so take extra precautions when you work on any part of the fuel system. See the* **Warning** *in Section 2.*
Note: *The following procedure is much easier to perform if the fuel tank is empty. Some tanks have a drain plug for this purpose. If the tank does not have a drain plug, drain the fuel into an approved fuel container using a commercially available siphoning kit (NEVER start a siphoning action by mouth) or wait until the fuel tank is nearly empty, if possible.*

1 Remove the fuel tank filler cap to relieve fuel tank pressure. Perform the fuel pressure release procedure (refer to Section 2).
2 Detach the cable from the negative terminal of the battery.
3 If the tank still has fuel in it, you can siphon it after raising the vehicle. If the tank has a drain plug remove it and allow the fuel to collect in an approved gasoline container. **Warning:** *Never attempt to siphon by mouth! Use a siphoning kit, which can be purchased at most auto parts stores. Store the fuel in an explosion-proof container.*
4 Raise the vehicle and place it securely on jackstands.
5 Disconnect the fuel filler tube from the body of the vehicle.
6 Disconnect the fuel lines and the vapor return line. **Note:** *The fuel feed and return lines and the vapor return line are three different diameters, so reattachment is simplified. If you have any doubts, however, clearly label the three lines and the fittings. Be sure to plug the hoses to prevent leakage and contamination of the fuel system.*
7 If there is still fuel in the tank, siphon it out from the fuel feed port. Remember - NEVER start the siphoning action by mouth! Use a siphoning kit, which can be purchased at most auto parts stores.
8 Support the fuel tank with a floor jack. Position a wood block between the jack head and the fuel tank to protect the tank.
9 If equipped, remove the bolts from the heat shield located on the front side of the fuel tank.
10 Disconnect the electrical connectors for the fuel pump and fuel level sending unit located under the rear seat (refer to Section 4).
11 Remove the bolts that retain the fuel tank to the chassis.
12 Remove the tank from the vehicle.
13 Installation is the reverse of removal.

7 Fuel tank cleaning and repair - general information

1 All repairs to the fuel tank or filler neck should be carried out by a professional who has experience in this critical and potentially dangerous work. Even after cleaning and flushing of the fuel system, explosive fumes can remain and ignite during repair of the tank.
2 If the fuel tank is removed from the vehicle, it should not be placed in an area where sparks or open flames could ignite the fumes coming out of the tank. Be especially careful inside garages where a natural gas-type appliance is located, because the pilot light could cause an explosion.

8 Fuel level sending unit - check and replacement

Refer to illustrations 8.4a and 8.4b
Warning: *Gasoline is extremely flammable, so take extra precautions when you work on any part of the fuel system. See the* **Warning** *in Section 2.*

Check

1 Before performing any tests on the fuel level sending unit, try to determine the actual fuel level in the tank. If this is not possible, either fill the tank with fuel or drain out any fuel in the tank into an approved gasoline container.
2 Raise the vehicle and support it securely with jackstands.
3 Lift the rear seat and remove the access cover plate to locate the fuel level sending unit electrical connector positioned on the fuel tank (refer to Section 4).
4 Position the ohmmeter probes into the electrical connector **(see illustrations)** and check the resistance. Use the 200-ohm or low scale on the ohmmeter.
5 With the fuel tank completely full, the resistance should be approximately 4.0 to 8.0 ohms. With the tank half full, the resistance should be approximately 27.0 to 35.0 ohms. With the fuel tank empty, the resistance of the sending unit should be approximately 73.0 to 88.0 ohms.
6 If the readings are incorrect, replace the sending unit. **Note:** *A more accurate check of the sending unit can be made by removing it (fuel level sending unit or assembly) from the fuel tank and checking its resistance while manually operating the float arm.*

4-6   Chapter 4 Fuel and exhaust systems

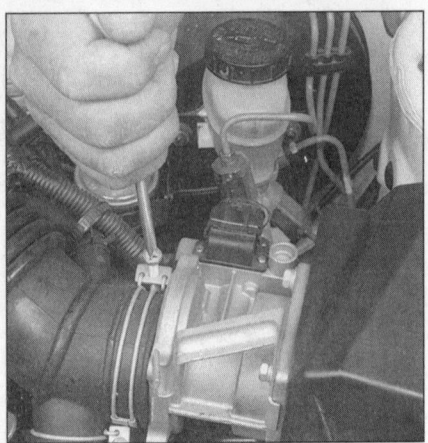

9.3a Loosen the clamp that retains the air intake duct to the air filter assembly

9.3b Disconnect the sensor wiring and then lift the air filter upper section from the engine compartment

9.5 Remove the mounting bolts from the air filter lower housing (arrows)

Replacement

7   Remove the rear seat from the vehicle (see Chapter 11).
8   If necessary, carefully peel back the insulation material from the top of the fuel tank without damaging it.
9   Refer to Section 4 and remove the fuel pump/fuel level sending unit from the tank.
10 Carefully angle the sending unit out of the opening without damaging the fuel level float located at the bottom of the assembly.
11 Disconnect the fuel pump from the sending unit.
12 Installation is the reverse of removal.

9   Air filter assembly - removal and installation

Refer to illustrations 9.3a, 9.3b and 9.5

1   Detach the cable from the negative terminal of the battery.
2   Remove the air filter from the air filter housing (see Chapter 1).
3   Loosen the clamp on the air intake duct and separate the duct from the air filter housing **(see illustrations)**.

4   Disconnect the intake air temperature sensor wiring harness. Remove the upper air filter housing.
5   Remove the mounting bolts from the lower air filter housing **(see illustration)**.
6   Remove the lower air filter housing from the engine compartment.
7   Installation is the reverse of removal.

10 Accelerator cable - removal, installation and adjustment

Removal

Refer to illustrations 10.1 and 10.3

1   Detach the accelerator cable from the throttle lever **(see illustration)**.
2   If equipped, remove the cruise control cable from the cable bracket.
3   Mark the exact position of the cable assembly in the bracket. Separate the accelerator cable from the cable bracket by loosening the lock-nut **(see illustration)**.
4   Detach the nylon collar from the upper end of the accelerator pedal arm.
5   Remove the cable through the firewall from the engine compartment side.

Installation

6   Installation is the reverse of removal. Be sure the cable is routed correctly.
7   If necessary, at the engine compartment side of the firewall, apply sealant around the accelerator cable to prevent water from entering the passenger compartment.

Adjustment

8   To adjust the cable:
a) Tighten the adjusting nut until there is zero freeplay at the throttle. Back off the adjusting nut 1-1/2 to 2 turns. Tighten the lock-nut.
b) After you have adjusted the throttle cable, have an assistant help you verify that the throttle valve opens all the way when you depress the accelerator pedal to the floor and that it returns to the idle position when you release the accelerator. Verify the cable operates smoothly. It must not bind or stick.

11 Electronic fuel injection system - general information

The fuel system on these vehicles is a

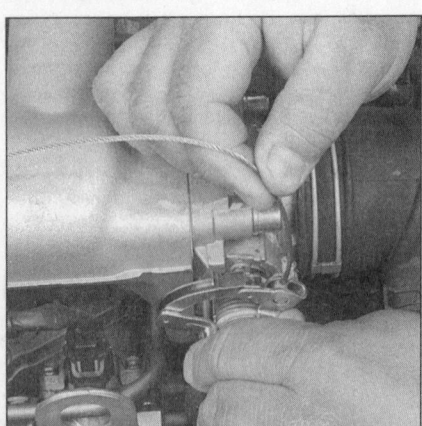

10.1 Rotate the throttle lever and pass the cable through the slot

10.3 Unscrew the nut nearest the exposed cable, then pass the cable through the slot in the bracket

12.6 Clean the interior of the throttle body using carburetor cleaner spray and shop towels or a toothbrush

Chapter 4 Fuel and exhaust systems

12.7 Use a stethoscope to determine if the injectors are working properly - they should make a steady clicking sound that rises and falls with engine speed changes

12.8 Install the noid light into the fuel injector electrical connector and confirm that it blinks when the engine is running

12.9 Measure the resistance of each injector. It should be within Specifications

Multi Port Fuel Injection (MPFI) system. On the MPFI system, fuel is metered into each intake port in accordance with engine demand through injectors mounted on the intake manifold. The two-piece air intake manifold includes an air intake plenum.

This system incorporates an on-board Engine Control Module (ECM) computer that accepts inputs from various engine sensors to compute the required fuel flow rate necessary to maintain a correct air/fuel ratio throughout the entire engine operational range. The computer then outputs a command to the fuel injectors to meter the approximate quantity of fuel. The system automatically senses and compensates for changes in altitude, load and speed. **Note:** *The Engine Control Module (ECM) is sometimes called the Powertrain Control Module (PCM) on later models due to standardization of the self-diagnosis system within the automotive industry.*

The fuel delivery system includes an electric in-tank fuel pump which delivers pressurized fuel through an inline fuel filter to the fuel charging manifold assembly.

A constant fuel pressure drop is maintained across the injector nozzles by a pressure regulator. The regulator is positioned downstream from the fuel injectors. Excess fuel passes through the regulator and returns to the fuel tank through a fuel return line.

12 Electronic fuel injection system - check

Warning: *Gasoline is extremely flammable, so take extra precautions when you work on any part of the fuel system. See the* **Warning** *in Section 2.*
Note: *The following procedure is based on the assumption that the fuel pump is working and the fuel pressure is adequate (see Section 3).*

Preliminary checks

1 Check all electrical connectors that are related to the system. Loose electrical connectors and poor grounds can cause many problems that resemble more serious malfunctions.
2 Check to see that the battery is fully charged, as the control unit and sensors depend on an accurate supply voltage in order to properly meter the fuel.
3 Check the air filter element - a dirty or partially blocked filter will severely impede performance and economy (see Chapter 1).
4 If a blown fuse is found, replace it and see if it blows again. If it does, search for a grounded wire in the harness to the fuel pump (see Chapter 12).

System checks

Refer to illustrations 12.6, 12.7, 12.8 and 12.9
5 Check the condition of the vacuum hoses connected to the intake manifold.
6 Remove the air intake duct from the throttle body and check for dirt, carbon or other residue build-up in the throttle body, particularly around the throttle plate. If it's dirty, clean it with aerosol carburetor cleaner, a rag and a toothbrush, if necessary **(see illustration)**.
7 With the engine running, place an automotive stethoscope against each injector, one at a time, and listen for a clicking sound, indicating operation **(see illustration)**. If you don't have a stethoscope, you can place the tip of a long screwdriver against the injector and listen through the handle.
8 If an injector isn't functioning (not clicking), purchase a special injector test light (sometimes called a "noid" light) and install it into the injector electrical connector **(see illustration)**. Start the engine and check to see if the noid light flashes. If it does, the injector is receiving proper voltage. If it doesn't flash, further diagnosis should be performed by a dealer service department or other repair shop.
9 With the engine OFF and the fuel injector electrical connectors disconnected, measure the resistance of each injector **(see illustration)**. Compare your measurements with the injector resistance listed in this Chapter's Specifications. If any injector is open or has an abnormally high resistance, replace it with a new one.

13 Multi Port Fuel Injection (MPFI) system components - check, removal and installation

Warning: *Gasoline is extremely flammable, so take extra precautions when you work on any part of the fuel system. See the* **Warning** *in Section 2.*

Air intake plenum with throttle body

Removal

1 Detach the cable from the negative terminal of the battery.
2 Disconnect the electrical connectors at the AAC valve, throttle position sensor (TPS), AIV control solenoid and EGR/EVAP canister control solenoid.
3 Detach the accelerator cable (see Section 6) and transmission linkage (see Chapter 7) from the throttle body assembly.
4 Remove the accelerator cable from the intake manifold (see Section 6).
5 Clearly label, then detach, the vacuum lines from air intake plenum, the EGR valve, the BPT valve and the fuel pressure regulator.
6 Detach the PCV system by disconnecting the hose from the fitting on the intake plenum.
7 If equipped, detach the canister purge line or lines from the throttle body.
8 Remove the upper air intake plenum retaining bolts and the lower (side) retaining bolts.
9 Remove the upper intake manifold and throttle body as an assembly from the lower intake manifold.

Installation

10 Be sure to clean and inspect the mounting faces of the lower intake manifold (see Chapter 2A) and the air intake plenum before positioning the new gasket(s) onto the lower intake mounting face. Install the air intake

Chapter 4 Fuel and exhaust systems

13.15 Remove the four bolts (arrows) which secure the throttle body

13.23 Remove the fuel rail mounting bolts (arrows)

13.24 Remove the fuel rail and fuel injector assembly from the engine as a single unit

plenum and throttle body assembly onto the intake manifold. Ensure the gasket remains in place. Install the upper intake manifold retaining bolts and tighten them to the torque listed in this Chapter's Specifications. The remainder of installation is the reverse of removal.

Throttle body

Removal

Refer to illustration 13.15

11 Detach the cable from the negative terminal of the battery.
12 Detach the Throttle Position Sensor (TPS) electrical connector.
13 Disconnect the accelerator cable (see Section 10) from the throttle body.
14 Carefully label and remove the hoses from the throttle body.
15 Remove the four throttle body mounting nuts **(see illustration)**.
16 Remove and discard the gasket between the throttle body air intake plenum.

Installation

17 Clean the gasket mating surfaces. If light scraping is necessary, be careful not to damage the gasket surfaces or allow material to drop into the manifold. Installation is the reverse of removal. Be sure to tighten the throttle body mounting nuts to the torque listed in this Chapter's Specifications. Do not use gasket sealant.

Throttle Position Sensor (TPS)

18 Refer to Chapter 6 for the check and replacement procedures for the TPS.

Fuel rail assembly

Refer to illustration 13.23 and 13.24

Removal

19 Relieve the fuel pressure (see Section 2).
20 Detach the cable from the negative terminal of the battery.
21 Remove the air intake plenum assembly (see Steps 1 through 9).
22 Disconnect the fuel feed and return lines from the fuel rail assembly. Disconnect the vacuum line from the fuel pressure regulator.
23 Remove the fuel rail assembly retaining bolts **(see illustration)**.

24 Disconnect each fuel injector electrical connector. Carefully pull the fuel rail and fuel injectors from the engine **(see illustration)**. **Note:** *Remove the injectors with the fuel rail as an assembly.*
25 If necessary, remove the injectors from the fuel rail (see Steps 43 and 44).

Installation

Note: *It's a good idea to replace the injector O-rings whenever the fuel rail is removed.*

26 Ensure that the injector caps are clean and free of contamination.
27 Install the fuel injectors on the fuel rail, if removed. Ensure that the injectors are well seated in the fuel rail assembly. Lubricate the O-rings with engine oil to ease assembly.
28 Install the fuel rail and injector assembly onto the engine and secure the fuel rail assembly with the retaining bolts. Tighten the bolts to the torque listed in this Chapter's Specifications.
29 The remainder of installation is the reverse of removal.

Fuel pressure regulator

Check

Note: *This procedure assumes the fuel filter is in good condition.*

30 Check the fuel pressure and perform the necessary steps to diagnose problems with the pressure regulator (see Section 3).
31 Start the engine and check for leakage around the fuel rail, fuel lines and the fuel pressure regulator.
32 If the fuel pressure regulator is faulty or leaking, replace it with a new part.

Replacement

Refer to illustration 13.37

33 Relieve the fuel pressure from the system (see Section 2).
34 Disconnect the cable from the negative terminal of the battery.
35 Remove the fuel rail from the engine (see Steps 19 through 24).
36 Clean any dirt from around the fuel pressure regulator.
37 Loosen the hose clamp and remove the screws from the fuel pressure regulator **(see illustration)**. Detach the regulator from the

fuel rail.
38 Install new O-rings on the pressure regulator and lubricate them with a light coat of engine oil.
39 Installation is the reverse of removal. Tighten the pressure regulator mounting bolts and the hose clamp securely.

Fuel injectors

Refer to illustrations 13.43, 13.44 and 13.45

Removal

40 Relieve the system fuel pressure (see Section 2).
41 Remove the air intake plenum assembly (see Steps 1 through 9).
42 Remove the fuel rail assembly (see Steps 19 through 24).
43 Remove the screws that retain the injector in the fuel rail **(see illustration)**.
44 Grasping the injector body, pull up while gently rocking the injector from side-to-side **(see illustration)**.
45 Inspect the injector O-rings (two per injector) for signs of deterioration **(see illustration)**. Replace as required. **Note:** *As long as you have the fuel rail off it's a good idea to replace the O-rings.*
46 Inspect the injector plastic "hat" (cover-

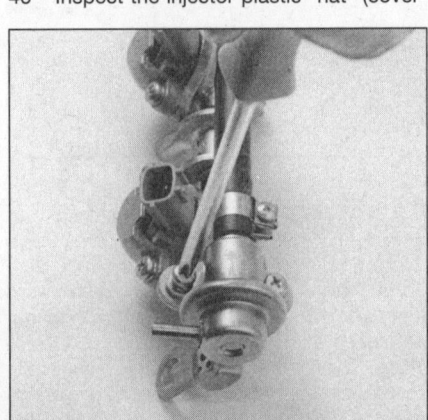

13.37 Remove the fuel pressure regulator mounting screws, loosen the hose clamp on the fuel return line, then detach the regulator from the fuel rail and the hose

Chapter 4 Fuel and exhaust systems

13.43 Remove the fuel injector retaining screws (arrows)

13.44 Carefully wiggle the fuel injector from the fuel rail recess

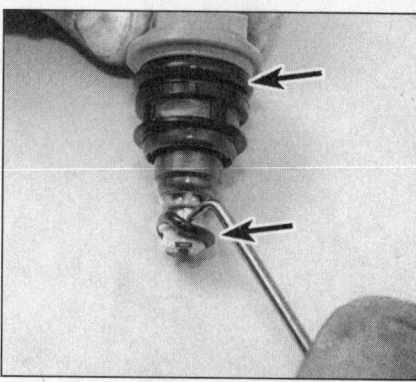

13.45 Remove the O-rings from the fuel injector (arrows)

ing the injector pintle) and washer for signs of deterioration. Replace as required. If the hat is missing, look for it in the intake manifold.

Installation

47 Lubricate the new O-rings with clean engine oil and install two on each injector. **Caution:** *Do not use silicone grease. It will clog the injectors.*
48 Using a light twisting motion, install the injector(s).
49 The remainder of installation is the reverse of removal.

Idle Air Control/Auxiliary Air Control (IAC/AAC) valve

Refer to illustrations 13.51a, 13.51b, 13.52, 13.54a and 13.54b

General Information

50 The IAC/AAC valve is attached to the air intake plenum. The ECM (computer) actuates the IAC/AAC valve by an on/off pulse signal. The longer the on-duty signal is left on, the larger the amount of air that will be allowed to flow through the IAC/AAC valve. This valve works in conjunction with the Fast Idle Control Device (FICD) solenoid.

Check

51 With the engine running, disconnect the

13.51a Disconnect the harness connector from the IAC/AAC valve - the engine rpm should decrease (GA16DE engine)

IAC/AAC valve electrical connector **(see illustrations)** and confirm that the idle speed drops immediately. If it does not, continue checking the idle control system.
52 Check the IAC/AAC valve resistance **(see illustration)**. Probe the terminals of the IAC/AAC valve with an ohmmeter and record the resistances. Compare your readings with those in the Specifications in this Chapter. If

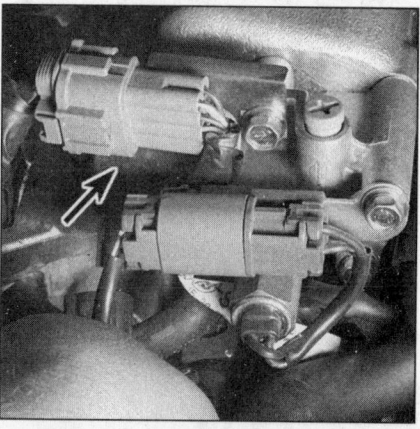

13.51b The IAC/AAC valve (arrow) is located behind the electrical connectors on the SR20DE engine

the IAC/AAC valve resistance is out of range, replace it.
53 Next, remove the valve and check the pintle for excessive carbon deposits. If necessary, clean it with carburetor cleaner spray. Also clean the valve housing to remove any deposits.
54 Also, check the FICD solenoid on SR20DE engines in the following manner.

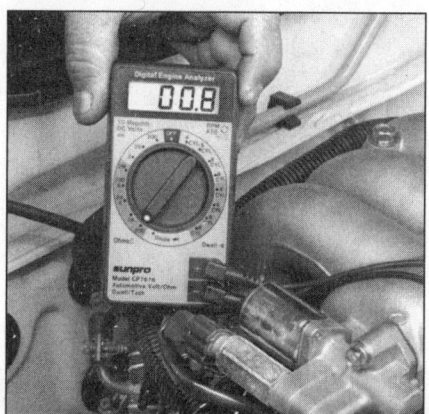

13.52 Check the resistance of the IAC/AAC valve - refer to the Specifications Section at the beginning of this Chapter

13.54a Apply battery voltage to the FICD solenoid and listen for a distinct "click" as the solenoid activates (GA16DE engine)

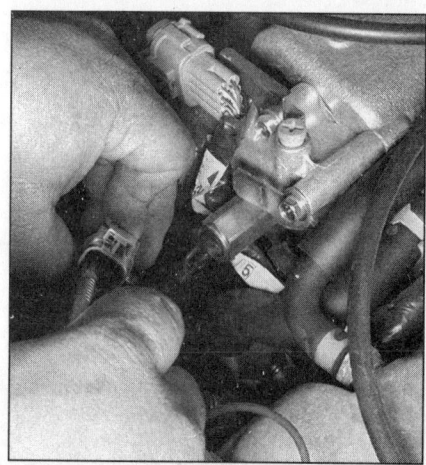

13.54b FICD location on the SR20DE engine

4-10   Chapter 4 Fuel and exhaust systems

13.66 Remove the two air regulator mounting bolts from the intake manifold (air intake plenum removed for clarity)

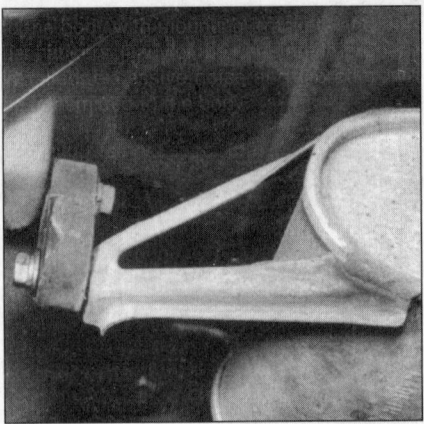

14.1 All rubber exhaust system components must be carefully checked. Also examine the welds on the brackets for deterioration

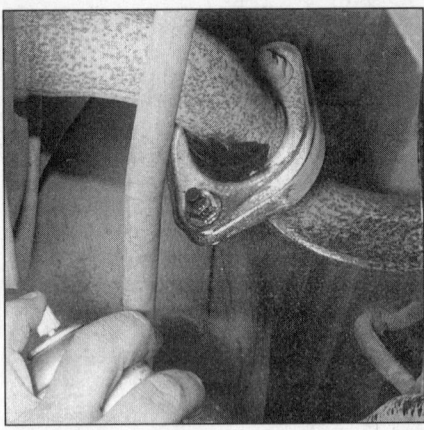

14.4 Liberally spray all exhaust fasteners with penetrating oil and allow it to soak in before attempting to loosen them

Using jumper wires from the battery terminals, apply battery voltage to the FICD solenoid. There should be a distinct "click" **(see illustrations)**. Listen carefully for the solenoid activation. If there is no sound, replace the FICD. On GA16DE engines, measure the resistance across the two end terminals (1 and 5) of the IACV-FICD connector and compare your reading to the figure in the Specifications in this Chapter.

Removal

55 Disconnect the electrical connector from the IAC/AAC valve and the FICD solenoid.
56 Remove the valve attaching screws and separate the assembly from the air intake plenum.
57 Check the condition of the gasket seal. Replace it with a new gasket.
58 Clean the sealing surface and the bore of the throttle body assembly to ensure a good seal. **Caution:** *The IAC/AAC valve itself is an electrical component and must not be soaked in any liquid cleaner, as damage may result.*

Installation

59 Install a new gasket on the IAC/AAC valve assembly.
60 Install the IAC/AAC valve and tighten the screws securely.
61 Plug in the electrical connector to the IAC/AAC valve assembly.

Air regulator (SR20DE engine only)

General information

62 The air regulator provides air bypass when the engine is cold for fast idle. It consists of a bimetal heater and rotary shutter. When the temperature of the bimetal is low, the shutter opens and allows air to bypass from the air intake duct into the intake manifold to create a fast idle. As the engine warms, the shutter closes and does not allow any air to circulate into the intake manifold. This condition continues until the engine is stopped and the temperature of the engine block cools down.

Check

63 With the engine completely cold, start the engine and check for the presence of battery voltage on the harness electrical connector. Battery voltage should be present.
64 Allow the engine to warm up to normal operating temperature, turn the ignition key OFF and remove the air regulator. Check to make sure the shutter has been activated and is completely closed (no air flow). If the shutter is open, replace the air regulator. Measure the resistance across the two terminals of the wiring connector. It must be between 70 and 80 ohms. Also check the regulator for clogging.

Replacement

Refer to illustration 13.66
65 Detach the air regulator electrical connector and disconnect the two hoses.
66 Remove the air regulator mounting bolts **(see illustration)**.
67 Lift the assembly from the intake manifold.
68 Installation is the reverse of removal.

14 Exhaust system - servicing and general information

Refer to illustrations 14.1 and 14.4
Warning: *Inspection and repair of exhaust system components should be done only after enough time has elapsed after driving the vehicle to allow the system components to cool completely. Also, when working under the vehicle, make sure it is securely supported on jackstands.*

1 The exhaust system consists of the exhaust manifold, the catalytic converters, the muffler, the tailpipe and all connecting pipes, brackets, hangers and clamps **(see illustration)**. The exhaust system is attached to the body with mounting brackets and rubber hangers. If any of the parts are improperly installed, excessive noise and vibration will be transmitted to the body.

2 Conduct regular inspections of the exhaust system to keep it safe and quiet. Look for any damaged or bent parts, open seams, holes, loose connections, excessive corrosion or other defects which could allow exhaust fumes to enter the vehicle. Deteriorated exhaust system components should not be repaired; they should be replaced with new parts.
3 If the exhaust system components are extremely corroded or rusted together, welding equipment will probably be required to remove them. The convenient way to accomplish this is to have a muffler repair shop remove the corroded sections with a cutting torch. If, however, you want to save money by doing it yourself (and you don't have a welding outfit with a cutting torch), simply cut off the old components with a hacksaw. If you have compressed air, special pneumatic cutting chisels can also be used. If you do decide to tackle the job at home, be sure to wear safety goggles to protect your eyes from metal chips and work gloves to protect your hands.
4 Here are some simple guidelines to follow when repairing the exhaust system:

 a) *Work from the back to the front when removing exhaust system components.*
 b) *Apply penetrating oil to the exhaust system component fasteners to make them easier to remove* **(see illustration)**.
 c) *Use new gaskets, hangers and clamps when installing exhaust system components.*
 d) *Apply anti-seize compound to the threads of all exhaust system fasteners during reassembly.*
 e) *Be sure to allow sufficient clearance between newly installed parts and all points on the underbody to avoid overheating the floor pan and possibly damaging the interior carpet and insulation. Pay particularly close attention to the catalytic converter and heat shield.*

Chapter 5
Engine electrical systems

Contents

	Section		Section
Alternator components - check and replacement	11	Drivebelt deflection	See Chapter 1
Alternator - removal and installation	10	General information	1
Battery cables - check and replacement	4	Ignition system - check	6
Battery check and maintenance	See Chapter 1	Ignition system - general information	5
Battery - emergency jump starting	2	Ignition timing - check and adjustment	See Chapter 1
Battery - removal and installation	3	Spark plug replacement	See Chapter 1
Camshaft position sensor - check	See Chapter 6	Spark plug wire, distributor cap and rotor - check	
Charging system - check	9	and replacement	See Chapter 1
Charging system - general information and precautions	8	Starter motor and circuit - in-vehicle check	13
CHECK ENGINE light	See Chapter 6	Starter motor - removal and installation	14
Crankshaft position sensor - check	See Chapter 6	Starter solenoid - replacement	15
Distributor - removal and installation	7	Starting system - general information and precautions	12
Drivebelt check, adjustment and replacement	See Chapter 1		

Specifications

General

Alternator brush length
  New .... 1/2 inch
  Minimum .... 1/4 inch
Battery voltage
  Engine off .... 12 volts
  Engine running .... 14.1 to 14.7 volts
Firing order 1-3-4-2

Ignition system

Ignition coil resistance
  Primary resistance
    SR20DE engine .... 0.5 to 1.0 ohms
    GA16DE engine .... 1.0 ohm (approximately)
  Secondary resistance
    SR20DE engine .... 25 K-ohms (approximately)
    GA16DE engine .... 10 K-ohms (approximately)
Ignition resistor resistance .... 2.2 K-ohms (approximately)
Ignition coil-to-distributor cap wire resistance .... 5,000 ohms per foot
Ignition timing .... See Chapter 1

Chapter 5 Engine electrical system

6.1a The distributor houses the camshaft position sensor, coil and power transistor. If any of these components are found to be faulty, the complete distributor assembly must be replaced

6.1b To use a calibrated ignition tester, simply disconnect a spark plug wire and connect it to the tester, clip the tester to a convenient ground (like a valve cover bolt) and operate the starter - if there is enough power to fire the plug, sparks will be visible between the electrode tip and the tester body

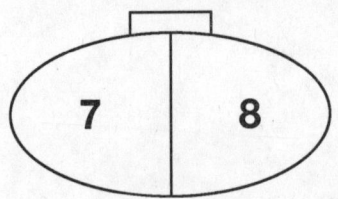

6.6 The chassis wire harness side of the ignition coil electrical connector looks like this

1 General information

The engine electrical systems include all ignition, charging and starting components. Because of their engine-related functions, these components are considered separately from chassis electrical devices like the lights, instruments, etc.

Be very careful when working on the engine electrical components. They are easily damaged if checked, connected or handled improperly. The alternator is driven by an engine drivebelt which could cause serious injury if your hands, hair or clothes become entangled in it with the engine running. Both the starter and alternator are connected directly to the battery and could arc or even cause a fire if mishandled, overloaded or shorted out.

Never leave the ignition switch on for long periods of time with the engine off. Don't disconnect the battery cables while the engine is running. Correct polarity must be maintained when connecting battery cables from another source, such as another vehicle, during jump starting. Always disconnect the negative cable first and hook it up last or the battery may be shorted by the tool being used to loosen the cable clamps.

Additional safety related information on the engine electrical systems can be found in *Safety first* near the front of this manual. It should be referred to before beginning any operation included in this Chapter.

2 Battery - emergency jump starting

Refer to the *Booster battery (jump) starting* procedure at the front of this manual.

3 Battery - removal and installation

1 Disconnect the negative battery cable, then the positive battery cable, from the battery. **Caution:** *Always disconnect the negative cable first and hook it up last or the battery may be shorted by the tool being used to loosen the cable clamps.*
2 Locate the battery hold-down clamp straddling the top of the battery. Remove the nuts and the hold-down clamp.
3 Lift out the battery. Special battery lifting straps that attach to the battery posts are available at auto parts stores; lifting and moving the battery is much easier if you use one.
4 Installation is the reverse of removal.

4 Battery cables - check and replacement

1 Periodically inspect the entire length of each battery cable for damage, cracked or burned insulation and corrosion. Poor battery cable connections can cause starting problems and decreased engine performance.
2 Check the cable-to-terminal connections at the ends of the cables for cracks, loose wire strands and corrosion. The presence of white, fluffy deposits under the insulation at the cable terminal connection is a sign that the cable is corroded and should be replaced. Check the terminals for distortion, missing mounting bolts and corrosion.
3 When replacing the cables, always disconnect the negative cable first and hook it up last or the battery may be shorted by the tool used to loosen the cable clamps. Even if only the positive cable is being replaced, be sure to disconnect the negative cable from the battery first.
4 Disconnect and remove the cable. Make sure the replacement cable is the same length and diameter.
5 Clean the threads of the relay or ground connection with a wire brush to remove rust and corrosion. Apply a light coat of petroleum jelly to the threads to prevent future corrosion.
6 Attach the cable to the relay or ground connection and tighten the mounting nut/bolt securely.
7 Before connecting the new cable to the battery, make sure that it reaches the battery post without having to be stretched. Clean the battery posts thoroughly and apply a light coat of petroleum jelly to prevent corrosion.
8 Connect the positive cable first, followed by the negative cable.
9 If only the terminal end of the battery cable is deteriorated, it is much easier and cheaper to simply replace the end instead of the entire cable. Follow the disconnection instructions above, then cut off the old terminal with pliers, strip back about an inch of cable insulation and attach the new terminal.

5 Ignition system - general information

1 The ignition system is designed to ignite the fuel/air charge entering each cylinder at just the right moment. It does this by producing a high voltage spark between the electrodes of each spark plug.
2 The vehicles covered by this manual are equipped with an electronic ignition system. This system consists of the camshaft position sensor, the power transistor, the ignition coil (all located in the distributor), an ignition circuit resistor and the primary and secondary wiring. **Note:** *If there is a problem with the ignition coil, the power transistor, the camshaft position sensor or the distributor itself, the entire distributor must be replaced.*
3 The Electronic Control Module (ECM) controls the spark timing advance characteristics. Vacuum advance and vacuum hoses have been eliminated from the design.
4 The camshaft position sensor is the basis of this computer controlled system. It monitors engine speed and piston position and relays this data to the computer which in turn controls the fuel injection duration (fuel injector on/off time) and ignition timing. The camshaft position sensor consists of a rotor plate and a wave forming circuit. The rotor plate has 360 slits for each degree, or one percent signal (engine speed signal) and four slits for the 180-degree camshaft position signal. Light Emitting Diodes (LED) and photo diodes are built into the wave forming circuit.

Chapter 5 Engine electrical system

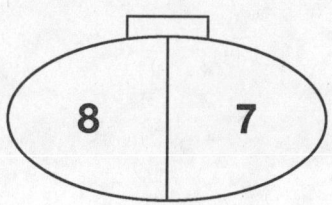

6.8 The coil wiring connector at the distributor side looks like this. The resistance between the two terminals is the primary coil resistance. Refer to the Specifications in this Chapter to see if your coil is within limits

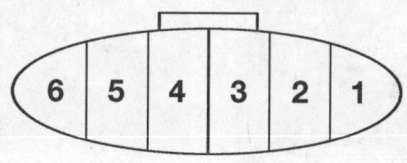

6.9 The distributor side of the camshaft position sensor electrical connector has its terminals numbered in this manner

7.5a Apply a small dab of white paint to the distributor directly below the tip of the rotor so the distributor can be reinstalled in exactly the same orientation

7.5b Mark the position of the distributor housing to the cylinder head. When the marks are precisely lined up during reassembly, ignition timing will be correct

When the rotor plate passes the space between the LED and the photo diode, the slits on the rotor plate continually cut the beam of light sent to the photo diode from the LED. They are then converted into on-off pulses by the wave forming circuit and then sent to the ECM.

5 The camshaft position sensor-type distributor must be replaced as a complete unit. There are no replacement parts available for the distributor except the cap, rotor and seal cover.

6 Ignition system - check

Refer to illustrations 6.1a, 6.1b, 6.6, 6.8 and 6.9

Warning: *Because of the high voltage generated by the ignition system, extreme care should be taken whenever an operation is performed involving ignition components.*

1 Any problem with the distributor, power transistor, ignition coil or camshaft position sensor will require the replacement of the entire distributor assembly **(see illustration)**. If the engine turns over but won't start, disconnect the spark plug wire from any spark plug and attach it to a calibrated ignition system tester (available at most auto parts stores). Connect the clip on the tester to a bolt or metal bracket on the engine **(see illustration)**. If you're unable to obtain a calibrated ignition tester, remove the wire from one of the spark plugs and using an insulated tool, pull back the boot and hold the end of the wire about 1/4-inch from a good ground.

2 Crank the engine and watch the end of the tester or spark plug wire to see if bright blue, well-defined sparks occur.

3 If sparks occur, sufficient voltage is reaching the plug to fire it. However, the plugs themselves may be fouled, so remove and check them as described in Chapter 1. Repeat the check at the remaining plug wires to verify that the distributor cap, rotor and all the spark plug wires are good.

4 If no sparks or intermittent sparks occur, remove the distributor cap and check the cap and rotor as described in Chapter 1. If moisture is present, dry out the cap and rotor, then reinstall the cap and repeat the spark test.

5 Check to make sure the ignition switch receives battery voltage. Also check for battery voltage to the Electronic Control Module (ECM). This can be observed by monitoring the red light on the ECM (see Chapter 6) after the ignition key is turned ON. If there is no power to the ECM, check the ECM relay which is located directly in front of the ECM (see Chapter 12).

6 Check to make sure the IGN SW fuse is not blown and there is voltage to the fuse. Disconnect the two-wire connector at the coil and turn the ignition ON **(see illustration)**. There should be voltage to the coil at terminal 7.

7 Refer to Chapter 6 for the camshaft position sensor testing procedure. If it does not test good, the entire distributor assembly will have to be replaced.

8 Disconnect the two-wire (coil) connector at the distributor and use an ohmmeter to check the coil (on the distributor side of the connector) **(see illustration)**. The resistance between the two terminals of the connector (the primary resistance) should be as listed in this Chapter's Specifications. The secondary resistance between terminal 7 and the secondary terminals on the distributor cap should also be as listed in the Specifications. If either does not check out, replace the entire distributor.

9 To check the power transistor, disconnect both connectors from the distributor and test the resistance between terminal 2 on the larger connector **(see illustration)** and terminal 8 on the coil connector. The reading should be more than zero ohms. If it is zero ohms, replace the entire distributor.

10 Disconnect the wiring at the ignition resistor and check the resistance with an ohmmeter. The correct reading is about 2.2 K-ohms. If your reading is not close to this, replace the resistor.

7 Distributor - removal and installation

Removal

Refer to illustrations 7.5a and 7.5b

1 Detach the cable from the negative terminal of the battery.

2 Disconnect the electrical connectors from the distributor. Follow the wires as they exit the distributor to find the connectors.

3 Note the raised "1" on the distributor cap. This marks the location for the number one cylinder spark plug wire terminal. **Note:** *Some distributor caps may have been replaced with aftermarket units that are not marked.*

4 Remove the distributor cap (see Chapter 1). Using a socket and breaker bar on the crankshaft pulley bolt, rotate the engine until the rotor is pointing toward the number one spark plug terminal (see the TDC locating procedure in Chapter 2).

5 Make a mark on the edge of the distributor base directly below the rotor tip and inline with it **(see illustration)**. Also, mark the distributor base and the cylinder head to ensure that the distributor can be reinstalled correctly **(see illustration)**.

6 Remove the distributor hold-down bolt, then pull the distributor straight out to remove it. **Caution:** *DO NOT turn the engine*

5-4   Chapter 5 Engine electrical system

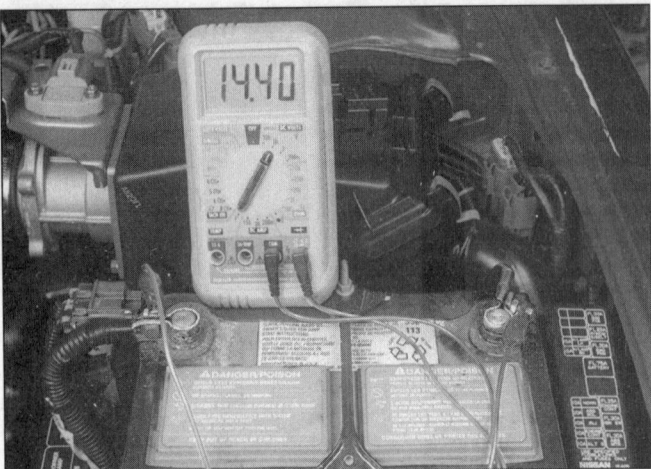

9.3 Once the engine has been started, confirm that the charging voltage increases to approximately 14 to 15 volts

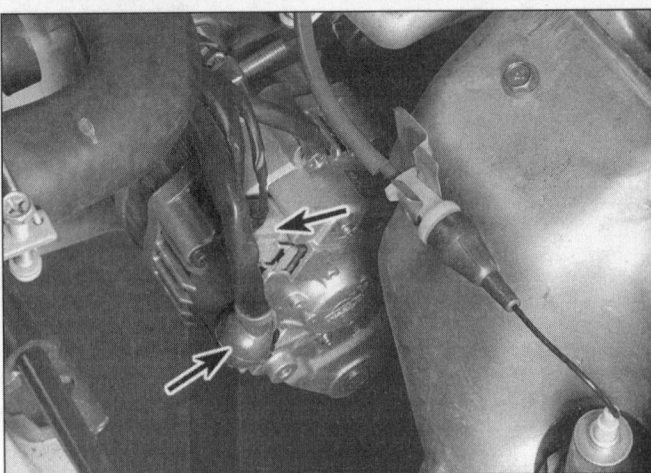

10.2 Disconnect the wiring (arrows) from the rear of the alternator

while the distributor is removed, or the alignment marks will be useless.

Installation

7   Insert the distributor into the engine in exactly the same relationship to the head that it was in when removed.

8   Make sure the distributor is seated completely and the alignment marks made previously are aligned. If not, remove the distributor and reposition it. **Note:** *If the crankshaft has been moved while the distributor is out, locate Top Dead Center (TDC) for the number one piston (see Chapter 2) and position the distributor and rotor accordingly.*

9   Loosely install the hold-down nuts.

10   Install the distributor cap and tighten the screws securely.

11   Plug in the electrical connectors.

12   Reattach the spark plug wires to the plugs (if removed).

13   Connect the cable to the negative terminal of the battery.

14   Check and, if necessary, adjust the ignition timing (see Chapter 1) and tighten the distributor hold-down bolt securely.

8   Charging system - general information and precautions

The charging system includes the alternator, an internal voltage regulator, a charge indicator or warning light, the battery and the wiring between all the components. The charging system supplies electrical power for the ignition system, the lights, the radio, etc. The alternator is driven by a drivebelt at the front of the engine. Refer to the wiring diagrams at the end of Chapter 12 for additional information.

The purpose of the voltage regulator is to limit the alternator's voltage to a preset value. This prevents power surges, circuit overloads, etc., during peak voltage output.

The charging system doesn't ordinarily require periodic maintenance. However, the drivebelt, battery and wires and connections should be inspected at the intervals outlined in Chapter 1.

Be very careful when making electrical circuit connections to a vehicle equipped with an alternator and note the following:

a) When reconnecting wires to the alternator from the battery, be sure to note the polarity.
b) Before using arc welding equipment to repair any part of the vehicle, disconnect the wires from the alternator and the battery terminals.
c) Never start the engine with a battery charger connected.
d) Always disconnect both battery cables before using a battery charger (negative cable first, positive cable last).

9   Charging system - check

Refer to illustration 9.3

1   If a malfunction occurs in the charging circuit, do not immediately assume that the alternator is causing the problem. First check the following items:

a) The battery cables where they connect to the battery. Make sure the connections are clean and tight.
b) The battery electrolyte specific gravity. If it is low, charge the battery.

10.3 The pivot bolts (arrows) are located on the bottom of the alternator

11.2 Make a mark across the front cover, stator and rear cover to aid in reassembly

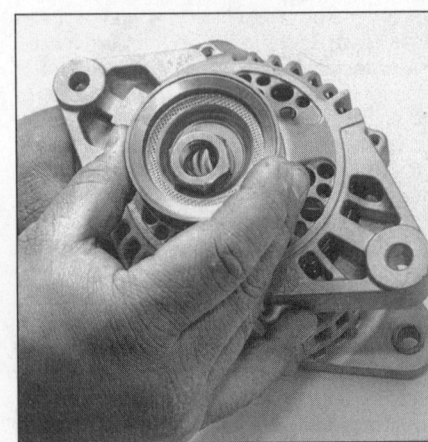

11.3 Remove the pulley nut and remove the pulley

Chapter 5 Engine electrical system

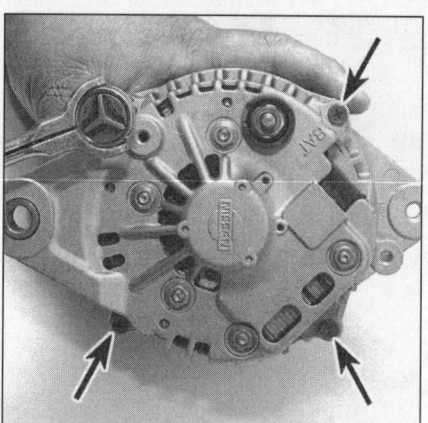

11.4a Remove the four bolts from the rear cover (arrows)

11.4b Separate the front cover from the alternator body

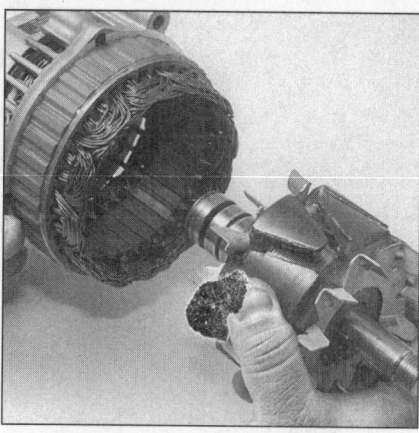

11.4c Remove the rotor from the alternator assembly

c) *Check the external alternator wiring and connections.*
d) *Check the fuses and fusible links (see Chapter 12).*
e) *Check the drivebelt condition and tension (see Chapter 1).*
f) *Check the alternator mounting bolts for tightness.*
g) *Run the engine and check the alternator for abnormal noise.*

2 Using a voltmeter, check the battery voltage with the engine off. It should be approximately 12-volts.
3 Start the engine and check the battery voltage again. It should now be approximately 14.1 to 14.7-volts **(see illustration)**.
4 If the indicated voltage reading is less or more than the specified charging voltage, the problem may be within the alternator.
5 Due to the special equipment necessary to test or service the alternator, it is recommended that if a fault is suspected the vehicle be taken to a shop with the proper equipment. But if you feel confident in the use of an ohmmeter, and in some cases a soldering iron, the component check and replacement procedures for the most common alternator type are included in Section 11. (It is generally faster and less expensive to install a rebuilt alternator than to attempt to repair one).
6 Some models are equipped with an ammeter on the instrument panel that indicates charge or discharge - current passing in or out of the battery. With all electrical equipment switched ON, and the engine idling, the gauge needle may show a discharge condition. At fast idle or normal driving speeds the needle should stay on the charge side of the gauge, with the charged state of the battery determining just how far over (the lower the battery state of charge, the farther the needle should swing toward the charge side).
7 Some models are equipped with a voltmeter on the instrument panel that indicates battery voltage with the key ON (engine not running), and alternator output when the engine is running.
8 The charge light on the instrument panel illuminates with the key ON and engine not running, and should go out when the engine runs.
9 If the gauge does not show a charge when it should or the alternator light (if equipped) remains on, there is a fault in the system. Before inspecting the brushes or replacing the alternator, the battery condition, alternator belt tension and electrical cable connections should be checked.

10 Alternator - removal and installation

Refer to illustrations 10.2 and 10.3
Note: *The Sentra models covered by this manual are equipped with several types of alternators. Be certain to take your old alternator with you when purchasing a new or rebuilt unit and compare them closely.*

1 Detach the cable from the negative terminal of the battery.
2 Disconnect the electrical connectors from the alternator **(see illustration)**.
3 Loosen the alternator adjustment and pivot bolts **(see illustration)** and detach the drivebelt.
4 Remove the adjustment and pivot bolts and separate the alternator from the engine.
5 Installation is the reverse of removal.
6 After the alternator is installed, adjust the drivebelt tension (see Chapter 1).

11 Alternator components - check and replacement

Disassembly

Refer to illustrations 11.2, 11.3, 11.4a, 11.4b, 11.4c, 11.5a, 11.5b, 11.6a, 11.6b, 11.6c and 11.7

1 Remove the alternator from the vehicle (see Section 10).
2 Scribe or paint marks on the front and rear end frame housings of the alternator to facilitate reassembly **(see illustration)**.
3 Remove the nut retaining the pulley to the rotor shaft and remove the pulley **(see illustration)**.
4 Remove the four through-bolts holding the front and rear covers together, then separate the rear cover assembly from the front cover. Remove the rotor **(see illustrations)**.
5 Remove the nuts retaining the stator, then separate the stator and diode assembly from the end frame **(see illustrations)**.

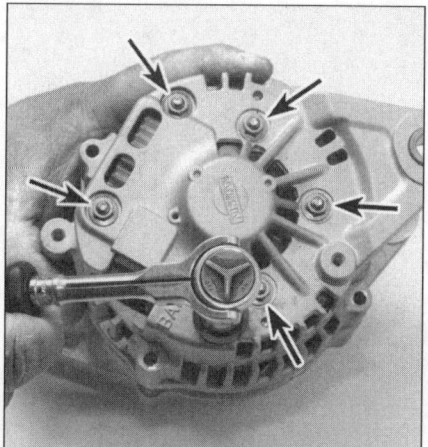

11.5a Remove the mounting nuts (arrows)

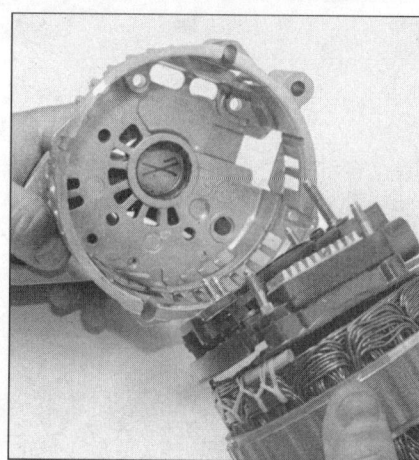

11.5b Separate the rear cover from the stator and diode assembly

5-6　Chapter 5　Engine electrical system

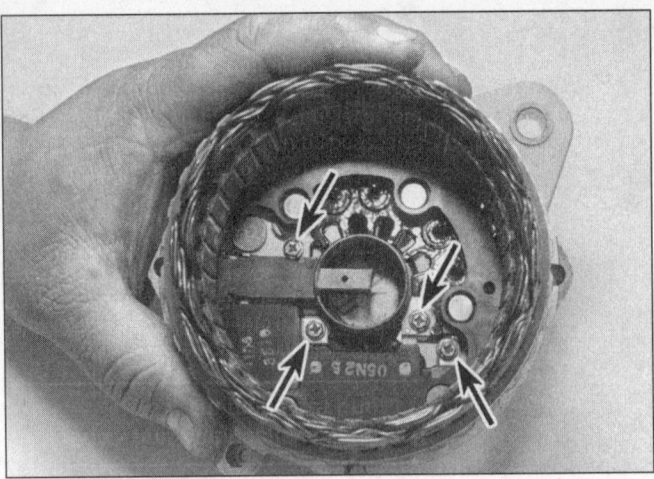

11.6a Remove the diode assembly mounting screws (arrows)

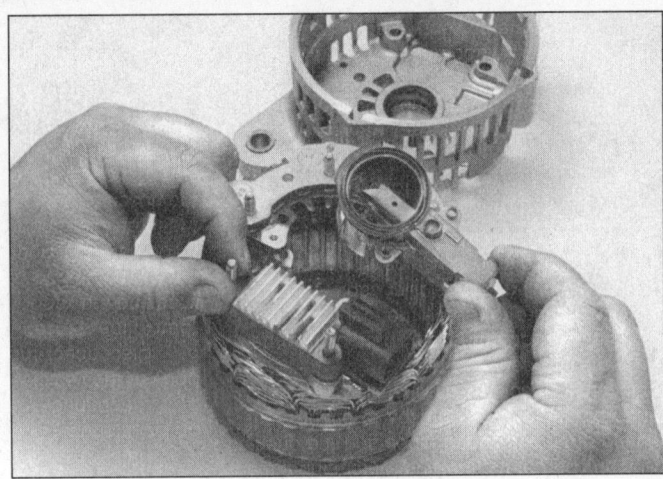

11.6b Remove the brush holder assembly from the diode assembly

6   Remove the screws attaching the regulator and the brush holder to the diode assembly and remove the brush holder and regulator **(see illustrations)**.

7   Remove the diode assembly from the stator **(see illustration)**. On some types of alternators it will be necessary to use a solder gun and heat sink to melt the solder joints that connect the stator leads to the diode assembly.

Component checks

Refer to illustrations 11.8a, 11.8b and 11.9

8   Check the rotor for an open between the two slip rings **(see illustration)**. There should be continuity between the slip rings. Check for grounds between each slip ring and the rotor shaft **(see illustration)**. There should be no continuity (infinite resistance) between the rotor shaft and either slip ring. If the rotor fails either test, or if the slip rings are excessively worn, the rotor is defective.

9   Check for opens between the center terminal and each end terminal of the stator windings **(see illustration)**. If either reading is high (infinite resistance), the stator is defec-

tive. Check for a grounded stator winding between each stator terminal and the frame. If there's continuity between any stator winding and the frame, the stator is defective.

10 Because the diode assembly varies between models and load requirements, have the diode assembly checked at a dealer service department or qualified automobile electric repair facility.

11 Measure the length of the brushes and replace them if they are at or near the minimum brush length found in this Chapter's Specifications. **Note:** *On some models the brush leads are soldered onto the voltage regulator assembly.*

Reassembly

Refer to illustration 11.13

12 Install the components in the reverse order of removal, noting the following:

13 Before installing the brush holder, push the brushes into the holder and slip a straightened paper clip or other suitable pin through the hole in the brush holder to hold the brushes in a retracted position. After the front and rear end frames have been bolted together, remove the paper clip **(see illustration)**.

12 Starting system - general information and precautions

1   The function of the starting system is to crank the engine quickly enough to allow it to start. The system is composed of the starter motor, starter solenoid, battery, ignition switch, clutch start switch (manual transaxle models), neutral start/back-up light switch (automatic transaxle models), diode box and connecting wires. Power is supplied through a fusible link, a fuse and through the theft-deterrent system, if used.

2   Turning the ignition key to the Start position actuates the starter relay through the starter control circuit. The starter solenoid then connects the battery to the starter. The battery supplies the electrical energy to the starter motor, which does the actual work of cranking the engine.

11.6c Remove the voltage regulator

11.7 Remove the four set screws and lift the diode assembly from the stator

Chapter 5 Engine electrical system

11.8a Continuity should exist between the rotor slip rings

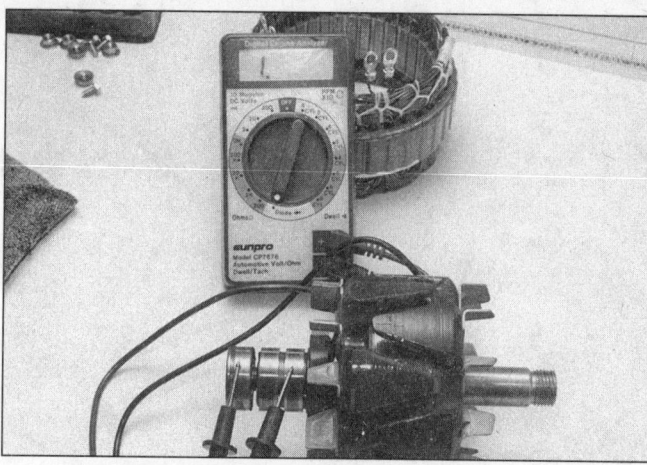

11.8b No continuity should exist between the slip ring(s) and rotor shaft

3   All vehicles are equipped with a starter/solenoid assembly that is mounted to the transmission bellhousing.
4   All vehicles are equipped with a clutch start switch or a neutral start switch in the starter control circuit, which prevents operation of the starter unless the shift lever is in Neutral or Park (automatic) or the clutch is depressed (manual).
5   Never operate the starter motor for more than 15 seconds at a time without pausing to allow it to cool for at least two minutes. Excessive cranking can cause overheating, which can seriously damage the starter.

13 Starter motor and circuit - in-vehicle check

Note: *Before diagnosing starter problems, make sure the battery is fully charged.*

General check

1   If the starter motor doesn't turn at all when the switch is operated, make sure the shift lever is in Neutral or Park or the clutch is fully depressed.
2   Make sure the battery is charged and that all cables at the battery and starter solenoid terminals are secure.
3   If the starter motor spins but the engine doesn't turn over, the drive assembly in the starter motor is slipping and the starter motor must be replaced (see Section 14).
4   If, when the switch is actuated, the starter motor doesn't operate at all but the starter solenoid operates (clicks), the problem lies with either the battery, the starter solenoid contacts or the starter motor connections.
5   If the starter solenoid doesn't click when the ignition switch is actuated, either the starter solenoid circuit is open or the solenoid itself is defective. Check the starter solenoid circuit (see the wiring diagrams at the end of Chapter 12) or replace the solenoid (see Section 15).
6   To check the starter solenoid circuit, remove the push-on connector from the solenoid wire. Make sure that the connection is clean and secure and the relay bracket is grounded. If the connections are good, check the operation of the solenoid with a jumper wire. To do this, place the transmission in Park or Neutral and apply the parking brake. Remove the push-on connector from the solenoid. Connect a jumper wire between the battery positive terminal and the exposed terminal on the solenoid. If the starter motor now operates, the starter solenoid is okay. The problem is in the ignition switch, Neutral start switch or in the starting circuit wiring (look for open or loose connections).
7   If the starter motor still doesn't operate, replace the starter solenoid (see Section 15).
8   If the starter motor cranks the engine at an abnormally slow speed, first make sure the battery is fully charged and all terminal connections are clean and tight. Also check the connections at the starter solenoid and battery ground. Eyelet terminals should not be easily rotated by hand. Also check for a short to ground. If the engine is partially seized, or has the wrong viscosity oil in it, it will crank slowly.

Starter cranking circuit test

Note: *To determine the location of excessive resistance in the starter circuit, perform the following simple series of tests.*

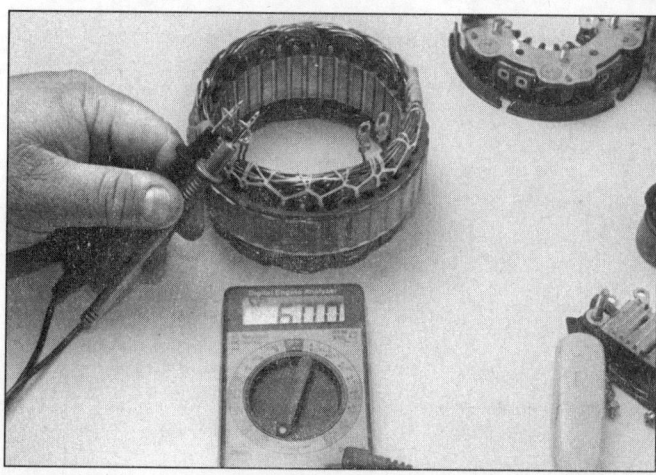

11.9 Check for continuity on each stator lead. There should be no breaks in the windings, therefore continuity should exist between each terminal

11.13 Install a paper clip into the brush holder to hold the brushes in place during reassembly - after installation, simply pull the paper clip out

5-8   Chapter 5 Engine electrical system

14.3 Disconnect the solenoid electrical connector from the harness and the cable from the starter terminal connection (arrows)

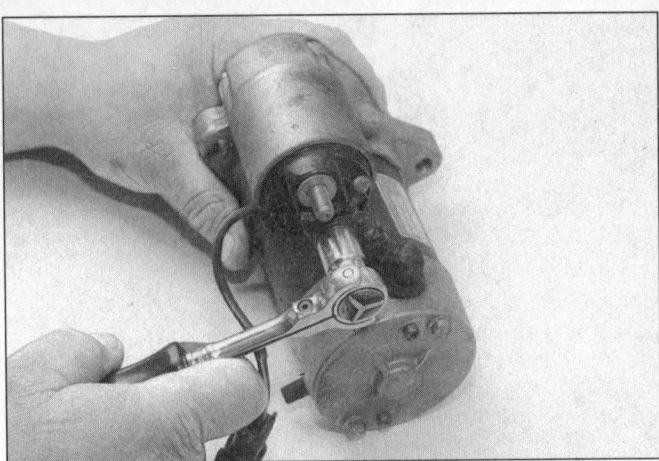

15.1 Remove the terminal nut from the solenoid

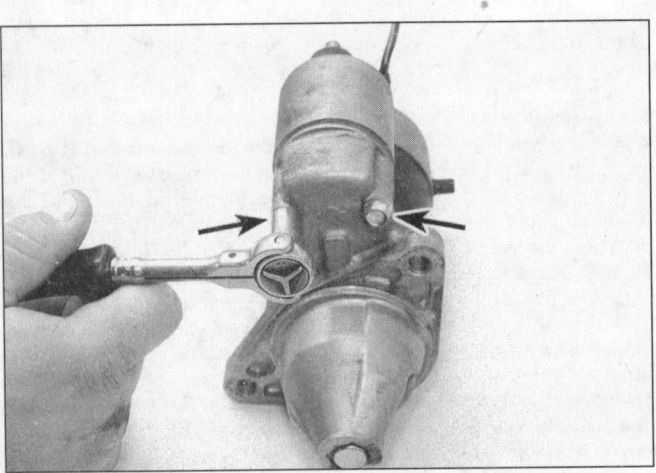

15.2a Remove the mounting bolts (arrows) that retain the solenoid to the starter housing

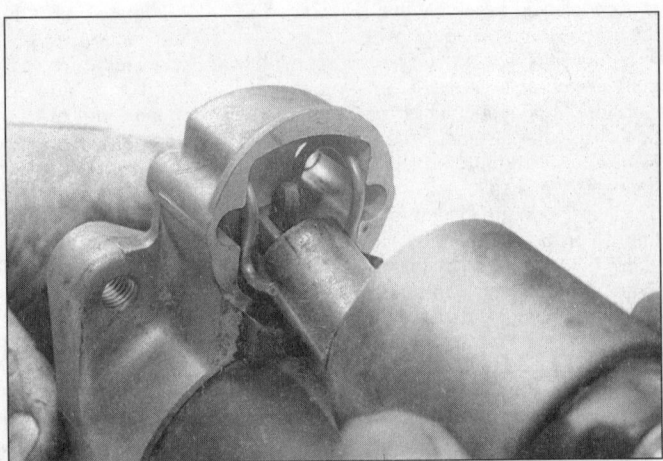

15.2b Remove the solenoid from the housing - note the position of the plunger, lever and the return spring

9 Disconnect the ignition coil wire from the distributor cap and ground it on the engine.
10 Connect a remote control starter switch from the battery terminal of the starter solenoid to the S terminal of the solenoid.
11 Connect a voltmeter positive lead to the starter motor terminal of the starter solenoid, then connect the negative lead to ground.
12 Actuate the ignition switch and take the voltmeter readings as soon as a steady figure is indicated. Do not allow the starter motor to turn for more than 15 seconds at a time. A reading of 9-volts or more, with the starter motor turning at normal cranking speed, is normal. If the reading is 9-volts or more but the cranking speed is slow, the motor is faulty. If the reading is less than 9-volts and the cranking speed is slow, the solenoid contacts are probably burned.

14 Starter motor - removal and installation

Refer to illustration 14.3
Note: *The models covered by this manual are equipped with several types of starters manufactured by both Hitachi and Mitsubishi. Be certain to take your old starter with you when purchasing a new or rebuilt unit and compare them closely.*
1 Detach the cable from the negative terminal of the battery.
2 Raise the vehicle and support it securely on jackstands.
3 Disconnect the battery cable and the solenoid connector from the starter motor **(see illustration)**.
4 Remove the starter motor mounting bolts and detach the starter from the engine.

Depending upon the year and engine type, the starter/solenoid assembly may be mounted above the transaxle or below the intake manifold.
5 If necessary, turn the wheels to one side to provide removal access.
6 Installation is the reverse of removal.

15 Starter solenoid - replacement

Refer to illustrations 15.1, 15.2a and 15.2b
1 Remove the nut and disconnect the wire from the solenoid electrical terminal **(see illustration)**.
2 Remove the two bolts from the solenoid and separate the solenoid from the starter assembly **(see illustrations)**.
3 Installation is the reverse of removal.

Chapter 6
Emissions and engine control systems

Contents

	Section		Section
Catalytic converter	8	General information	1
CHECK ENGINE light	See Section 2	Information sensors	4
Engine Control Module (ECM) - removal and installation	3	On Board Diagnosis (OBD) system and trouble codes	2
Evaporative Emissions Control (EVAP) system	6	Positive Crankcase Ventilation (PCV) system	7
Exhaust Gas Recirculation (EGR) system	5		

1 General information

Refer to illustrations 1.1, 1.7a and 1.7b

To prevent pollution of the atmosphere from incompletely burned and evaporating gases, and to maintain good driveability and fuel economy, a number of emission control systems are incorporated **(see illustration)**. They include the:

  EGR with backpressure transducer (BPT)
  Evaporative Emission Control (EVAP) system
  Multi Port Fuel Injection (MPFI) system
  Positive Crankcase Ventilation (PCV) system
  Exhaust Gas Recirculation (EGR) system
  Catalytic converter

All of these systems are linked, directly or indirectly, to the emission control system.

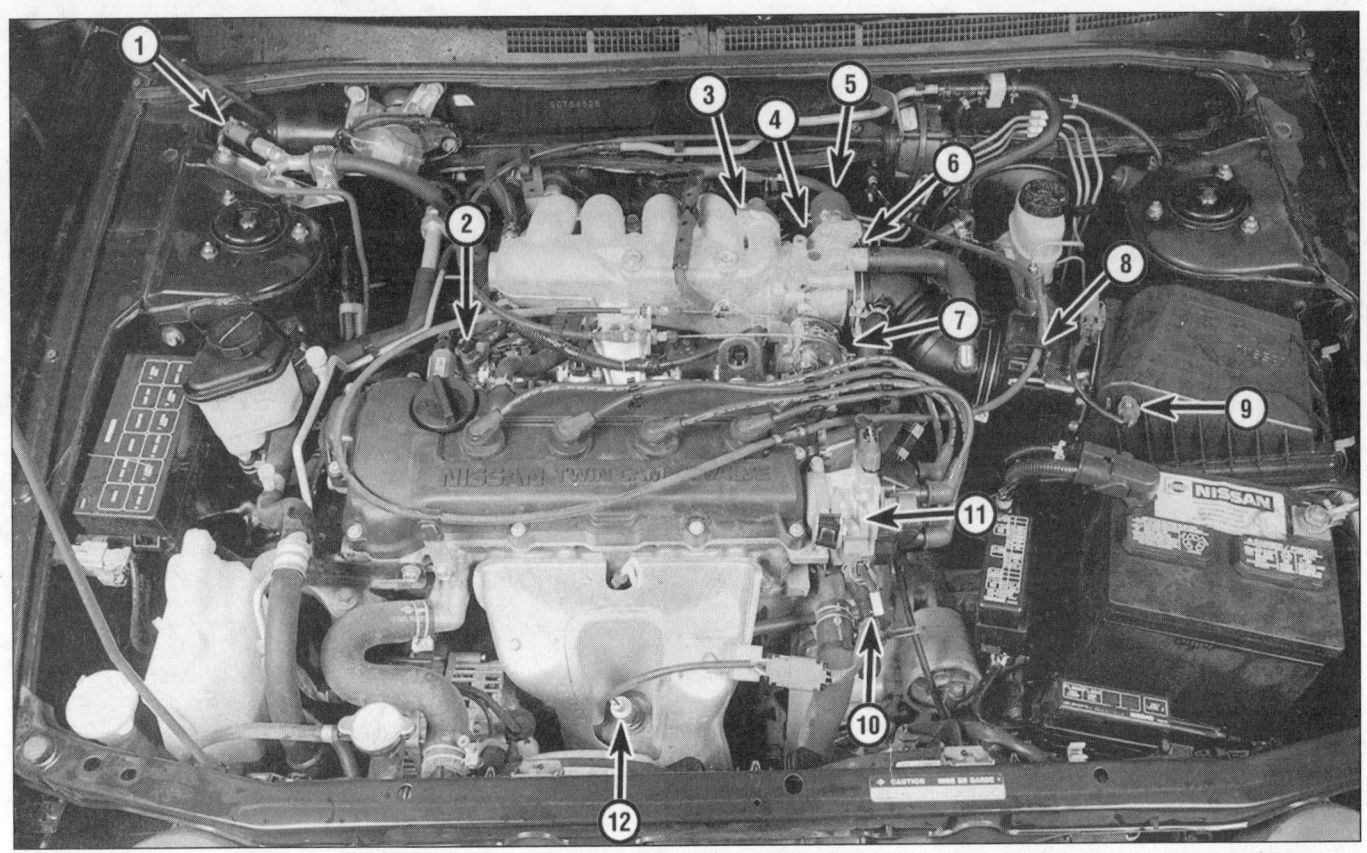

1.1 Emission and engine control system component locations for the GA16DE engine (the SR20DE engine has its sensors located in the same areas)

1   Power steering pressure sensor
2   Engine Coolant Temperature (ECT) sensor
3   EGR temperature sensor (not visible)
4   Throttle Position Sensor (TPS) (not visible)
5   IACV-AAC valve and IACV-FICD valve
6   EGR valve and canister control solenoid valve (not visible)
7   EGRC-BPT valve
8   Mass Air Flow (MAF) sensor
9   Intake Air Temperature (IAT) sensor
10 Crankshaft position sensor (not visible)
11 Ignition coil, power transistor and camshaft position sensor (in distributor)
12 Front oxygen sensor

Chapter 6 Emissions and engine control systems

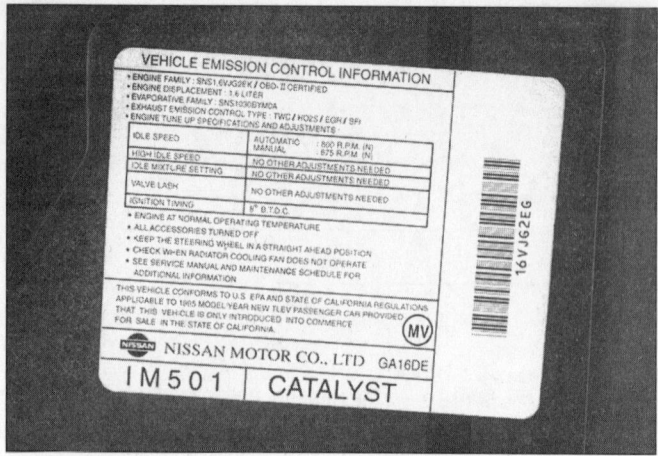

1.7a The Vehicle Emission Control Information (VECI) label is located under the hood and contains tune-up and emissions control information on your vehicle

1.7b Typical vacuum hose routing diagram (1995 model shown)

The Sections in this Chapter include general descriptions, checking procedures within the scope of the home mechanic and component replacement procedures (when possible) for each of the systems listed above.

Before assuming that an emissions control system is malfunctioning, check the fuel and ignition systems carefully. The diagnosis of some emission control devices requires specialized tools, equipment and training. If checking and servicing become too difficult or if a procedure is beyond your ability, consult a dealer service department or other qualified repair facility. Remember, the most frequent cause of emissions problems is simply a loose or broken vacuum hose or wire, so always check the hose and wiring connections first.

This doesn't mean, however, that emission control systems are particularly difficult to maintain and repair. You can quickly and easily perform many checks and do most of the regular maintenance at home with common tune-up and hand tools. *Note: Because of a Federally mandated extended warranty which covers the emission control system components, check with your dealer about warranty coverage before working on any emissions-related systems. Once the warranty has expired, you may wish to perform some of the component checks and/or replacement procedures in this Chapter to save money.*

Pay close attention to any special precautions outlined in this Chapter. It should be noted that the illustrations of the various systems may not exactly match the system installed on the vehicle you're working on because of changes made by the manufacturer during production or from year to year.

A Vehicle Emissions Control Information (VECI) label and a vacuum hose routing decal are located in the engine compartment **(see illustrations)**. The VECI label contains important emissions specifications and adjustment information. The vacuum hose decal has a vacuum hose schematic with the emissions components identified. When servicing the engine or emissions systems, the VECI label in your particular vehicle should always be checked for up-to-date information.

2 On Board Diagnosis (OBD) system and trouble codes

General information

1 The Engine Control Module (ECM), also known as the ECCS control module, controls the fuel injection system, the spark advance system, the self-diagnosis system, the cooling fans, etc. The ECM is located beneath the center console (refer to Section 3).

2 The ECM receives signals from various sensors which monitor changing engine operations such as intake air volume, intake air temperature, coolant temperature, engine RPM, acceleration/deceleration, exhaust temperature, etc. These signals are utilized by the ECM to determine the correct injection pulse duration and ignition timing.

3 The Sections in this Chapter include general descriptions and checking procedures, within the scope of the home mechanic and component replacement procedures (when possible). Before assuming the fuel and ignition systems are malfunctioning, check the emission control system thoroughly. The emission system and the fuel system are closely interrelated, but can be checked separately. The diagnosis of some of the fuel and emission control devices requires specialized tools, equipment and training. If checking and servicing become too difficult or if a procedure is beyond your ability, consult a dealer service department or other qualified repair facility. Remember, the most frequent cause of fuel and emissions problems is simply a loose or broken vacuum hose or wire, so always check the hose and wiring connections first. *Note: Because of a Federally mandated extended warranty which covers the emission control system components (and any other components which have a primary purpose other than emission control but have significant effects on emissions), check with your dealer about warranty coverage before working on any emission related systems. Once the warranty has expired, you may wish to perform some of the component checks and/or replacement procedures in this Chapter to save you money.*

Precautions

4 Always disconnect the power by either turning off the ignition switch or disconnecting the battery terminals before disconnecting engine control systems wiring connectors.

5 When installing a battery, be particularly careful to avoid reversing the positive and negative cables.

6 Do not subject engine control systems or emission related components to severe impact during removal or installation.

7 Do not be careless during troubleshooting. Even slight terminal contact can invalidate a testing procedure and even damage one of the numerous transistor circuits.

8 Never attempt to work on the ECM or open the ECM cover. The ECM is protected by a government mandated extended warranty that will be nullified if you tamper with it.

9 If you are inspecting electronic control system components during rainy weather, make sure water does not enter any part. When washing the engine compartment, do not spray these parts or their connectors with water.

Self diagnosis

Refer to illustration 2.12

10 The self-diagnosis mode is useful to diagnose malfunctions in the sensors and actuators of the ECCS system. There are two different modes available for diagnosing driveability problems.

11 Mode I is the initial test mode of the vehicle with the ignition key ON and the engine not running. The CHECK ENGINE light on the dash will light and remain lit without pulsing. This indicates that the ECM is receiv-

Chapter 6 Emissions and engine control systems

ing power and the bulb is working. **Note:** *The CHECK ENGINE light will remain **on when the engine is running** in the event of a malfunction or a stored trouble code. However, there are a few codes that will not illuminate the CHECK ENGINE light but the trouble code is stored in the ECM memory. It is a good idea to check for trouble codes if you experience driveability problems.*

12 Mode II is used to access the self-diagnosis system. To access the correct mode for self diagnosis, turn the ignition switch to the ON (engine not running) position. Turn the diagnostic mode selector on the ECM fully clockwise **(see illustration)**. Wait a minimum of two seconds and then turn the mode selector fully counterclockwise. Wait until the inspection lamp flashes (at least two seconds).

13 Carefully observe the CHECK ENGINE light on the dash. If everything in the self-diagnosis system is functioning properly, the computer will flash a code 55 on 1995 models or a code 0505 on 1996 and later models. This code will be represented by five long flashes on the CHECK ENGINE light followed by five quick flashes. If the computer has actual trouble codes stored, carefully observe the flashes and record the exact number onto paper. For example, code 43 (throttle position sensor or TPS circuit) is indi-

2.12 To start the self-diagnosis system inspection, with the ignition key ON (engine not running), turn the mode selector clockwise and hold it there for two seconds then completely counterclockwise and observe the flashing CHECK ENGINE light on the dash

cated by four long flashes followed by three quick flashes. Refer to the accompanying trouble code chart for the exact codes and component failures. **Note:** *1996 and later models are equipped with the OBD II self-diagnosis system. However, translation codes are available for the home mechanic. Consult the code chart for the code number and component failure. For example, code 1005, EGRC solenoid, will be 10 long flashes followed by five short flashes. Code 0205 would be 2 long flashes followed by 5 short flashes. Observe the flashing light carefully and write down the digits as they are flashed to avoid confusion.*

14 If the ignition key is turned OFF during the code extraction process and turned back on, the diagnosis will automatically return to Mode I. Restart the procedure to extract the codes. **Note:** *Switching the diagnostic modes is not possible if the engine is running.*

Clearing codes

15 After the tests have been performed and the repairs completed, erase the memory. With the ignition key ON (engine not running), turn the diagnostic mode selector on the ECM fully clockwise, wait at least two seconds and then fully counterclockwise. This will erase any memory the ECM has stored concerning a particular component. **Note:** *Be sure to turn the mode selector fully counterclockwise for normal vehicle operation if the codes were not erased from the memory.*

Trouble code chart - 1995 models

Trouble codes	Circuit or system	Probable cause
Code 11* (1 long flash, 1 short flash)	Camshaft position sensor/circuit	Check the camshaft position sensor or circuit (see Section 4).
Code 12 (1 long flash, 2 short flashes)	Mass airflow sensor/circuit	The mass airflow sensor source or ground circuit(s) may be shorted or open. Check the mass airflow sensor or circuit (see Section 4).
Code 13 (1 long flash, 3 short flashes)	Coolant temperature sensor	The coolant sensor or circuit may be shorted or open. Check the coolant temperature sensor and circuit (see Section 4).
Code 14 (1 long flash, 4 short flashes)	Vehicle speed sensor	The vehicle speed sensor signal circuit is open (see Section 4).
Code 21* (2 long flashes, 1 short flash)	Ignition signal	The ignition signal in the primary circuit is not entered during engine cranking or running (see a dealer service department or other qualified repair shop).
Code 28 (2 long flashes, 8 short flashes)	Overheat condition	Check the cooling fans and circuits.
Code 31 (3 long flashes, 1 short flash)	ECM control unit	The ECM input signal is beyond "normal" range (see a dealer service department or other qualified repair shop).
Code 32 (3 long flashes, 2 short flashes)	EGR function	The EGR control valve does not operate (see Section 5).
Code 33 (3 long flashes, 3 short flashes)	Oxygen sensor	The oxygen sensor circuit is open (see Section 4).
Code 34* (3 long flashes, 4 short flashes)	Knock sensor	The knock sensor circuit is open or shorted (see Section 4).
Code 35 (3 long flashes, 5 short flashes)	Exhaust gas temperature sensor	The exhaust gas temperature sensor circuit is open or shorted (see Section 4).
Code 36 (3 long flashes, 6 short flashes)	EGRC-BPT valve problem	Check this component and its circuit.
Code 37 (3 long flashes, 7 short flashes)	Closed loop operation	Front oxygen sensor malfunction, or circuit malfunction shorted or open.

Trouble code chart - 1995 models (continued)

Trouble codes	Circuit or system	Probable cause
Code 41 (4 long flashes, 1 short flash)	Intake air temperature	Check sensor and sensor malfunction circuit.
Code 43 (4 long flashes, 3 short flashes)	Throttle position sensor (TPS)	The TPS circuit is open or shorted (see Section 4).
Code 55* (5 long flashes, 5 short flashes)	EFI system	Normal operation. No problems are detected.
Codes 65 through 68 (six long flashes, five to eight short flashes)	Misfire on cylinders 4, 3, 2 and 1 respectively	Check the spark plug wire and cap terminal for the cylinder in question.
Code 71 (7 long flashes, 1 short flash)	Random misfire	Check ignition system components.
Code 72 (7 long flashes, 2 short flashes)	Catalytic converter system problem	Refer the vehicle to a specialist.
Code 76 (7 long flashes, 6 short flashes)	Fuel injection system problem	Refer the vehicle to a specialist.
Code 77 (7 long flashes, 7 short flashes)	Rear oxygen sensor malfunction	Check the component and its circuit.
Code 82 (8 long flashes, 2 short flashes)	Crankshaft position sensor signal problem	Check the component and its circuit.
Code 84 (8 long flashes, 4 short flashes)	A/T diagnostic communications line	Refer vehicle to an automatic transaxle specialist.
Code 91 (9 long flashes, 1 short flash)	Front oxygen sensor heater problem	Check the component and its circuit.
Code 95 (9 long flashes, 5 short flashes)	Crank P/S COG	Refer the vehicle to a specialist.
Code 98 (9 long flashes, 8 short flashes)	Coolant temperature sensor out of range	Check the component and its circuit.
Code 103 (10 long flashes, 3 short flashes)	Park/neutral switch malfunction	Refer vehicle to an automatic transaxle specialist.
Code 105 (10 long flashes, 5 short flashes)	EGRC solenoid	Check the component and its circuit.
Code 111 (11 long flashes, 1 short flash)	Inhibitor switch	Refer vehicle to an automatic transaxle specialist.
Code 112 (11 long flashes, 2 short flashes)	Vehicle speed sensor, A/T	Refer vehicle to an automatic transaxle specialist.
Code 113 (11 long flashes, 3 short flashes)	A/T first signal	Refer vehicle to an automatic transaxle specialist.
Code 114 (11 long flashes, 4 short flashes)	A/T second signal	Refer vehicle to an automatic transaxle specialist.
Code 115 (11 long flashes, 5 short flashes)	A/T third signal	Refer vehicle to an automatic transaxle specialist.
Code 116 (11 long flashes, 6 short flashes)	A/T fourth signal or TCC	Refer vehicle to an automatic transaxle specialist.
Code 118 (11 long flashes, 8 short flashes)	Shift solenoid, A/T	Refer vehicle to an automatic transaxle specialist.
Code 121 (12 long flashes, 1 short flash)	Shift solenoid, A/T	Refer vehicle to an automatic transaxle specialist.
Code 123 (12 long flashes, 3 short flashes)	Overrun clutch, A/T	Refer vehicle to an automatic transaxle specialist.
Code 124 (12 long flashes, 4 short flashes)	Torque converter clutch	Refer vehicle to an automatic transaxle specialist.
Code 125 (12 long flashes, 5 short flashes)	Line pressure, A/T	Refer vehicle to an automatic transaxle specialist.
Code 126 (12 long flashes, 6 short flashes)	Throttle position sensor, A/T	Refer vehicle to an automatic transaxle specialist.
Code 127 (12 long flashes, 7 short flashes)	Engine speed signal to A/T	Refer vehicle to an automatic transaxle specialist.
Code 128 (12 long flashes, 8 short flashes)	A/T fluid temperature sensor	Refer vehicle to an automatic transaxle specialist.

*These codes do not illuminate the CHECK ENGINE light on the dash, but it is possible to access the OBD system and read the codes on the flashing light on the computer

Chapter 6 Emissions and engine control systems

Trouble code chart - 1996 and later models

Trouble codes	Circuit or system	Probable cause
Code 0505 (5 long flashes, 5 short flashes)	No self diagnostic codes present	
Code 0101 (1 long flash, 1 short flash)	Camshaft position sensor	Check the camshaft sensor or the circuit (see Section 4).
Code 0102 (1 long flash, 2 short flashes)	Mass airflow sensor/circuit	The mass airflow sensor source or ground circuit(s) may be shorted or open. Check the mass airflow sensor or circuit (see Section 4).
Code 0103 (1 long flash, 3 short flashes)	Coolant temperature sensor	The coolant sensor or circuit may be shorted or open. Check the coolant temperature sensor and circuit (see Section 4).
Code 0104 (1 long flash, 4 short flashes)	Vehicle speed sensor	The vehicle speed sensor signal circuit is open (see Section 4).
Code 0111 (1 long flash, 11 short flashes)	EVAP purge flow monitoring system	Have the vehicle diagnosed by a dealer service department.
Code 0114 (1 long flash, 14 short flashes)	Fuel system rich	Check the injectors, fuel pressure, fuel pressure regulator, oxygen sensor and/or the MAF sensor (see Section 4).
Code 0115 (1 long flash, 15 short flashes)	Fuel system lean	Check the injectors, fuel pressure, fuel pressure regulator, oxygen sensor and/or the MAF sensor (see Section 4).
Code 0201 (2 long flashes, 1 short flash)	Ignition signal	The ignition signal in the primary circuit is not entered during engine cranking or running (see a dealer service department or other qualified repair shop).
Code 0203 (2 long flashes, 3 short flashes)	Closed throttle switch	Check throttle position sensor.
Code 0205 (2 long flashes, 5 short flashes)	IACV/AAC valve	Check the IACV/AAC valve and circuit (see Chapter 4).
Code 0208 (2 long flashes, 8 short flashes)	Overheating	Cooling fan or system is not operating correctly.
Code 0213 (2 long flashes, 13 short flashes)	EVAP system	Check for small leak in system.
Code 0214 (2 long flashes, 14 short flashes)	Purge volume control valve	Have the vehicle diagnosed by a dealer service department or other qualified repair shop.
Code 0301 (3 long flashes, 1 short flash)	ECM control unit	The ECM input signal is beyond "normal" range (see a dealer service department or other qualified repair shop).
Code 0302 (3 long flashes, 2 short flashes)	EGR Function	The EGR control valve does not operate (see Section 5).
Code 0303 (3 long flashes, 3 short flashes)	Front heated oxygen sensor malfunction	Check the component and its circuit.
Code 0306 (3 long flashes, 6 short flashes)	EGRC BPT valve	Check the EGRC BPT valve (see Section 5).
Code 0304 (3 long flashes, 4 short flashes)	Knock sensor	The knock sensor circuit is open or shorted (see Section 4).
Code 0305 (3 long flashes, 5 short flashes)	Exhaust gas temperature sensor	The exhaust gas temperature sensor circuit is open or shorted (see Section 4).
Code 0307 (3 long flashes, 3 short flashes)	Closed loop operation	ECM does not enter closed loop operation. Check front oxygen sensor and/or heater.
Code 0309 (3 long flashes, 9 short flashes)	Vent control valve	Check EVAP canister vent control valve.
Code 0311 (3 long flashes, 11 short flashes)	Vacuum cut valve bypass valve	Have the vehicle diagnosed by a dealer service department.
Code 0312 (3 long flashes, 12 short flashes)	EVAP system purge control valve	Have the vehicle diagnosed by a dealer service department or other qualified repair shop.
Code 0401 (4 long flashes, 1 short flash)	Intake air temperature sensor	IAT circuit is open or shorted (see Section 4).

Trouble code chart - 1996 and later models (continued)

Trouble codes	Circuit or system	Probable cause
Code 0402 (4 long flashes, 2 short flashes)	Fuel temperature sensor or circuit	Check the component and its circuit.
Code 0403 (4 long flashes, 3 short flashes)	Throttle position sensor (TPS)	The TPS circuit is open or shorted (see Section 4).
Code 0409 (4 long flashes, 9 short flashes)	Front oxygen sensor	Check the component and its circuit.
Code 0410 (4 long flashes, 10 short flashes)	Front oxygen sensor	Check the component and its circuit.
Code 0411 (4 long flashes, 11 short flashes)	Front oxygen sensor	Check the component and its circuit.
Code 0412 (4 long flashes, 12 short flashes)	Front oxygen sensor	Check the component and its circuit.
Code 0503 (5 long flashes, 3 short flashes)	Front oxygen sensor	Defective oxygen sensor, shorted oxygen sensor circuit, defective injectors or incorrect fuel pressure.
Code 0510 (5 long flashes, 10 short flashes)	Rear oxygen sensor	Check the rear oxygen sensor and circuit (see Section 4).
Code 0511 (5 long flashes, 11 short flashes)	Rear oxygen sensor	Check the rear oxygen sensor and circuit (see Section 4).
Code 0512 (5 long flashes, 12 short flashes)	Rear oxygen sensor	Check the rear oxygen sensor and circuit (see Section 4).
Code 0514 (5 long flashes, 14 short flashes)	EGR system	Have the vehicle diagnosed by a dealer service department.
Code 0605 (6 long flashes, 5 short flashes)	Cylinder 4 misfire	Check number 4 spark plug, ignition wire, compression or injector.
Code 0606 (6 long flashes, 6 short flashes)	Cylinder 3 misfire	Check number 3 spark plug, ignition wire, compression or injector.
Code 0607 (6 long flashes, 7 short flashes)	Cylinder 2 misfire	Check number 2 spark plug, ignition wire, compression or injector.
Code 0608 (6 long flashes, 8 short flashes)	Cylinder 1 misfire	Check number 1 spark plug, ignition wire, compression or injector.
Code 0701 (7 long flashes, 1 short flash)	Multiple cylinder misfire	Check the spark plugs, ignition wires, injectors and compression readings for each cylinder.
Code 0702 (7 long flashes, 2 short flashes)	Catalytic converter	Possible three way catalytic converter defective. Check for injector problems, damaged exhaust tube or intake manifold leak.
Code 0704 (7 long flashes, 4 short flashes)	EVAP system	Check the EVAP control system pressure sensor.
Code 0705 (7 long flashes, 5 short flashes)	EVAP system	Check for a small leak.
Code 0706 (7 long flashes, 6 short flashes)	Fuel injection system malfunction	Mixture is too lean or too rich.
Code 0707 (7 long flashes, 7 short flashes)	Rear oxygen sensor	Check the rear oxygen sensor and circuit (see Section 4).
Code 0801 (8 long flashes, 1 short flash)	Vacuum cut valve bypass valve or circuit	Have the vehicle diagnosed by a dealer service department or other qualified repair shop.
Code 0802 (8 long flashes, 2 short flashes)	Crankshaft sensor	Check the crankshaft sensor and circuit (see Section 4).
Code 0803 (8 long flashes, 3 short flashes)	Absolute pressure sensor	Check the component and its circuit.
Code 0804 (8 long flashes, 4 short flashes)	A/T diagnosis comm line	Possible dead battery. Harness connectors between the ECM and automatic transaxle damaged.
Code 0807 (8 long flashes, 7 short flashes)	EVAP canister purge control	Have the vehicle diagnosed by a dealer service department or other qualified repair shop.
Code 0901 (9 long flashes, 1 short flash)	Front oxygen sensor heater	Check the front oxygen sensor heater and circuit (see Section 4).

Chapter 6 Emissions and engine control systems

Trouble codes	Circuit or system	Probable cause
Code 0902 (9 long flashes, 2 short flashes)	Rear oxygen sensor heater	Check the rear oxygen sensor heater and circuit (see Section 4).
Code 0903 (9 long flashes, 3 short flashes)	Vent control valve	Check the EVAP canister vent control valve.
Code 0905 (9 long flashes, 5 short flashes)	Crankshaft position sensor	Possible damaged flywheel. Check the crank sensor and/or circuit (see Section 4).
Code 0908 (9 long flashes, 8 short flashes)	Coolant temperature sensor	Check the ECT sensor and circuit (see Section 4).
Code 1003 (10 long flashes, 3 short flashes)	Park/neutral switch	Check the park/neutral switch circuit (see Chapter 7).
Code 1005 (10 long flashes, 5 short flashes)	EGR solenoid valve	The EGR valve and EVAP canister purge control solenoid valve circuit is open or shorted (see Sections 5 and 6).
Code 1008 (10 long flashes, 8 short flashes)	EVAP system	Check the EVAP canister purge control valve.
Code 1101 (11 long flashes, 1 short flash)	Inhibitor switch	Check the Inhibitor switch and circuit (see Chapter 7).
Code 1102 (11 long flashes, 2 short flashes)	A/T vehicle speed sensor	Check the vehicle speed sensor and circuit for the automatic transaxle (see Chapter 7).
Code 1103 (11 long flashes, 3 short flashes)	A/T first signal	Have the vehicle diagnosed by a dealer service department or other qualified repair shop.
Code 1104 (11 long flashes, 4 short flashes)	A/T second signal	Have the vehicle diagnosed by a dealer service department or other qualified repair shop.
Code 1105 (11 long flashes, 5 short flashes)	A/T third signal	Have the vehicle diagnosed by a dealer service department or other qualified repair shop.
Code 1106 (11 long flashes, 6 short flashes)	A/T fourth signal	Have the vehicle diagnosed by a dealer service department or other qualified repair shop.
Code 1107 (11 long flashes, 7 short flashes)	A/T torque converter clutch	Have the vehicle diagnosed by a dealer service department or other qualified repair shop.
Code 1108 (11 long flashes, 8 short flashes)	Shift solenoid A	Have the vehicle diagnosed by a dealer service department or other qualified repair shop.
Code 1201 (12 long flashes, 1 short flash)	Shift solenoid B	Have the vehicle diagnosed by a dealer service department or other qualified repair shop.
Code 1203 (12 long flashes, 3 short flashes)	Overrun clutch solenoid	Have the vehicle diagnosed by a dealer service department or other qualified repair shop.
Code 1204 (12 long flashes, 4 short flashes)	TCC solenoid	Have the vehicle diagnosed by a dealer service department or other qualified repair shop.
Code 1205 (12 long flashes, 5 short flashes)	Line pressure solenoid	Have the vehicle diagnosed by a dealer service department or other qualified repair shop.
Code 1206 (12 long flashes, 6 short flashes)	TPS for A/T	Have the vehicle diagnosed by a dealer service department or other qualified repair shop.
Code 1207 (12 long flashes, 7 short flashes)	Speed signal A/T	Have the vehicle diagnosed by a dealer service department or other qualified repair shop.
Code 1208 (12 long flashes, 8 short flashes)	A/T fluid temperature sensor	Have the vehicle diagnosed by a dealer service department or other qualified repair shop.
Code 1302 (13 long flashes, 2 short flashes)	MAP/BAR switch solenoid valve or circuit	Have the vehicle diagnosed by a dealer service department or other qualified repair shop.
Code 1305 (13 long flashes, 5 short flashes)	Fuel pump control module or circuit	Have the vehicle diagnosed by a dealer service department or other qualified repair shop.
Code 1308 (13 long flashes, 8 short flashes)	Cooling fan	Have the vehicle diagnosed by a dealer service department or other qualified repair shop.

6-8   Chapter 6   Emissions and engine control systems

3.1 Center console side covers must be removed to gain access to the ECM

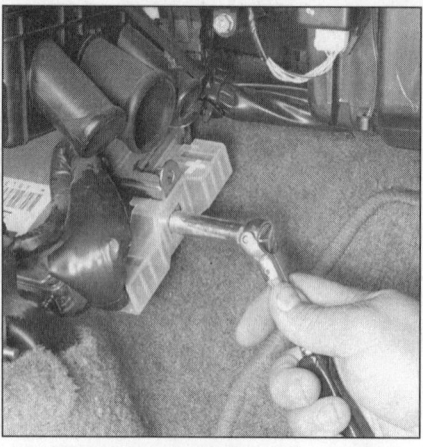

3.4 Remove the bolt and unplug the harness connector from the ECM

3.6 Pull the ECM out and detach the relay (if equipped) from its bracket

4.2a Check the resistance of the coolant temperature sensor (arrow) with the engine completely cold and then with the engine at operating temperature; resistance should decrease as temperature increases (GA16DE engine shown)

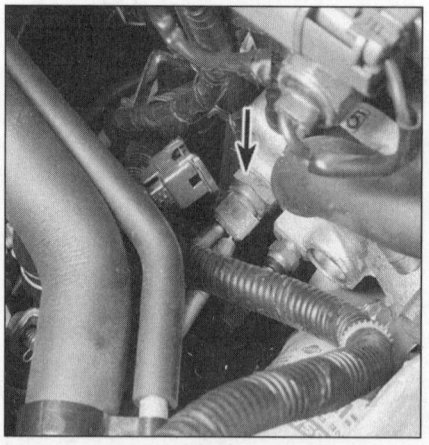

4.2b Coolant temperature sensor location here on a SR20DE engine

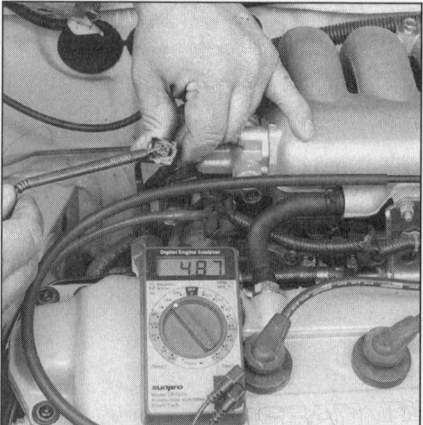

4.3 Working on the harness side, check the reference voltage from the ECM to the coolant temperature sensor with the ignition key ON and the engine not running; it should be approximately 5.0 volts

3   Engine Control Module (ECM) - removal and installation

Refer to illustration 3.1, 3.4 and 3.6

1   The Engine Control Module (ECM) is located under the instrument panel, in front of the center console, to the right of the accelerator pedal **(see illustration)**.

2   Disconnect the negative battery cable from the battery. **Warning:** *The models covered by this manual have airbags. Always disconnect the negative battery cable, then the positive cable and wait 10 minutes before working in the vicinity of the impact sensors, steering column or instrument panel to avoid the possibility of accidental deployment of the airbag, which could cause personal injury (see Chapter 12). The airbag circuits are easily identified by yellow insulation covering the entire wiring harness or just prior to the wire harness connectors. Do not use electrical test equipment on any of these wires or tamper with them in any way.*

3   Remove the side trim panels to expose the ECM on both sides of the center console.
4   Disconnect the harness connector from the ECM **(see illustration)**.
5   Remove the retaining nuts from the ECM bracket.
6   Carefully slide the ECM out **(see illustration)**. **Note:** *It is a good idea to avoid any static electricity damage to the computer by using special gloves or a technician's personal body grounding connector. If not, ground yourself to the vehicle before touching the ECM. If one is available, use a special anti-static pad to store the ECM on once it is removed.*

4   Information sensors

Engine Coolant Temperature (ECT) sensor

General description

1   The coolant temperature sensor is a thermistor (a resistor which varies the value of its resistance in accordance with temperature changes). The change in the resistance values will directly affect the voltage signal from the coolant sensor to the ECM. As the sensor temperature DECREASES, the resistance values will INCREASE. As the sensor temperature INCREASES, the resistance values will DECREASE. A failure in the coolant sensor circuit should set a code. This code indicates a failure in the coolant temperature circuit, so in most cases the appropriate solution to the problem will be either repair of a wire or replacement of the sensor.

Check

Refer to illustrations 4.2a, 4.2b and 4.3

2   Check the resistance value of the coolant temperature sensor while it is completely cold (65 to 70-degrees F = 2.1 to 2.9 K-ohms). Next, start the engine and warm it up until it reaches operating temperature **(see illustrations)**. The resistance should be lower (194-degrees F = 236 to 260 ohms). **Note:** *Access to the coolant temperature sensor makes it difficult to position test probes on the terminals. If necessary, remove the sensor and perform the tests in a pan of heated water to simulate the conditions.*

Chapter 6 Emissions and engine control systems

4.6 The rear heated oxygen sensor is located in the exhaust pipe, downstream of the catalytic converter

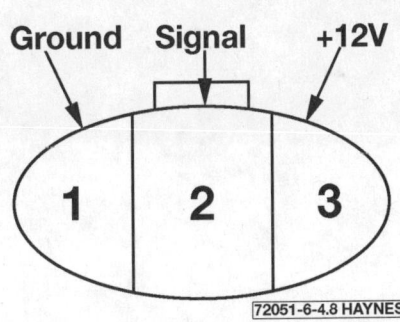

4.8 The wiring harness side of the oxygen sensor connector looks like this. Terminal 1 is ground, terminal 2 is signal and terminal 3 is the 12 volt supply voltage

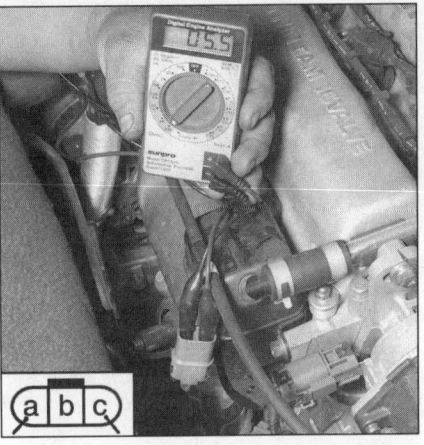

4.10 Probe terminals A and C to check the resistance of the oxygen sensor heater

3   If the resistance values on the sensor are correct, check the reference voltage to the sensor from the ECM **(see illustration)**. It should be approximately 5.0 volts.

Replacement

Warning: *Wait until the engine is completely cool before removing the sensor.*

4   Before installing the new sensor, wrap the threads with Teflon sealing tape to prevent leakage and thread corrosion. Remove the radiator cap to relieve any residual pressure in the system, then reinstall the cap.
5   Disconnect the electrical connector and remove the sensor. Be prepared for coolant leakage. **Caution:** *Handle the coolant sensor with care. Damage to this sensor will affect the operation of the entire fuel injection system.* Install the sensor and tighten it securely. Check the level of the coolant and add some, if necessary (see Chapter 1).

Heated oxygen sensor (HO2S) (front and rear)

General information

Refer to illustration 4.6

6   There are two heated (three wire) oxygen sensors used in these models. One sensor is located upstream of the catalytic converter in the exhaust manifold, the other downstream **(see illustration)**. They measure the oxygen content of the exhaust gas stream. The oxygen content in the exhaust reacts with the oxygen sensor to produce a voltage output which varies from 0.1-volt (high oxygen, lean mixture) to 0.9-volts (low oxygen, rich mixture). The ECM constantly monitors this variable voltage output to determine the ratio of oxygen to fuel in the mixture. The ECM alters the air/fuel mixture ratio by controlling the pulse width (open time) of the fuel injectors. A mixture ratio of 14.7 parts air to 1 part fuel is the ideal mixture ratio for minimizing exhaust emissions, thus allowing the catalytic converter to operate at maximum efficiency. It is this ratio of 14.7 to 1 which the ECM and the oxygen sensor attempt to maintain under normal driving conditions.
7   The oxygen sensor produces no voltage when it is below its normal operating temperature of about 600-degrees F. During this initial period before warm-up, the ECM operates in open loop mode.

Check

Refer to illustrations 4.8 and 4.10

8   Locate the oxygen sensor electrical connector and backprobe the connector with a straight-pin **(see illustration)**. Connect the positive probe of a voltmeter onto the pin and the negative probe to ground. Backprobe the center wire **(see illustration)** and observe the sensor voltage signal. **Note:** *The signal voltage on rear heated oxygen sensors will not display the wide range of change in the voltage values. The rear oxygen sensor detects very small amounts of oxygen because the emissions have been catalyzed at this point. If the rear heated oxygen sensor displays the same voltage values as the front heated oxygen sensor, then either the catalytic converter is defective or the oxygen sensor is malfunctioning.*
9   Monitor the voltage signal as the engine goes from cold to warm. The oxygen sensor will produce a steady voltage signal at first (open loop) of approximately 0.1 to 0.2 volts with the engine cold. After a period of approximately two minutes, the engine will reach operating temperature and the oxygen sensor will start to fluctuate between 0.1 to 0.9 volts (closed loop). If the oxygen sensor fails to reach the closed loop mode or there is a very long period of time until it does switch into closed loop mode, replace the oxygen sensor with a new part.
10   Check the resistance of the oxygen sensor heater by connecting a voltmeter across the two outermost terminals **(see illustration)**. **Note:** *You must remove the rubber grommet from the floorpan of the vehicle near the catalytic converter to remove the rear heated oxygen sensor harness connector.*
11   The resistance of the front sensor heater should be between 3.3 and 6.3 ohms when it is at room temperature. The rear oxygen sensor heater resistance should be 5.2 to 8.2 ohms.
12   Check for proper supply voltage to the heater. Measure the voltage on the harness side of the oxygen sensor electrical connector, terminal C. There should be battery voltage with the ignition key ON (engine not running). If there is no voltage, check the circuit between the ignition switch, fuse box and the sensor (see Chapter 12).
13   The proper operation of the oxygen sensor depends on four conditions:

a) *Electrical* - The low voltages generated by the sensor depend upon good, clean connections which should be checked whenever a malfunction of the sensor is suspected or indicated.
b) *Outside air supply* - The sensor is designed to allow air circulation to the internal portion of the sensor. Whenever the sensor is removed and installed or replaced, make sure the air passages are not restricted.
c) *Proper operating temperature* - The ECM will not react to the sensor signal until the sensor reaches approximately 600-degrees F. This factor must be taken into consideration when evaluating the performance of the sensor.
d) *Unleaded fuel* - The use of unleaded fuel is essential for proper operation of the sensor. Make sure the fuel you are using is of this type.

14   In addition to observing the above conditions, special care must be taken whenever the sensor is serviced.

a) The oxygen sensor has a permanently attached pigtail and electrical connector which should not be removed from the sensor. Damage or removal of the pigtail or electrical connector can adversely affect operation of the sensor.
b) Grease, dirt and other contaminants should be kept away from the electrical connector and the louvered end of the sensor.

c) Do not use cleaning solvents of any kind on the oxygen sensor.
d) Do not drop or roughly handle the sensor.
e) The silicone boot must be installed in the correct position to prevent the boot from being melted and to allow the sensor to operate properly.

15 If the oxygen sensor fails any of these tests, replace it with a new part.

Replacement

Refer to illustrations 4.18 and 4.19

Note: *Because it is installed in the exhaust manifold or pipe, which contracts when cool, the oxygen sensor may be very difficult to loosen when the engine is cold. Rather than risk damage to the sensor (assuming you are planning to reuse it in another manifold or pipe), start and run the engine for a minute or two, then shut it off. Be careful not to burn yourself during the following procedure.*

16 Disconnect the cable from the negative terminal of the battery.
17 If necessary, raise the vehicle and place it securely on jackstands.
18 Carefully disconnect the electrical connector from the sensor **(see illustration)**.
19 Carefully unscrew the sensor **(see illustration)**. **Caution:** *Excessive force may damage the threads.*
20 Anti-seize compound must be used on the threads of the sensor to facilitate future removal. The threads of new sensors will already be coated with this compound, but if an old sensor is removed and reinstalled, recoat the threads.
21 Install the sensor and tighten it securely.
22 Reconnect the electrical connector of the pigtail lead to the main engine wiring harness.
23 Lower the vehicle and reconnect the cable to the negative terminal of the battery.

Throttle Position Sensor (TPS)

Refer to illustrations 4.24a and 4.24b

General description

24 The Throttle Position Sensor (TPS) is located on the end of the throttle shaft on the throttle body **(see illustration)**. By monitoring the output voltage from the TPS, the ECM can determine fuel delivery based on throttle valve angle (driver demand). A broken or loose TPS can cause intermittent bursts of fuel from the injector and an unstable idle because the ECM thinks the throttle is moving. Any problems in the TPS or circuit will set a code. **Note:** *SR20DE models equipped with an automatic transaxle use a wide open throttle and closed throttle position switch as part of the throttle angle detection system. This switch shuts on/off only at these particular points in the throttle angle range. The switch is wired directly to the automatic transaxle control module and is not used for engine control. On automatic transaxle models, there are two harnesses which connect to the TPS. The flat connector is for the TPS, the other is the closed-and-open throttle switch.*

4.18 Disconnect the wiring harness to the oxygen sensor (rear one shown) and squirt a little penetrating oil on the threads prior to removal

4.19 Remove the oxygen sensor using a special slotted socket, if available

4.24a The throttle position sensor (arrow) is mounted on the end of the throttle shaft on all models

4.24b On SR20DE engines with an automatic transaxle, the TPS electrical connector (left arrow) is mounted on the air intake plenum, away from the sensor (right arrow)

Check

Refer to illustrations 4.25, 4.26a, 4.26b and 4.27

25 To check the TPS, turn the ignition switch to ON (engine not running) and probe the ends of the voltmeter into the ground wire and signal (center) wire on the (connected) electrical connector. This test checks for the proper signal voltage from the TPS **(see illustrations)**. **Note 1:** *Use the TPS harness connector mounted on the bracket for all of the tests. Be careful when backprobing the electrical connector. Do not damage the wiring harness or pull on any connectors to make clean contact.* **Note 2:** *Automatics equipped with a wide open and closed throttle position switch can also have this switch checked using an ohmmeter* **(see illustration)**.
26 The center terminal of the sensor should read about 0.35 to 0.65-volts at closed throttle. Open the throttle. The sensor should increase voltage to about 4-volts **(see illustration)**. If the TPS voltage readings are incorrect, replace it with a new unit.

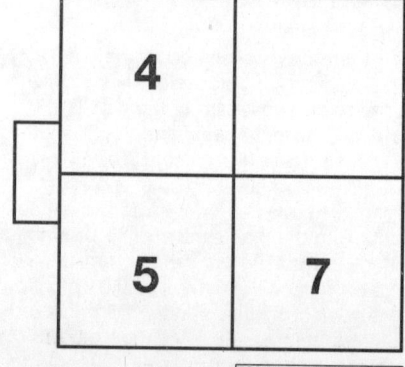

4.25 This is the additional wiring connector used on automatic transaxle models. Terminals 5 and 7 respond to a closed throttle condition and terminals 4 and 7 respond to wide open throttle

Chapter 6 Emissions and engine control systems   6-11

4.26a Backprobe between terminals A and B. This will give the output voltage of the TPS. At closed throttle it should be about 0.5 volts

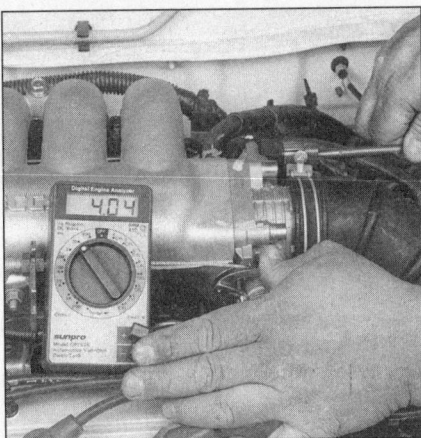

4.26b Open the throttle to wide open and check that the signal voltage increases to approximately 4.0 volts

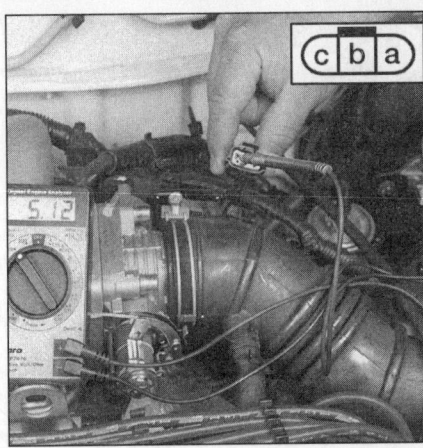

4.27 Check the reference voltage to the TPS with a voltmeter. Probe terminal C and make sure the reference voltage is approximately 5.0 volts

27 Also, check the TPS reference voltage. With the ignition key ON (engine not running), connect the positive (+) probe of the voltmeter onto the reference voltage wire **(see illustration)**. There should be approximately 5.0 volts sent from the ECM to the TPS. Check for a good ground at the opposite terminal using an ohmmeter. There should be good continuity between that terminal and ground. As a final check, disconnect the wiring harness and use an ohmmeter to check the resistance between terminals A and B **(see illustration 4.26a)** as you slowly operates the throttle. The resistance should smoothly and evenly change from about 1 K-ohm at closed throttle to about 10 K-ohms at full throttle.

Adjustment

Refer to illustration 4.28

28 With the TPS installed, loosen the two mounting screws **(see illustration)**. Connect the voltmeter and check signal voltage at idle (ignition key ON - engine not running).
29 Adjust the TPS as follows:
  a) *Manual transaxle models - rotate the TPS body and observe the signal voltage. It should be 0.3 to 0.5 volts. Rotate*

the TPS until the correct voltage is attained and tighten the mounting screws
  b) *Automatic transaxle models - with the engine warmed up and idling, check the closed throttle position switch circuit using an ohmmeter. Probe terminal numbers 5 and 7 (see illustration 4.25) and observe that at 900 +/- 150 rpm the switch continuity goes OFF to ON. This is the base setting.*

Replacement

30 Remove the two retaining screws **(see illustration 4.28)** and separate the TPS from the throttle body.
31 Install the new TPS leaving the mounting screws loose, then adjust the switch.
32 Tighten the screws securely.

Mass Airflow (MAF) sensor

Refer to illustrations 4.33, 4.34 and 4.41

General Information

33 The Mass Airflow Sensor (MAF) is located on the air intake duct at the air filter housing **(see illustration)**. This sensor uses a hot wire sensing element to measure the

amount of air entering the engine. The air passing over the hot wire causes it to cool. Consequently, this change in temperature can be converted into an analog voltage signal to the ECM which in turn calculates the required fuel injector pulse width.

Check

34 Check for power to the MAF sensor. Disconnect the MAF sensor electrical connector and connect a voltmeter to terminal 3 and ground **(see illustration)**. There should be battery voltage.
35 Reconnect the wiring and start the engine. Allow it to warm up. Using a straight pin, backprobe terminal 1.
36 At idle, the voltage should be between 1.3 and 1.7 volts. When the engine is revved to 4,000 rpm, the voltage should increase to about 4.0 volts. When the engine is stopped but the ignition is ON, the voltage should be less than 1.0 volts.
37 If the MAF sensor voltage does not vary or if the sensor causes the engine to stumble when rapped with your knuckles, replace the sensor.
38 If the voltage readings are correct, check the wiring harness for open circuits or other damage.

Replacement

39 Disconnect the electrical connector from the MAF sensor.
40 Remove the air cleaner assembly (see Chapter 4).

4.28 The TPS is secured by two screws

4.33 The mass air flow sensor is attached to the air cleaner housing

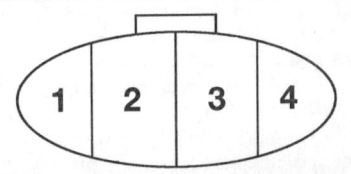

4.34 The chassis harness side of the MAF sensor wiring is numbered in this manner. Terminal 3 should have twelve volts when the ignition is ON

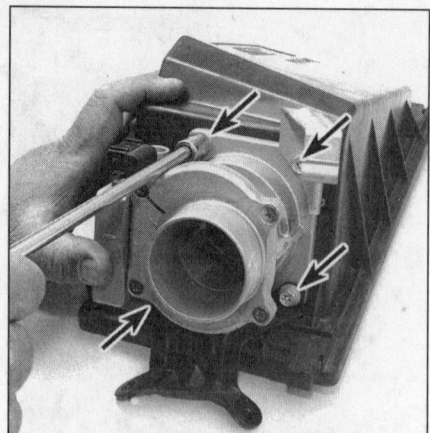

4.41 Remove the four bolts (arrows) and separate the MAF sensor from the air cleaner housing

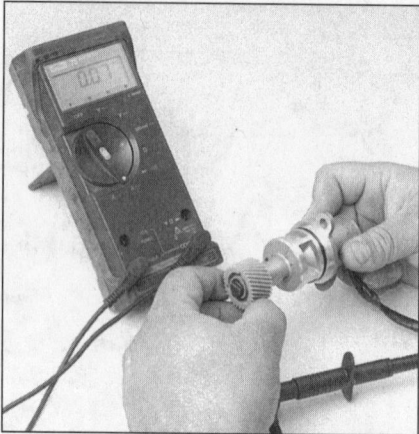

4.45 Using a voltmeter set on the AC scale, backprobe the VSS harness connector and check for voltage pulses that range from 0 to 0.5 volts as the gear is turned

4.48 The IAT sensor (arrow) is mounted on the air cleaner housing

41 Remove the four bolts **(see illustration)** and detach the MAF sensor from the air cleaner housing.
42 Installation is the reverse of removal.

Vehicle Speed Sensor (VSS)

Refer to illustration 4.45

General description

43 The Vehicle Speed Sensor (VSS) is located in the transaxle (see Chapter 7B). This sensor works in conjunction with a reed switch which is installed into the speedometer unit. The reed switch transforms vehicle speed into a pulsing voltage signal that is translated by the ECM and provided as information for other systems for fuel and transmission shift control. Any problems with the VSS will usually set a code.

Check

44 It is best to refer the vehicle to well-equipped shop if you suspect trouble with the VSS. However, for those wishing to attempt this procedure, the following checks are included. To check the vehicle speed sensor, unplug the electrical connector in the wiring harness near the sensor. Using an ohmmeter, check the resistance across the two terminals of the Vehicle Speed Sensor wiring harness connector. The resistance of the VSS should be approximately 250 ohms. If the resistance is incorrect, remove the VSS from the transaxle.
45 Place the VSS on a bench and check for a pulsing voltage. Use the AC scale on the voltmeter. Backprobe the electrical connector using two pins and, while slowly spinning the VSS gear, confirm that the voltage signal pulses from 0 to 0.5 volts **(see illustration)**. If the VSS doesn't respond as described, replace the VSS. The VSS is dependent on proper speedometer circuit operation. If there is a problem with the speedometer circuit, repair it before attempting to service the VSS.

Replacement

46 To replace the VSS, disconnect the electrical connector from the VSS and remove the retaining bolt and lift the VSS from the transaxle.
47 Installation is the reverse of removal.

Intake air temperature (IAT) sensor

Refer to illustration 4.48

General description

48 The air temperature sensor is mounted on the air cleaner housing **(see illustration)**. This sensor is a resistor which changes value according to the temperature of the air entering the engine. The ECM supplies approximately 5-volts (reference voltage) to the air temperature sensor. The voltage will change according to the temperature of the incoming air. The resistance will be high when the air temperature is low, and low when the air temperature is high. Any problems with the air temperature sensor will usually set a code.

Check

49 To check the air temperature sensor, disconnect the electrical connector and turn the ignition key ON, but do not start the engine.
50 Measure the reference voltage. The voltmeter should read approximately 5-volts. If the voltage signal is not correct, have the ECM diagnosed by a dealer service department or other repair shop. Also check the quality of the ground circuit (the other terminal) using an ohmmeter. There should be good continuity between terminal 1 and ground.
51 Measure the resistance across the sensor terminals. The resistance should be HIGH when the air temperature is LOW. Next, start the engine and let it idle. Allow the engine to reach normal operating temperature. Turn the ignition OFF, disconnect the air temperature sensor and measure the resistance across the terminals. The resistance should be LOW when the air temperature is HIGH. If the sensor does not exhibit this change in resistance, replace it with a new part. The sensor can also be removed and checked by blowing on it with a hair dryer. Low temperatures produce a high resistance value (for example, at 68-degrees F the resistance is 2.1 to 2.9 K-ohms) while high temperatures produce low resistance values (at 122-degrees F the resistance is 680 to 1,000 ohms and at 176-degrees F the resistance is 270 to 380 ohms).

Replacement

52 Disconnect the electrical connector, remove the screws and separate the sensor from the air cleaner housing. Installation is the reverse of removal.

EGR gas temperature sensor

General description

53 Most models are equipped with an EGR gas temperature sensor mounted near the EGR valve, installed into the EGR tube. This sensor detects the temperature of the exhaust as it moves through the EGR valve. The information is sent to the ECM and in turn the EGR on/off time is regulated more efficiently. Any malfunction with the EGR gas temperature sensor will set a code.

Check

54 Disconnect the harness connector for the EGR gas temperature sensor and measure the resistance of the sensor. Resistance should decrease as temperature increases. At 122-degrees F, the resistance of the sensor should be 570 to 700 K-ohms and at 212-degrees F should measure 80 to 98 K-ohms. It will probably be necessary to remove the sensor and place it into a pan of water which can then be heated to develop these temperatures.

Removal and installation

55 Disconnect the harness connector for the EGR gas temperature sensor and remove the sensor from the intake manifold.
56 Installation is the reverse of removal.

Chapter 6 Emissions and engine control systems

Knock sensor (KS)
Refer to illustration 4.58

General description
57 The knock sensor is threaded into the back side of the engine block. It detects abnormal vibration (shock waves due to detonation, or "knocking") in the engine. The sensor produces an AC output voltage which increases with the severity of the knock. The signal is fed into the ECM and the timing is retarded up to 10 degrees to compensate for severe detonation. Any problems with the knock sensor will set a code.

Check
58 Disconnect the electrical connector at the knock sensor. Using an ohmmeter set on the high scale, check for resistance between terminal 2 and ground **(see illustration)**. It should be 500 to 620 K-ohms at 77 degrees F (use the 10M scale on the ohmmeter).
59 Also check for a loose knock sensor. Tighten it if necessary.

Camshaft position sensor (CMPS)
Refer to illustration 4.65

Check
60 Remove the distributor cap (see Chapter 1).
61 Remove the rotor retaining screw and pull the rotor off the shaft (see Chapter 5).
62 Remove the sensor dust cover retaining screws and lift off the cover.
63 Inspect the camshaft signal plate for damage and dirt intrusion. Blow any accumulated dust out of the distributor and reinstall the cover, rotor and cap.
64 Inspect the electrical connections at the six wire connector for damage and corrosion and correct any defects.
65 Disconnect the six-wire connector **(see illustration)** from the distributor and use a voltmeter to check the voltage at terminal 5 with the ignition ON. If battery voltage is not present, check the circuit from the camshaft sensor to the ECM main relay for an open circuit or a defective relay.

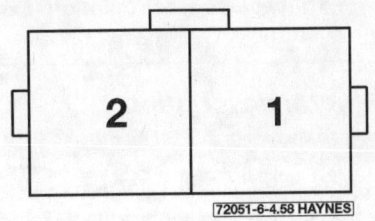

4.58 Check the resistance between terminal 2 of the knock sensor and ground

66 Reconnect the electrical connector. Start the engine and allow it to idle.
67 On 1995 and 1996 models and also 1997 SR20DE models, backprobe terminals 3 and then 4 with a voltmeter set to the AC volts scale. There should be about 2.7 volts AC at both terminals.
68 On 1997 GA16DE engines and all 1998 models, probe the terminals in the same manner, but set the voltmeter to the DC volts scale. Terminal 3 should read 0.1 to 0.4 volts. Terminal 4 should read about 2.5 volts.
69 If the camshaft sensor signal test results are incorrect, replace the distributor as a complete unit.

Crankshaft position sensor (CKPS)

General information
Refer to illustrations 4.70 and 4.71

70 All models are equipped with a crankshaft position sensor that is mounted on the transaxle bellhousing **(see illustration)**. It detects changes in crankshaft speed using a permanent magnet, core and coil. The changing gap causes the magnetic field near the sensor to change this in turn varies the voltage signal to the ECM. This sensor is used only as the on-board diagnostic device for detecting engine misfire.
71 Working on the harness side of the electrical connector, check for reference voltage to the sensor with the ignition key ON (engine not running). Voltage should be approximately 5.0 volts. Also check for a good ground at terminal 2 **(see illustration)**.

72 Next, check the resistance of the crankshaft sensor. It should be 432 to 528 ohms for GA16DE engines with manual transaxles. All other models should read from 166 to 204 ohms.

Replacement
73 Disconnect the electrical connector from the crankshaft sensor.
74 Remove the crankshaft sensor bolt.
75 Remove the crankshaft sensor from the engine. Installation is the reverse of removal.

Absolute Pressure Sensor
Refer to illustration 4.77

General information
76 1997 models equipped with the GA16DE engine and all later models use an absolute pressure sensor. This sensor is located on the upper firewall, slightly towards the center of the vehicle from the cruise control servo. It informs the ECM of ambient barometric pressure and also the pressure (vacuum) in the intake manifold. In this way it serves as both a BARO sensor and a manifold absolute pressure (MAP) sensor.

Check
77 Disconnect the electrical connector and check for 5-volts at terminal 3 with the ignition ON **(see illustration)**.
78 Use an ohmmeter to verify that there is good continuity between terminal 1 and ground.
79 Remove the sensor and disconnect the vacuum hose from it. Leave the wiring con-

4.70 The crankshaft position sensor

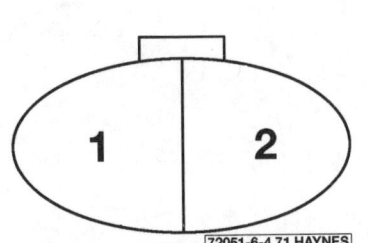

4.71 Disconnect the harness from the crankshaft position sensor. There should be 5 volts at terminal 1 and a good ground at terminal 2 when the ignition is ON

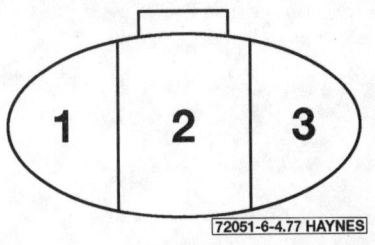

4.77 The wiring harness leading to the absolute pressure sensor (not used on early models) has its terminals numbered in this way. Terminal 1 must be a good ground and terminal 3 should be 5 volts when the ignition is ON

4.65 This is the numbering of the distributor electrical connector on the chassis side. Terminal 5 should have 12 volts when the ignition is ON

nected. Turn the ignition ON and check the voltage at (the center) terminal 2. There should be 3.2 to 4.8 volts. **Note:** *On 1998 vehicles, this check can be made by measuring the voltage at terminal 66 on the ECM in order to get a more accurate reading.*

Replacement

80 Disconnect the wiring and the vacuum hose from the sensor. Remove the two mounting screws.
81 Installation is the reverse of removal.

Park/Neutral Back-up light switch

Refer to Chapter 7 for diagnosis and replacement procedures.

Power steering pressure switch

Refer to illustration 4.82

General information

82 This switch senses loads on the steering system which would normally result in a demand on the power steering pump and a resulting decrease in idle speed **(see illustration)**. It signals the ECM to increase idle speed.

Check

83 Disconnect the wiring harness at the switch. Use an ohmmeter to verify that terminal 2 has a good continuity with ground.
84 With the wiring still disconnected, start the engine and use the ohmmeter to test the resistance between the two terminals on the switch. There should be infinite resistance (no continuity).
85 Have an assistant turn the steering wheel while you observe the ohmmeter. When the steering wheel is being turned, there should be near zero resistance (continuity).
86 If the switch does not operate as described, replace it.

5 Exhaust Gas Recirculation (EGR) system

General description

Refer to illustrations 5.1, 5.5, 5.6 and 5.12
1 The EGR system is used to lower NOx (oxides of nitrogen) emission levels caused by high combustion temperatures. The EGR recirculates a small amount of exhaust gases into the intake manifold **(see illustration)**. The additional mixture lowers the temperature of combustion thereby reducing the formation of NOx compounds.
2 The EGR systems are equipped with an EGR valve and also an "EGR and canister control solenoid valve" which receives ported and manifold vacuum. The operation of the system is controlled by the ECM which operates this control solenoid valve. The manifold vacuum system utilizes a vacuum tap in the air intake system positioned after the throttle valve. The ported vacuum control system uses a vacuum tap in the throttle body which is exposed to an increasing percentage of manifold vacuum as the throttle valve is opened during acceleration.
3 A backpressure transducer (EGRC-BPT) valve monitors the exhaust backpressure as the engine rpm increases or decreases to aid in controlling the EGR vacuum signal. An EGR temperature sensor is also used to inform the ECM of the heat in the exhaust gas which is being supplied to the engine.

Check

4 Check all hoses for cracks, kinks, broken sections and proper connection. Inspect all system connections for damage, cracks and leaks.
5 To check the EGR system operation, bring the engine up to operating temperature and, with the transmission in Neutral (parking brake set and tires blocked to prevent movement), allow it to idle for 70 seconds. Open the throttle so the engine speed is between 2,000 and 4,000 rpm and then allow it to close. The EGR valve stem should move if the control system is working properly. The test should be repeated several times. Movement of the stem indicates the control system is functioning correctly **(see illustration)**.
6 If the EGR valve stem does not move, check all of the hose connections to make sure they are not leaking or clogged. Disconnect the vacuum hose and apply ten inches of vacuum with a hand-held vacuum pump **(see illustration)**. If the stem still does not move, replace the EGR valve with a new one. If the valve does open, measure the valve travel to make sure it is approximately 1/8-inch. Also, the engine should idle roughly when the valve is open. If it doesn't, the passages are probably clogged.
7 Apply vacuum with the pump and then clamp the hose shut. The valve should stay open for 30 seconds or longer. If it does not, the diaphragm is leaking and the valve should be replaced with a new one.
8 If the EGR valve is not receiving manifold or ported vacuum, the throttle body unit must be removed to check and clean the slotted port in the throttle bore and the vacuum passages and orifices in the throttle body. Use solvent to remove deposits and check for flow with light air pressure.
9 If the engine idles roughly and it is suspected the EGR valve is not closing, remove the EGR valve and inspect the poppet and seat area for deposits.
10 If the deposits are more than a thin film of carbon, the valve should be cleaned. To clean the valve, apply solvent and allow it to penetrate and soften the deposits, making sure that none gets on the valve diaphragm, as it could be damaged.
11 Use a vacuum pump to hold the valve open and carefully scrape the deposits from the seat and poppet area with a tool. Inspect the poppet and stem for wear and replace the valve with a new one if wear is found.
12 Locate the EGRC-BPT valve and plug one of the ports with a finger **(see illustra-**

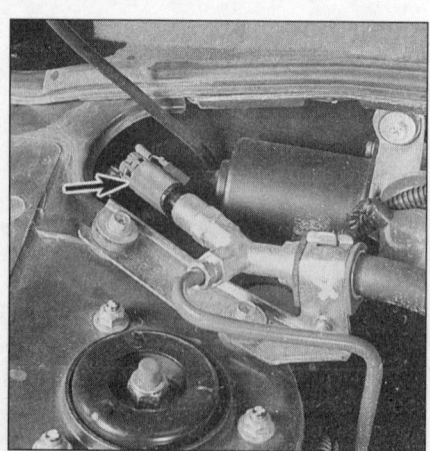

4.82 The power steering pressure switch (arrow) is located in the right rear corner of the engine compartment

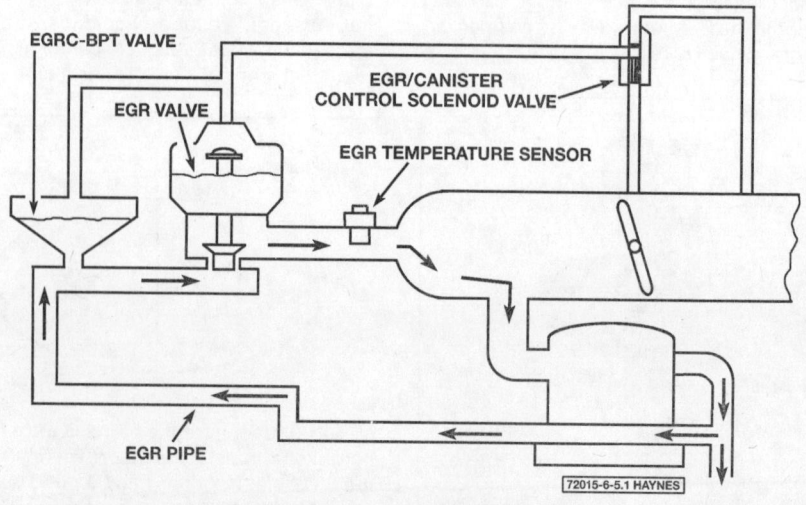

5.1 Schematic of the EGR system

Chapter 6 Emissions and engine control systems

5.5 Be careful not to burn your hand when checking for EGR valve diaphragm movement. The diaphragm and rod beneath it should move when the engine is revved

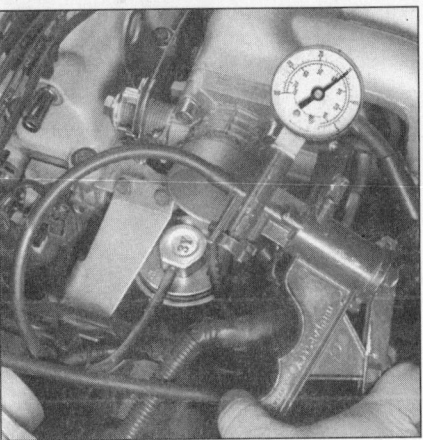

5.6 Use a hand-held vacuum pump to apply vacuum to the EGR valve. As vacuum increases, the diaphragm in its base should lift

5.12 Connect the vacuum pump hose to one port of the EGRC-BPT valve and apply vacuum. There should be a moderate amount of leakage

tion). Use a hand-held vacuum pump and apply vacuum to the valve. The valve should leak. **Note:** *It should stop leaking when pressure greater than 4 in-Hg is applied to the port under the valve. (The valve must be removed from the engine to apply pressure at this point).*

13 With the engine running, check for a vacuum signal from the intake manifold. This test will determine if manifold vacuum is reaching the various valves and components of the EGR system.

14 Also apply vacuum to the port (vacuum hose) that is routed to the EGR valve. The gauge should hold vacuum. If not, check for broken hoses or ruptured diaphragms in the EGR valve.

15 Refer to Section 4 for the testing procedure for the EGR gas temperature sensor.

16 The unit which controls the operation of both the EGR valve and the canister solenoid is located on top of the engine behind the EGR valve and to the right of the EGRC-BPT valve. To check the operation of this "EGR and canister control solenoid valve", disconnect the wiring from it and then connect a ground wire to one of its terminals and fused battery voltage to the other.

17 It should be possible to blow air through the two vacuum ports that are adjacent to one another when there is battery power applied, but impossible to do so when it is not energized with battery voltage.

Component replacement

EGR valve

Refer to illustration 5.21 and 5.22

18 When buying a new EGR valve, make sure that you have the right EGR valve. Use the stamped code located on the top of the EGR valve.

19 Detach the cable from the negative terminal of the battery.

20 Remove the air cleaner housing assembly (see Chapter 4).

5.21 Remove the EGR flange nut from the EGR valve using a large open end wrench

21 Detach the vacuum line from the EGR valve. Disconnect the EGR pipe nut **(see illustration)**.
22 Remove the EGR valve mounting fasteners **(see illustration)**.
23 Remove the EGR valve and gasket from the manifold. Discard the gasket.
24 With a wire wheel, buff the exhaust deposits from the EGR valve mounting surface on the manifold and, if you plan to use the same valve, the mounting surface of the valve itself. Look for exhaust deposits in the valve outlet. Remove deposit build-up with a screwdriver. **Caution:** *Never wash the valve in solvents or degreaser - both agents will permanently damage the diaphragm. Sandblasting is also not recommended because it will affect the operation of the valve.*
25 If the EGR passage contains an excessive build-up of deposits, clean it out with a wire wheel. Make sure that all loose particles are completely removed to prevent them from clogging the EGR valve or from being ingested into the engine.

5.22 Remove the EGR mounting nuts (arrows) and remove the EGR valve from the intake manifold

26 If there are large amounts of deposits within the EGR valve, remove the EGR pipe from the exhaust manifold and clean out the deposits inside the tube. **Note:** *Remove the EGRC-BPT valve and pipe and clean any deposits from the passages.*
27 Installation is the reverse of removal.

EGR and canister control solenoid valve

28 Unplug the electrical connector from the solenoid.
29 Clearly label and detach both vacuum hoses.
30 Remove the solenoid mounting screw and remove the solenoid.
31 Installation is the reverse of removal.

EGR temperature sensor

32 Disconnect the wiring from the sensor.
33 Unscrew the sensor from the engine.
34 Apply anti-seize to the threads of the sensor before installing it.
35 Installation is the reverse of removal.

6 Evaporative Emissions Control (EVAP) System

General description

Refer to illustration 6.1

1 The Evaporative Emissions Control (EVAP) system absorbs fuel vapors and, during engine operation, releases them into the engine intake where they mix with the incoming air-fuel mixture. The EVAP system consists of a charcoal-filled canister and the lines connecting the canister to the fuel tank, ported vacuum and intake manifold vacuum **(see illustration)**.
2 When the engine is not operating, fuel vapors are transferred from the fuel tank, throttle body and intake manifold to the charcoal canister where they are stored. When the engine is running, the fuel vapors are purged from the canister by the purge control valve. The gasses are consumed in the normal combustion process.
3 On 1997 and earlier models, a vacuum operated purge control valve is mounted on top of the canister. On 1998 and later models, an electronic purge control valve, controlled by the PCM is used. The electronic system is also equipped with a canister purge cut valve to close the purge line during deceleration and idling conditions.

Check

4 Poor idle, stalling and poor driveability can be caused by a defective canister purge control valve, a damaged canister, split or cracked hoses or hoses connected to the wrong tubes.
5 Evidence of fuel loss or fuel odor can be caused by the following items: an improperly functioning fuel cap relief valve, a damaged fuel cap gasket, fuel leaking from fuel lines or a cracked or damaged canister, an inoperative purge valve, disconnected, misrouted, kinked, deteriorated or damaged vapor or control hoses or an improperly seated air filter or air filter gasket.
6 Inspect each hose attached to the canister for kinks, leaks and breaks along its entire length. Repair or replace as necessary.
7 Inspect the canister. If it is cracked or damaged, replace it.
8 Look for fuel leaking from the bottom of the canister. If fuel is leaking, replace the canister and check the hoses and hose routing.

1997 and earlier models only

Refer to illustration 6.10

Note: *On 1998 and later models, on which the EVAP system is computer controlled, further diagnosis should be left to a dealer service department or other qualified shop since special equipment is required.*

9 Attach a short length of hose to the lower tube on the purge control valve and attempt to blow through it. Little or no air should pass into the canister (a small amount of air will pass because the canister has a constant purge hole).
10 With a hand-held vacuum pump, apply vacuum to the purge control valve signal tube (upper tube) **(see illustration)**.
11 If the purge control valve does not hold vacuum for at least 20 seconds, the purge control valve is leaking and must be replaced.
12 If the diaphragm holds vacuum, apply battery voltage to the EGR and canister control solenoid valve and see if vacuum (vapors) are allowed to pass through to the intake system. **Note:** *Follow the testing procedure in Section 5 to test the EGR valve and canister control solenoid valve.*

Canister replacement

1997 and earlier models

Refer to illustration 6.14

13 Clearly label, then detach, all vacuum lines from the canister.
14 Remove the canister mounting bolts and pull the canister out of the mounting bracket **(see illustration)**.
15 Installation is the reverse of removal.

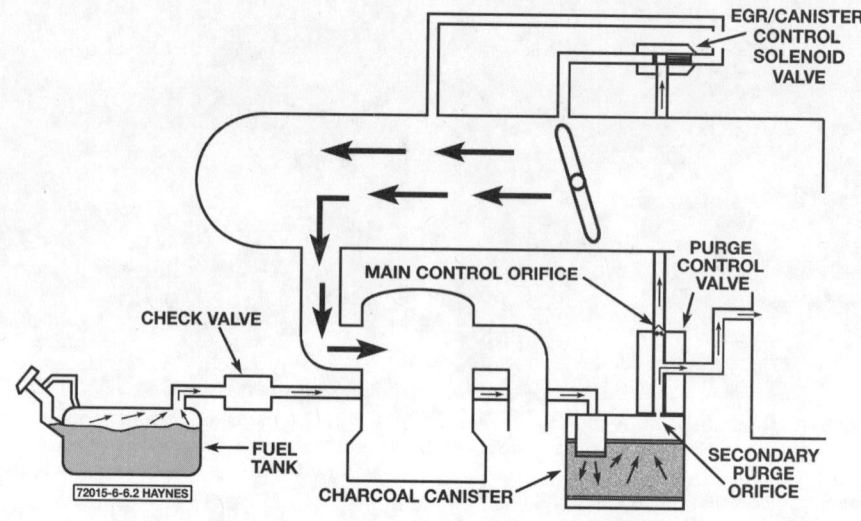

6.1 Schematic of a typical EVAP system

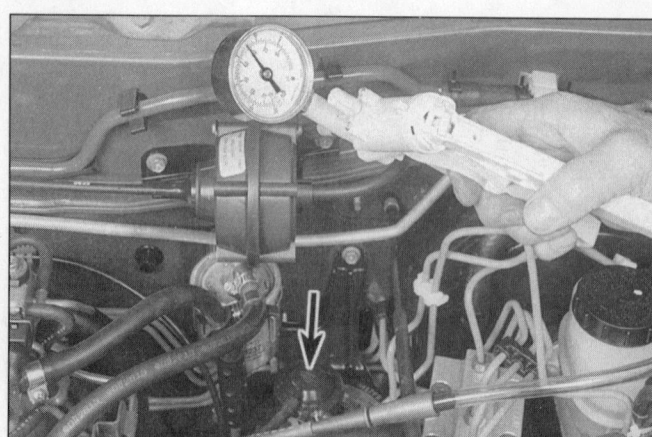

6.10 Apply vacuum to the purge control valve (arrow) and make sure that it holds vacuum for at least 20 seconds

6.14 The charcoal canister (arrow) is located underneath the master cylinder on 1997 and earlier models

Chapter 6 Emissions and engine control systems

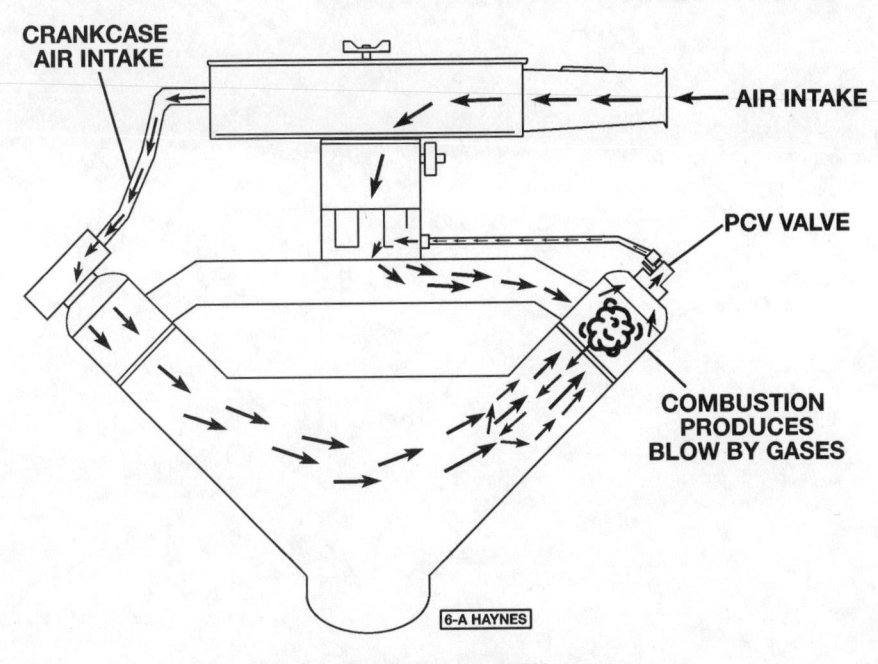

7.1 Gas flow in a typical PCV system

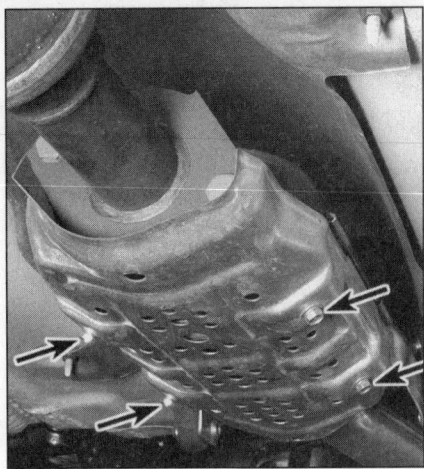

8.1 Remove the heat shield mounting nuts (arrows) to gain access to the catalytic converter for inspection

1998 and later models

Note: *On 1995 California emissions models and all 1996 and later models, the EVAP canister is located on the left side of the vehicle, behind the fuel tank.*

16 Raise the vehicle and support it securely on jackstands.
17 Clearly label, then detach, all vacuum lines from the canister.
18 Remove the canister mounting bolts and remove the canister.
19 Installation is the reverse of removal.

7 Positive Crankcase Ventilation (PCV) system

Refer to illustration 7.1

1 The Positive Crankcase Ventilation (PCV) system reduces hydrocarbon emissions by scavenging crankcase vapors. It does this by circulating fresh air from the air cleaner through the crankcase, where it mixes with blow-by gases and is then rerouted through a PCV valve to the intake manifold **(see illustration)**.
2 The main components of the PCV system are the PCV valve, a breather separator, a fresh air filtered inlet and the vacuum hoses connecting these two components with the engine.
3 To maintain idle quality, the PCV valve restricts the flow when the intake manifold vacuum is high. If abnormal operating conditions arise, the system is designed to allow excessive amounts of blow-by gases to flow back through the crankcase vent tube into the air cleaner to be consumed by normal combustion.
4 Checking and replacement of the PCV valve and filter is covered in Chapter 1.

8 Catalytic converter

General description

Refer to illustration 8.1

1 The catalytic converter **(see illustration)** is an emission control device added to the exhaust system to reduce pollutants from the exhaust gas stream. A single-bed converter design is used in combination with a three-way (reduction) catalyst. The catalytic coating on the three-way catalyst contains platinum and rhodium, which lowers the levels of oxides of nitrogen (NOx) as well as hydrocarbons (HC) and carbon monoxide (CO).

Check

2 The test equipment for a catalytic converter is expensive and highly sophisticated. If you suspect that the converter on your vehicle is malfunctioning, take it to a dealer or authorized emissions inspection facility for diagnosis and repair.
3 Whenever the vehicle is raised for servicing of underbody components, check the converter for leaks, corrosion and other damage. If damage is discovered, the converter should be replaced.

Replacement

4 Because the converter part of the exhaust system, converter replacement requires removal of the exhaust pipe assembly (see Chapter 4). Take the vehicle, or the exhaust system, to a dealer or a muffler shop.

Notes

Chapter 7 Part A
Manual transaxle

Contents

	Section		Section
Back-up light and neutral position switch - check and replacement	5	Oil seal replacement	2
General information	1	Shift linkage - removal and installation	4
Manual transaxle lubricant change	See Chapter 1	Transaxle - removal and installation	6
Manual transaxle lubricant level check	See Chapter 1	Transaxle mount - check and replacement	3
		Transaxle overhaul - general information	7

Specifications

Torque specifications

Ft-lbs (unless otherwise indicated)

Shift linkage
  Control rod
    Control rod-to-transaxle bolt/nut .... 120 to 156 in-lbs
    Control rod-to-shift lever bolt/nut .... 156 to 204 in-lbs
  Support rod
    Support rod-to-support rod bracket nut .... 26 to 35
    Support rod-to-shift lever socket nuts .... 108 to 132 in-lbs
    Support rod-to-mass damper nut .... 192 to 264 in-lbs
  Holder bracket nuts .... 156 to 204 in-lbs
Engine-to-transaxle bolts
  GA16DE engine
    Upper bolts .... 22 to 30
    Two lower bolts .... 12 to 15
  SR20DE engine
    Upper bolts .... 51 to 59
    Two lower bolts .... 22 to 30

1 General information

The vehicles covered by this manual are equipped with a 5-speed manual transaxle, or an automatic transaxle. Information on the manual transaxle is included in this Part of Chapter 7. Service procedures for the automatic transaxle are contained in Chapter 7, Part B.

The manual transaxle is a compact, two-piece, lightweight aluminum alloy housing containing both the transmission and differential assemblies.

Because of the complexity, difficulty in obtaining replacement parts and special tools necessary, internal repair procedures for the manual transaxle are beyond the scope of this manual. For readers who wish to tackle a transaxle rebuild, a brief *Manual transaxle overhaul - general information* Section is provided. The bulk of information in this Chapter is devoted to removal and installation procedures.

7A-2 Chapter 7 Part A Manual transaxle

2.4 Carefully pry out the driveaxle oil seal with a seal removal tool or a screwdriver; make sure you don't damage the seal bore or the new seal may leak

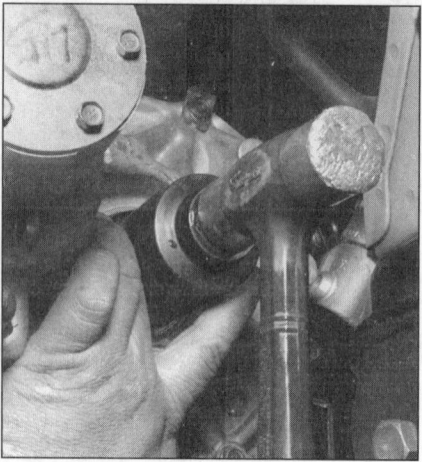

2.6 Use a seal installer, a large socket or a piece of pipe to install the new seal

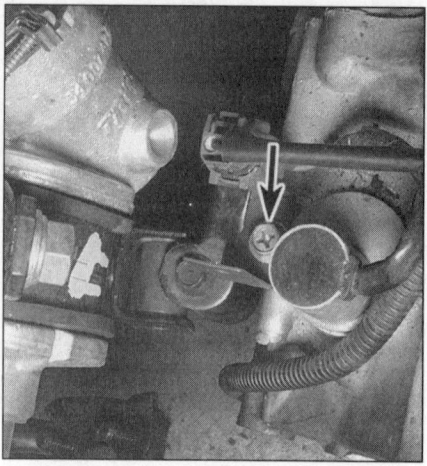

2.9 To remove the vehicle speed sensor, disconnect the electrical connector, remove the hold-down bolt (arrow) and remove the sensor from the transaxle

2.10 Remove the oil seal O-ring from the sensor pinion gear assembly with a small hook tool or a small screwdriver; make sure you don't gouge the groove

2 Oil seal replacement

1 Oil leaks frequently occur due to wear of the driveaxle oil seals and/or the vehicle speed sensor oil seal and O-rings. Replacement of these seals is relatively easy, since the repairs can usually be performed without removing the transaxle from the vehicle.

Driveaxle oil seals

Refer to illustrations 2.4 and 2.6

2 The driveaxle oil seals are located on the sides of the transaxle, where the inner ends of the driveaxles are splined into the differential side gears. If you suspect that a driveaxle oil seal is leaking, raise the vehicle and support it securely on jackstands. If the seal is leaking, you'll see lubricant on the side of the transaxle, below the seal.
3 Remove the driveaxle(s) (see Chapter 8).
4 Using a screwdriver or prybar, carefully pry the oil seal out of the transaxle bore **(see illustration)**.
5 If the oil seal cannot be removed with a screwdriver or prybar, a special oil seal removal tool (available at auto parts stores) will be required.
6 Using a large section of pipe or a large deep socket as a drift, install the new oil seal. Drive it into the bore squarely and make sure that it is completely seated **(see illustration)**. Lubricate the lip of the new seal with multi-purpose grease.
7 Install the driveaxle(s). Be careful not to damage the lip of the new seal.

Vehicle speed sensor O-ring

Refer to illustrations 2.9 and 2.10

8 The vehicle speed sensor is located on the transaxle housing. Look for lubricant around the housing to determine if the O-ring is leaking.
9 Disconnect the electrical connector, remove the hold-down bolt **(see illustrations)** and remove the vehicle speed sensor from the transaxle.
10 Using a scribe or a small screwdriver, remove the O-ring seal **(see illustration)**.
11 Install a new O-ring on the driven gear housing.
12 Installation is the reverse of removal.

Control rod seal

Refer to illustrations 2.14a, 2.14b, 2.15, 2.16, 2.17 and 2.18

13 If the seal for the control rod is obviously

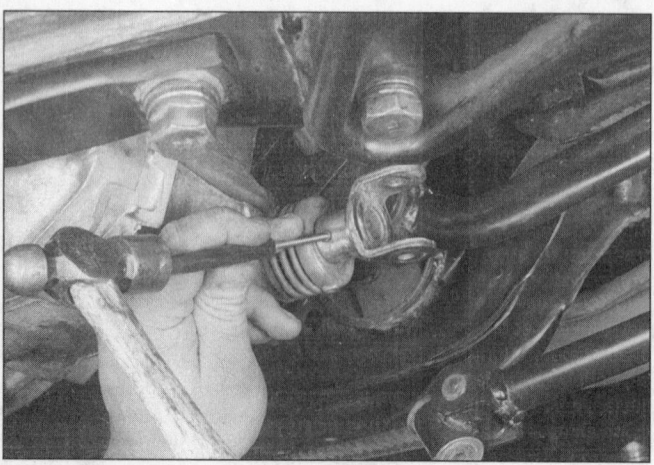

2.14a Drive out the yoke retaining pin with a hammer and punch . . .

2.14b . . . then remove the yoke

Chapter 7 Part A Manual transaxle

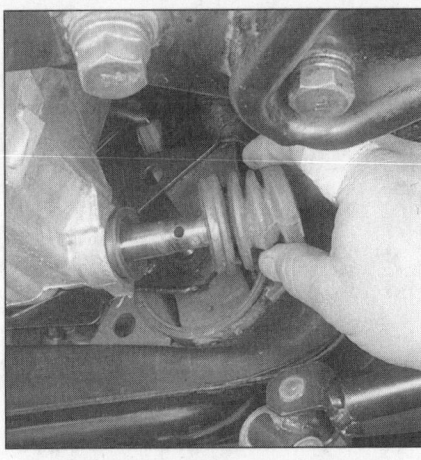

2.15 Slide the dust boot off the shaft

2.16 Pry out the control rod seal

2.17 A hammer and a socket of the proper size are used to install the new seal

leaking, refer to Section 4 and remove the control rod from the transaxle.
14 Remove the yoke from the shaft with a hammer and punch (see illustration). Remove the yoke (see illustration).
15 Remove the dust boot from the shaft (see illustration).
16 Pry out the old seal (see illustration).
17 Lubricate the lip of the new seal with multi-purpose grease. Install the seal using a socket of the correct diameter and a hammer (see illustration).
18 Installation is the reverse of removal (see illustration).

3 Transaxle mount - check and replacement

Check

Refer to illustration 3.2

1 Raise the vehicle and place it securely on jackstands.
2 Insert a large screwdriver or prybar between the mount and the bracket and pry up (see illustration).

3 The transaxle should not move excessively away from the mount. If it does, replace the mount.

Replacement

4 To replace a mount, support the transaxle with a jack, remove the nuts and bolts and remove the mount. It may be necessary to raise the transaxle slightly to provide enough clearance to remove the mount.
5 Installation is the reverse of removal.

4 Shift linkage - removal and installation

Removal

Refer to illustrations 4.3, 4.4a, 4.4b, 4.4c and 4.5

1 Remove the shift lever knob. If the knob is difficult to loosen, break it loose with a pair of adjustable pliers, but be sure to protect the knob with a clean shop rag. Remove the dust boot.
2 Raise the vehicle and place it securely on jackstands.

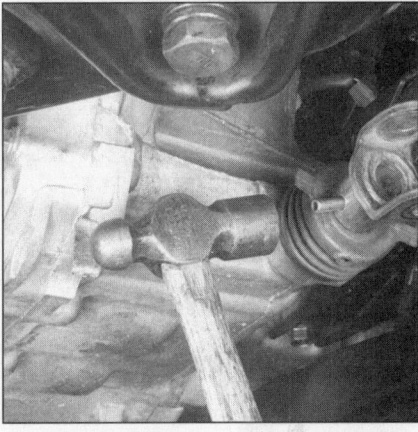

2.18 Use a new roll pin when assembling the yoke to the shaft

3 Disconnect the support and control rod from the transaxle (see illustrations).
4 Remove the nut and bolt that attach the control rod to the shift lever (see illustration). Disconnect the return spring, remove the nuts that attach the support rod

3.2 To check a transaxle mount, place a large screwdriver or prybar between the mount and the bracket and try to lever the transaxle up and down; if it moves excessively, replace the mount

4.3 To disconnect the control and support rods from the transaxle, remove these fasteners (arrows)

4.4a To disconnect the control rod from the shift lever, remove this nut (arrow) and bolt

7A-4   Chapter 7 Part A   Manual transaxle

4.4b To disconnect the support rod from the shift lever socket, remove these nuts

4.4c To disconnect the support rod from the mass damper, remove this nut (center arrow); to replace the rubber mount, remove the two outer nuts (arrows)

4.5 To replace the shift lever dust boot, remove the two front nuts (upper arrows) and the two rear nuts from below (not visible in this photo)

to the shift lever socket and mass damper **(see illustrations)** and remove the shift linkage assembly. Inspect the condition of all the bushings at both ends of the support and control rods. If they're cracked or worn, replace them. Inspect the return spring rubber; replace it if it's cracked or torn. Also inspect the condition of the rubber insulator portion of the mass damper holder bracket. If the rubber is cracked or torn, replace the holder bracket.

5   Pull out the shift lever and slide off the dust boot, seat, insulator and shift lever socket. Inspect these parts for cracks and tears. Replace as necessary. The condition of the shift lever socket and the seat are especially important; if either of these parts is damaged, shifting will be difficult. Also inspect the condition of the shift lever boot **(see illustration)**. If it's damaged, replace it.

Installation

6   Installation is the reverse of removal. Make sure you lubricate all friction surfaces - the spherical bearing surface of the shift lever, the inside of the shift lever socket and the inside of the seat - with silicone grease. Also be sure to lubricate all support and control rod bushings and collars with silicone grease. Finally, make sure you tighten all fasteners to the torque listed in this Chapter's Specifications.

5   Back-up light and neutral position switch - check and replacement

Check

Refer to illustration 5.3

1   Raise the vehicle and place it securely on jackstands. The back-up light switch is located on the lower rear side of the transaxle. The neutral position switch is located on the opposite side of the transaxle. The harness connector for the back-up light switch is on the left side of the transaxle. The harness connector for the neutral switch is at the base of the bellhousing near the rear of the engine.
2   Disconnect the switch electrical connectors.
3   Place the transaxle into Reverse and verify (using an ohmmeter) that there's continuity between the two connector terminals of the back-up light switch harness **(see illustration)**.
4   Verify that there's no continuity between the two terminals in any other gear.
5   If continuity isn't as specified, replace the back up light switch (see below).
6   Put the transaxle into Neutral and verify (using an ohmmeter) that there's continuity between the two connector terminals of the neutral switch harness.
7   Verify that there's no continuity between the two terminals in any other gear.
8   If continuity isn't as specified, replace the neutral position switch (see below).

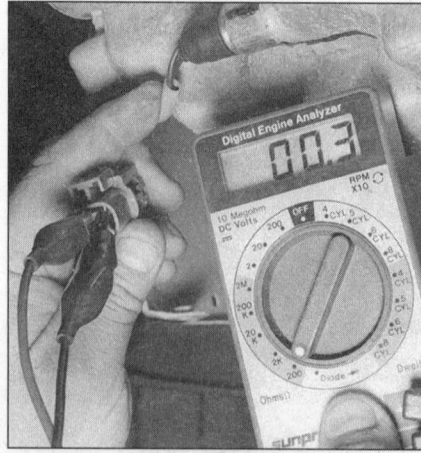

5.3 To check the back-up light switch, disconnect the connector, probe the terminals with an ohmmeter and verify that there's continuity when the transaxle is in Reverse, but no continuity in any other gear (the neutral switch, on the other side of the transaxle, is checked in the same way, except that continuity should exist only in Neutral)

Replacement

9   Drain the transaxle lubricant (see Chapter 1).
10   Disconnect the switch electrical connector. Unscrew the switch.
11   Install the new switch and tighten it securely.
12   Connect the electrical connector.
13   Remove the jackstands and lower the vehicle. Refill the transaxle with lubricant (see Chapter 1).

6   Transaxle - removal and installation

Removal

Refer to illustrations 6.3, 6.7, 6.11a, 6.11b, 6.11c, 6.14, 6.15, 6.17a, 6.17b, 6.18, 6.20a and 6.20b

1   Disconnect the negative cable from the battery.
2   Remove the air intake duct (see Chapter 4).
3   Disconnect the clutch cable **(see illustration)**.

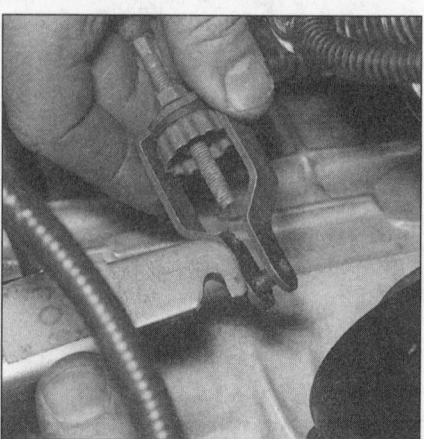

6.3 Free the clutch cable from the release lever and secure it out of the way

Chapter 7 Part A Manual transaxle

7A-5

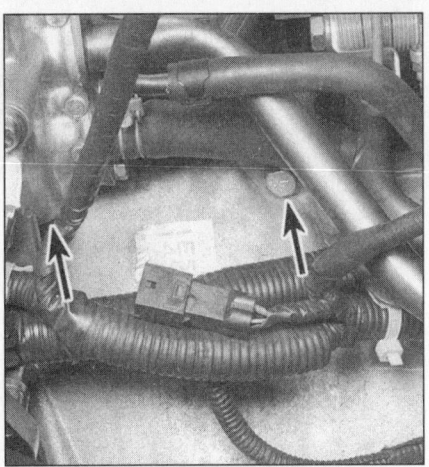

6.7 Upper transaxle-to-engine bolts (arrows)

6.11a Remove the bolt and the washer securing the gear shift linkage support rod to the mounting bracket

6.11b Unscrew the nut, then remove the pivot bolt . . .

4 Disconnect the electrical connector for the vehicle speed sensor.
5 Disconnect the electrical connectors for the back-up light and neutral position switch (see Section 5). Disconnect all ground wires.
6 If the starter motor is easier to remove from above, remove it now (see Chapter 5).
7 Remove the upper transaxle-to-engine bolts **(see illustration)**.
8 Loosen the wheel lug nuts. Raise the vehicle and support it securely on jackstands. Remove the wheels.
9 Remove the section of exhaust pipe underneath the engine/transaxle (see Chapter 4).
10 Drain the transaxle fluid (see Chapter 1).
11 Disconnect the shift support and shift control rods from the transaxle **(see illustration)** (see Section 4).
12 Remove the driveaxles (see Chapter 8).
13 Remove the starter motor, if you haven't already done so (see Chapter 5).
14 Remove the bolts from the engine-to-transaxle braces **(see illustration)**.
15 Support the engine. This can be done from above by using an engine hoist or sup-

6.11c . . . and free the selector rod from the transaxle

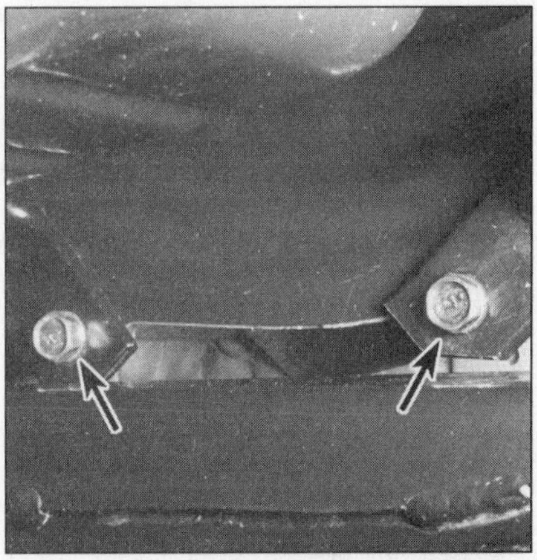

port fixture **(see illustration)**, or by placing a jack (with a block of wood as an insulator) under the engine oil pan. The engine must remain supported at all times while the transaxle is out of the vehicle.
16 Support the transaxle with a jack

6.14 Remove these bolts (arrows) from the engine-to-transaxle braces

(preferably a special jack made for this purpose). Safety chains will help steady the transaxle on the jack.
17 Raise the engine and transaxle slightly and disconnect the transaxle mounts **(see illustrations)**.

7A

6.15 This type of support is the preferred method of supporting the engine. A conventional engine hoist can also be used

6.17a Remove the left-side mount from the transaxle

6.17b The rear mounting bracket must also be removed

6.18 Remove the bolts from around the bellhousing

6.20a After supporting the engine securely, carefully pull the transaxle away from the engine until the input shaft clears the clutch hub

18 Remove the remainder of the bolts securing the transaxle to the engine **(see illustration)**.
19 Make a final check that all wires and hoses have been disconnected from the transaxle.
20 Lower the left (driver's) end of the engine, then roll the transaxle and jack toward the side of the vehicle **(see illustration)**. Once the input shaft is clear of the splines in the clutch hub, lower the transaxle and remove it from under the vehicle **(see illustration)**. Try to keep the transaxle as level as possible.
21 The clutch components can now be inspected (see Chapter 8). In most cases, new clutch components should be routinely installed whenever the transaxle is removed.

Installation

22 If removed, install the clutch components (see Chapter 8).
23 With the transaxle secured to the jack as on removal, raise it into position and then carefully slide it forward, engaging the input shaft with the splines in the clutch hub. Do not use excessive force to install the transaxle - if the input shaft does not slide into place, readjust the angle of the transaxle so it is level and/or turn the input shaft so the splines engage properly with the clutch.
24 Install the transaxle-to-engine bolts. Tighten all engine-to-transaxle bolts to the torque listed in this Chapter's Specifications.
25 Install the transaxle mount nuts and bolts. Tighten all nuts and bolts securely.
26 Remove the jacks supporting the transaxle and the engine.
27 Install the various items removed previously. Refer to Chapter 8 for the installation of the driveaxles and Chapter 4 for information regarding the exhaust system components.
28 Make a final check that all wires and hoses have been connected and that the transaxle has been filled with the specified lubricant to the proper level (see Chapter 1). Lower the vehicle.
29 Connect the negative battery cable. Road test the vehicle to check for proper transaxle operation and check for leakage.

7 Transaxle overhaul - general information

1 Overhauling a manual transaxle is a difficult job for the do-it-yourselfer. It involves the disassembly and reassembly of many small parts. Numerous clearances must be precisely measured and, if necessary, changed with select fit spacers and snap-rings. As a result, if transaxle problems arise, it can be removed and installed by a competent do-it-yourselfer, but overhaul should be left to a transmission repair shop. Rebuilt transaxles may be available - check with your dealer parts department and auto parts stores. At any rate, the time and money involved in an overhaul is almost sure to exceed the cost of a rebuilt unit.
2 Nevertheless, it's not impossible for an inexperienced mechanic to rebuild a transaxle if the special tools are available and the job is done in a deliberate step-by-step manner so nothing is overlooked.
3 The tools necessary for an overhaul include internal and external snap-ring pliers, a bearing puller, a slide hammer, a set of pin punches, a dial indicator and a hydraulic press. In addition, a large, sturdy workbench and a vise or transaxle stand will be required.
4 During disassembly of the transaxle, make careful notes of how each piece comes off, where it fits in relation to other pieces and what holds it in place. Your notes plus the manufacturer's shop manual which contains exploded views will make it much easier to get the transaxle back together.
5 Before taking the transaxle apart for repair, it will help if you have some idea what area of the transaxle is malfunctioning. Certain problems can be closely tied to specific areas in the transaxle, which can make component examination and replacement easier. Refer to the *Troubleshooting* section at the front of this manual for information regarding possible sources of trouble.

6.20b Remove the transaxle from beneath the vehicle. Take care to avoid dropping it - the case can be easily cracked

Chapter 7 Part B
Automatic transaxle

Contents

Section		Section
Automatic transaxle fluid change .... See Chapter 1		Oil seal - replacement .... See Chapter 7A
Automatic transaxle fluid level check .... See Chapter 1		Shift cable - check, adjustment and replacement 3
Automatic transaxle - removal and installation 6		Shift lock system - description, check and component
Diagnosis - general 2		replacement 5
General information 1		
Neutral start/back-up light (inhibitor) switch - check, adjustment and replacement 4		

Specifications

Torque specifications

Ft-lbs (unless otherwise indicated)

Driveplate-to-torque converter bolts
    Bolt with flange.. 33 to 43
    Bolt without flange .... 29 to 36
Transaxle-to-engine bolts
    RL4F03A
        Upper bolts .... 22 to 30
        Lower two bolts .... 144 to 180 in-lbs
        Front gusset-to-engine bolts .... 22 to 30
        Rear gusset-to-engine bolts .... 144 to 180 in-lbs
    RL4F03V and RE4F03V
        Upper bolts .... 51 to 59
        Lower two bolts .... 144 to 180 in-lbs

1 General information

All models covered by this manual are equipped with either a 5-speed manual transaxle or a 4-speed automatic transaxle. All information on the automatic transaxle is included in this Part of Chapter 7. Information for the manual transaxle can be found in Part A of this Chapter.

Because of the complexity of the automatic transaxles and the specialized equipment necessary to perform most service operations, this Chapter contains only those procedures related to general diagnosis, routine maintenance, adjustment and removal and installation. If the transaxle requires major repair work, it should be left to a dealer service department or an automotive or transmission repair shop. You can, however, remove and install the transaxle yourself and save the expense, even if the repair work is done by a transmission shop.

These models use either a RL4F03A or RE4F03V transaxle. These transaxles are very similar and all external procedures are the same for both units.

2 Diagnosis - general

Note: *Automatic transaxle malfunctions may be caused by five general conditions: poor engine performance, improper adjustments, hydraulic malfunctions, mechanical malfunctions or malfunctions in the computer or its signal network. Diagnosis of these problems should always begin with a check of the easily repaired items: fluid level and condition (see Chapter 1), shift linkage adjustment and throttle linkage adjustment. Next, perform a road test to determine if the problem has been corrected or if more diagnosis is necessary. If the problem persists after the preliminary tests and corrections are completed, additional diagnosis should be done by a dealer service department or transmission repair shop. Refer to the Troubleshooting section at the front of this manual for information on symptoms of transaxle problems.*

Preliminary checks

1   Drive the vehicle to warm the transaxle to normal operating temperature.
2   Check the fluid level as described in Chapter 1:
   a) *If the fluid level is unusually low, add enough fluid to bring the level within the designated area of the dipstick, then check for external leaks (see below).*
   b) *If the fluid level is abnormally high, drain off the excess, then check the drained fluid for contamination by coolant. The presence of engine coolant in the automatic transmission fluid indicates that a failure has occurred in the internal radiator walls that separate the coolant from the transmission fluid (see Chapter 3).*
   c) *If the fluid is foaming, drain it and refill the transaxle, then check for coolant in the fluid, or a high fluid level.*

Chapter 7 Part B Automatic transaxle

3.3 Loosen the shift cable locknut

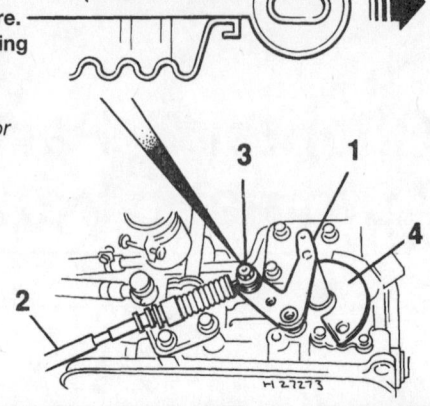

3.4 Shift cable adjustment procedure. Pull the cable end fitting in the direction of the arrow

1. Transaxle selector lever
2. Shift cable
3. Nut
4. Inhibitor switch

3 Check the engine idle speed. **Note:** *If the engine is malfunctioning, do not proceed with the preliminary checks until it has been repaired and runs normally.*
4 Inspect the shift cable (see Section 4). Make sure that it's properly adjusted and operates smoothly.

Fluid leak diagnosis

5 Most fluid leaks are easy to locate visually. Repair usually consists of replacing a seal or gasket. If a leak is difficult to find, the following procedure may help.
6 Identify the fluid. Make sure it's transmission fluid and not engine oil or brake fluid (automatic transmission fluid is a deep red color).
7 Try to pinpoint the source of the leak. Drive the vehicle several miles, then park it over a large sheet of cardboard. After a minute or two, you should be able to locate the leak by determining the source of the fluid dripping onto the cardboard.
8 Make a careful visual inspection of the suspected component and the area immediately around it. Pay particular attention to gasket mating surfaces.
9 A mirror is often helpful for finding leaks in areas that are hard to see.
10 If the leak still cannot be found, clean the suspected area thoroughly with a degreaser or solvent, then dry it.
11 Drive the vehicle for several miles at normal operating temperature and varying speeds. After driving the vehicle, visually inspect the suspected component again.
12 Once the leak has been located, the cause must be determined before it can be properly repaired. If a gasket is replaced but the sealing flange is bent, the new gasket will not stop the leak. The bent flange must be straightened.
13 Before attempting to repair a leak, check to make sure that the following conditions are corrected or they may cause another leak. **Note:** *Some of the following conditions cannot be fixed without highly specialized tools and expertise. Such problems must be referred to a transmission shop or a dealer service department.*

Gasket leaks

14 Check the pan periodically. Make sure the bolts are snug, no bolts are missing, the gasket is in good condition and the pan is flat (dents in the pan may indicate damage to the valve body inside).
15 If the pan gasket is leaking, the fluid level or the vent may be plugged, the pan bolts may be too tight, the pan sealing flange may be warped, the sealing surface of the transaxle housing may be damaged, the gasket may be damaged or the transaxle casting may be cracked or porous. If sealant instead of gasket material has been used to form a seal between the pan and the transaxle housing, it may be the wrong sealant.

Seal leaks

16 If a transaxle seal is leaking, the fluid level may be too high, the vent may be plugged, the seal bore may be damaged, the seal itself may be damaged or improperly installed, the surface of the shaft protruding through the seal may be damaged or a loose bearing may be causing excessive shaft movement.
17 Make sure the dipstick tube seal is in good condition and the tube is properly seated. Periodically check the area around the speedometer gear or sensor for leakage. If transmission fluid is evident, check the O-ring for damage.

Case leaks

18 If the case itself appears to be leaking, the casting is porous and will have to be repaired or replaced.
19 Make sure the oil cooler hose fittings are tight and in good condition.

Fluid comes out vent pipe or fill tube

20 If this condition occurs, the transaxle is overfilled, there is coolant in the fluid, the case is porous, the dipstick is incorrect, the vent is plugged or the drain-back holes are plugged.

3 Shift cable - check, adjustment and replacement

Check

1 Move the shift lever from the "P" position to the "1" position. You should be able to

3.9 To disconnect the shift cable from the manual lever, remove this nut (right arrow); to disconnect the cable from the transaxle, remove these two bracket bolts (arrows)

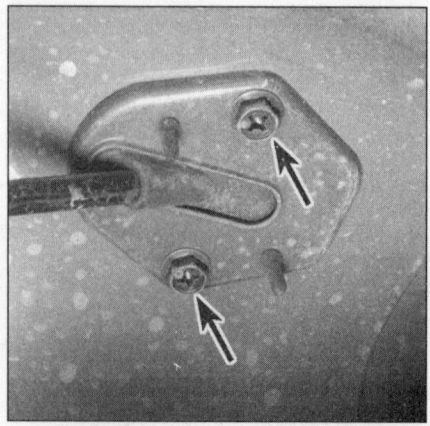

3.10 To disconnect the cable housing from the floorpan, remove these two bolts (arrows)

Chapter 7 Part B Automatic transaxle

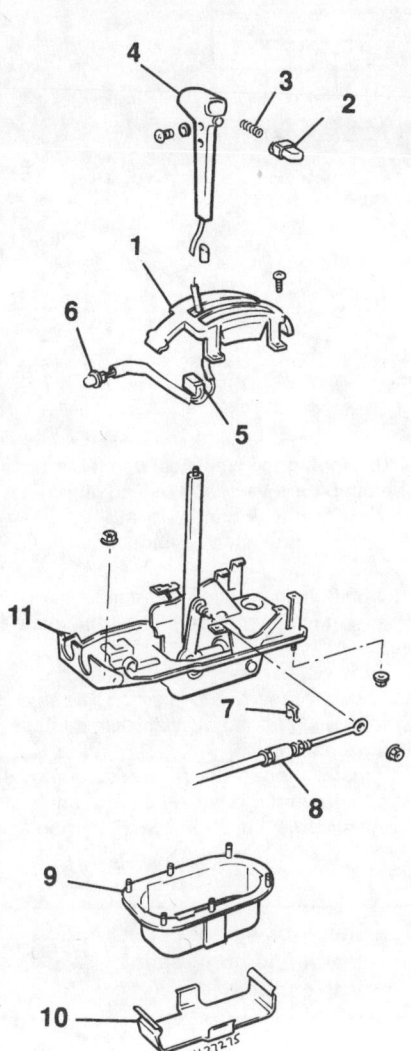

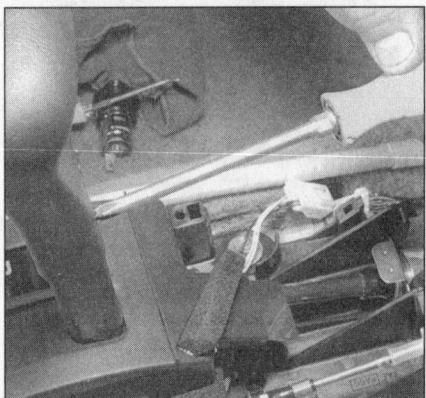

3.11b Remove the Phillips screw in the front of the shift lever handle . . .

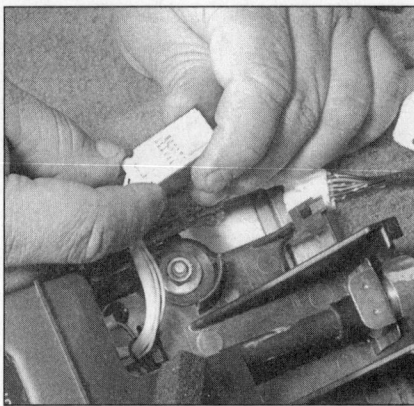

3.11c . . . unplug the electrical connector for the overdrive switch . . .

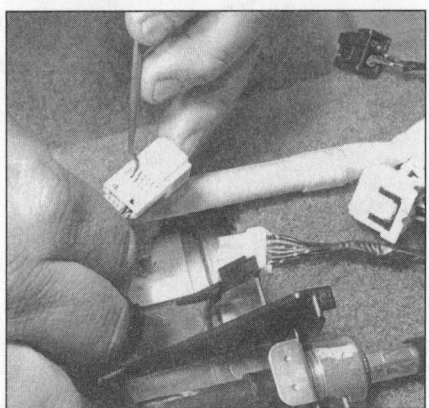

3.11d . . . disengage the white overdrive switch leads from the connector with an awl . . .

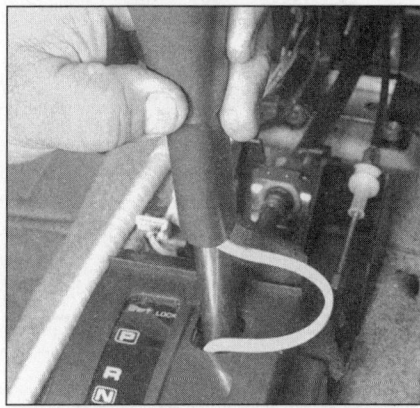

3.11e . . . and carefully pull the shift lever handle straight up to remove it

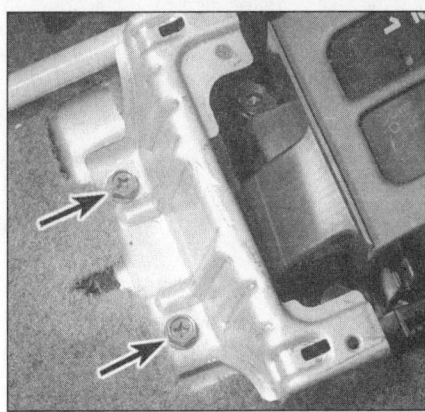

3.11f To remove the shift lever indicator panel, remove these two screws (arrows) and this small console bracket . . .

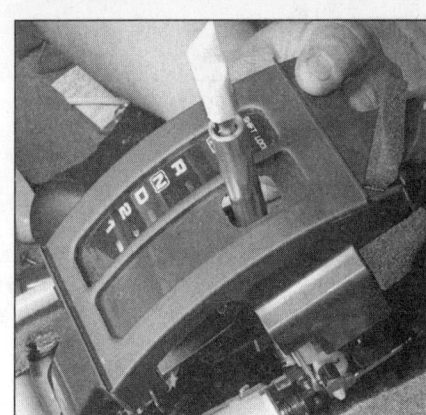

3.11g . . . then pop the plastic retainers loose and lift the indicator panel straight up

3.11a Typical shift lever components

1	Indicator panel	7	Clip
2	Detent button	8	Cable
3	Spring	9	Dust cover
4	Shift lever handle	10	Dust cover support
5	Wiring connector	11	Shift lever assembly
6	Gear indicator bulb		

feel the detents in each range. If you can't feel the detents, or if the pointer indicating the ranges is incorrectly aligned, adjust the shift cable.

Adjustment

Refer to illustrations 3.3 and 3.4

2 Raise the vehicle and support it securely on jackstands.
3 Loosen the shift cable locknut **(see illustration)** on the transaxle and place the manual lever in the "P" position.
4 Adjust the cable by pulling outwards as far as possible **(see illustration)**.

5 Return the cable in the opposite direction 1/32-inch to 3/64-inch.
6 Tighten the shift cable locknut.
7 Move the shift lever from the "P" position to the "1" position and make sure that the shift lever moves smoothly and without any sliding noise.
8 Apply silicone grease to the contact areas of the shift lever and the shift cable.

Replacement

Refer to illustrations 3.9, 3.10, 3.11a through 3.11g, 3.12a and 3.12b

9 At the transaxle end, loosen the locknut and disconnect the cable from the manual lever and disconnect the cable bracket from the transaxle **(see illustration)**.
10 Disconnect the cable housing from the floorpan **(see illustration)**.
11 Remove the shift lever housing **(see illustrations)**.

7B-4 Chapter 7 Part B Automatic transaxle

3.12a To disconnect the cable from the shift lever, remove this nut

3.12b To detach the cable from the shift lever base, pry off this clip

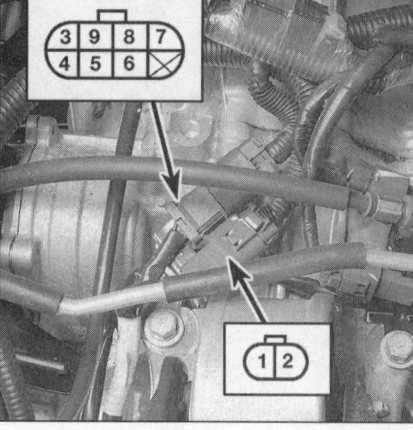

4.1a Unplug the electrical connectors for the inhibitor switch and use an ohmmeter to check continuity at the indicated terminals

12 Disconnect the cable from the shift lever **(see illustrations)**.
13 Installation is the reverse of removal.
14 Be sure to adjust the cable when you're done.

4 Neutral start/back-up light (inhibitor) switch - check, adjustment and replacement

Check
Refer to illustrations 4.1a and 4.1b

1 Disconnect the switch electrical connector and check continuity between terminals 1 and 2, and between terminals 3 and 4, 5, 6, 7, 8 and 9 while moving the shift lever through each range. Compare your results to the accompanying table **(see illustrations)**.
  If the switch fails any of these continuity checks, disconnect the shift cable from the manual lever and retest the switch.

  a) If the switch passes all the continuity checks this time, reconnect the shift cable and adjust it, then retest the switch.
  b) If the switch still fails any of the continuity tests, remove it from the transaxle and try testing it again on the bench.
  c) If the switch passes all the continuity tests on the bench, install it and adjust it (see below).
  d) If the switch still fails any of the continuity tests, replace it (see below).

Adjustment
Refer to illustration 4.2

2 Disconnect the shift cable from the manual lever, loosen the switch retaining screws, set the manual lever at the Neutral position and insert a 5/32-inch (4 mm) pin or drill bit into the adjustment holes in both the switch and the lever **(see illustration)**. Make sure the pin is perpendicular to the switch and the lever. Tighten the screws securely.
3 Recheck the switch continuity. If switch continuity is not as specified, replace the switch.

Replacement
Refer to illustration 4.8

4 Disconnect the negative cable from the battery.
5 Shift the transaxle into Neutral.
6 Remove the nut and lift off the manual lever.
7 Disconnect the electrical connector.
8 Remove the switch retaining screws and lift the switch off the manual lever shaft **(see illustration)**.
9 Installation is the reverse of removal. Don't tighten the retaining screws until you have adjusted the switch as described in Step 5.

5 Shift lock system - description, check and component replacement

Description

1 The shift lock system prevents the shift lever from being shifted out of Park or Neutral until the brake pedal is applied. The system consists of a shift lock control unit, a shift lock solenoid, a detention switch and a brake light switch. Other than the following simple

Inhibitor switch terminals

	1	2	3	4	5	6	7	8	9
P	x—x		x—x						
R			x		x				
N	x—x		x			x			
D			x				x		
2			x					x	
1			x						x

4.1b Inhibitor switch continuity chart

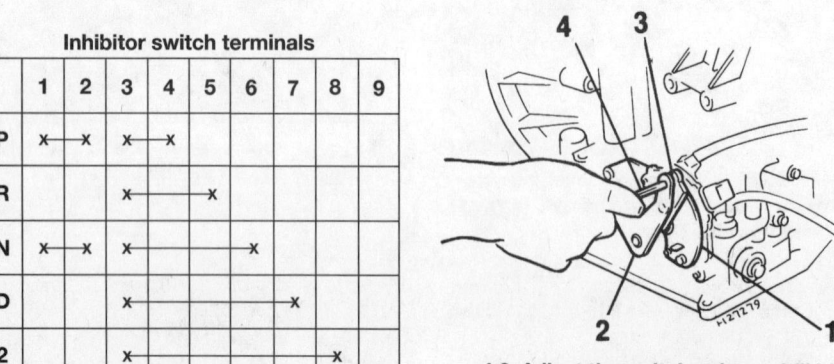

4.2 Adjust the switch using a drill bit as described in the text
1 Inhibitor switch
2 Shift lever
3 Adjustment holes
4 Drill bit

4.8 To remove the inhibitor switch, remove the nut that attaches the manual lever to the switch (arrow), then remove the switch retaining screws (arrows)

Chapter 7 Part B Automatic transaxle   7B-5

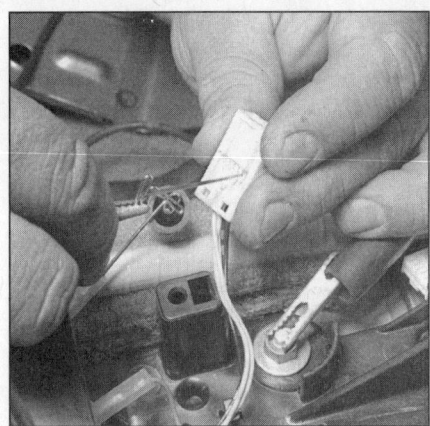

5.3a To test the shift lock solenoid, apply battery voltage to the shift lock solenoid and verify that there's an audible "click" from the solenoid

component checks, diagnosis of the shift lock system should be left to a dealer service department or other qualified repair shop.

Check

Shift lock solenoid

Refer to illustrations 5.3a and 5.3b

2   Remove the floor console (see Chapter 11).
3   Disconnect the shift lock solenoid connector. Apply battery voltage to the shift lock solenoid **(see illustrations)** and verify that there's an audible "click" from the solenoid.
4   If the shift lock solenoid doesn't click when energized by the battery, replace it.

Park position switch

Refer to illustration 5.5

5   Check continuity between the park position switch terminal of the shift lock electrical connector **(see illustration)** while moving the shift lever through the following positions:
  a) When the shift lever is in the Park position with the shift lever button is released, there should be no continuity.
  b) When the shift lever is in any position except the above, there should be continuity.

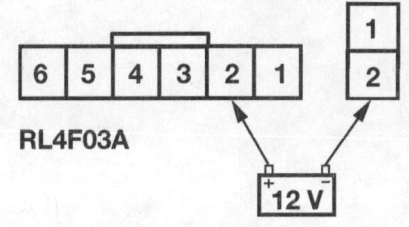

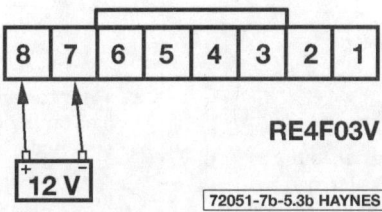

5.3b Depending on model and connector type, apply battery voltage to the shift lock solenoid at the indicated terminals

Brake light switch

6   To test the brake light switch, check continuity between terminals 1 and 2 of the brake light switch electrical connector. When the brake pedal is depressed, there should be continuity; when the brake pedal is released, there should be no continuity.

Component replacement

Shift lock solenoid/park position switch

7   Remove the shift lever knob (refer to Section 3).
8   Remove the center console (see Chapter 11).
9   Remove the position indicator (refer to Section 3).
10 Disconnect the electrical connector for the shift lock solenoid/park position switch.
11 Remove the shift lock solenoid/park position switch.
12 Installation is the reverse of removal.

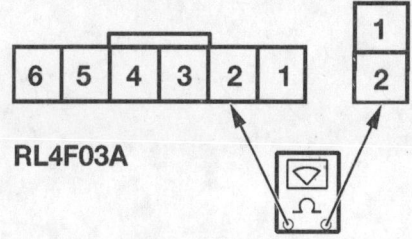

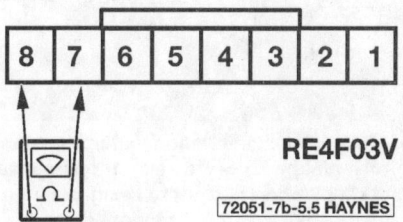

5.5 Depending on model and connector type, check for continuity at the indicated terminals to test the park position switch

Brake light switch

13 Refer to Chapter 9.

6   Automatic transaxle - removal and installation

Removal

Refer to illustrations 6.3a, 6.3b and 6.14

1   Detach the cable from the negative battery terminal.
2   Disconnect the inhibitor switch electrical connector (see Section 4).
3   Disconnect the vehicle speed sensor electrical connector and all other interfering wiring **(see illustrations)**.
4   Remove the starter (see Chapter 5).
5   Remove the upper transaxle-to-engine bolts.
6   Loosen the wheel lug nuts.

6.3a Unplug all of the electrical connectors (arrows) on top of the transaxle; also detach the vent hose (far right arrow)

6.3b Unplug the electrical connector and remove the crankshaft position sensor so that it doesn't become damaged during the removal of the transaxle

7B

Chapter 7 Part B Automatic transaxle

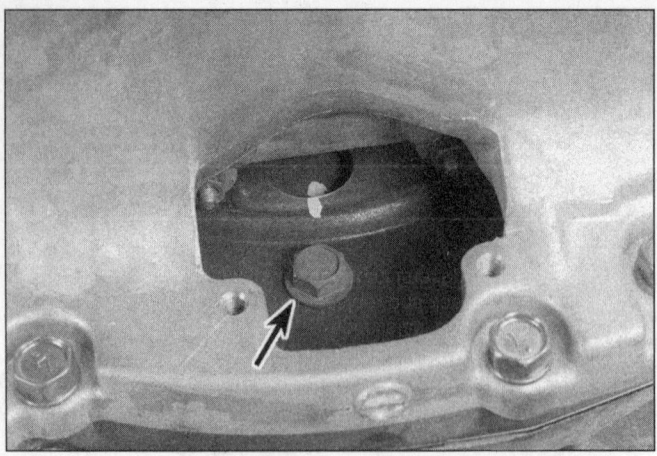

6.14 Mark the relationship of the torque converter to the driveplate with white paint, then remove the bolts (arrow) by rotating the crankshaft to bring all of the bolts into the access window in turn

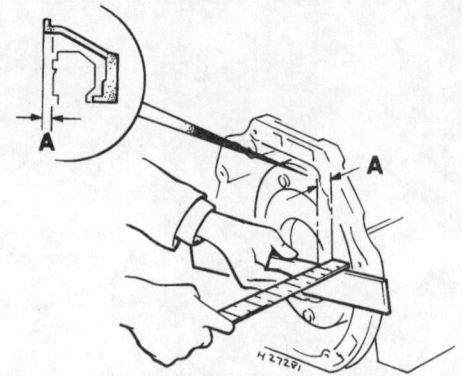

6.22 Prior to installing the transaxle, make sure that the torque converter is fully seated into the transaxle by checking the distance from the converter bolt hole to the transaxle mating surface (A)

GA16 models should be at least 7/8 inch
SR20 models should be at least 5/8 inch

7   Raise the vehicle and support it securely on jackstands. Remove the wheels.
8   Drain the transaxle fluid (see Chapter 1).
9   Disconnect the oil cooler hoses. Plug the hoses to keep out dirt and moisture.
10 Remove the dipstick tube.
11 Disconnect the shift cable from the transaxle (see Section 3).
12 Remove the driveaxles (see Chapter 8).
13 Remove the torque converter cover.
14 Mark the relationship of the torque converter to the driveplate so they can be installed in the same position **(see illustration)**.
15 Remove the torque converter mounting bolts. Turn the crankshaft for access to each one in turn.
16 Support the engine using a hoist from above, or a jack and a wood block under the oil pan to spread the load (refer to Chapter 7A).
17 Support the transaxle with a jack - preferably a special jack made for this purpose. Safety chains will help steady the transaxle on the jack.
18 Remove the mounting bolts from the transaxle mounting brackets.

19 Remove the lower engine-to-transaxle bolts.
20 Lower the transaxle slightly and disconnect and plug the transaxle cooler lines.
21 Move the transaxle to the side to disengage it from the engine block dowel pins and make sure the torque converter is detached from the driveplate. Secure the torque converter to the transaxle so that it will not fall out during removal. Lower the transaxle from the vehicle.

Installation

Refer to illustration 6.22

22 Make sure that the torque converter is completely engaged in the transaxle prior to installation **(see illustration)**.
23 With the transaxle secured to the jack, raise it into position. Be sure to keep it level so the torque converter does not slide forward. Connect the cooler lines.
24 Move the transaxle carefully into place until the dowel pins are engaged and the torque converter is engaged.
25 Turn the torque converter to align the bolt holes with the holes in the driveplate.

The match marks on the torque converter and driveplate, made during step 14, must line up.
26 Install the transaxle-to-engine bolts and nuts. Tighten them to the torque listed in this Chapter's Specifications.
27 Install the torque converter-to-driveplate bolts. Tighten them to the torque listed in this Chapter's Specifications.
28 Install the transaxle/engine mounting bracket bolts and nuts. Tighten them securely.
29 Remove the jacks supporting the transaxle and the engine.
30 Install the dipstick tube.
31 Install the starter.
32 Connect the shift cable.
33 Plug in the electrical connectors for the inhibitor switch and the vehicle speed sensor.
34 Install the torque converter cover.
35 Connect the driveaxles to the transaxle (see Chapter 8).
36 Adjust the shift cable (see Section 3).
37 Lower the vehicle.
38 Fill the transaxle with the proper type and amount of fluid (see Chapter 1). Run the vehicle and check for fluid leaks.

Chapter 8
Clutch and driveaxles

Contents

	Section		Section
Clutch cable - removal and installation	3	Driveaxle boot check	See Chapter 1
Clutch components - removal, inspection and installation	4	Driveaxle boot replacement	10
Clutch - description and check	2	Driveaxles - general information and inspection	7
Clutch pedal and release lever freeplay check and adjustment	See Chapter 1	Driveaxles - removal and installation	8
		Flywheel - removal and installation	See Chapter 2
Clutch release bearing and lever - removal, inspection and installation	5	General information	1
		Oil seal - replacement	See Chapter 7A
Clutch start switch - check and replacement	6	Support bearing assembly - removal, overhaul and installation	9

Specifications

CV joint length (dimension "L" in illustration 10.10a)
Inner
    Tripod type .... 4.0 inches
    Ball-and-cage type .... 3-7/8 inches
Outer .... 3-11/16 inches

Torque specifications   **Ft-lbs** (unless otherwise indicated)
Clutch pressure plate bolts
    Step 1 .... 84 to 168 in-lbs
    Step 2 .... 16 to 22
Driveaxle/hub nut .... 145 to 202
Support bearing bolts
    Inner and middle bolts .... 19 to 26
    Outer bolt .... 22 to 30

8-2 Chapter 8 Clutch and driveaxles

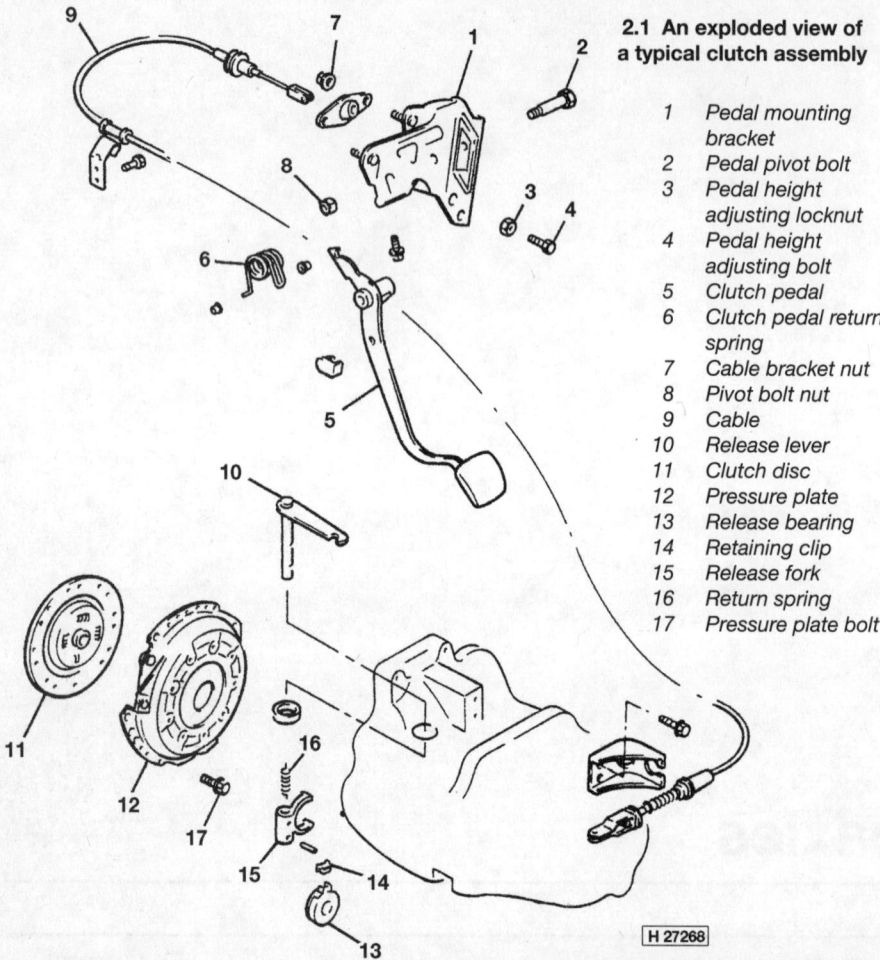

2.1 An exploded view of a typical clutch assembly

1 Pedal mounting bracket
2 Pedal pivot bolt
3 Pedal height adjusting locknut
4 Pedal height adjusting bolt
5 Clutch pedal
6 Clutch pedal return spring
7 Cable bracket nut
8 Pivot bolt nut
9 Cable
10 Release lever
11 Clutch disc
12 Pressure plate
13 Release bearing
14 Retaining clip
15 Release fork
16 Return spring
17 Pressure plate bolt

1 General information

The information in this Chapter deals with the components from the rear of the engine to the front wheels, except for the transaxle, which is dealt with in Chapter 7A and 7B. For the purposes of this Chapter, these components are grouped into two categories: Clutch and driveaxles. Separate Sections within this Chapter offer general descriptions and checking procedures for both groups.

Since nearly all the procedures covered in this Chapter involve working under the vehicle, make sure it's securely supported on sturdy jackstands or a hoist where the vehicle can be easily raised and lowered.

2 Clutch - description and check

Refer to illustration 2.1

1 All vehicles with a manual transaxle use a single dry plate, diaphragm spring type clutch **(see illustration)**. The clutch disc has a splined hub which allows it to slide along the splines of the transaxle input shaft. The clutch and pressure plate are held in contact by spring pressure exerted by the diaphragm in the pressure plate.

2 The clutch release system is cable-operated. The release system consists of the clutch pedal, the clutch cable, the clutch release lever and the clutch release (or throw-out) bearing **(see illustration 2.1)**.

3 When pressure is applied to the clutch pedal to release the clutch, the cable transmits this movement to the clutch release lever. As the lever pivots, the shaft fingers push against the release bearing. The bearing pushes against the fingers of the diaphragm spring of the pressure plate assembly, which in turn releases the clutch plate.

4 Terminology can be a problem regarding the clutch components because common names have in some cases changed from that used by the manufacturer. For example, the driven plate is also called the clutch plate or disc, the pressure plate assembly is sometimes referred to as the clutch cover, and the clutch release bearing is sometimes called a throw-out bearing.

5 Other than replacing components that have obvious damage, some preliminary checks should be performed to diagnose a clutch system failure.

a) The first check should be the freeplay of the clutch pedal (see Chapter 1). After adjusting the cable, retest the clutch operation.

b) To check "clutch spin down time," run the engine at normal idle speed with the transaxle in Neutral (clutch pedal up - engaged). Disengage the clutch (pedal down), wait several seconds and shift the transaxle into Reverse. No grinding noise should be heard. A grinding noise would most likely indicate a problem in the pressure plate or the clutch disc.

c) To check for complete clutch release, run the engine (with the parking brake applied to prevent movement) and hold the clutch pedal approximately 1/2-inch from the floor. Shift the transaxle between 1st gear and Reverse several times. If the shift is not smooth, component failure is indicated.

d) Visually inspect the clutch pedal bushing at the top of the clutch pedal to make sure there is no sticking or excessive wear.

e) Under the vehicle, check that the clutch release lever is solidly mounted on the ball stud.

3 Clutch cable - removal and installation

Refer to illustrations 3.1, 3.2 and 3.3

1 Open the hood and loosen the clutch cable at the release lever by backing off the cable adjuster knob **(see illustration)**.
2 Disconnect the cable from the release lever **(see illustration)**.
3 Disconnect the cable housing from the firewall **(see illustration)**.
4 Remove the dash lower trim panel (see Chapter 11).
5 Disconnect the cable from the clutch pedal.
6 Installation is the reverse of removal. Adjust the clutch cable by following the procedure in Chapter 1, Section 10.

4 Clutch components - removal, inspection and installation

Warning: *Dust produced by clutch wear and deposited on clutch components may contain asbestos, which is hazardous to your health. DO NOT blow it out with compressed air and DO NOT inhale it. DO NOT use gasoline or petroleum-based solvents to remove the dust. Brake system cleaner should be used to flush the dust into a drain pan. After the clutch components are wiped clean with a rag, dispose of the contaminated rags and cleaner in a labeled, covered container.*

Removal

Refer to illustrations 4.5, 4.7a and 4.7b

1 Access to the clutch components is normally accomplished by removing the transaxle, leaving the engine in the vehicle. If, of course, the engine is being removed for major overhaul, then the opportunity should always be taken to check the clutch for wear

Chapter 8 Clutch and driveaxles   8-3

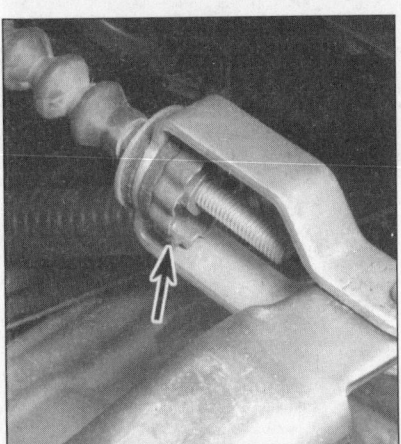

3.1 Loosen the clutch cable adjuster at the release lever

3.2 Disconnect the clutch cable from the release lever

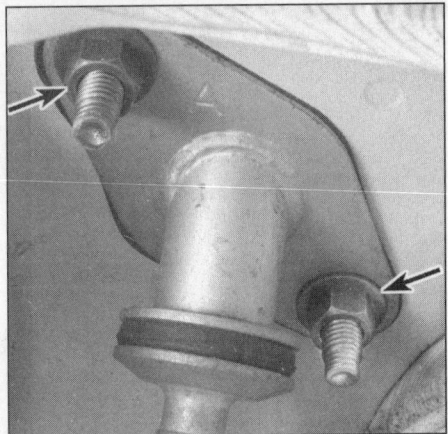

3.3 To disconnect the clutch cable from the firewall, remove these nuts (arrows)

4.5 Mark the relationship of the pressure plate to the flywheel (in case you're going to re-use the same pressure plate)

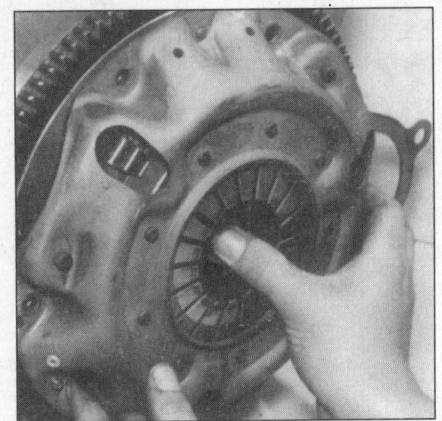

4.7a Remove the pressure plate . . .

4.7b . . . and the clutch disc

and replace worn components as necessary. However, the relatively low cost of the clutch components compared to the time and labor involved in gaining access to them warrants their replacement any time the engine or transaxle is removed, unless they are new or in near-perfect condition. The following procedures assume that the engine will stay in place.

2   Remove the transaxle from the vehicle (see Chapter 7A). Support the engine while the transaxle is out. Preferably, an engine hoist should be used to support it from above. However, if a jack is used underneath the engine, make sure a piece of wood is used between the jack and oil pan to spread the load. **Caution:** *The pick-up for the oil pump is very close to the bottom of the oil pan. If the pan is bent or distorted in any way, engine oil starvation could occur.*

3   The release fork and release bearing can remain attached to the transaxle for the time being.

4   To support the clutch disc during removal, install a clutch alignment tool through the clutch disc hub.

5   Carefully inspect the flywheel and pressure plate for indexing marks. The marks are usually an X, an O or a white letter. If they cannot be found, scribe marks yourself so the pressure plate and the flywheel will be in the same alignment during installation **(see illustration)**.

6   Slowly loosen the pressure plate-to-flywheel bolts. Work in a diagonal pattern and loosen each bolt a little at a time until all spring pressure is relieved.

7   Hold the pressure plate securely and completely remove the bolts, followed by the pressure plate and clutch disc **(see illustrations)**.

Inspection

Refer to illustrations 4.10, 4.12a and 4.12b

8   Ordinarily, when a problem occurs in the clutch, it can be attributed to wear of the clutch driven plate assembly (clutch disc). However, all components should be inspected at this time.

9   Inspect the flywheel for cracks, heat checking, score marks and other damage. If the imperfections are slight, a machine shop can resurface it to make it flat and smooth. Refer to Chapter 2 for the flywheel removal procedure.

10   Inspect the lining on the clutch disc. There should be at least 1/16-inch of lining above the rivet heads. Check for loose rivets,

4.10 Examine the clutch disc for evidence of excessive wear, such as smeared friction material, chewed-up rivets, worn hub splines and distorted damper cushions or springs

distortion, cracks, broken springs and other obvious damage **(see illustration)**. As mentioned above, ordinarily the clutch disc is replaced as a matter of course, so if in doubt about the condition, replace it with a new one.

11   The release bearing should be replaced along with the clutch disc (see Section 5).

8-4   Chapter 8 Clutch and driveaxles

NORMAL FINGER WEAR

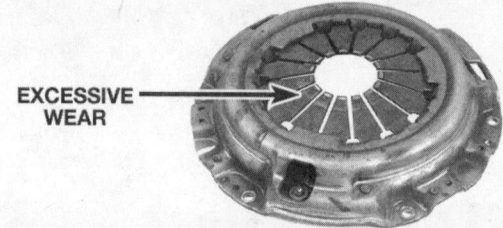

EXCESSIVE WEAR

EXCESSIVE FINGER WEAR

BROKEN OR BENT FINGERS

4.12a Replace the pressure plate if any of these conditions are noted

4.12b Examine the pressure plate friction surface for score marks, cracks and evidence of overheating (blue spots)

4.14 Center the clutch disc in the pressure plate with a clutch alignment tool, then tighten the pressure plate-to-flywheel bolts

12   Check the machined surface and the diaphragm spring fingers of the pressure plate **(see illustrations)**. If the surface is grooved or otherwise damaged, replace the pressure plate assembly. Also check for obvious damage, distortion, cracking, etc. Light glazing can be removed with emery cloth or sandpaper. If a new pressure plate is indicated, new or factory rebuilt units are available.

Installation

Refer to illustration 4.14

13   Before installation, carefully wipe the flywheel and pressure plate machined surfaces clean. It's important that no oil or grease is on these surfaces or the lining of the clutch disc. Handle these parts only with clean hands.
14   Position the clutch disc and pressure plate with the clutch held in place with an alignment tool **(see illustration)**. Make sure it's installed properly (most replacement clutch plates will be marked "flywheel side" or something similar - if not marked, install the clutch disc with the damper springs or cushion toward the transaxle).
15   Tighten the pressure plate-to-flywheel bolts only finger tight, working around the pressure plate.
16   Center the clutch disc by ensuring the alignment tool is through the splined hub and into the recess in the crankshaft. Wiggle the tool up, down or side-to-side as needed to bottom the tool. Tighten the pressure plate-to-flywheel bolts a little at a time, working in a crisscross pattern to prevent distortion of the cover. After all of the bolts are snug, tighten them to the torque listed in this Chapter's Specifications. Remove the alignment tool.
17   Using high-temperature grease, lubricate the inner groove of the release bearing (see Section 5). Also place grease on the release lever contact areas and the transaxle input shaft bearing retainer.
18   Install the clutch release bearing (see Section 5).
19   Install the transaxle, clutch release cable and all components removed previously, tightening all fasteners to the proper torque specifications.

5   Clutch release bearing and lever - removal, inspection and installation

Warning: *Dust produced by clutch wear and deposited on clutch components may contain asbestos, which is hazardous to your health. DO NOT blow it out with compressed air and DO NOT inhale it. DO NOT use gasoline or petroleum-based solvents to remove the dust. Brake system cleaner should be used to flush it into a drain pan. After the clutch components are wiped clean with a rag, dispose of the contaminated rags and cleaner in a labeled, covered container.*

Removal

Refer to illustrations 5.2a, 5.2b and 5.3

1   Remove the transaxle (see Chapter 7).
2   Note how the return spring engages the release fork. Disconnect the spring **(see illustration)**. Grasp the release bearing firmly and pull it off the input shaft **(see illustration)**.
3   Rotate the release lever and release fork so that the roll pins are pointing toward the cavity in the transaxle housing behind the release fork, then drive out the roll pins with a small punch **(see illustration)**. Separate the clutch release lever from the release fork.

Inspection

Refer to illustration 5.4

4   Hold the bearing by the outer race and rotate the inner race while applying pressure **(see illustration)**. If the bearing doesn't turn smoothly or if it's noisy, replace the bearing/hub assembly with a new one. Wipe the bearing with a clean rag and inspect it for

Chapter 8 Clutch and driveaxles

5.2a Disengage the return spring from the release bearing

5.2b To remove the release bearing, grasp it firmly and pull it straight off the input shaft

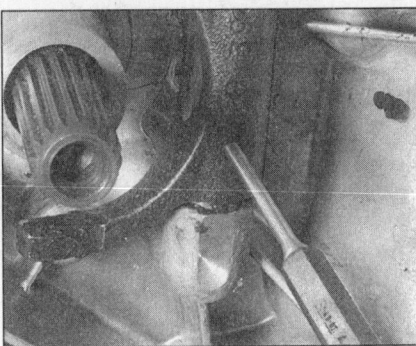

5.3 Before you can remove the release lever and the release fork, you'll have to knock out the two roll pins that lock the release lever and fork together; make sure you angle the pins so they're pointing toward the cavity in the clutch housing

5.4 To check the condition of the bearing, hold it by the outer race and rotate the inner race while applying pressure - the bearing should turn smoothly - if it doesn't, replace it

5.7 Install the two roll pins to lock the release fork to the release lever; make sure the return spring is installed properly as shown

5.8 This is how the release bearing should look when the retainers (spring clips) and the return spring are properly engaged

damage, wear and cracks. Don't immerse the bearing in solvent - it's sealed for life and to do so would ruin it. Also check the release lever for cracks and bends.

Installation

Refer to illustrations 5.7 and 5.8

5   Fill the inner groove of the release bearing with high-temperature grease. Also apply a light coat of the same grease to the transaxle input shaft splines.

6   Lubricate the release lever bearing surfaces (where the lever seats and where it goes through the transaxle bellhousing), the tips of the release fork (where it contacts the release bearing) and the bearing retainers with high-temperature grease.

7   Install the release lever and slide on the return spring and release fork **(see illustration)**. Line up the holes in the release fork and the release lever and install the roll pins.

8   Slide the release bearing onto the transaxle input shaft and engage the retainers (spring clips) and return spring as shown **(see illustration)**.

9   Apply a light coat of high-temperature grease to the face of the release bearing where it contacts the pressure plate diaphragm fingers.

10   The remainder of installation is the reverse of the removal procedure.

6   Clutch start switch - check and replacement

Refer to illustration 6.1

1   The clutch start switch **(see illustration)** is located at the top of the clutch pedal.

2   Verify that the engine will not start when the clutch pedal is released.

3   Verify that the engine will start when the clutch pedal is depressed all the way.

4   If the clutch start switch doesn't perform as described above, adjust the clutch pedal (see Chapter 1). The switch should now operate properly. If it doesn't, check switch continuity.

5   Verify that there is continuity between the clutch start switch terminals when the switch is On (pedal depressed).

6   Verify that no continuity exists between the switch terminals when the switch is Off (pedal released).

6.1 The clutch start switch is located at the top of the clutch pedal; to replace it, unplug the electrical connector (arrow), loosen the locknut (above the bracket) and unscrew the switch

7   If the switch fails either of these continuity tests, replace it: loosen the nut near the body of the switch, then unscrew the switch. Unplug the electrical connector. Installation is the reverse of removal.

Chapter 8 Clutch and driveaxles

1. Outer CV joint assembly
2. Boot clamps
3. CV joint boot
4. Circlip
5. Axleshaft
6. Boot clamps
7. Inner CV joint boot
8. Retaining ring
9. Snap-ring
10. Inner race
11. Cage
12. Ball
13. Snap-ring
14. Inner CV joint housing (left driveaxle)
15. Dust shield (left driveaxle)
16. Circlip (left driveaxle)
17. Inner CV joint housing and extension shaft (right driveaxle)
18. Snap-ring (right driveaxle)
19. Dust shield (right driveaxle)
20. Support bearing (right driveaxle)
21. Support bearing retainer (right driveaxle)
22. Bracket (right driveaxle)
23. Snap-ring
24. Dust shield

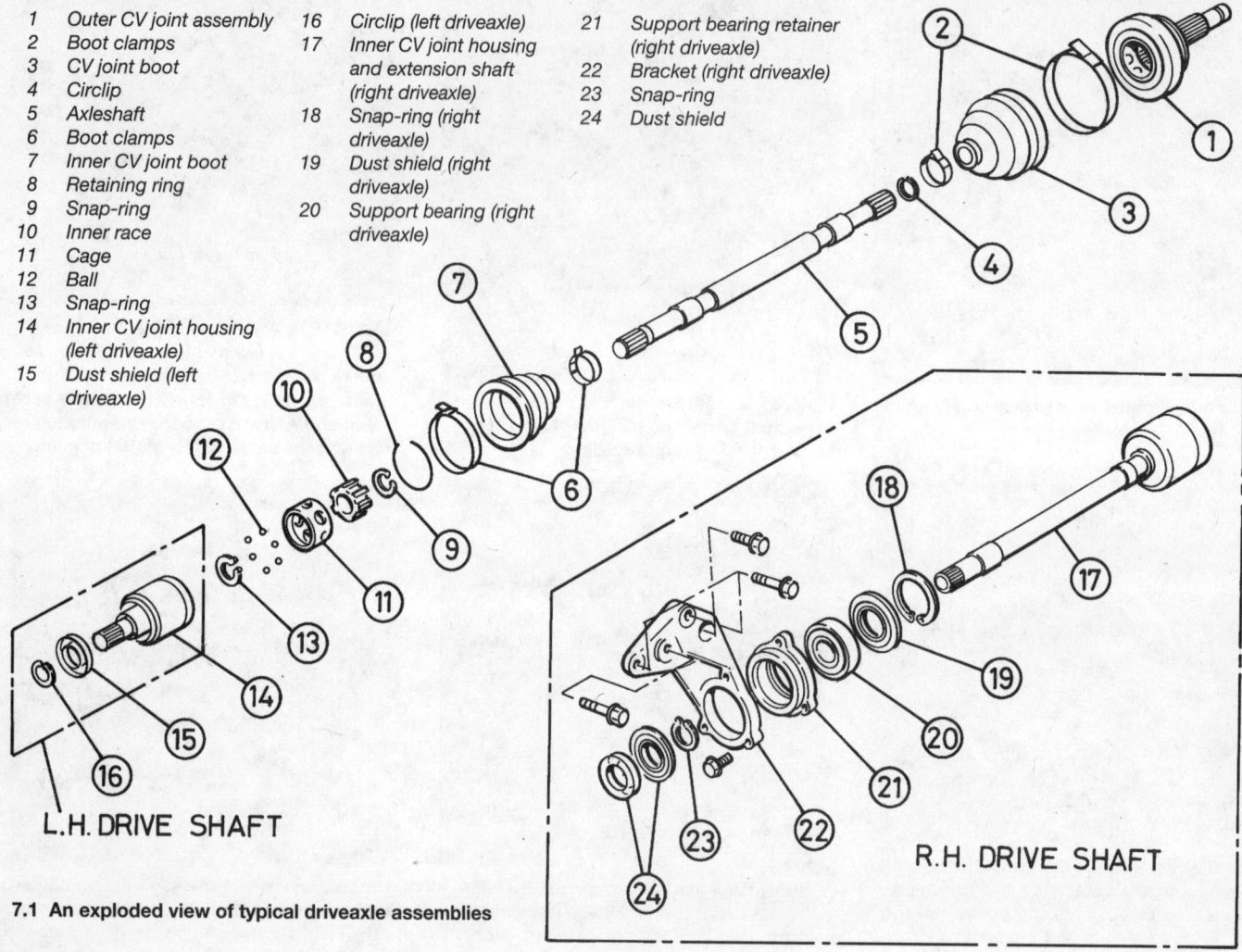

7.1 An exploded view of typical driveaxle assemblies

8 Adjust the clutch pedal (see Chapter 1).
9 Verify that the engine doesn't start when the clutch pedal is released.

7 Driveaxles - general information and inspection

Refer to illustration 7.1

1 Power is transmitted from the transaxle to the wheels through a pair of driveaxles **(see illustration)**. The inner end of each driveaxle is splined into the differential side gears. The outer ends of the driveaxles are splined to the axle hubs and locked in place by a large nut.
2 The inner ends of the driveaxles are equipped with sliding constant velocity joints, which are capable of both angular and axial motion. Some inner joints consist of a tripod bearing and a joint housing (outer race) in which the joint is free to slide in and out as the driveaxle moves up and down with the wheel. These joints can be disassembled and cleaned in the event of a boot failure (see Section 10), but if any parts are damaged, the joints must be replaced as a unit. Other inner joints are the "double-offset" type, which consists of ball bearings running between an inner race and an outer cage. They can also be disassembled, cleaned and inspected, but they too must be replaced as a single unit if defective.
3 The outer CV joints are the "Rzeppa" (pronounced "sheppa") or "Birfield" type, which also consist of ball bearings running between an inner race and an outer cage. However, Rzeppa/Birfield joints are capable of angular - but not axial - movement. These outer joints should be cleaned, inspected and repacked whenever replacing an outer CV joint boot, but they cannot be disassembled. If an outer joint is damaged, it must be replaced.
4 The boots should be inspected periodically for damage and leaking lubricant. Torn CV joint boots must be replaced immediately or the joints can be damaged. Boot replacement involves removal of the driveaxle. **Note:** *Some auto parts stores carry "split" type replacement boots, which can be installed without removing the driveaxle from the vehicle. This is a convenient alternative; however,* the driveaxle should be removed and the CV joint disassembled and cleaned to ensure the joint is free from contaminants such as moisture and dirt which will accelerate CV joint wear. The most common symptom of worn or damaged CV joints, besides lubricant leaks, is a clicking noise in turns, a clunk when accelerating after coasting and vibration at highway speeds. To check for wear in the CV joints and driveaxle shafts, grasp each axle (one at a time) and rotate it in both directions while holding the CV joint housings, feeling for play indicating worn splines or sloppy CV joints. Also check the driveaxle shafts for cracks, dents and distortion.

8 Driveaxles - removal and installation

Removal

Refer to illustrations 8.4, 8.5, 8.9, 8.10a and 8.10b

1 Disconnect the cable from the negative terminal of the battery.
2 Set the parking brake.

Chapter 8 Clutch and driveaxles   8-7

8.4 Remove the cotter pin and the nut lock

8.5 Use a large prybar to immobilize the hub while loosening the driveaxle/hub nut

8.9 Pull the steering knuckle out and slide the end of the driveaxle out of the hub

8.10a To separate the inner end of the driveaxle from the transaxle, pry on the CV joint housing like this with a large screwdriver or prybar (you may need to give the prybar a sharp rap with a hammer)

8.10b If you can't pry the left driveaxle out on a model equipped with an automatic transaxle, remove the right driveaxle, insert a screwdriver through the differential side gears and tap the left driveaxle out

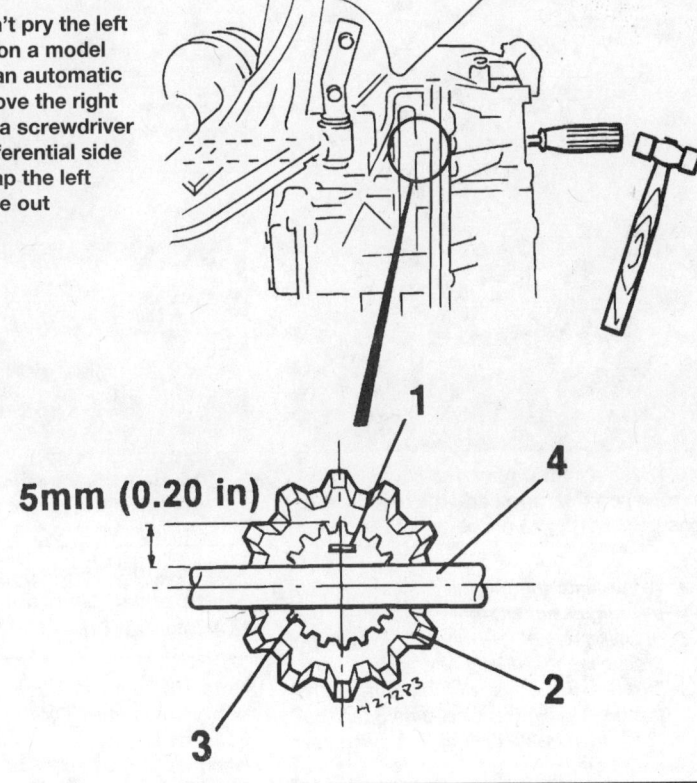

3   Loosen the front wheel lug nuts, raise the vehicle and support it securely on jackstands. Remove the wheel.
4   Remove the cotter pin and the bearing nut lock from the driveaxle/hub nut (see illustrations).
5   Remove the driveaxle/hub nut and washer. To prevent the hub from turning, wedge a prybar between two of the wheel studs and allow the prybar to rest against the ground or the floorpan of the vehicle (see illustration).
6   To loosen the driveaxle from the hub splines, tap the end of the driveaxle with a soft-faced hammer or a hammer and a brass punch.
7   Remove the engine splash shields (see Chapter 1). Place a drain pan underneath the transaxle to catch the lubricant that may spill out when the driveaxles are removed.
8   Disconnect the control arm from the steering knuckle (see Chapter 10)
9   Pull out on the steering knuckle and detach the driveaxle from the hub (see illustration). If the driveaxle sticks in the hub, remove the brake disc (see Chapter 9) and push the driveaxle through the hub with a puller.

10   Carefully pry the inner CV joint out of the transaxle (see illustration) and remove the driveaxle assembly. On some models, the inner CV joint housing on the right (passenger side) driveaxle terminates at a support bearing instead of the transaxle. The inner side of the housing has an extended shaft - instead of a short stub axle - that goes through the support bearing assembly and engages the right differential side gear. On these vehicles, insert the prybar between the inner CV joint housing and the bearing support assembly and pop the inner end of the extension shaft loose from the side gear the same way you would any conventional inner CV joint, then unbolt the bearing support assembly from the block. Do not try to separate the driveaxle from the CV joint housing until you have the entire assembly on the bench (see Section 10). Note: *When removing the left (driver's) side driveaxle on models with an automatic transaxle, it may not be possible to pry the inner CV joint out of the transaxle. In this case, it will be necessary to remove the right side driveaxle and insert a screwdriver through the differential side gears and knock the left side shaft free* (see illustration).
11   Refer to Chapter 7 for the driveaxle oil seal replacement procedure.

Installation

12   Installation is the reverse of the removal procedure, but with the following additional points:

8

8-8 Chapter 8 Clutch and driveaxles

10.2a Pry up the retaining tabs on the boot clamps . . .

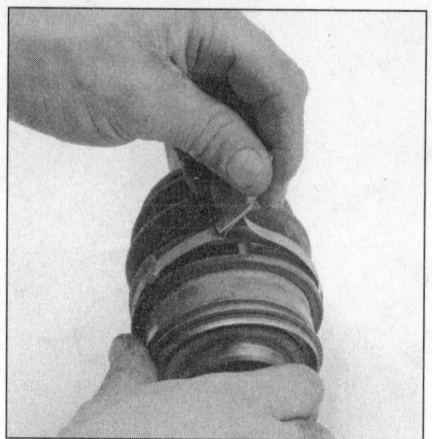

10.2b . . . then open the clamps and remove them from the boot

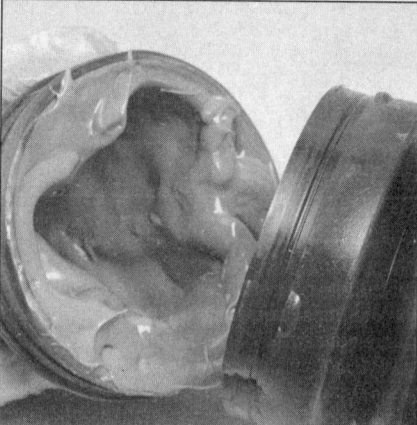

10.3 Once the boot is detached from the inner CV joint housing, the housing can be removed

10.4 Use a center punch to place marks (arrows) on the tripod and the driveaxle to ensure that they're properly reassembled

10.5 Remove the snap-ring from the groove in the end of the axleshaft

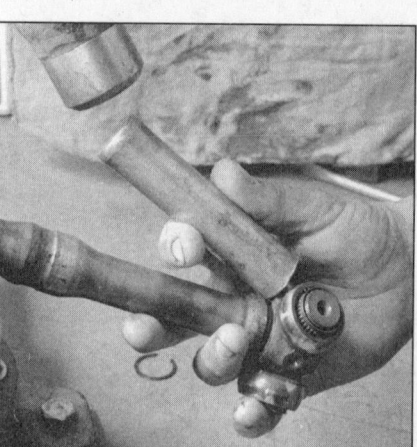

10.6 Drive the tripod joint from the axleshaft with a brass punch and hammer - make sure you don't damage the bearing surfaces or the splines on the shaft

a) When installing the left driveaxle, push the driveaxle sharply in to seat the retaining ring on the inner CV joint in the groove in the differential side gear.
b) Tighten the driveaxle/hub nut to the torque listed in this Chapter's Specifications, then install the nut lock and a new cotter pin.
c) Install the wheel and lug nuts, lower the vehicle and tighten the lug nuts to the torque listed in the Chapter 1 Specifications.
d) Check the transaxle or differential lubricant and add, if necessary, to bring it to the proper level (see Chapter 1).
e) If you're installing the right driveaxle assembly on a vehicle using a support bearing and extension shaft, reattach the outer driveaxle assembly to the housing for the inner CV joint on the bench and install the outer driveaxle/support bearing/extension shaft as a single assembly. Be sure to tighten the support bearing bolts to the torque listed in this Chapter's Specifications.

9 Support bearing assembly - removal, overhaul and installation

Note: Some models use an extension shaft between the inner CV joint housing and the transaxle to give the right driveaxle the same length as the left unit. This extension shaft is hung from a support bearing assembly bolted to the block. The following procedure applies only to these models.

1 Remove the right driveaxle/support bearing/extension shaft assembly (see Section 8).
2 Remove the boot clamps for the inner CV joint and slide the boot back out of the way.
3 Remove the large snap-ring (see illustration 7.1) and withdraw the axleshaft and bearing assembly from the outer housing.
4 Rotate the extension shaft and listen to the bearing. It should feel smooth. If the bearing is rough or noisy, take the support bearing assembly, a new bearing and three new dust shields (see illustration 7.1) to an automotive machine shop and have the old bearing pressed off and a new bearing installed.

10.8a Wrap the splined area of the axleshaft with tape to prevent damage to the boots when installing them

5 Reattach the right driveaxle to its inner CV joint housing.
6 Installation is the reverse of removal (see Section 8).

Chapter 8 Clutch and driveaxles

8-9

10.8b Install the tripod with the chamfered (tapered) ends of the splines facing toward the axleshaft

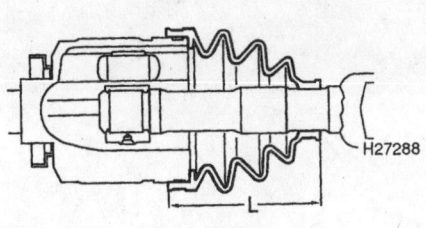

10.10a Adjust the joint to the length indicated in the Specifications at the beginning of this Chapter, making sure the boot isn't distorted

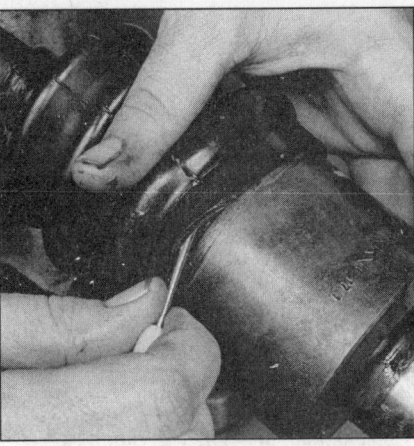

10.10b Equalize the pressure inside the boot by inserting a small, DULL screwdriver between the boot and the CV joint housing

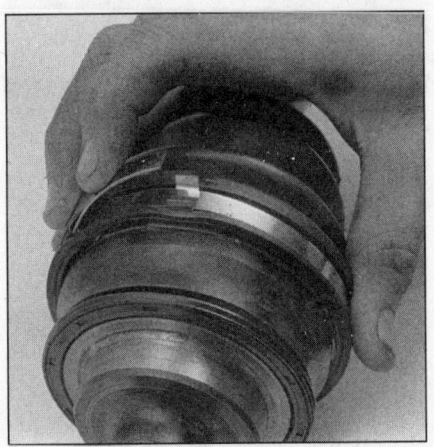

10.10c To install the new clamps, bend the tang down . . .

10.10d . . . and tap the tabs down to hold it in place

10.11 Remove both boot clamps, then slide the boot down the axleshaft so it's out of the way

10 Driveaxle boot replacement

Note: *If the CV joints must be overhauled (usually due to torn boots), explore all options before beginning the job. Complete rebuilt driveaxles are available on an exchange basis, which eliminates much time and work. Whichever route you choose to take, check on the cost and availability of parts before disassembling the vehicle.*

All units

1 Remove the driveaxle (see Section 8).

Inner CV joint

Tripod type

Disassembly

Refer to illustrations 10.2a, 10.2b, 10.3, 10.4, 10.5 and 10.6

2 Remove the boot clamps **(see illustrations)**.
3 Pull the boot back from the inner CV joint and slide the joint housing off **(see illustration)**.
4 Use a center punch to mark the tripod and axleshaft to ensure that they are reassembled properly **(see illustration)**.
5 Remove the snap-ring from the end of the axleshaft with a pair of snap-ring pliers **(see illustration)**.
6 Use a hammer and a brass punch to drive the tripod joint from the driveaxle **(see illustration)**. Some driveaxles are equipped with a rubber dynamic balancer that is retained by clamps. If it is necessary to remove the balancer, be sure to first mark the location of the balancer and clamps so they can be reinstalled in the same location.

Check

7 Clean all components with solvent to remove the grease, and check for cracks, pitting, scoring and other signs of wear.

Reassembly

Refer to illustrations 10.8a, 10.8b, 10.10a, 10.10b, 10.10c and 10.10d

8 Slide the clamps and boot onto the axleshaft. It's a good idea to wrap the axleshaft splines with tape to prevent damaging the boot **(see illustration)**. Place the tripod on the shaft **(see illustration)** and install the snap-ring. Apply grease to the tripod assembly, the inside of the joint housing and the inside of the boot.
9 Slide the boot into place, making sure both ends seat in their grooves.
10 Adjust the length of the joint **(see illustration)**, equalize the pressure in the boot **(see illustration)**, then tighten and secure the boot clamps **(see illustrations)**. Proceed to Step 30.

Double-offset (ball-and-cage) type

Disassembly

Refer to illustrations 10.11, 10.12, 10.13, 10.14 and 10.15

11 Remove both boot clamps **(see illustrations 10.2a and 10.2b)** and discard them. Slide the boot out of the way **(see illustration)**.
12 Mark the shaft, the inner race, the cage and the outer race (housing) so they can be

Chapter 8 Clutch and driveaxles

10.12 Mark the shaft, inner race, cage and outer race (housing) so they can be reassembled in the same relationship to each other

10.13 Pry the retainer from the housing with a small screwdriver

10.14 Slide the housing off the bearing assembly - some of the ball bearings may fall out when the race is removed, so be ready to catch them

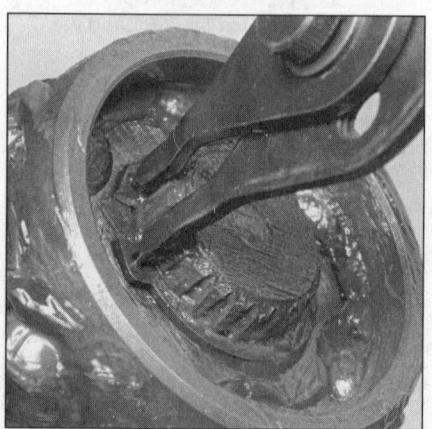

10.15 Remove the snap-ring from the groove in the axleshaft with a pair of snap-ring pliers

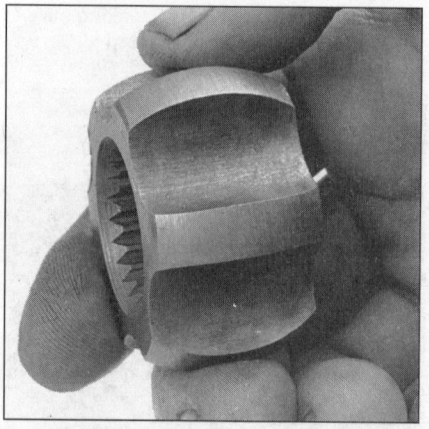

10.19a Inspect the inner race lands and grooves for pitting, score marks, cracks and other signs of wear and damage

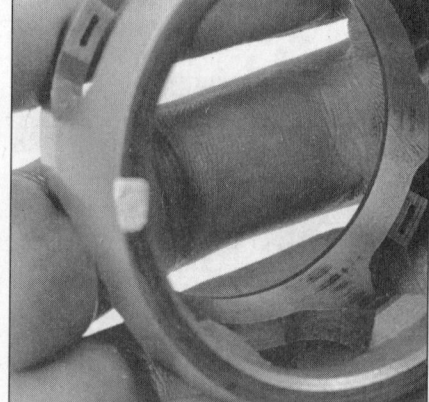

10.19b Inspect the cage for cracks, pitting and score marks (shiny, polished spots are normal and will not adversely affect CV joint performance)

reassembled in the same way **(see illustration)**.

13 Pry the wire ring bearing retainer from the housing **(see illustration)**.

14 Pull the housing off the inner bearing assembly **(see illustration)**.

15 Remove the snap-ring from the groove in the axleshaft with a pair of snap-ring pliers **(see illustration)**.

16 Slide the inner race off the axleshaft.

17 Using a screwdriver or piece of wood, pry the ball bearings from the cage. Be careful not to scratch the inner race, the ball bearings or the cage.

18 Remove the cage.

Inspection

Refer to illustrations 10.19a and 10.19b

19 Clean the components with solvent to remove all traces of grease. Inspect the cage and races for pitting, score marks, cracks and other signs of wear and damage. Shiny, polished spots are normal and will not adversely affect CV joint performance **(see illustrations)**.

Reassembly

Refer to illustration 10.25

20 Wrap the axleshaft splines with tape to avoid damaging the boot. Slide the small boot clamp and boot onto the axleshaft **(see illustration 10.8a)**, then remove the tape. Slide the large boot clamp over the boot.

21 Install the cage on the axleshaft with the smaller diameter side of the cage facing toward the boot.

22 Install the inner race onto the axleshaft with the matchmark on the race (or the larger diameter side) aligned with the mark on the end of the axleshaft.

23 Install the snap-ring in the groove. Make sure it's completely seated by pushing on the inner race.

24 Move the cage up over the inner race, aligning the match marks. Press the ball bearings into the cage windows with your thumbs. If they won't stay in place, apply CV joint grease to hold them.

25 Fill the outer race and boot with CV joint grease (normally included with the new boot kit). Pack the inner race and cage assembly with grease, by hand, until grease is worked completely into the assembly **(see illustration)**.

26 Slide the inner race, balls and cage into the CV joint housing and install the wire ring bearing retainer.

27 Wipe any excess grease from the axle boot groove on the outer race. Seat the small

10.25 Pack the inner race and cage assembly with grease, by hand, until grease is worked completely into the assembly (also note that the larger diameter side, or "bulge," is facing out)

diameter of the boot in the recessed area on the axleshaft. Push the other end of the boot onto the CV joint housing and move the race in or out until there's no deformation (distortion or dents) in the boot.

Chapter 8 Clutch and driveaxles

10.34 After the old grease has been rinsed away and the cleaning solvent has been blown out with compressed air, rotate the outer joint through its full range of motion and inspect the bearing surfaces for wear or damage - if any of the balls, the race or the cage look damaged, replace the outer joint assembly

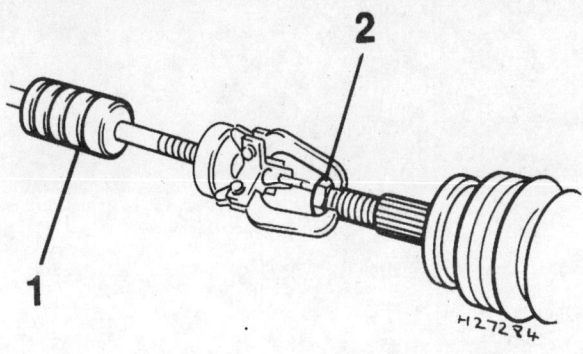

10.38a The outer CV joint can be removed from the driveaxle shaft with a slide hammer

28 Adjust the length of the joint (**see illustration 10.10a**), then equalize the pressure in the boot by inserting a dull screwdriver between the boot and the outer race (**see illustration 10.10b**). Don't damage the boot with the tool.

29 Install the boot clamps (**see illustrations 10.10c and 10.10d**).

All inner CV joints

30 Install a new circlip on the inner CV joint stub axle.

31 Install the driveaxle (see Section 8).

Outer CV joint

Refer to illustrations 10.34, 10.38a and 10.38b

32 Remove the boot clamps (**see illustrations 10.2a and 10.2b**) and slide the boot back far enough to inspect the joint. If the axleshaft is rough and you're planning to re-use the same boot, wrap the axleshaft with tape to protect the boot from damage (**see illustration 10.8a**).

33 Thoroughly wash the outer CV joint in clean solvent and blow dry it with compressed air, if available. The outer joint can't be disassembled, so it's difficult to wash away all the old grease and to rid the bearing of solvent once it's clean. But it's imperative that the job be done thoroughly, so take your time and do it right.

34 Bend the outer CV joint housing at an angle to the driveaxle to expose the bearings, inner race and cage (**see illustration**). Inspect the bearing surfaces for signs of wear. If the joint is worn, replace it (see step 38).

35 If the boot is damaged but the joint is OK, remove the inner CV joint and boot (see Steps 2 through 6 for tripod and Steps 11 through 18 for double-offset). **Caution:** *The outer CV joint should only be removed if it is going to be replaced.* Proceed to the next step.

36 Slide the new outer boot onto the driveaxle. It's a good idea to wrap vinyl tape around the shaft splines to prevent damage to the boot (**see illustration 10.8a**). When the boot is in position, add the specified amount of grease (included in the boot replacement kit) to the outer joint and the boot (pack the joint with as much grease as it will hold and put the rest into the boot). Slide the boot on the rest of the way, equalize the pressure inside the boot and install the new clamps (**see illustrations 10.10b, 10.10c and 10.10d**).

37 Slide on the inner boot and install the inner CV joint (see Steps 8 through 10 for tripod and 20 through 29 for double-offset).

38 If the bearings are damaged or worn, replace the joint. This can be done by knocking the joint from the shaft using a hammer and punch, or by pulling the joint off with a slide hammer (**see illustration**). Install the new joint by first sliding the new boot and clamps onto the axleshaft. Then thread the old driveaxle/hub nut onto the end of the

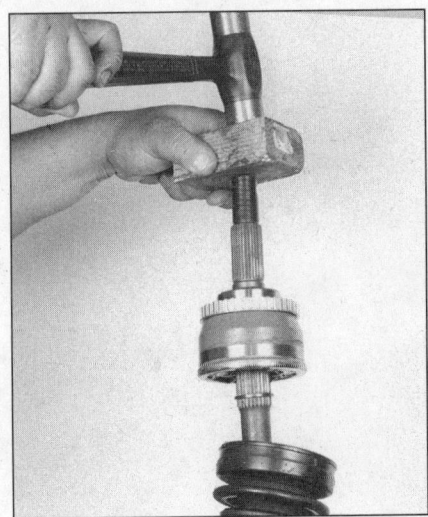

10.38b To install the outer CV joint, put the axleshaft into a bench vise (equipped with protective jaws and tap the CV joint with a hammer and a block of wood; drive the joint onto the driveaxle shaft until the retaining ring in the joint seats in the groove in the shaft

shaft and assemble the joint onto the shaft splines. Drive the joint into place (**see illustration**).

All units

39 Install the driveaxle (see Section 8).

Notes

Chapter 9 Brakes

Contents

	Section		Section
Anti-Lock Brake System (ABS) - general information	2	Drum brake shoes - replacement	6
Brake check	See Chapter 1	General information	1
Brake disc - inspection, removal and installation	5	Master cylinder - removal, overhaul and installation	8
Brake hydraulic system - bleeding	11	Parking brake - check and adjustment	13
Brake light switch - check and replacement	15	Parking brake cables - replacement	14
Brakes hoses and lines - inspection and replacement	10	Power brake booster - check, removal and installation	12
Disc brake caliper - removal, overhaul and installation	4	Proportioning valve - replacement	9
Disc brake pads - replacement	3	Wheel cylinder - removal, overhaul and installation	7

Specifications

General
Brake fluid type .... See Chapter 1

Disc brakes
Minimum brake pad thickness .... See Chapter 1
Brake disc minimum thickness .... Refer to the marks cast into the disc
Maximum disc runout .... 0.0028 inch
Maximum disc thickness variation .... 0.0008 inch

Rear drum brakes
Maximum inside diameter .... Refer to the marks cast into the drum
Maximum radial runout .... 0.0020 inch
Maximum out-of-roundness .... 0.0012 inch
Shoe friction material minimum thickness .... See Chapter 1

Brake pedal
Freeplay .... 3/64 to 1/8 inch
Depressed height
    Manual transaxle .... 3 inches
    Automatic transaxle .... 3-1/3 inches
Free height
    Manual transaxle .... 5-3/4 to 6-1/4 inch
    Automatic transaxle .... 6-1/4 to 6-1/2 inch
Clearance between pedal stopper and brake light switch .... 1/64 inch

ABS sensor wheel-to-sensor pickup clearance
Front .... 0.067 to 0.071 inch
Rear .... 0.008 to 0.043 inch

Power brake booster
Output rod length "A"	13/32 inch (see illustration 12.16)
Pushrod length "L"	4-59/64 inches (see illustration 12.14)

Torque specifications
Ft-lbs (unless otherwise indicated)

Brake booster-to-body mounting nuts	108 to 144 in-lbs
Brake caliper pin bolts	16 to 23
Brake caliper mounting bracket (torque plate) bolts	
Front	40 to 47
Rear	16 to 23
Brake hose-to-caliper banjo bolt	144 to 168 in-lbs
Master cylinder-to-brake booster nuts	108 to 120 in-lbs
Brake booster mounting nuts	108 to 144 in-lbs
Wheel cylinder bolts	52 to 95 in-lbs

1 General information

The vehicles covered by this manual are equipped with hydraulically operated front and rear brake systems. The front brakes are disc type and the rear brakes are drum or disc type. Both the front and rear brakes are self adjusting. The disc brakes automatically compensate for pad wear, while the drum brakes incorporate an adjustment mechanism which is activated as the parking brake is applied.

Hydraulic system

The hydraulic system consists of two separate circuits. The master cylinder has separate reservoirs for the two circuits, and, in the event of a leak or failure in one hydraulic circuit, the other circuit will remain operative. A dual proportioning valve, either built into the master cylinder or mounted on the firewall, provides brake balance between the front and rear brakes.

Power brake booster

The power brake booster, utilizing engine manifold vacuum and atmospheric pressure to provide assistance to the hydraulically operated brakes, is mounted on the firewall in the engine compartment.

Parking brake

The parking brake operates the rear brakes only, through cable actuation. It's activated by a lever mounted in the center console.

Service

After completing any operation involving disassembly of any part of the brake system, always test drive the vehicle to check for proper braking performance before resuming normal driving. When testing the brakes, perform the tests on a clean, dry, flat surface. Conditions other than these can lead to inaccurate test results.

Test the brakes at various speeds with both light and heavy pedal pressure. The vehicle should stop evenly without pulling to one side or the other. Avoid locking the brakes, because this slides the tires and diminishes braking efficiency and control of the vehicle.

Tires, vehicle load and wheel alignment are factors which also affect braking performance.

2 Anti-lock Brake System (ABS) - general information

1 The Anti-lock Brake System (ABS), is designed to maintain vehicle steerability, directional stability and optimum deceleration under severe braking conditions and on most road surfaces. It does so by monitoring the rotational speed of each wheel and controlling the brake line pressure to each wheel during braking. This prevents the wheels from locking up.

Components
Actuator assembly

2 The actuator assembly consists of an electric hydraulic pump and four solenoid valves. The electric pump provides hydraulic pressure to charge the reservoirs in the actuator, which supplies pressure to the braking system. The solenoid valves modulate brake line pressure during ABS operation. The body contains four valves - one for each wheel. The pump, the reservoirs and the solenoid valves are all housed in the actuator assembly.

Speed sensors

3 The speed sensors, which are located at each wheel, generate small electrical pulsations when the toothed sensor wheels are turning, sending a variable voltage signal indicating wheel rotational speed to the ABS control unit.

4 The front speed sensors are mounted on the steering knuckles in close relationship to the toothed sensor wheels, which are pressed onto the outer constant velocity (CV) joints.

5 The rear wheel sensor pickups are bolted to the axle hubs. The sensor wheels are pressed onto the rear hub assemblies.

ABS control unit

6 The ABS control unit, which is mounted behind the right (passenger side) kick panel, is the "brain" of the ABS system. The function of the control unit is to accept and process information received from the wheel speed sensors to control the hydraulic line pressure, avoiding wheel lock up. The control unit also constantly monitors the system, even under normal driving conditions, to find faults within the system.

7 If a problem develops within the system, an "ABS" light will glow on the dashboard. A diagnostic code will also be stored in the control unit, which indicates the problem area or component.

Diagnosis and repair
Refer to illustrations 2.8a and 2.8b

8 If the ABS warning light on the dash comes on and stays on while the vehicle is in operation, the ABS system requires attention. Diagnosis is quite complex, involving a number of lengthy diagnostic procedures, so we don't recommend attempting to fix the ABS system at home. However, if you're willing to do a little work, you can obtain a diagnostic trouble code - which will indicate the general area of the problem - as follows:

a) *Drive the vehicle above 20 mph for at least one minute.*
b) *Stop the vehicle and turn the engine off. Turn the ignition switch on while grounding terminal no. 4 of the data link connector located in the fuse block.*
c) *With the terminal grounded, turn the ignition switch On and wait 3.6 seconds.*
d) *Count the number of flashes of the ABS light located in the instrument cluster, then refer to the table below.*
e) *After checking the codes and making any repairs, erase the malfunction codes by ungrounding the check connector, then within 1.25 seconds, grounding it again three times for one second each. The ABS light should now go out.*

Chapter 9 Brakes

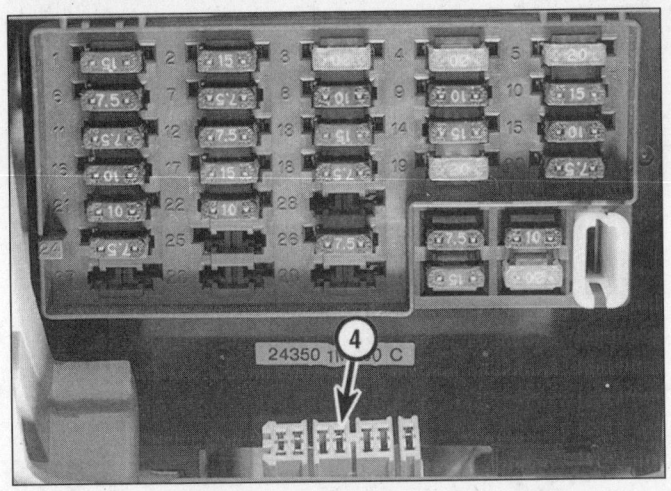

2.8a Locate terminal 4 of the diagnostic connector below the fuse block

2.8b Use a jumper wire to ground the number 4 terminal of the diagnostic connector, then turn the ignition key On to read the ABS system codes

Code or symptom	Component
12	Normal self diagnosis (no malfunctions)
45	Left front outlet solenoid valve actuator
46	Left front inlet solenoid valve actuator
41	Right front outlet solenoid valve actuator
42	Right front inlet solenoid valve actuator
51	Right rear outlet solenoid valve actuator
52	Right rear inlet solenoid valve actuator
55	Left rear outlet solenoid valve actuator
56	Left rear inlet solenoid valve actuator
25	Left front sensor - open circuit
26	Left front sensor - short circuit
21	Right front sensor - open circuit
22	Right front sensor - short circuit
35	Left rear sensor - open circuit
36	Right rear sensor - short circuit
31	Right rear sensor - open circuit
32	Right rear sensor - short circuit
18	Sensor wheel - Incorrect clearance between sensor and sensor wheel, damaged sensor wheel teeth, faulty wheel bearing or wrong tire pressure
61	Actuator motor or motor relay malfunction
63	Solenoid valve relay malfunction
57	Power supply voltage low
71	Control unit malfunction
ABS light remains On when ignition switch is turned On	Solenoid valve relay coil power supply malfunction Warning lamp bulb circuit malfunction Control unit power supply circuit malfunction Solenoid valve relay stuck Control unit or connector malfunction
ABS light remains lit during self-diagnosis	Control unit malfunction
ABS light does not light when ignition	Faulty fuse, warning lamp bulb or circuit switch is turned on; control unit malfunction
Brake pedal noise or vibration	Control unit malfunction
Excessive stopping distance	Control unit malfunction
Unexpected brake pedal action	Control unit malfunction
ABS doesn't operate	Control unit malfunction
ABS operates frequently	Control unit malfunction
Brake pedal action is inconsistent	Control unit malfunction

Chapter 9 Brakes

3.5 When replacing the front brake pads, use a C-clamp to depress the piston into the caliper before removing the caliper and pads

9 You can perform the following preliminary checks yourself:
   a) Check the brake fluid level in the reservoir.
   b) Check that all electrical connectors are securely connected.
   c) Check the fuses.
   d) Check the brake system (see Chapter 1).
   e) Check the brake pedal (see Chapter 1).
   f) Check the brake pads (see Section 3).

10 Start the engine again by slowly turning the ignition key (turning the key too quickly may cause the ABS warning light to stay on when there is nothing wrong).
   a) If the light stays on, take the vehicle to a dealer service department or other qualified repair shop and have the ABS system repaired.
   b) If the light doesn't stay on, drive the vehicle above 20 mph for at least one minute and verify that the warning light on the dash doesn't come on again. If it does, take the vehicle to a dealer service department or other qualified repair shop and have the ABS system repaired.

Front brakes

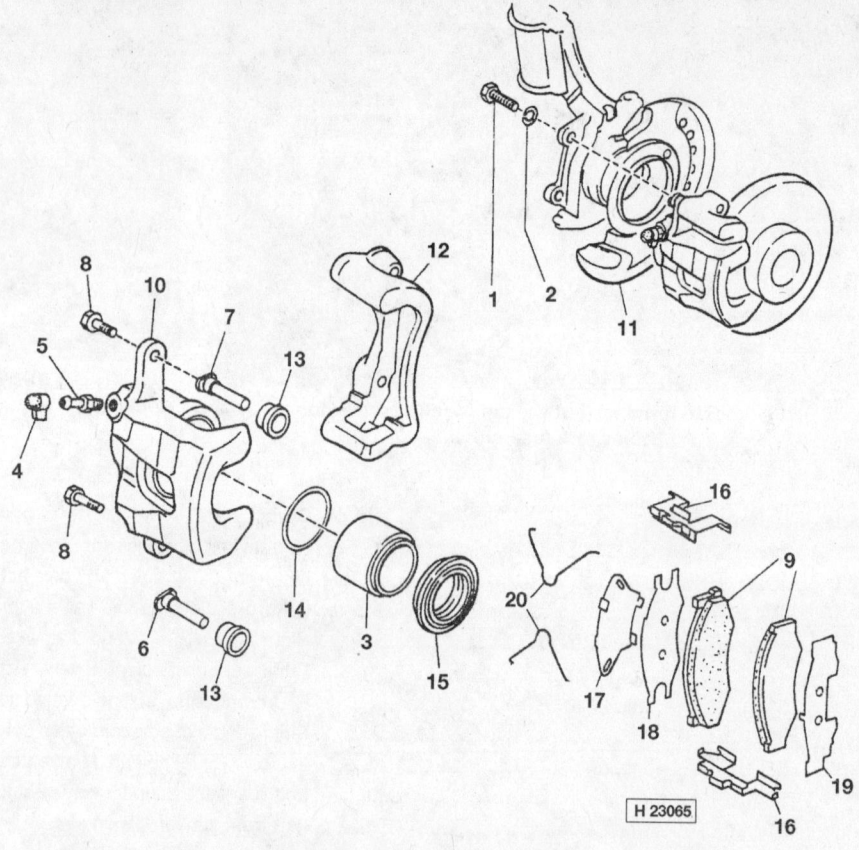

3.6a Exploded view of the front brake caliper and pads

1	Caliper mounting bracket bolt	7	Upper caliper pin	15	Piston boot
2	Washer	8	Caliper pin bolt	16	Pad retainers
3	Piston	9	Inner and outer pads	17	Inner pad shim cover
4	Bleeder valve dust cap	10	Caliper body	18	Inner pad shim (some models)
5	Bleeder valve	11	Splash shield	19	Outer pad shim
6	Lower caliper pin	12	Mounting bracket	20	Anti-rattle spring
		13	Pin boots		
		14	Piston seal		

3 Disc brake pads - replacement

Refer to illustrations 3.5, 3.6a through 3.6s and 3.6t through 3.6oo

Warning: *Disc brake pads must be replaced on both front or both rear wheels at the same time - never replace the pads on only one wheel. Also, the dust created by the brake system may contain asbestos, which is harmful to your health. Never blow it out with compressed air and don't inhale any of it. An approved filtering mask should be worn when working on the brakes. Do not, under any circumstances, use petroleum-based solvents to clean brake parts. Use brake system cleaner only!*

1 Remove the cap from the brake fluid reservoir.
2 Loosen the front or rear wheel lug nuts, raise the front or rear of the vehicle and support it securely on jackstands. Block the wheels at the opposite end.
3 Remove the wheels. Work on one brake assembly at a time, using the assembled brake for reference if necessary. Wash the brake assembly thoroughly with brake system cleaner.
4 Inspect the brake disc carefully as outlined in Section 5. If machining is necessary, follow the information in that Section to remove the disc, at which time the pads can be removed as well.
5 If you're replacing the front pads, push the piston back into its bore to provide room for the new brake pads. A C-clamp can be used to accomplish this **(see illustration)**. As the piston is depressed to the bottom of the caliper bore, the fluid in the master cylinder will rise. Make sure that it doesn't overflow. If necessary, siphon off some of the fluid. **Note:** *This step does not apply to rear caliper pistons; rear pistons use a ratcheting mechanism inside the caliper which will be damaged if you try to depress the piston with a C-clamp. Rear pistons are retracted with a pair of needle-nose pliers, as shown in the photo sequence for the rear pad replacement procedure.*
6 Follow **illustrations 3.6a through 3.6s** for the front pad replacement procedure. Be sure to stay in order and read the caption under each illustration. If you're replacing rear brake pads, follow the second photo sequence **(see illustrations 3.6t through 3.6oo)**.
7 When reinstalling the caliper, be sure to tighten the mounting bolts to the torque listed in this Chapter's Specifications.
8 After the job has been completed, firmly depress the brake pedal a few times to bring the pads into contact with the disc. Check the level of the brake fluid, adding some if necessary. Check the operation of the brakes carefully before placing the vehicle into normal service.

Chapter 9 Brakes

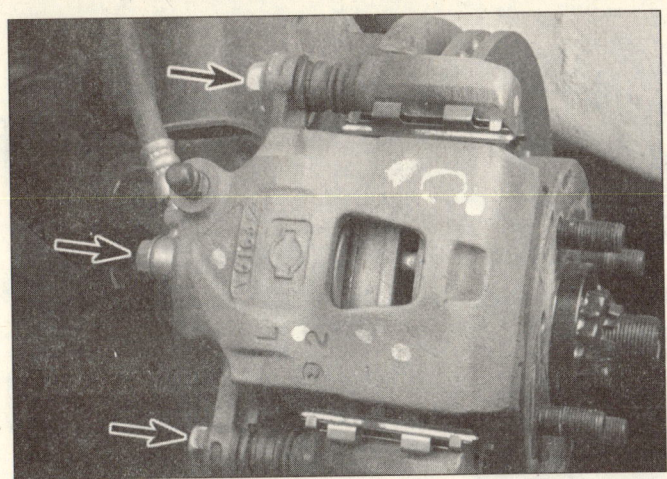

3.6b Remove these two bolts (upper and lower arrows) to detach the caliper from the mounting bracket; the middle arrow points to the brake hose banjo bolt (which shouldn't be removed unless the caliper requires service)

3.6c Lift the caliper off the pads

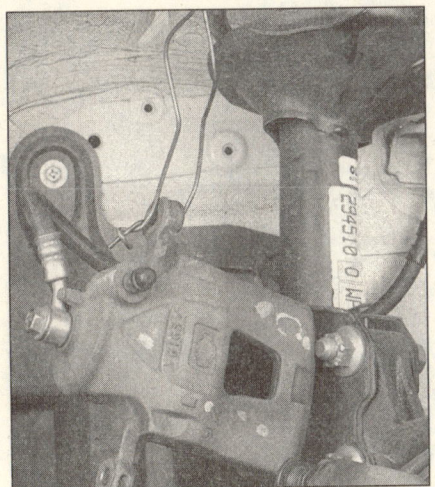

3.6d Hang the caliper out of the way with a piece of coat hanger or wire

3.6e Remove anti-rattle springs, followed by the shim from the outer brake pad

3.6f Remove the outer pad

3.6g Remove the shim(s) from the inner brake pad

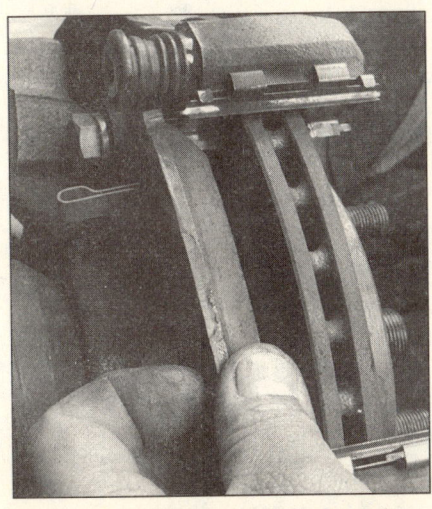

3.6h Remove the inner brake pad

3.6i Remove the lower pad retainer

Chapter 9 Brakes

3.6j Remove the upper pad retainer

3.6k Remove the caliper pins and the dust boots; inspect the boots for cracks and tears and replace them if they're damaged

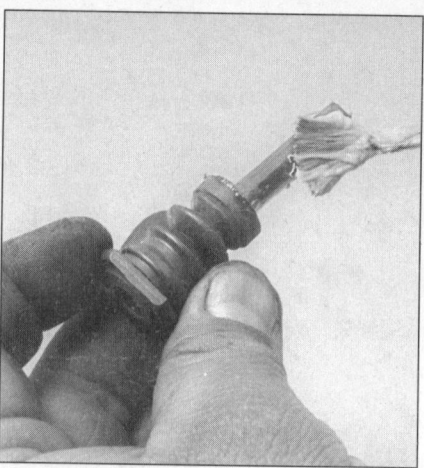

3.6l Clean the caliper pins, inspect them for scoring and corrosion, and replace them if necessary; coat the pins with high-temperature grease, then install them

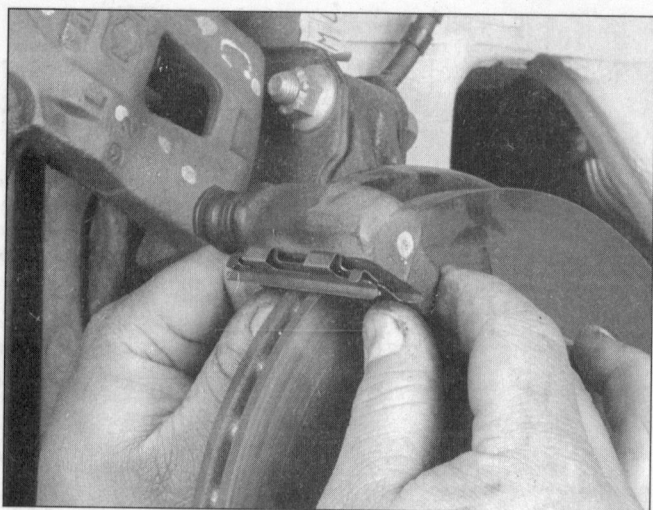

3.6m Install the upper pad retainer

3.6n Install the lower pad retainer

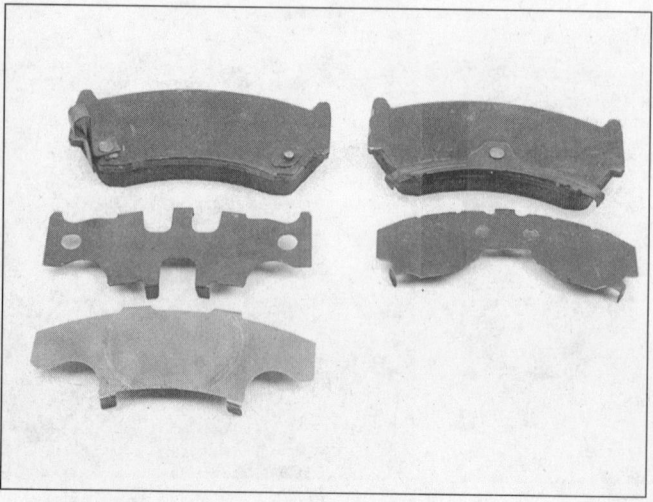

3.6o Inner and outer pads and shim details

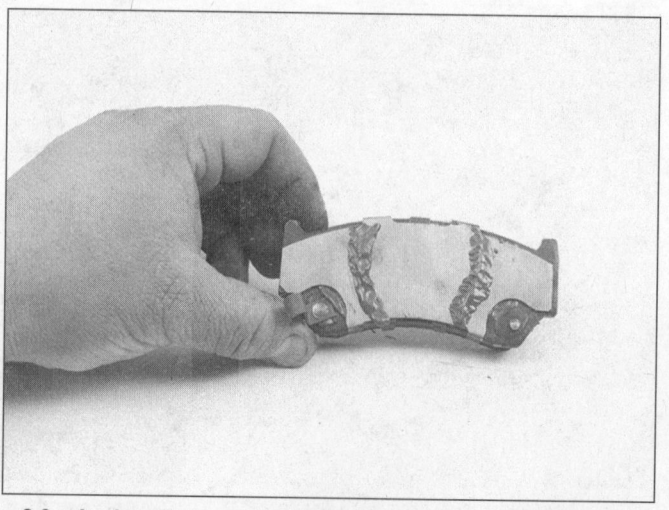

3.6p Apply anti-squeal compound to the backing plates of the new pads

Chapter 9 Brakes

3.6q Install the inner brake pad and shim

3.6r Install the outer brake pad and shim

3.6s Install the caliper and the caliper bolts, then tighten the caliper bolts to the torque listed in this Chapter's Specifications. **Note:** *If the caliper won't fit over the pads, use a C-clamp to push the piston into the caliper a little further*

Rear brakes

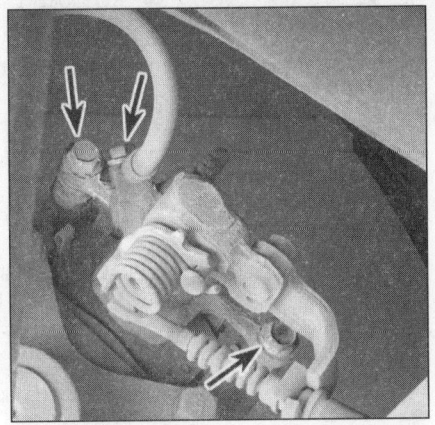

3.6t Remove the rear caliper pin bolts (left and right arrows); the center arrow points to the banjo bolt for the brake hose - it isn't necessary to remove this bolt unless you're planning to remove the caliper for overhaul

3.6u Remove the caliper . . .

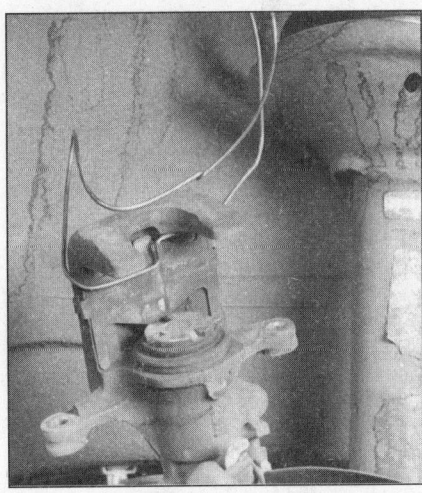

3.6v . . . and hang it out of the way with a piece of wire

3.6w Remove the outer brake pad

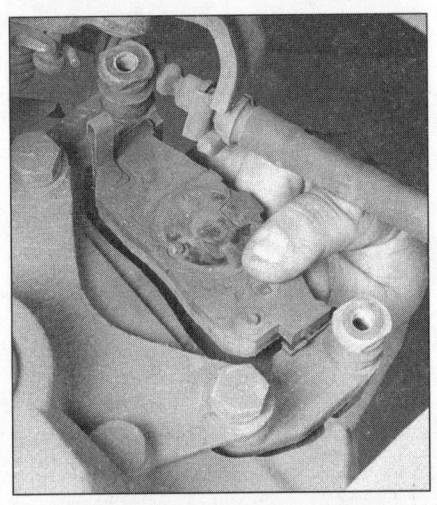

3.6x Remove the inner brake pad

3.6y Remove the upper pad retainer

9-8 Chapter 9 Brakes

3.6z Remove the lower pad retainer

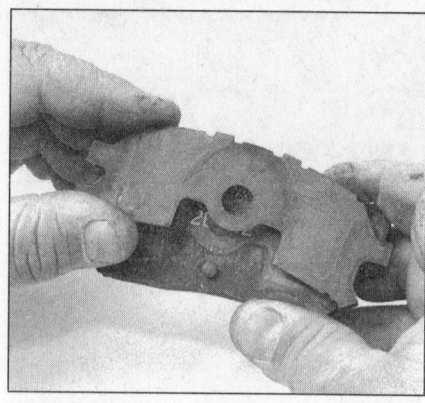
3.6aa Remove the shim from the outer pad

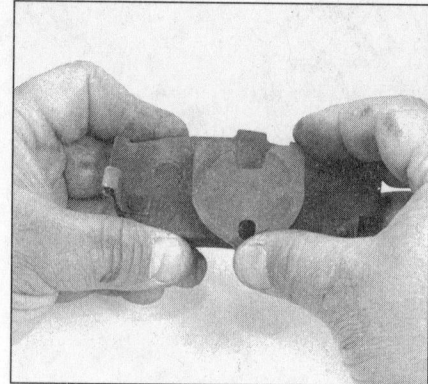
3.6bb Remove the shim from the inner pad

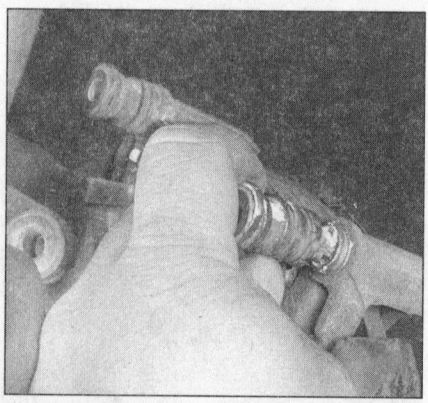
3.6cc Remove the caliper pins and boots

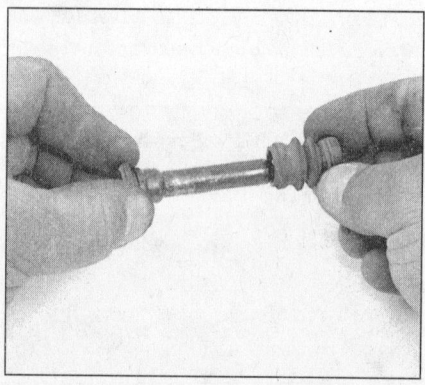
3.6dd Inspect the pins for scoring and the boots for cracks and tears

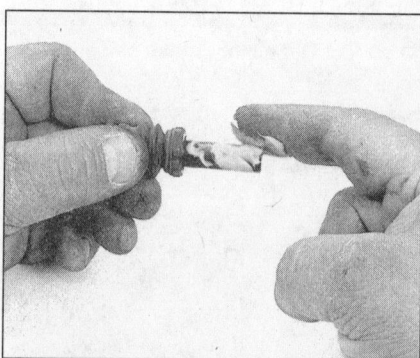
3.6ee Wipe off the pins and lubricate them with high-temperature grease before installing them

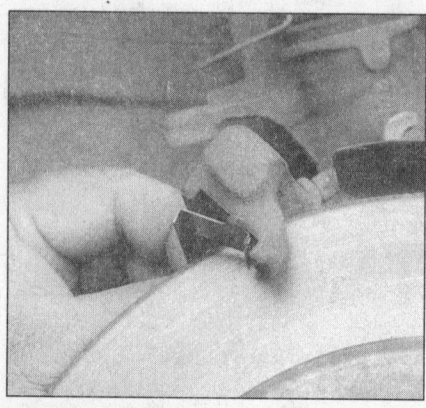
3.6ff Install the upper pad retainer

3.6gg Install the lower pad retainer

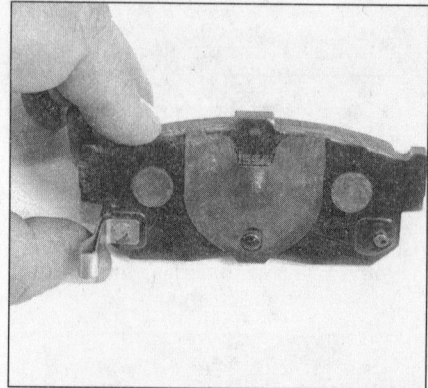
3.6hh Install the shim on the inner pad

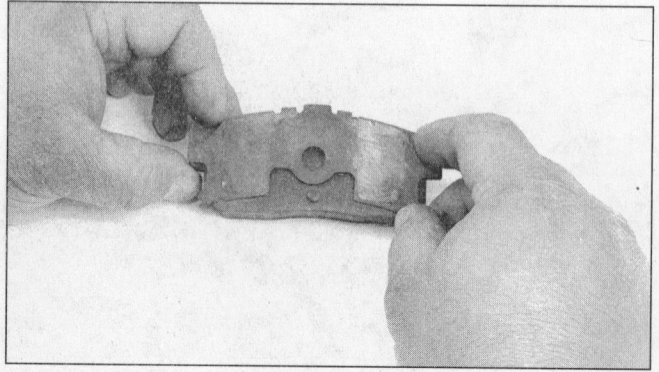
3.6ii Install the shim on the outer pad

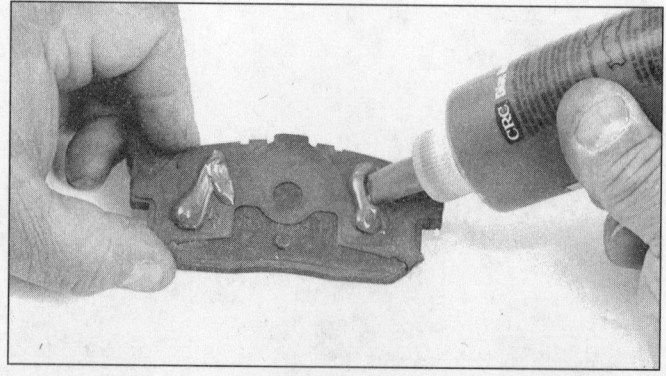
3.6jj Apply anti-squeal compound to the backs of both pads

Chapter 9 Brakes

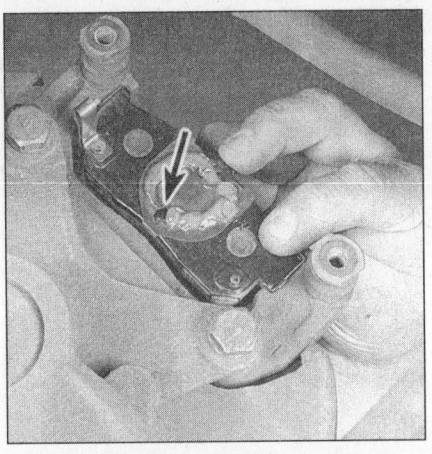

3.6kk Install the inner pad - notice the small projection on the pad backing plate (arrow) this must fit into a notch in the face of the caliper piston

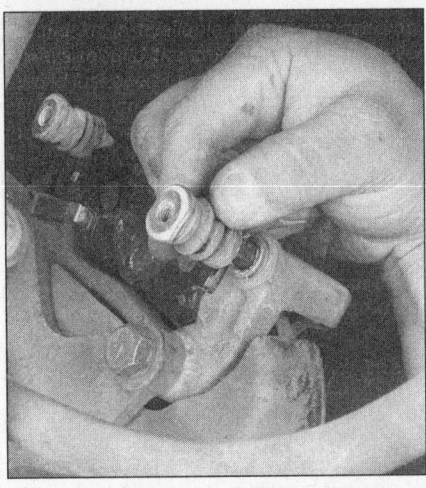

3.6ll Install the caliper boots and pins

3.6mm Install the outer pad

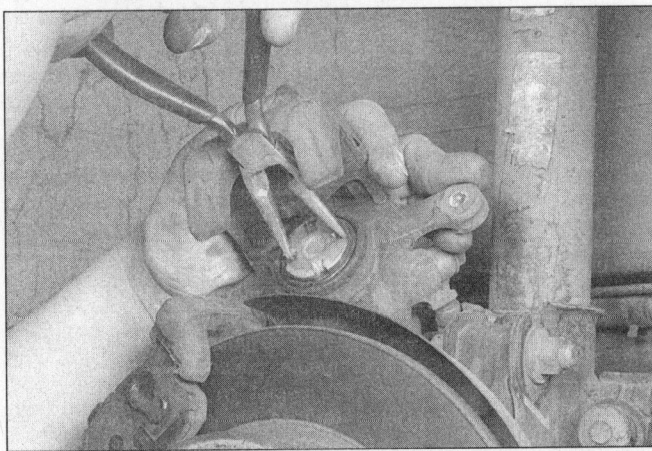

3.6nn Using a pair of needle-nose pliers, turn the piston clockwise to retract it into the caliper bore - make sure the piston is positioned so that one of the notches in the piston face will engage with the projection on the pad backing plate

3.6oo Install the caliper and caliper bolts and tighten the bolts to the torque listed in this Chapter's Specifications

4 Disc brake caliper - removal, overhaul and installation

Warning: *The dust created by the brake system may contain asbestos, which is harmful to your health. Never blow it out with compressed air and don't inhale any of it. An approved filtering mask should be worn when working on the brakes. Do not, under any circumstances, use petroleum-based solvents to clean brake parts. Use brake system cleaner only!*

Removal

Refer to illustration 4.2

1 Loosen the front or rear wheel lug nuts, raise the front or rear of the vehicle and place it securely on jackstands. Block the wheels at the opposite end. Remove the front or rear wheel.

2 Remove the banjo bolt and discard the old copper washers (new ones should be used during reassembly). Disconnect the brake hose from the caliper. Plug the brake hose to keep contaminants out of the brake system and to prevent losing any more brake fluid than is necessary **(see illustration)**.

3 Refer to Section 3 for the caliper removal procedure (it's part of brake pad replacement).

Overhaul

Refer to illustrations 4.5, 4.7 and 4.10

Note: *If an overhaul is indicated (usually because of fluid leakage), explore all options before beginning the job. New and factory rebuilt calipers are available on an exchange basis, which makes this job quite easy. If it's decided to rebuild the calipers, make sure a rebuild kit is available before proceeding. Always rebuild the calipers in pairs - never rebuild just one of them. Overhauling the rear caliper is extremely difficult without special tools, so if the rear caliper is leaking or otherwise defective, obtain a rebuilt unit or have it*

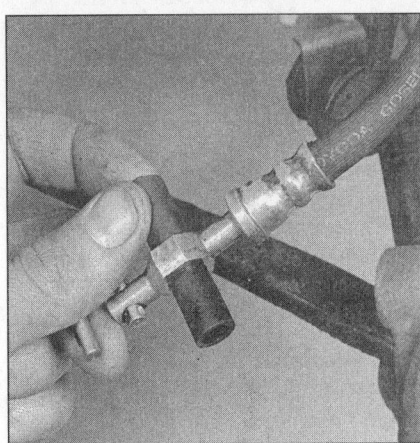

4.2 Using a piece of rubber hose of the appropriate size, plug the brake line

overhauled by a shop equipped with the proper tools.

4 Place a wood block or a bundle of rags between the piston and the caliper frame to

Chapter 9 Brakes

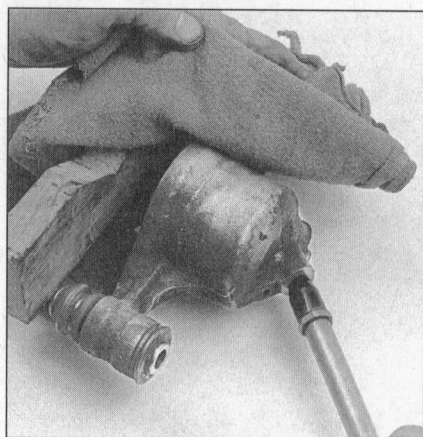

4.5 With the caliper padded to catch the piston, use compressed air to force the piston out of its bore - make sure your hands or fingers are not between the piston and caliper

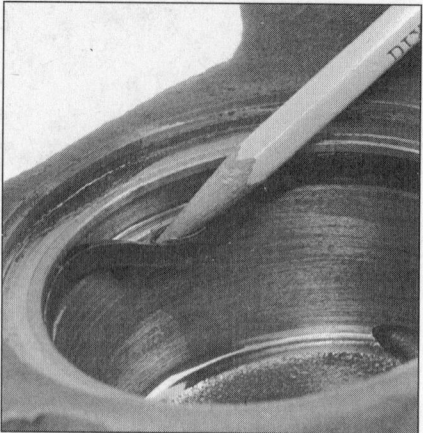

4.7 To remove the seal from the caliper bore, use a plastic or wooden tool, such as a pencil

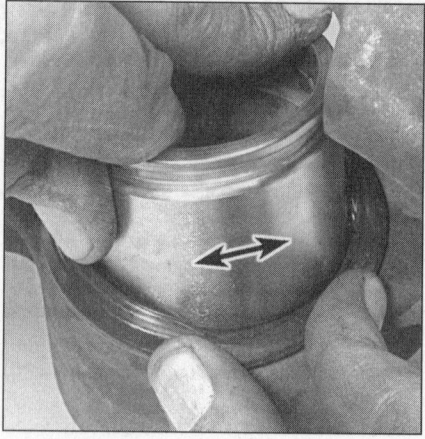

4.10 Install the dust boot flange in the upper groove in the bore, then insert the piston into the boot (NOT the bore) at an angle and, using a rotating motion, work the piston completely into the dust boot

prevent damage as it is ejected.

5 To remove the piston from the caliper, apply compressed air to the brake fluid hose connection on the caliper body (see illustration). Use only enough pressure to ease the piston out of its bore. **Warning:** *Be careful not to place your fingers between the piston and the caliper, as the piston may come out with some force.*

6 Inspect the mating surfaces of the piston and caliper wall. If there is any scoring, rust, pitting or bright areas, replace the complete caliper unit with a new one.

7 If these components are in good condition, remove the piston seal from the caliper bore using a wooden or plastic tool (see illustration). Metal tools may damage the cylinder bore.

8 Wash all the components with brake system cleaner.

9 Submerge the new piston in clean brake fluid and install it into the lower groove in the caliper bore, making sure it isn't twisted.

10 Install the flange of the dust boot in the upper groove of the caliper bore. Make sure it seats completely. Lubricate the piston with clean brake fluid, carefully slide it into the new boot (see illustration), then position it squarely in the caliper bore and apply firm (but not excessive) pressure to install it. Make sure the piston boot seats in the groove in the piston.

11 Install the caliper pin dust boots.

Installation

12 Install the caliper by reversing the removal procedure. Don't forget to replace the copper washers at the brake hose-to-caliper banjo bolt (new washers normally come with the rebuild kit).

13 Bleed the brake circuit according to the procedure in Section 11. Make sure there are no leaks from the hose connections. Test the brakes carefully before returning the vehicle to normal service.

5 Brake disc - inspection, removal and installation

Inspection

Refer to illustrations 5.3, 5.4a, 5.4 and 5.5

1 Loosen the wheel lug nuts, raise the vehicle and support it securely on jackstands. Remove the wheel and install the lug nuts to hold the disc in place (washers may have to be installed under the nuts in order for pressure to be applied to the disc).

2 Remove the brake caliper as outlined in Section 3 (it's part of the brake pad replacement procedure). It isn't necessary to disconnect the brake hose. After removing the caliper bolts, suspend the caliper out of the way with a piece of wire. Remove the two mounting bracket bolts and detach the mounting bracket.

3 Visually inspect the disc surface for score marks and other damage. Light scratches and shallow grooves are normal after use and may not always be detrimental to brake operation, but deep scoring - over 0.039-inch (1.0 mm) - requires disc removal

5.3 The brake pads on this vehicle were obviously neglected, as they wore down to the rivets and cut deep grooves into the disc - wear this severe means the disc must be replaced

and refinishing by an automotive machine shop (see illustration). Be sure to check both sides of the disc. If pulsating has been noticed during application of the brakes, suspect disc runout.

4 To check disc runout, place a dial indicator at a point about 1/2-inch from the outer edge of the disc (see illustration). Set the indicator to zero and turn the disc. The indicator reading should not exceed the specified allowable runout limit. If it does, the disc should be refinished by an automotive machine shop. **Note:** *The discs should be resurfaced regardless of the dial indicator reading, as this will impart a smooth finish and ensure a perfectly flat surface, eliminating any brake pedal pulsation or other undesirable symptoms related to questionable discs. At the very least, if you elect not to have the discs resurfaced, remove the glaze from the surface with emery cloth using a swirling motion* (see illustration).

5 It's absolutely critical that the disc not be machined to a thickness under the specified minimum thickness. The minimum wear

5.4a To check disc runout, mount a dial indicator as shown and rotate the disc

Chapter 9 Brakes

9-11

5.4b Using a swirling motion, remove the glaze from the disc surface with sandpaper or emery cloth

5.5 Use a micrometer to measure disc thickness

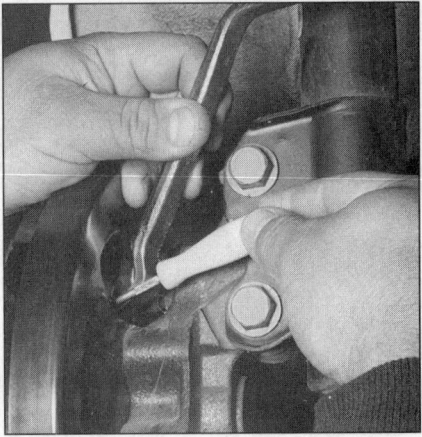

6.2a If the brake drums are hanging up on the brake shoes because of excessive wear, remove the access plug from the backing plate, insert a pair of small screwdrivers (or a screwdriver and a brake adjuster tool, as shown) and retract the shoes

(or discard) thickness is cast into the disc. The disc thickness can be checked with a micrometer **(see illustration)**.

Removal

6   Remove the lug nuts which you installed to hold the disc in place, remove the brake caliper mounting bracket and remove the disc from the hub. If the disc is stuck to the hub and won't come off, install bolts of the proper diameter and thread pitch into the threaded holes between the wheel studs and tighten them, which will force the disc off the hub.

Installation

7   Place the disc in position over the threaded studs.
8   Install the mounting bracket over the disc and position it on the steering knuckle. Tighten the mounting bracket bolts to the torque listed in this Chapter's Specifications. Install the brake caliper, tightening the caliper pin bolts to the torque listed in this Chapter's Specifications.
9   Install the wheel, then lower the vehicle to the ground. Tighten the lug nuts to the torque listed in the Chapter 1 Specifications. Depress the brake pedal a few times to bring the brake pads into contact with the disc. Bleeding won't be necessary unless the brake hose was disconnected from the caliper. Check the operation of the brakes carefully before driving the vehicle.

6 Drum brake shoes - replacement

Refer to illustrations 6.2a, 6.2b, 6.3, 6.4a through 6.4t and 6.5
Warning: *Drum brake shoes must be replaced on both wheels at the same time - never replace the shoes on only one wheel. Also, the dust created by the brake system may contain asbestos, which is harmful to your health. Never blow it out with compressed air and don't inhale any of it. An approved filtering mask should be worn when working on the brakes. Do not, under any circumstances, use petroleum-based solvents to clean brake parts. Use brake system cleaner only!*
Caution: *Whenever the brake shoes are replaced, the return and hold-down springs should also be replaced. Due to the continuous heating/cooling cycle the springs are subjected to, they lose tension over a period of time and may allow the shoes to drag on the drum and wear at a much faster rate than normal.*

1   Loosen the wheel lug nuts, raise the rear of the vehicle and support it securely on jackstands. Block the front wheels to keep the vehicle from rolling. Remove the wheels.
2   Release the parking brake and remove the brake drums. If the brake drum is difficult to remove, remove the access plug from the backing plate, insert a small screwdriver through the hole, lift the adjuster lever off the adjusting wheel and turn the wheel with another screwdriver to back off the brake shoes **(see illustrations)**. The drum should now come off. Remove the rear hubs (Chapter 10).
3   All four rear brake shoes must be replaced at the same time, but to avoid mixing up parts, work on only one brake assembly at a time **(see illustration)**.

6.2b Lift the adjuster lever off the adjuster wheel with the screwdriver and rotate the adjuster wheel until the shoes are retracted sufficiently to allow drum removal (brake drum removed for clarity)

6.3 Drum brake details (left side shown)

9

Chapter 9 Brakes

6.4a Before disassembling the brake, wash it thoroughly with brake system cleaner and allow it to dry - position a drain pan under the brake to catch the residue - DO NOT use compressed air to blow the brake dust off!

6.4b Remove the brake hold-down springs and retainers, then pull the hold-down pins through the rear of the backing plate

4   Before disassembling anything, wash off the brake assembly with brake system cleaner **(see illustration)**. Follow the accompanying illustrations for the brake shoe replacement procedures **(see illustrations)**. Be sure to stay in order and read the caption under each illustration.

5   Before reinstalling the drum, it should be checked for cracks, score marks, deep scratches and hard spots, which will appear as small discolored areas. If the hard spots cannot be removed with fine emery cloth or if any of the other conditions listed above exist, the drum must be taken to an automotive machine shop to have it resurfaced. **Caution:** *Professionals recommend resurfacing the drums each time a brake job is done. Resurfacing will eliminate the possibility of out-of-*

6.4c Lift the brake assembly off the backing plate as a single assembly, then remove the upper and lower return springs

6.4d Disengage the adjuster and the adjuster spring (the one *behind* the adjuster) and remove the leading shoe

6.4e Disengage and remove the adjuster from the trailing shoe

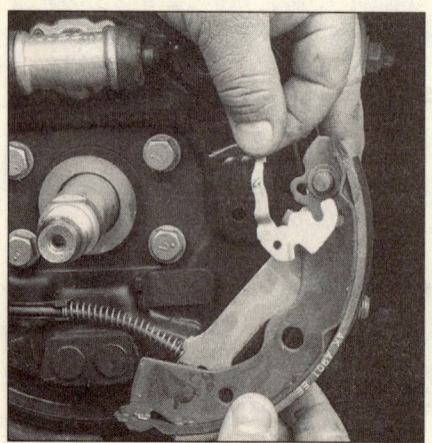

6.4f Disengage and remove the adjuster lever from the trailing shoe

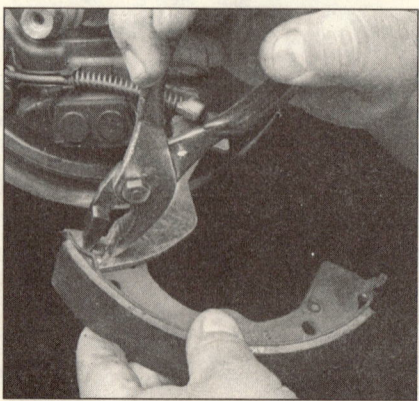

6.4g Remove the C-clip retainer from the parking brake lever pivot pin and disconnect the parking brake lever from the trailing shoe

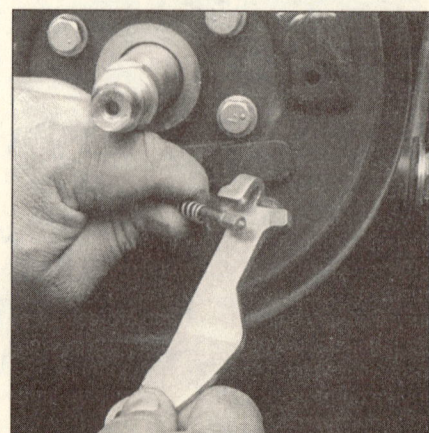

6.4h Disengage the parking brake cable from the parking brake lever

Chapter 9 Brakes 9-13

6.4i Apply a thin film of high-temperature grease to the friction surfaces on the brake backing plate

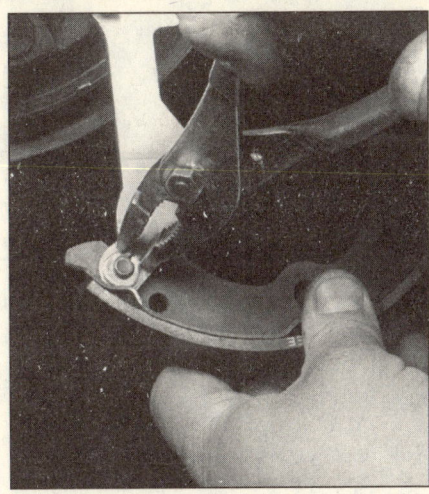

6.4j Install the parking brake lever on the new trailing shoe and secure it with a new C-clip retainer

6.4k Connect the parking brake cable and install the new trailing shoe; make sure it's properly engaged with the wheel cylinder

6.4l Insert the hold-down pin through the rear of the backing plate and install the trailing shoe hold-down spring and retainer

6.4m Attach the adjuster spring to the trailing shoe as shown; note that the spring is hooked into the hole from the *rear*

6.4n Attach the lower return spring to the trailing shoe as shown; note that the spring is hooked into the hole from the *front*

6.4o Attach the adjuster spring to the post on the *backside* of the leading shoe

6.4p Insert the adjuster between the two shoes

6.4q Make sure the adjuster is properly engaged with both shoes

9-14   Chapter 9 Brakes

6.4r Insert the hold-down pin through the rear of the backing plate and install the hold-down spring and retainer for the leading shoe

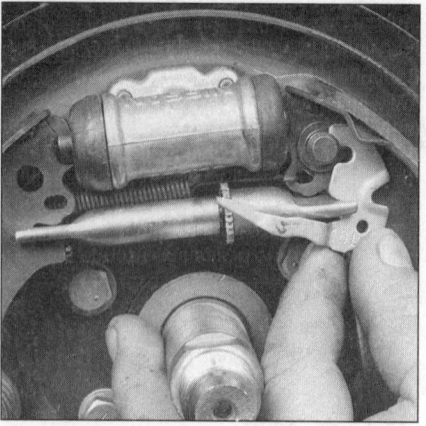

6.4s Install the adjuster lever - make sure it's properly engaged with the pivot pin and the adjuster

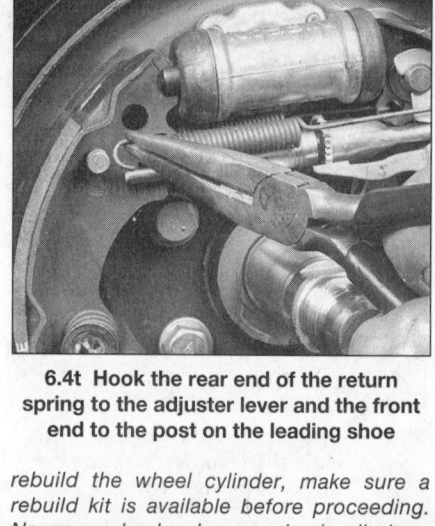

6.4t Hook the rear end of the return spring to the adjuster lever and the front end to the post on the leading shoe

6   Install the brake drum.
7   To make a preliminary adjustment of the brake, turn the adjuster star wheel until the brakes just begin to drag on the drum as it is turned, then back-off the star wheel until no dragging can be heard when you rotate the drum. Depress the brake pedal firmly several times, then rotate the drum to ensure that the brakes are not dragging. If they are, back off the star wheel a little more.
8   Mount the wheel, install the lug nuts, then lower the vehicle. Tighten the wheel lug nuts to the torque listed in the Chapter 1 Specifications.
9   Make a number of forward and reverse stops and operate the parking brake to adjust the brakes until satisfactory pedal action is obtained.
10   Check the operation of the brakes carefully before driving the vehicle.

7   Wheel cylinder - removal, overhaul and installation

Caution: *If an overhaul is indicated (usually because of fluid leaks or sticky operation), explore all options before beginning the job. New wheel cylinders are available, which makes this job quite easy. If it's decided to*

rebuild the wheel cylinder, make sure a rebuild kit is available before proceeding. Never overhaul only one wheel cylinder - always rebuild both of them at the same time.

Removal
Refer to illustration 7.4
1   Raise the rear of the vehicle and support it securely on jackstands. Block the front wheels to keep the vehicle from rolling.
2   Remove the brake shoe assembly (see Section 6).
3   Remove all dirt and foreign material from around the wheel cylinder.
4   Disconnect the brake line with a flare-nut wrench, if available **(see illustration)**. Don't pull the brake line away from the wheel cylinder.
5   Remove the wheel cylinder mounting bolts.
6   Detach the wheel cylinder from the brake backing plate and place it on a clean workbench. Immediately plug the brake line to prevent fluid loss and contamination.

Overhaul
Refer to illustration 7.7
7   Remove the bleeder screw, cups, pistons, boots and spring assembly from the wheel cylinder body **(see illustration)**.

6.5   The brake drum maximum allowable inside diameter is cast inside the drum

round drums. If the drums are worn so much that they can't be resurfaced without exceeding the maximum allowable diameter, which is stamped into the drum **(see illustration)**, then new ones will be required. At the very least, if you elect not to have the drums resurfaced, remove the glaze from the surface with emery cloth using a swirling motion.

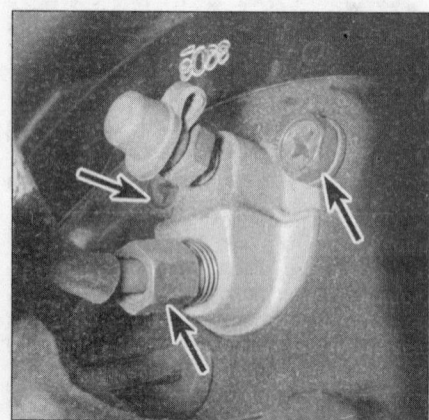

7.4   To remove the wheel cylinder, disconnect the brake line fitting (lower arrow), then remove the two wheel cylinder bolts (upper arrows)

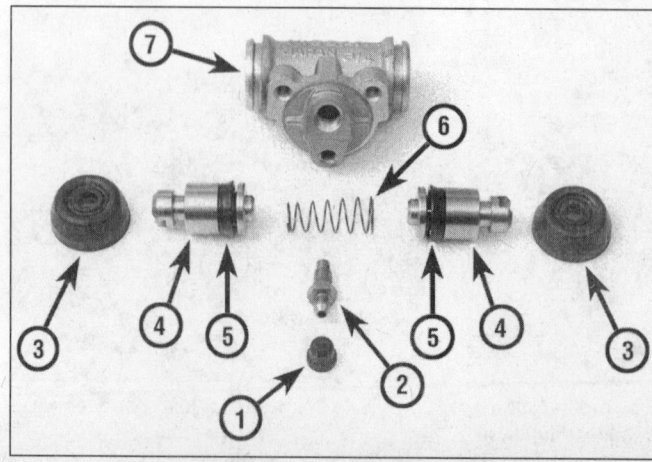

7.7   Exploded view of the rear wheel cylinder

1   Bleeder screw cap
2   Bleeder screw
3   Dust boot
4   Piston
5   Piston cup
6   Spring
7   Wheel cylinder housing

Chapter 9 Brakes

9-15

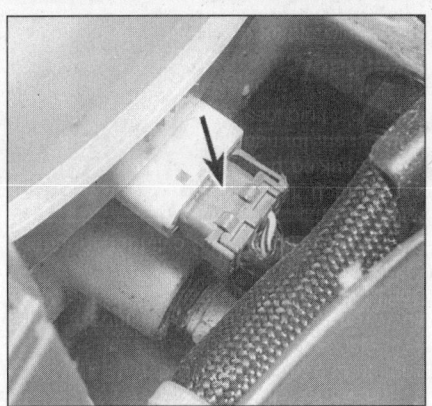

8.2 Unplug the electrical connector (arrow) for the fluid level warning switch

8.4 Loosen the brake line fittings with a flare-nut wrench

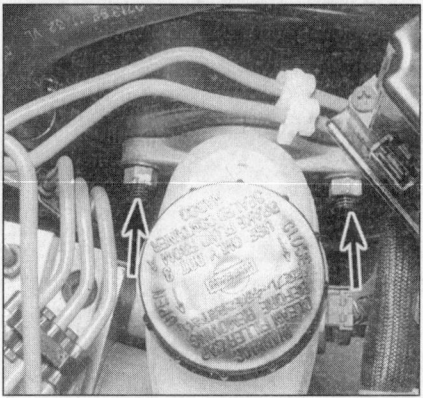

8.6 Remove the master cylinder mounting nuts (arrows)

8 Clean the parts with brake system cleaner. Do not, under any circumstances, use petroleum-based solvents to clean brake parts!
9 Use filtered, unlubricated compressed air to dry the wheel cylinder and blow out the passages.
10 Check the bore for corrosion and score marks. Replace the wheel cylinder with a new one if either of these conditions are found.
11 Lubricate the new cups with brake fluid.
12 Assemble the wheel cylinder components. Make sure the cup lips face in.

Installation

13 Place the wheel cylinder in position and install the bolts finger tight. Connect the brake line to the cylinder, being careful not to cross-thread the fitting. Tighten the wheel cylinder bolts to the torque listed in this Chapter's Specifications.
14 Tighten the brake line securely and install the brake shoe assembly.
15 Bleed the brakes (see Section 11).
16 Check the operation of the brakes carefully before driving the vehicle.

8 Master cylinder - removal, overhaul and installation

Note: *Before deciding to overhaul the master cylinder, check on the availability and cost of a new or factory rebuilt unit and also the availability of a rebuild kit.*

Removal

Refer to illustrations 8.2, 8.4 and 8.6
1 Disconnect the cable from the negative battery terminal.
2 Unplug the electrical connector for the fluid level warning switch **(see illustration)**.
3 Remove as much fluid as possible from the reservoir with a syringe.
4 Place rags under the fittings and prepare caps or plastic bags to cover the ends of the lines once they're disconnected. **Caution:** *Brake fluid will damage paint. Cover all body parts and be careful not to spill fluid during this procedure.* Loosen the fittings at the ends of the brake lines where they enter the master cylinder **(see illustration)**. To prevent rounding off the flats, use a flare-nut wrench, which wraps around the fitting hex.
5 Pull the brake lines away from the master cylinder and plug the ends to prevent contamination.
6 Remove the nuts attaching the master cylinder to the power booster **(see illustration)**. Pull the master cylinder off the studs to remove it. Again, be careful not to spill the fluid as this is done. Remove and discard the old gasket between the master cylinder and the power brake booster.

Overhaul

Refer to illustrations 8.8a, 8.8b, 8.8c, 8.9a, 8.9b, 8.10a, 8.10b, 8.11, 8.14a, 8.14b, 8.14c, 8.15a, 8.15b, 8.16a, 8.16b, 8.16c, 8.16d and 8.17
7 Before attempting the overhaul of the master cylinder, obtain the proper rebuild kit.
8 Pull off the reservoir and remove the grommets **(see illustrations)**.
9 Remove the stopper cap **(see illustrations)**.

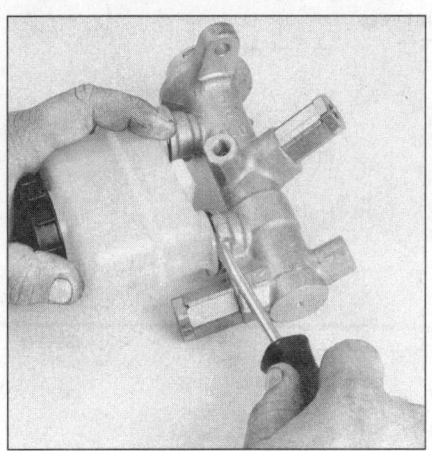

8.8a Pry the reservoir loose from the grommets . . .

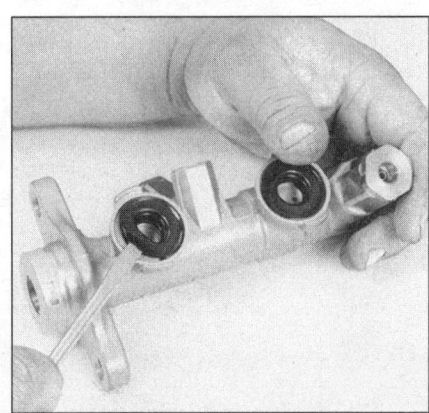

8.8c . . . and pry out the grommets

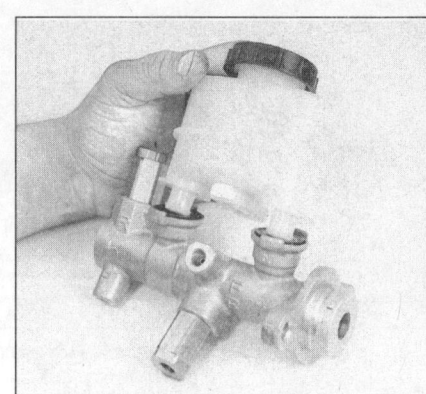

8.8b . . . pull off the reservoir . . .

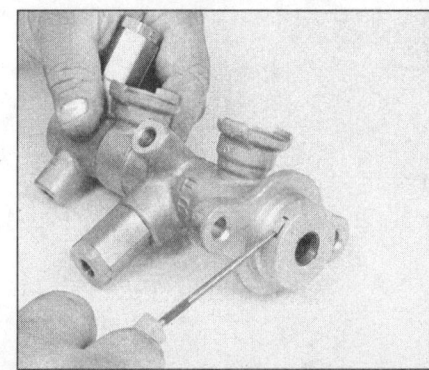

8.9a To remove the stopper cap, pry open the two small tangs on the side with a small screwdriver . . .

9

Chapter 9 Brakes

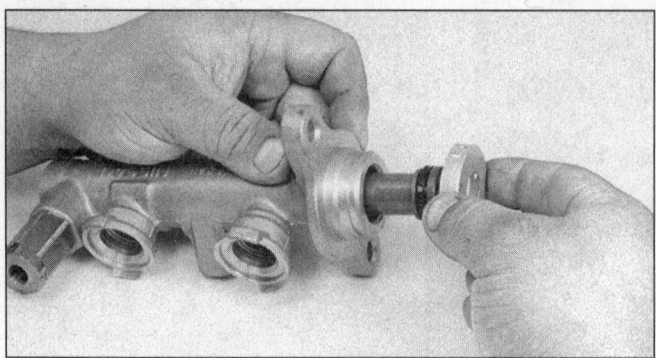

8.9b ... and remove the stopper cap

8.10b ... tap the master cylinder on a block of wood and remove the secondary piston assembly

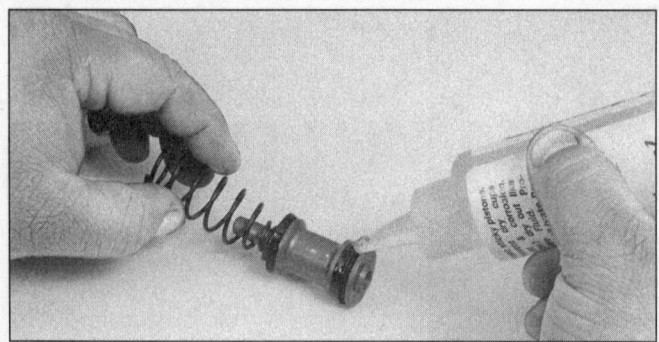

8.14a Lubricate the piston cups with brake assembly lube or clean brake fluid

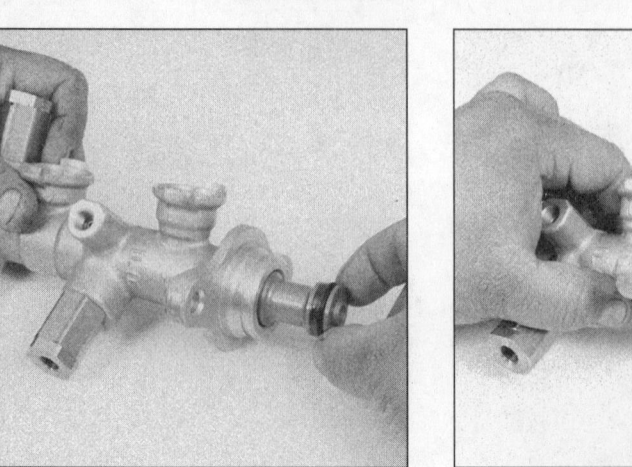

8.14b Insert the secondary piston assembly ...

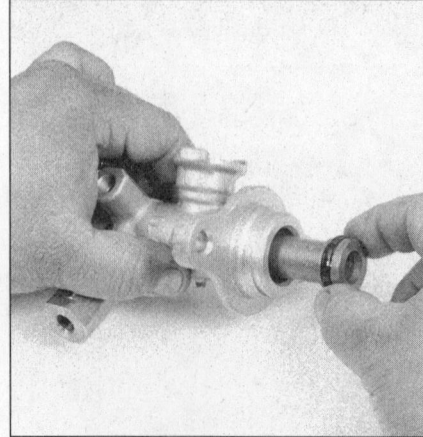

8.14c ... and the primary piston assembly into the master cylinder

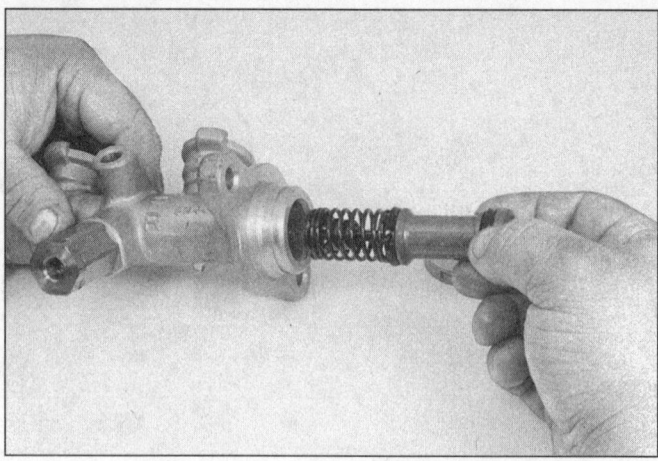

8.10a Remove the primary piston assembly ...

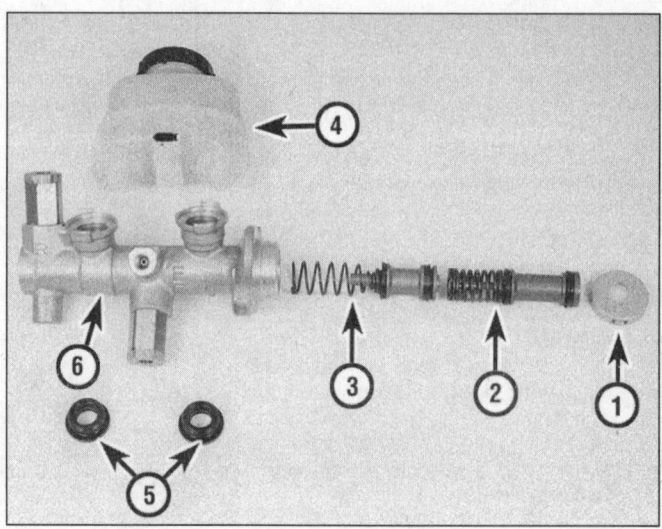

8.11 Lay out the parts like this to ensure proper reassembly - note the direction that the piston cups face

1 Stopper cap
2 Primary piston assembly
3 Secondary piston assembly
4 Reservoir
5 Grommets
6 Master cylinder body

10 The internal components can now be removed from the bore (see illustrations).
11 Note the sequence in which the parts were disassembled so they can be returned to their original locations. Laying all the parts out in order is one way to ensure everything is reassembled correctly (see illustration).
12 Carefully inspect the bore of the master cylinder. Any score marks, corrosion or other damage will mean a new master cylinder is required. DO NOT attempt to hone the bore.
13 Replace all parts included in the rebuild kit, following any instructions in the kit. Clean all re-used parts with brake system cleaner. **Warning:** *Do not use any petroleum-based solvents. During reassembly, lubricate all parts liberally with clean brake fluid.*
14 Lubricate the assembled secondary and primary piston assemblies with clean brake fluid or brake assembly lube and insert them

Chapter 9 Brakes

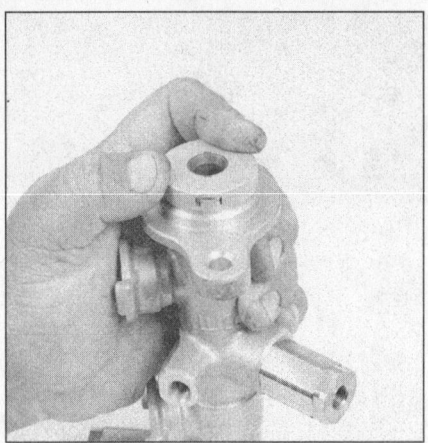

8.15a Install the new stopper cap . . .

8.15b . . . and bend the tangs inward

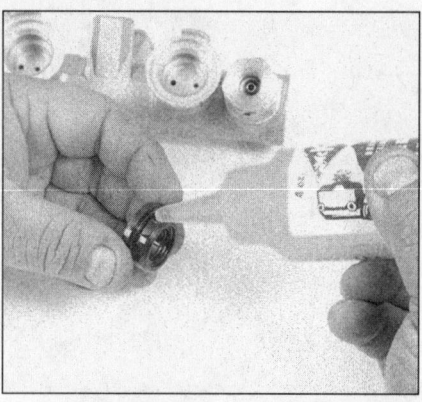

8.16a Lubricate the outer surfaces of the new grommets with brake assembly lube or clean brake fluid to make them easier to push into the master cylinder . . .

into the bore, bottoming them against the end of the master cylinder **(see illustrations)**.
15 Place the cylinder in a bench vise and install the stopper cap **(see illustrations)**.
16 Install the reservoir grommets and reservoir **(see illustrations)**.
17 If you'd like to make the fluid level in the reservoir easier to read, use a laundry marker to highlight the raised "MAX" and "MIN" marks on the side of the reservoir **(see illustration)**.
18 Bench bleed the master cylinder before installing it. You'll have to apply pressure to the master cylinder piston and, at the same time, control flow from the brake line outlets, so put the master cylinder in a vise, with the jaws of the vise clamping on the mounting flange.
19 If you can find them, insert threaded plugs into the brake line outlet holes and snug them down so no air will leak past them, but not so tight that they can't be easily loosened. If you can't find plugs that will fit, you can use your fingers to block the holes (see Step 24).
20 Fill the reservoir with brake fluid of the recommended type (see Chapter 1).
21 Remove one plug and push the piston

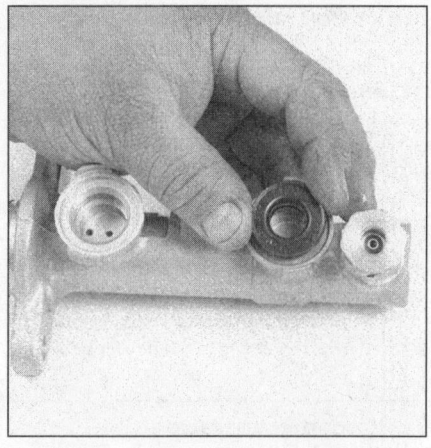

8.16b . . . install the grommets . . .

assembly into the bore to expel the air from the master cylinder. A large Phillips screwdriver can be used to push on the piston assembly.
22 To prevent air from being drawn back into the master cylinder, the plug must be replaced and snugged down before releasing

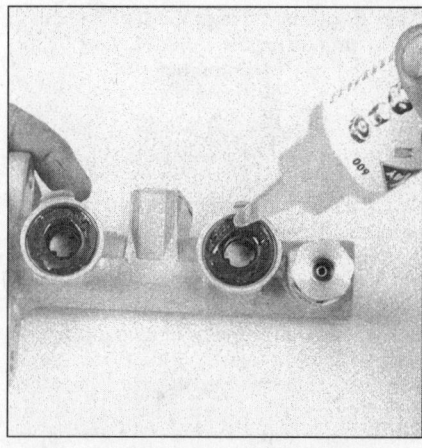

8.16c . . . lubricate the inner surface of the grommets to make the reservoir easier to push into the grommets . . .

the pressure on the piston.
23 Repeat this procedure until brake fluid, free of air bubbles, is expelled from the brake line outlet hole. When only brake fluid is expelled, repeat the procedure at the other outlet hole and plug. Be sure to keep the

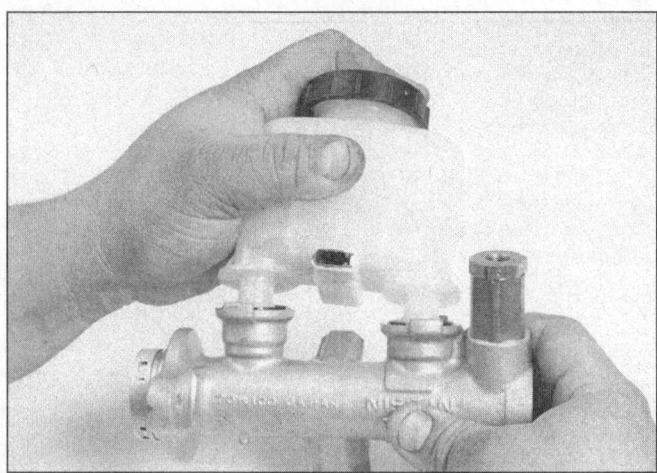

8.16d . . . and install the reservoir by pushing it firmly into the grommets

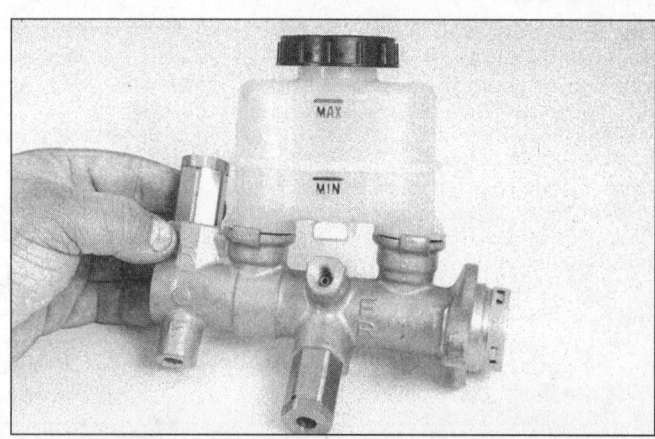

8.17 While you've got the master cylinder out of the vehicle, this is an excellent time to highlight the "MIN" and "MAX" level markings with a laundry marker; this will make them much easier to read

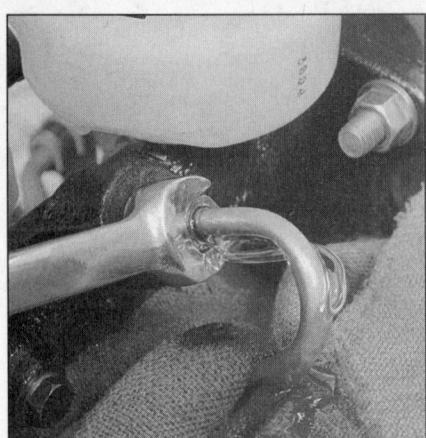

8.28 Have an assistant depress the brake pedal and hold it down, then loosen the fitting nut, allowing the air and fluid to escape; repeat this procedure on both fittings until the fluid is clear of air bubbles

9.1 A typical proportioning valve; to replace it, simply unscrew all the threaded fittings and disconnect and plug the brake lines, then remove the mounting bolt

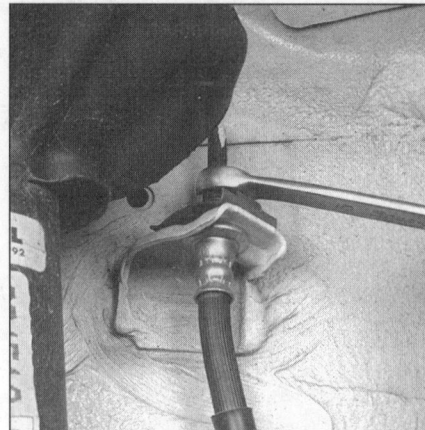

10.3 Loosen the threaded fitting on the brake line; use a flare-nut wrench, if available, to protect the corners of the nut

master cylinder reservoir filled with brake fluid to prevent the introduction of air into the system.

24   Since high pressure isn't involved in the bench bleeding procedure, an alternative to the removal and replacement of the plugs with each stroke of the piston assembly is available. Before pushing in on the piston assembly, remove the plug as described in Step 21. Before releasing the piston, however, instead of replacing the plug, simply put your finger tightly over the hole to keep air from being drawn back into the master cylinder. Wait several seconds for brake fluid to be drawn from the reservoir into the bore, then depress the piston again, removing your finger as brake fluid is expelled. Be sure to put your finger back over the hole each time before releasing the piston, and when the bleeding procedure is complete for that outlet, replace the plug and tighten it before going on to the other port.

Installation

Refer to illustration 8.28

25   Install the master cylinder over the studs on the power brake booster and tighten the nuts only finger-tight at this time. Don't forget to use a new gasket.
26   Thread the brake line fittings into the master cylinder. Since the master cylinder is still a bit loose, it can be moved slightly so the fittings thread in easily. Don't strip the threads as the fittings are tightened.
27   Tighten the mounting nuts to the torque listed in this Chapter's Specifications. Tighten the brake line fittings securely.
28   Fill the master cylinder reservoir with fluid, then bleed the the lines at the master cylinder, followed by bleeding the remainder of the brake system (see Section 11). To bleed the lines at the master cylinder, have an assistant depress the brake pedal and hold it down. Loosen the fitting to allow air and fluid to escape **(see illustration)**. Tighten the fitting, then allow your assistant to return the pedal to its rest position. Repeat this procedure on both fittings until the fluid is free of air bubbles, then bleed the rest of the system. Check the operation of the brake system carefully before driving the vehicle.
Warning: *If you do not have a firm brake pedal at the end of the bleeding procedure, or have any doubts as to the effectiveness of the brake system, DO NOT drive the vehicle. Have it towed to a dealer service department or other qualified repair shop for diagnosis.*

9   Proportioning valve - replacement

Refer to illustration 9.1

1   On some models the proportioning valve is mounted on the firewall **(see illustration)**. On other models it is built into the master cylinder. Its purpose is to limit hydraulic pressure to the rear brakes under heavy braking conditions to prevent rear wheel lockup.
2   The valve is not serviceable; if you suspect it's malfunctioning, have it checked by a dealer service department or repair shop equipped with the necessary pressure gauges.
3   If the valve is leaking or has been determined to be defective, replace it by unscrewing the brake lines (using a flare-nut wrench, if available) and unbolting the valve from its mounting bracket. After the new valve is installed, bleed the complete brake system as described in Section 11.

10   Brake hoses and lines - inspection and replacement

Inspection

1   About every six months, with the vehicle raised and supported securely on jackstands, the rubber hoses which connect the steel brake lines with the front and rear brake assemblies should be inspected for cracks, chafing of the outer cover, leaks, blisters and other damage. These are important and vulnerable parts of the brake system and inspection should be complete. A light and mirror will be helpful for a thorough check. If a hose exhibits any of the above conditions, replace it with a new one.

Replacement

Front brake hose

Refer to illustrations 10.3 and 10.4

2   Loosen the wheel lug nuts, raise the vehicle and support it securely on jackstands. Remove the wheel.
3   At the bracket, unscrew the brake line fitting from the hose. Use a flare-nut wrench to prevent rounding off the corners **(see illustration)**.
4   Remove the U-clip from the female fitting at the bracket with a pair of pliers **(see illustration)**, then pass the hose through the bracket.
5   At the caliper end of the hose, remove the banjo bolt, then separate the hose from the caliper. Note that there are two copper

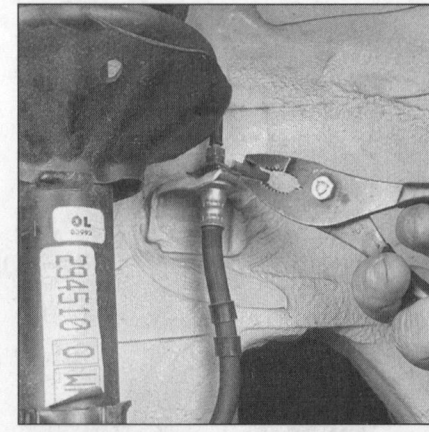

10.4 Pull off the U-clip with a pair of pliers

Chapter 9 Brakes

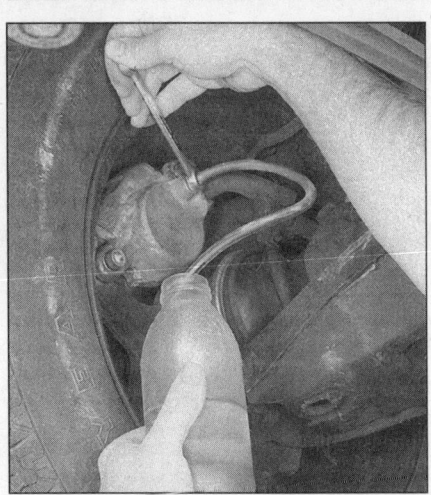

11.8 When bleeding the brakes, a hose is connected to the bleed screw at the caliper or wheel cylinder and then submerged in brake fluid - air will be seen as bubbles in the tube and container (all air must be expelled before moving to the next wheel)

sealing washers on either side of the banjo fitting - they should be replaced with new ones during installation.
6   Remove the U-clip from the strut bracket, then detach the hose from the bracket.
7   To install the hose, pass the caliper fitting end through the strut bracket, then connect the fitting to the caliper with the banjo bolt and copper washers.
8   Make sure the hose isn't twisted between the caliper and the strut bracket.
9   Route the hose into the frame bracket, again making sure it isn't twisted, then connect the brake line fitting, starting the threads by hand. Install the U-clip, then tighten the fitting securely.
10 Bleed the caliper (see Section 11).
11 Install the wheel and lug nuts, lower the vehicle and tighten the lug nuts to the torque listed in the Chapter 1 Specifications.

Rear brake hose

12 The rear brake hose serves as the flexible connection between two rigid metal lines, one on the body and the other on the axle. Both ends of the hose are attached to these metal lines with threaded fittings and U-clips. Refer to Steps 2, 3 and 4. Be sure to bleed the wheel cylinder when you're done (see Section 11).

Metal brake lines

13 When replacing brake lines, be sure to use the correct parts. Don't use copper tubing for any brake system components. Purchase steel brake lines from a dealer or auto parts store.
14 Prefabricated brake lines, with the ends already flared and fittings installed, are available at auto parts stores and dealer service departments. If necessary, carefully bend the line to the proper shape. A tube bender is recommended for this. **Caution:** *Don't crimp or damage the line.*
15 When installing the new line, make sure it's securely supported in the brackets and has plenty of clearance between moving or hot components.
16 After installation, check the master cylinder fluid level and add fluid as necessary. Bleed the brake system (see Section 11) and test the brakes carefully before driving the vehicle in traffic.

11   Brake hydraulic system - bleeding

Refer to illustration 11.8
Warning: *Wear eye protection when bleeding the brake system. If the fluid comes in contact with your eyes, immediately rinse them with water and seek medical attention.*
Note: *Bleeding the hydraulic system is necessary to remove any air that manages to find its way into the system when it's been opened during removal and installation of a hose, line, caliper or master cylinder.*
1   You'll probably have to bleed the system at all four brakes if air has entered it due to low fluid level, or if the brake lines have been disconnected at the master cylinder.
2   If a brake line was disconnected only at a wheel, then only that caliper or wheel cylinder must be bled.
3   If a brake line is disconnected at a fitting located between the master cylinder and any of the brakes, that part of the system served by the disconnected line must be bled.
4   Remove any residual vacuum from the brake power booster by applying the brake several times with the engine off.
5   Remove the master cylinder reservoir cover and fill the reservoir with brake fluid. Reinstall the cover. Check the fluid level often during the bleeding operation and add fluid as necessary to prevent the fluid level from falling low enough to allow air bubbles into the master cylinder. If you're working on a model equipped with ABS, turn the ignition switch off and disconnect the electrical connector from the ABS actuator.
6   Have an assistant on hand, as well as a supply of new brake fluid, a clear plastic container partially filled with clean brake fluid, a length of plastic, rubber or vinyl tubing to fit over the bleeder valve and a wrench to open and close the bleeder valve.
7   Beginning at the right rear wheel, loosen the bleeder valve slightly, then tighten it to a point where it's snug but can still be loosened quickly and easily.
8   Place one end of the tubing over the bleeder valve and submerge the other end in brake fluid in the container **(see illustration)**.
9   Have the assistant depress the brake pedal and hold it down firmly.
10 While the pedal is held down, open the bleeder valve just enough to allow a flow of fluid to leave the valve. Watch for air bubbles to exit the submerged end of the tube. When the fluid flow slows after a couple of seconds, close the valve and have your assistant release the pedal.
11 Repeat Steps 9 and 10 until no more air is seen leaving the tube, then tighten the bleeder valve and proceed to the left front wheel, the left rear wheel and the right front wheel, in that order, and perform the same procedure. Be sure to check the fluid in the master cylinder reservoir frequently.
12 Never use old brake fluid. It contains moisture which will deteriorate the brake system components and could cause the fluid to boil, which could render the brake system inoperative.
13 Refill the master cylinder with fluid at the end of the operation. If you're working on a model with ABS, be sure to reconnect the electrical connector to the ABS actuator.
14 Check the operation of the brakes. The pedal should feel solid when depressed, with no sponginess. If necessary, repeat the entire process. **Warning:** *Do not operate the vehicle if you're in doubt about the effectiveness of the brake system.*

12   Power brake booster - check, removal and installation

Operating check

1   Depress the brake pedal several times with the engine off and make sure there's no change in the pedal reserve distance.
2   Depress the pedal and start the engine. If the pedal goes down slightly, operation is normal.

Airtightness check

3   Start the engine and turn it off after one or two minutes. Depress the brake pedal slowly several times. If the pedal depresses less each time, the booster is airtight.
4   Depress the brake pedal while the engine is running, then stop the engine with the pedal depressed. If there's no change in the pedal reserve travel after holding the pedal for 30 seconds, the booster is airtight.

Removal

5   Power brake booster units shouldn't be disassembled. They require special tools not normally found in most automotive repair stations or shops. They're fairly complex and, because of their critical relationship to brake performance, should be replaced with a new or rebuilt one.
6   To remove the booster, first remove the brake master cylinder (see Section 8).
7   Remove the steering column lower finish panel. Locate the pushrod clevis connecting the booster to the brake pedal. It's accessible from inside the vehicle, under the dash on the driver's side.
8   Remove the clevis pin retaining clip with pliers and pull out the clevis pin.
9   Disconnect the hose leading from the engine to the booster. Be careful not to dam-

9-20　Chapter 9　Brakes

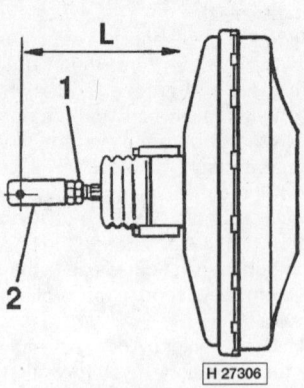

12.14 Measure the power brake booster input rod length to make sure it is as specified

L The pushrod length
1 Locknut
2 Pushrod clevis

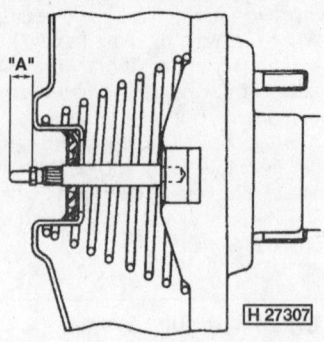

12.16 Measure the power brake booster output rod length (A) and check the measurement against the one in the Specifications Section

13.4 To adjust parking brake lever travel, turn the adjusting nut until the specified travel is obtained

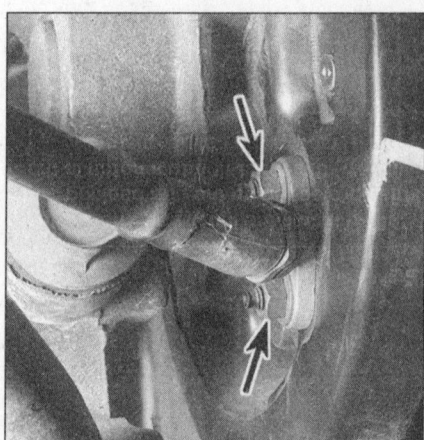

14.4a The parking brake cable retainer is attached to the backing plate by these two nuts; to detach the cable housing from the backing plate, remove the nuts and pull the cable out of the plate

age the hose when removing it from the booster fitting.

10 Remove the four nuts and washers holding the brake booster to the firewall; you may need a light to see them.

11 Slide the booster straight out from the firewall until the studs clear the holes.

Installation

12 Installation procedures are basically the reverse of removal. However, be sure to make the following adjustments where necessary.

Adjustment

13 Measure and - where possible - adjust the following dimensions: The length of the input rod between the clevis and the booster, and the length of the booster output rod (pushrod) between the booster and the master cylinder primary piston.

Input rod length

Refer to illustration 12.14

14 Measure the length of the input rod and compare your measurement to the dimension listed in this Chapter's Specifications **(see illustration)**.

15 If the input rod length isn't within specification, loosen the locknut and turn the clevis until the dimension is correct, then tighten the locknut securely.

Output rod length

Refer to illustration 12.16

16 Apply approximately 20 in-Hg of vacuum to the brake booster with a hand-operated vacuum pump and measure the length of the output rod, comparing your measurement to the dimensions listed in this Chapter's Specifications **(see illustration)**. If the rod length is outside specifications, replace the booster.

13 Parking brake - check and adjustment

Check

1 The parking brake, when properly adjusted, should travel seven to eight clicks on models with rear drum brakes and eight to nine clicks on models with rear disc brakes.

2 If the parking brake lever travels less than the specified minimum number of clicks, it might not be releasing completely and the shoes or pads could even be dragging against the drum or disc. If the lever can be pulled up more than the specified maximum number of clicks, the parking brake may not hold adequately on an incline, allowing the car to roll.

Adjustment

Refer to illustration 13.4

3 To gain access to the parking brake cable adjuster, remove the center console (see Chapter 11).

4 Loosen or tighten the adjusting nut until the desired travel is attained **(see illustration)**. Turn the nut clockwise to tighten the cable and decrease the number of clicks at the parking brake lever, or turn it counter-clockwise to loosen the cable and increase the number of clicks at the lever.

5 Install the console (see Chapter 11).

14 Parking brake cables - replacement

1 Make sure the parking brake is completely released.

2 Loosen the rear wheel lug nuts, raise the rear of the vehicle and support it securely on jackstands. Block the front wheels. Remove the wheel.

Rear cables

Refer to illustrations 14.4a, 14.4b, 14.4c and 14.7

3 On models with rear drum brakes, remove the brake drums and the brake shoes, and disconnect the cable from the parking brake levers on both trailing shoes (see Section 6).

4 On models with rear drum brakes, disconnect the cable retainers from the backing plates and pull the cables through the plates **(see illustration)**. On models with rear disc brakes, remove the large lock plates that attach the cables to the brackets on the calipers, then disconnect the cables from the rear calipers **(see illustrations)**.

5 Unbolt any cable brackets from the frame or the suspension pieces.

6 Remove the exhaust pipe and catalytic converter heat shields.

7 To disconnect the rear cables, remove the two cable housing nuts just behind the equalizer and disengage the cables from the equalizer **(see illustration)**.

8 Installation is the reverse of removal. Apply a light coat of grease to the portion of the cable end that engages with the equalizer.

9 Adjust the parking brake when you're

Chapter 9 Brakes 9-21

14.4b On models with rear disc brakes, the parking brake cable is attached to a bracket on the caliper by a lock plate which can be removed with a prybar . . .

14.4c . . . then unhook the cable end from the parking brake lever on the caliper

done (see Section 13).
10 Installation is the reverse of removal. Apply a light coat of grease to the portion of the cable end that engages with the equalizer.
11 Adjust the parking brake lever when you're done (see Section 13).

15 Brake light switch - check and replacement

Check

Refer to illustration 15.1
1 The brake light switch is located on a bracket at the top of the brake pedal **(see illustration)**. The switch activates the brake lights at the rear of the vehicle when the pedal is depressed.
2 To check the brake light switch, simply note whether the brake lights come on when the pedal is depressed and go off when the pedal is released. If they don't, adjust the switch as described in Chapter 1, Section 16 (adjusting the switch is part of brake pedal adjustment).
3 If the switch still doesn't work properly, either it's not getting voltage, or the switch itself is defective.
4 Use a voltmeter or test light to verify that there's voltage at the switch connector. With the pedal at rest, voltage should be present at one of the terminals of the switch. With the pedal depressed, voltage should be present at both terminals. If voltage isn't present at both terminals when the pedal is depressed, replace the switch. If voltage is present on both terminals with the pedal depressed, trace the circuit between the switch and the brake lights for an open-circuit condition.

Replacement

5 Unplug the electrical connector for the brake light switch.
6 Loosen the locknuts and unscrew the switch from the pedal bracket.
7 Installation is the reverse of removal.
8 Adjust the brake pedal and brake light switch (see Chapter 1, Section 16).

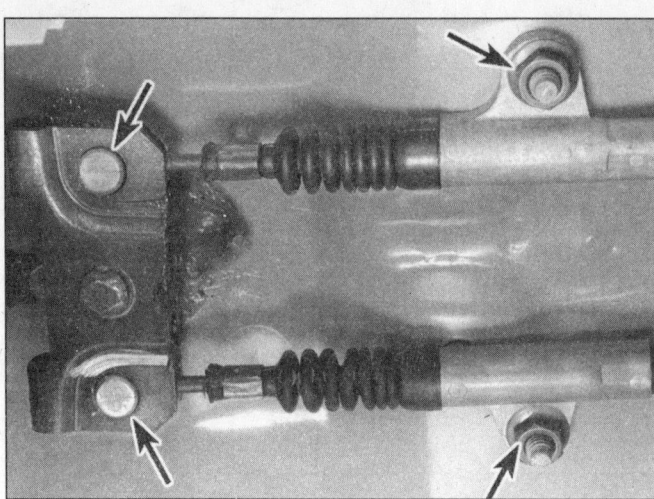

14.7 To disconnect the rear parking brake cables from the equalizer, simply remove these bracket nuts (right arrows), slide the cables forward slightly and disengage the cable ends (left arrows) from the equalizer

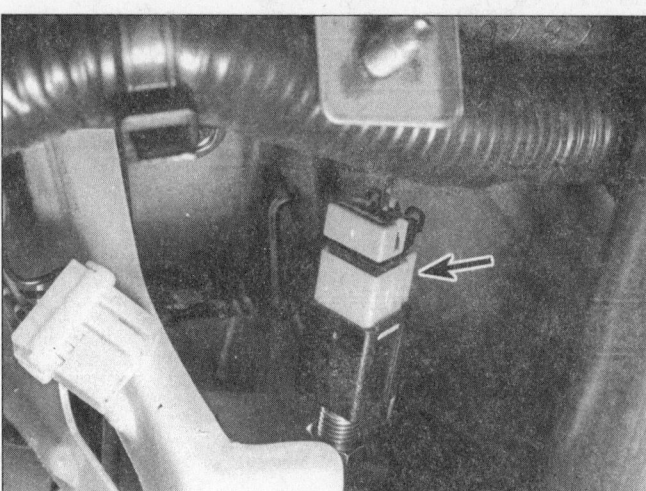

15.1 The brake light switch (arrow) is located at the top of the brake pedal

Notes

Chapter 10
Suspension and steering systems

Contents

	Section		Section
Balljoints - replacement	6	Steering gear boots - replacement	14
Control arm - removal, inspection and installation	5	Steering knuckle and hub - removal and installation	7
General information	1	Steering wheel - removal and installation	12
Hub and bearing assembly (front) - removal and installation	8	Strut assembly (front) - removal, inspection and installation	2
Hub and bearing assembly (rear) - removal and installation	11	Shock absorber/coil spring assembly (rear) - removal, inspection and installation	10
Power steering pump - removal and installation	16	Strut/coil spring assembly (front) - replacement	3
Power steering system - bleeding	17	Tie-rod ends - removal and installation	13
Rear axle assembly - removal and installation	9	Wheel alignment - general information	19
Stabilizer bar (front) - removal and installation	4	Wheels and tires - general information	18
Steering gear - removal and installation	15		

Specifications

Torque specifications   Ft-lbs

Front suspension
Strut piston rod self-locking nut	43 to 54
Strut upper mounting nuts	18 to 22
Strut-to-steering knuckle bolts/nuts	84 to 98
Balljoint-to-steering knuckle nut	43 to 54
Control arm	
Pivot bolt/nut	76 to 90
Bushing clamp bolts	58 to 72
Stabilizer bar	
Link-to-control arm nut	12 to 16
Link-to-stabilizer bar nut	34 to 38
Bushing clamp bolts	23 to 31

Rear suspension
Shock absorber assembly	
Piston rod nut	13 to 17
Upper mounting nuts	14 to 16
Lower mounting bolts/nuts	72 to 87
Trailing arm-to-chassis	72 to 87
Lateral link	
Lateral link-to-axle nut	72 to 87
Control rod-to-lateral link bolt/nut	43 to 58
Control rod-to-axle nut	72 to 87
Lateral link-to-chassis nut/bolt	72 to 87
Rear hub/bearing retainer nut	137 to 188

Steering
Airbag module Torx bolts (T50H)	11 to 18
High pressure line-to-steering gear fitting	14 to 20
High pressure line-to-pump banjo bolt	51 to 58
Pump pulley nut	40 to 50
Steering gear mounting clamp bolts	54 to 72
Steering shaft lower U-joint pinch-bolts	17 to 22
Steering wheel nut	22 to 29
Tie-rod-to-steering knuckle nut	22 to 29

Chapter 10 Suspension and steering systems

1.1 Front suspension components

1 Control arm front pivot bolt	4 Steering gear assembly	6 Steering knuckle
2 Control arm	5 Tie-rod end	7 Strut/coil spring assembly
3 Control arm rear bushing/clamp		

1 General information

Refer to illustrations 1.1 and 1.2

The front suspension is a MacPherson strut design. The upper end of each strut is attached to the vehicle body. The lower end of the strut is connected to the upper end of the steering knuckle **(see illustration)**. The steering knuckle is attached to a balljoint mounted on the outer end of the control arm. On some models a stabilizer bar is attached to the body with a pair of clamps and to the control arms with link bolts.

The rear suspension consists of a beam axle located by trailing arms and a lateral link, and is suspended by shock absorber/coil spring assemblies **(see illustration)**. The lateral link is fastened to the chassis and the axle and pivots on an internal control link.

The rack-and-pinion steering gear is located behind the engine/transaxle assembly at the bottom of the firewall (it's bolted to the lower rear crossmember and actuates the tie-rods, which are attached to the steering knuckles. The inner ends of the tie-rods are protected by rubber boots which should be inspected periodically for secure attachment, tears and leaking lubricant.

The power assist system consists of a belt-driven pump and associated lines and hoses. The fluid level in the power steering pump reservoir should be checked periodically (see Chapter 1).

The steering wheel operates the steering shaft, which actuates the steering gear through universal joints. Looseness in the steering can be caused by wear in the steering shaft universal joints, the steering gear, the tie-rod ends and loose retaining bolts.

Frequently, when working on the suspension or steering system components, you may come across fasteners which seem impossible to loosen. These fasteners on the underside of the vehicle are continually subjected to water, road grime, mud, etc., and can become rusted or "frozen," making them extremely difficult to remove. In order to unscrew these stubborn fasteners without damaging them (or other components), be sure to use lots of penetrating oil and allow it to soak in for a while. Using a wire brush to clean exposed threads will also ease removal of the nut or bolt and prevent damage to the threads. Sometimes a sharp blow with a hammer and punch will break the bond between a nut and bolt threads, but care must be taken to prevent the punch from slipping off the fastener and ruining the threads. Heating the stuck fastener and surrounding area with a torch sometimes helps too, but isn't recommended because of the obvious dangers associated with fire. Long breaker bars and extension, or "cheater," pipes will increase leverage, but never use an extension pipe on a ratchet - the ratcheting mechanism could be damaged. Sometimes tightening the nut or bolt first will help to break it loose. Fasteners that require drastic measures to remove should always be replaced with new ones.

Since most of the procedures dealt with in this Chapter involve jacking up the vehicle and working underneath it, a good pair of jackstands will be needed. A hydraulic floor jack is the preferred type of jack to lift the vehicle, and it can also be used to support certain components during various operations. **Warning:** *Never, under any circumstances, rely on a jack to support the vehicle while working on it. Whenever any of the sus-*

Chapter 10 Suspension and steering systems

1.2 Rear suspension components

1. Lateral link
2. Trailing arm
3. Shock absorber/coil spring assembly
4. Axle beam
5. Control link

pension or steering fasteners are loosened or removed they must be inspected and, if necessary, replaced with new ones of the same part number or of original equipment quality and design. Torque specifications must be followed for proper reassembly and component retention. Never attempt to heat or straighten any suspension or steering components. Instead, replace any bent or damaged part with a new one.

2 Strut assembly (front) - removal, inspection and installation

Removal

Refer to illustrations 2.3 and 2.5

1 Loosen the front wheel lug nuts, raise the front of the vehicle and support it securely on jackstands. Remove the wheels.
2 Unclip the brake hose from the strut bracket and detach it from the bracket. If the vehicle is equipped with ABS, detach the speed sensor wiring harness from the strut by removing the clamp bracket bolt.
3 Remove the strut-to-knuckle nuts **(see illustration)** and knock the bolts out with a hammer and punch.

4 Separate the strut from the steering knuckle. Be careful not to overextend the inner CV joint.
5 Support the strut and spring assembly with one hand and remove the strut upper mounting nuts **(see illustration)**. Remove the assembly out from the fenderwell.

Inspection

6 Check the strut body for leaking fluid, dents, cracks and other obvious damage which would warrant repair or replacement.
7 Check the coil spring for chips or cracks in the spring coating (this will cause premature spring failure due to corrosion). Inspect

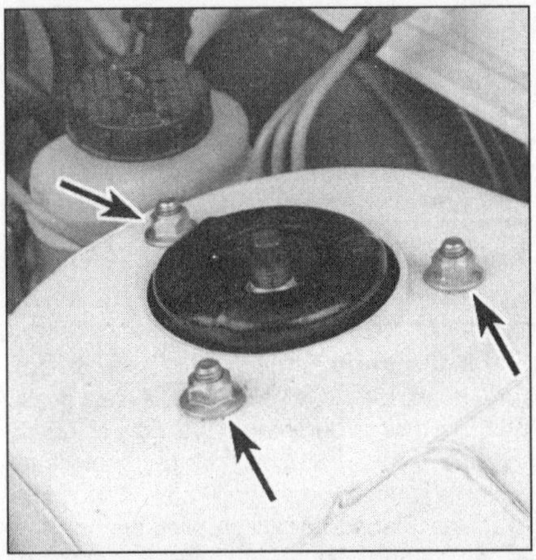

2.3 To detach the strut from the steering knuckle, remove the two nuts (arrows), then knock out the bolts with a hammer and punch

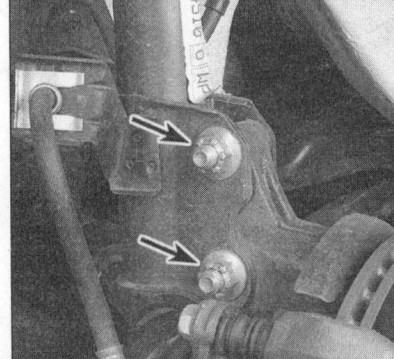

2.5 To detach the upper end of the strut assembly from the body, remove the upper mounting nuts (arrows). Warning: *Don't remove the center nut!*

Chapter 10 Suspension and steering systems

3.3 Install the spring compressor according to the tool manufacturer's instructions and compress the spring until all pressure is relieved from the upper spring seat

3.4 Remove the damper shaft nut

3.5 Lift the suspension support off the damper shaft

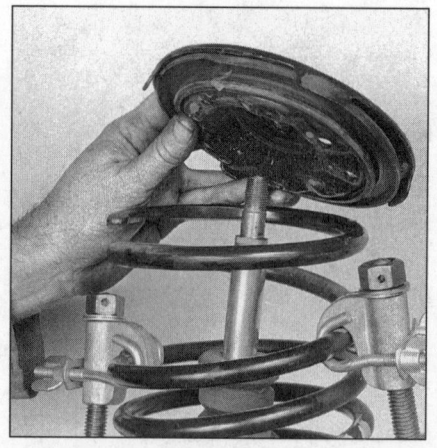

3.6 Remove the spring seat from the damper shaft

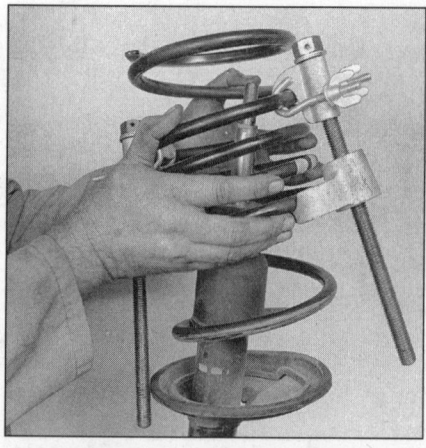

3.7 Remove the compressed spring assembly - keep the ends of the spring pointed away from your body

the spring seat for cuts, hardness and general deterioration.
8 If any undesirable conditions exist, proceed to the strut disassembly procedure (see Section 3).

Installation

9 Guide the strut assembly up into the fenderwell and insert the upper mounting studs through the holes in the shock tower. Once the studs protrude from the shock tower, install the nuts so the strut won't fall back through. This is most easily accomplished with the help of an assistant, as the strut is quite heavy and awkward.
10 Slide the steering knuckle into the strut flange and insert the two bolts. Install the nuts and tighten them to the torque listed in this Chapter's Specifications.
11 Guide the brake hose through its bracket in the strut and install the clip.
12 Install the wheel and lug nuts, then lower the vehicle and tighten the lug nuts to the torque listed in the Chapter 1 Specifications.
13 Tighten the upper mounting nuts to the torque listed in this Chapter's Specifications.
14 Drive the vehicle to an alignment shop to have the front end alignment checked, and if necessary, adjusted.

3 Strut/shock absorber or coil spring - replacement

1 If the struts/shock absorbers or coil springs exhibit the telltale signs of wear (leaking fluid, loss of damping capability, chipped, sagging or cracked coil springs) explore all options before beginning any work. The strut/shock absorber assemblies are not serviceable and must be replaced if a problem develops. However, strut/shock absorber assemblies complete with springs may be available on an exchange basis, which eliminates much time and work. Whichever route you choose to take, check on the cost and availability of parts before disassembling your vehicle. **Warning:** *Disassembling a strut or coil-over shock absorber is potentially dangerous and utmost attention must be directed to the job, or serious injury may result. Use only a high-quality spring compressor and carefully follow the manufacturer's instructions furnished with the tool. After removing the coil spring from the strut or shock absorber, set it aside in a safe, isolated area.*

Disassembly

Refer to illustrations 3.3, 3.4, 3.5, 3.6 and 3.7
2 Remove the strut or shock absorber assembly following the procedure described in Section 2 (front) or Section 10 (rear). Mount the assembly in a vise. Line the vise jaws with wood or rags to prevent damage to the unit and don't tighten the vise excessively.
3 Following the tool manufacturer's instructions, install the spring compressor (which can be obtained at most auto parts stores or equipment yards on a daily rental basis) on the spring and compress it sufficiently to relieve all pressure from the upper spring seat **(see illustration)**. This can be verified by wiggling the spring.

4 Loosen the damper shaft nut **(see illustration)**.
5 Remove the nut and suspension support **(see illustration)**. Inspect the bearing in the suspension support for smooth operation. If it doesn't turn smoothly, replace the suspension support. Check the rubber portion of the suspension support for cracking and general deterioration. If there is any separation of the rubber, replace it.
6 Lift the spring seat and upper insulator from the damper shaft **(see illustration)**. Check the rubber spring seat for cracking and hardness, replacing it if necessary.
7 Carefully lift the compressed spring from the assembly **(see illustration)** and set it in a safe place. **Warning:** *Never place your head near the end of the spring!*
8 Slide the rubber bumper off the damper shaft.
9 Check the lower insulator (if equipped) for wear, cracking and hardness and replace it if necessary.

Reassembly

Refer to illustration 3.11
10 If the lower insulator is being replaced,

Chapter 10 Suspension and steering systems

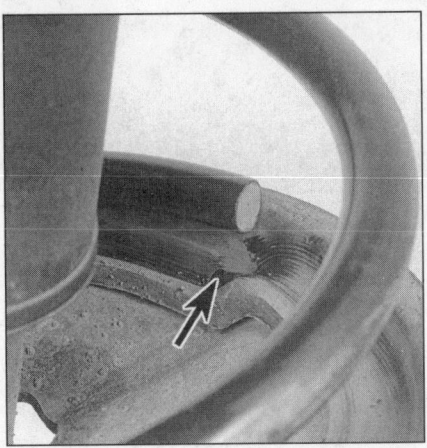

3.11 When installing the spring, make sure the end fits into the recessed portion of the lower seat (arrow)

4.2 To disconnect the stabilizer link from the control arm, remove this nut (arrow)

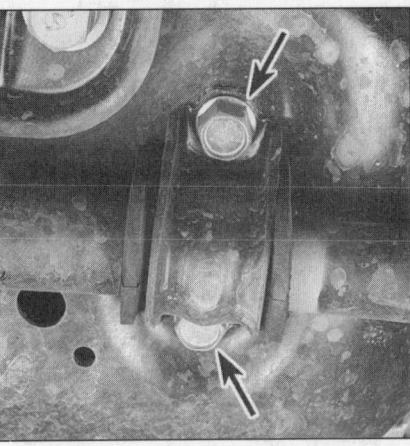

4.3 To disconnect the stabilizer bar from the body, remove these two bushing clamp bolts (arrows) (left clamp shown, right clamp identical)

set it into position with the dropped portion seated in the lowest part of the seat. Extend the damper rod to its full length and install the rubber bumper.

11 Carefully place the coil spring onto the lower insulator, with the end of the spring resting in the lowest part of the insulator **(see illustration)**.

12 Install the upper insulator and spring seat.

13 Install the dust seal and suspension support to the damper shaft.

14 Install the nut and tighten it to the torque listed in this Chapter's Specifications.

15 Install the strut/shock absorber and coil spring assembly following the procedure described in Section 2 (front) or Section 10 (rear).

4 Stabilizer bar (front) - removal and installation

Refer to illustrations 4.2 and 4.3

1 Loosen the wheel lug nuts, raise the front of the vehicle, support it securely on jackstands and remove the wheels.

2 Disconnect the stabilizer link nuts from both control arms **(see illustration)**.

3 Remove the stabilizer bar clamp bolts **(see illustration)**.

4 Remove the stabilizer assembly.

5 Inspect the clamp bushings and the link bushings. If they're cracked or torn, replace them.

6 Installation is the reverse of removal. Be sure to tighten all fasteners to the torque listed in this Chapter's Specifications.

5 Control arm - removal, inspection and installation

Removal

Refer to illustrations 5.3a, 5.3b, 5.3c, 5.3d, 5.4a and 5.4b

1 Loosen the wheel lug nuts on the side to be dismantled, raise the front of the vehicle, support it securely on jackstands and remove the wheel.

2 Disconnect the stabilizer link from the control arm (see Section 4).

3 Remove the cotter pin **(see illustration)**,

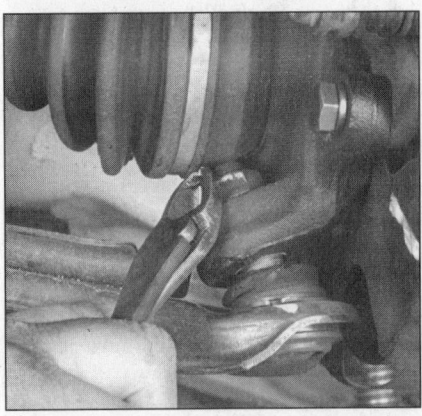

5.3a Remove the cotter pin . . .

loosen the balljoint stud nut **(see illustration)** and separate the balljoint stud from the steering knuckle with a "picklefork"-type balljoint separator **(see illustration)**. Be sure to grease the picklefork to protect the balljoint dust boot. Remove the nut and separate the arm from the steering knuckle **(see illustration)**.

5.3b . . . loosen the balljoint stud nut and back it off as far as it will go (without actually removing it) . . .

5.3c . . . then pop the balljoint stud loose with a "picklefork" (be sure to grease the picklefork to protect the balljoint dust boot)

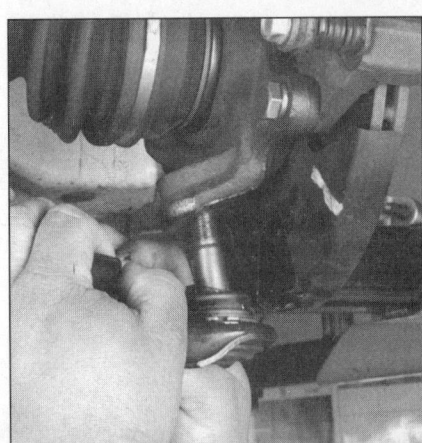

5.3d Remove the nut and separate the control arm from the steering knuckle by pulling down firmly

10-6 Chapter 10 Suspension and steering systems

5.4a Remove the front pivot bolt nut and the pivot bolt (arrows)

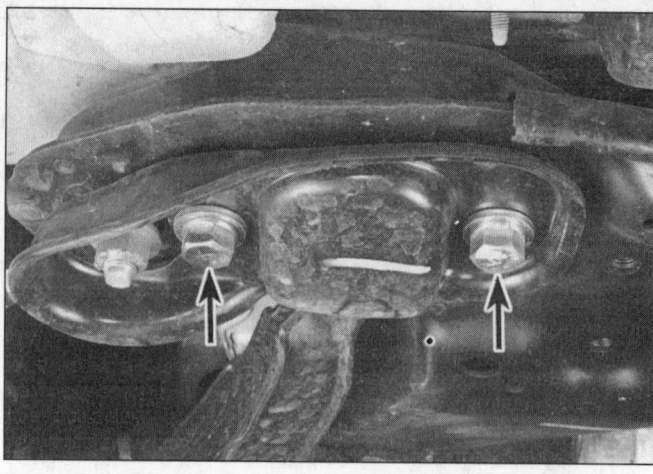

5.4b Remove the rear bushing clamp bolts (arrows)

4   Remove the front pivot bolt nut and/or bolt. Remove the rear bushing clamp bolts **(see illustrations)**. Remove the control arm.

Inspection

5   Inspect the front and rear bushings for cracks and tears. If the rear bushing is damaged, you can replace it yourself. The front bushing is not replaceable; if it is worn or damaged, replace the control arm.
6   Inspect the control arm for straightness. If it's bent, replace it. Do not attempt to straighten a bent control arm.

Installation

7   Installation is the reverse of removal. Tighten all of the fasteners to the torque listed in this Chapter's Specifications. Also be sure to install a new cotter pin. If the cotter pin hole doesn't line up with the slots on the nut, tighten the nut until the next hole in the nut lines up with the hole in the balljoint stud. **Note:** *The control arm pivot bolt nut and bushing clamp bolts should be tightened with the suspension at normal ride height. This can be simulated by raising the outer end of the arm with a floor jack.*
8   Install the wheel and lug nuts, lower the vehicle and tighten the lug nuts to the torque listed in the Chapter 1 Specifications.

6   Balljoints - replacement

1   Loosen the wheel lug nuts, raise the vehicle and support it securely on jackstands. Remove the wheel.
2   Remove the control arm (see Section 5).
3   On these models, the balljoint is an integral part of the control arm and cannot be replaced by itself, so if there is any wear or damage, replace the control arm.
4   Installation is the reverse of removal. Be sure to tighten all fasteners to the torque listed in this Chapter's Specifications.
5   Install the wheel and lug nuts. Lower the vehicle and tighten the lug nuts to the torque listed in the Chapter 1 Specifications.

6   It's a good idea to have the front wheel alignment checked, and if necessary, adjusted after this job has been performed.

7   Steering knuckle and hub - removal and installation

Warning: *Dust created by the brake system may contain asbestos, which is harmful to your health. Never blow it out with compressed air and don't inhale any of it. Do not, under any circumstances, use petroleum-based solvents to clean brake parts. Use brake system cleaner only.*

Removal

1   Loosen the wheel lug nuts, raise the vehicle and support it securely on jackstands. Remove the wheel.
2   Remove the brake caliper and the brake disc, then disconnect the brake hose from the strut (see Chapter 9).
3   If the vehicle is equipped with ABS, disconnect and remove the wheel speed sensor.
4   Remove the strut-to-steering knuckle nuts, but don't remove the bolts yet (see Section 2).
5   Separate the tie-rod end from the steering knuckle arm (see Section 13).
6   Separate the balljoint from the steering knuckle (see Section 5).
7   Push the driveaxle from the hub as described in Chapter 8. Support the end of the driveaxle with a piece of wire.
8   Remove the strut-to-knuckle bolts and separate the knuckle from the strut.

Installation

9   Guide the knuckle and hub assembly into position, inserting the driveaxle into the hub.
10 Push the knuckle into the strut flange and install the bolts and nuts, but don't tighten them yet.
11 Attach the control arm to the steering knuckle (see Section 5).
12 Attach the tie-rod end to the steering

knuckle arm (see Section 13). Tighten the strut-to-knuckle nuts to the torque listed in this Chapter's Specifications.
13 Place the brake disc on the hub and install the caliper as outlined in Chapter 9.
14 Install the driveaxle/hub nut and tighten it to the torque listed in the Chapter 8 Specifications.
15 Install the wheel and tighten the lug nuts but don't torque them yet.
16 Lower the vehicle and tighten the wheel lug nuts to the torque listed in the Chapter 1 Specifications.

8   Hub and bearing assembly (front) - removal and installation

Due to the special tools and expertise required to press the hub and bearing from the steering knuckle, this job should be left to a professional mechanic. However, the steering knuckle and hub may be removed and the assembly taken to a dealer service department or other repair shop. See Section 7 for the steering knuckle and hub removal procedure.

9   Rear axle assembly - removal and installation

Refer to illustrations 9.7 and 9.8

1   Loosen the wheel lug nuts, raise the vehicle and support it securely on jackstands. Remove the wheels.
2   On vehicles with an Anti-lock Brake System (ABS), remove the rear wheel speed sensors.
3   Disconnect the parking brake cables and remove the brake calipers and brake discs (see Chapter 9).
4   Disconnect the brake hose from the wheel cylinder or caliper (see Chapter 9). Have some rags and a container handy to catch the brake fluid. Unclip the hose from the strut bracket and push it through. If the vehicle is equipped with ABS, detach the

Chapter 10 Suspension and steering systems   10-7

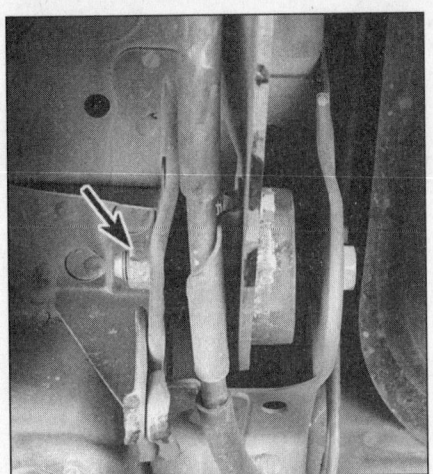

9.7 Remove the trailing arm front pivot nut and bolt (arrow)

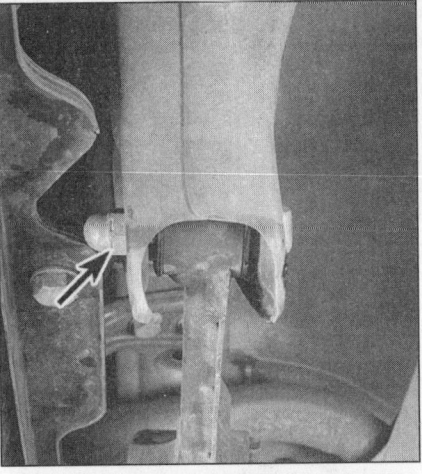

9.8 Disconnect the lateral rod from the chassis

10.2 To disconnect the lower end of the shock absorber from the axle, remove this nut and bolt (arrow)

speed sensor wiring harness by removing the clamp bracket bolt.
5   Remove the rear hub and wheel bearing assemblies (see Section 11).
6   Place a floor jack underneath the center of the rear axle assembly. Make sure the jack doesn't contact the lateral rod. Raise the jack just enough to support the axle assembly. **Note:** *If two floor jacks are available, place one at each end of the axle beam.*
7   Remove the pivot bolts for the trailing arms and, with an assistant helping to balance the axle assembly, carefully lower the axle/lateral link assembly on the jack **(see illustration)**.
8   Disconnect the lateral rod from the chassis **(see illustration)**.
9   Disconnect the lower ends of both shock absorbers from the axle (see Section 10).
10 Installation is the reverse of removal. Make sure you tighten all fasteners to the torque listed in this Chapter's Specifications. **Note:** *The lateral link and shock absorber fasteners should be tightened with the suspension at normal ride height (this can be simulated by raising the rear suspension with a floor jack). The rear axle bolts (at the forward ends of the trailing arms) should be tightened with the rear suspension in the unloaded (fully extended) position.*
11 Connect the brake hose and reconnect it to the wheel cylinder or caliper and bleed the brakes (see Chapter 9). If the vehicle is equipped with ABS, install the speed sensor wiring harness bracket.

10 Shock absorber/coil spring assembly (rear) - removal, inspection and installation

Removal

Refer to illustrations 10.2 and 10.3

1   Loosen the rear wheel lug nuts, raise the rear of the vehicle and support it securely on

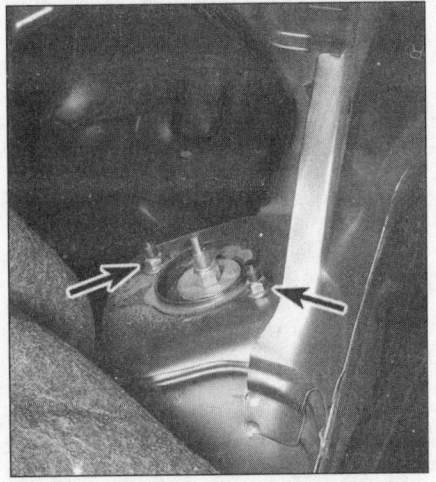

10.3 The shock absorber upper mounting nuts (arrows) are located in the trunk

jackstands. Remove the wheels.
2   Remove the lower shock absorber-to-axle nut, remove the bolt and separate the shock from the axle **(see illustration)**.
3   The upper mounting nuts are located inside the trunk. Have an assistant support the shock and spring assembly while you remove the two mounting nuts **(see illustration)**. Remove the assembly out from the fenderwell.

Inspection

4   Refer to Steps 6 and 7 in Section 2.
5   If any undesirable conditions exist, proceed to the disassembly procedure (see Section 3).

Installation

6   Guide the assembly up into the fenderwell and insert the upper mounting studs through the holes in the shock tower. Make sure the upper spring seat is aligned. Once the studs protrude from the shock tower, install the nuts so the shock absorber won't

11.3 Remove the hub/bearing grease cap with a hammer and chisel

fall back through. This is most easily accomplished with the help of an assistant, as the assembly is quite heavy and awkward.
7   Slide the lower end of the shock into the axle mount and insert the bolt. Raise the suspension with a floor jack to simulate normal ride height,, install the nut and tighten it to the torque listed in this Chapter's Specifications.
8   Tighten the upper mounting nuts to the torque listed in this Chapter's Specifications.

11 Hub and bearing assembly (rear) - removal and installation

Refer to illustrations 11.3, 11.4, 11.5, 11.6 and 11.7

1   Loosen the rear wheel lug nuts, raise the rear of the vehicle, support it securely on jackstands and remove the wheels.
2   Remove the brake drum or disc (see Chapter 9).
3   Pry off the grease cap **(see illustration)**.

10-8   Chapter 10 Suspension and steering systems

11.4 Remove the cotter pin

11.5 Remove the bearing/hub retaining nut

11.6 Remove the hub/bearing assembly

4   Remove the cotter pin **(see illustration)**.
5   Remove the bearing retainer nut **(see illustration)**.
6   Remove the hub and bearing assembly **(see illustration)**.
7   If it is necessary to remove the brake backing plate, disconnect the brake line fitting and parking brake cable, then remove the bolts **(see illustration)**.
8   Installation is the reverse of removal. Be sure to tighten all fasteners to the torque listed in this Chapter's Specifications.

12 Steering wheel - removal and installation

Warning: These models are equipped with airbags. The airbag is armed and can deploy (inflate) anytime the battery is connected. To prevent accidental deployment (and possible injury), disconnect the negative battery cable, followed by the positive battery cable, whenever working near airbag components. After the battery is disconnected, wait at least ten minutes before beginning work (the system has a back-up capacitor that must fully discharge).

Removal

Refer to illustrations 12.3, 12.4a, 12.4b, 12.6 and 12.7

1   Turn the steering wheel so that the front wheels are pointing straight ahead. Turn the ignition key to the lock position.
2   Disconnect the cable from the negative battery terminal, followed by the positive cable, and wait at least ten minutes before proceeding.
3   Remove the cover from the underside of the steering wheel and unplug the airbag module connector **(see illustration)**.
4   Remove the side covers and remove the Torx bolts behind them **(see illustrations)** with a T50H Torx bit. The Torx bolts are coated with a special bonding agent, so they must be discarded; be sure to replace them with new ones during reassembly.
5   Lift the airbag module off the steering wheel. **Warning:** Handle the airbag module with care and store it in a safe location with the trim side facing up. See the precautions in Chapter 12.

11.7 To remove the brake backing plate from the rear knuckle, remove these bolts (arrows); it's not necessary to remove the wheel cylinder, but you'll have to disconnect the brake line fitting and the parking brake cable

12.3 The airbag module connector is located inside a recess in the underside of the steering wheel; it can be accessed by removing the cover with a small screwdriver

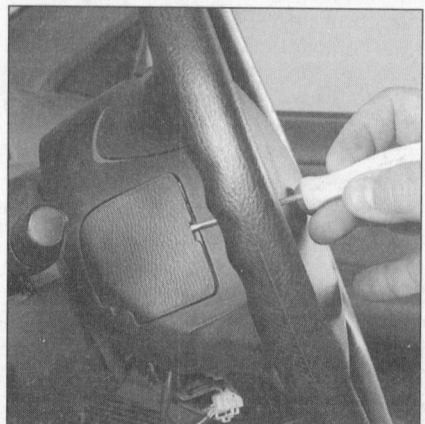

12.4a Use a small screwdriver to remove the plastic covers for access to the airbag module bolts

12.4b The airbag module is attached to the steering wheel with a pair of Torx T50H bolts located on either side of the steering wheel

Chapter 10 Suspension and steering systems

12.6 Remove the steering wheel retaining nut

12.7 Mark the relationship of the steering wheel to the shaft before removing the wheel

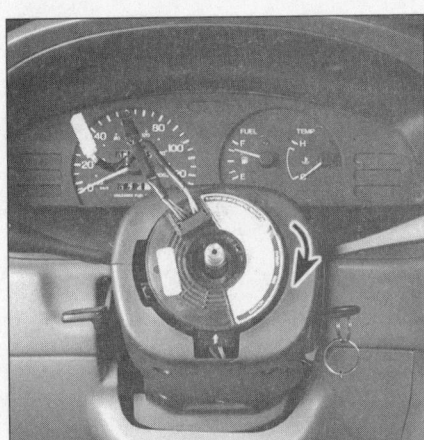

12.8a To align the spiral cable with the neutral position, turn the cable clockwise until it becomes hard to turn . . .

12.8b . . . then back it off two and a half turns until the white arrow lines up with the white mark on the spiral cable housing

6 Remove the steering wheel retaining nut **(see illustration)**.

7 Mark the relationship of the steering shaft to the hub (if the marks don't already exist or don't line up) to simplify installation and ensure steering wheel alignment **(see illustration)**. Remove the steering wheel from the shaft (use a puller if necessary - do not hammer on the shaft!). **Warning:** *Do not turn the steering shaft while the steering wheel is removed.*

Installation

Refer to illustrations 12.8a and 12.8b

8 Verify that the front wheels are pointing straight ahead, then turn the spiral cable clockwise by hand until it becomes hard to turn, then rotate the spiral cable counter-clockwise two and a half turns until the white arrow on the housing lines up with the white alignment mark on the cable assembly **(see illustrations)**.

9 Install the steering wheel. Make sure the airbag spiral cable pin guides are properly engaged with their corresponding holes in the back of the steering wheel and pull the spiral cable through.

10 Install the steering wheel retaining nut and tighten it to the torque listed in this Chapter's Specifications.

11 Plug in the horn connector and engage the spiral cable with the pawls in the steering wheel.

12 Install the airbag module and secure it with new Torx bolts. Do not reuse the old bolts. Install the side covers.

13 Plug in the airbag module connector. Install the cover.

14 Verify the airbag circuit is operational by turning the ignition key to the On or Start position. The "AIRBAG" warning light should illuminate for about seven seconds, then turn off.

13 Tie-rod ends - removal and installation

Removal

Refer to illustrations 13.2a, 13.2b and 13.4

1 Loosen the wheel lug nuts. Raise the front of the vehicle, support it securely on jackstands, block the rear wheels and set the parking brake. Remove the front wheel.

2 Loosen the jam nut enough to mark the position of the tie-rod end in relation to the threads **(see illustrations)**.

3 Remove the cotter pin and loosen, but don't remove, the nut on the tie-rod end stud.

4 Disconnect the tie-rod end from the

13.2a Loosen the jam nut . . .

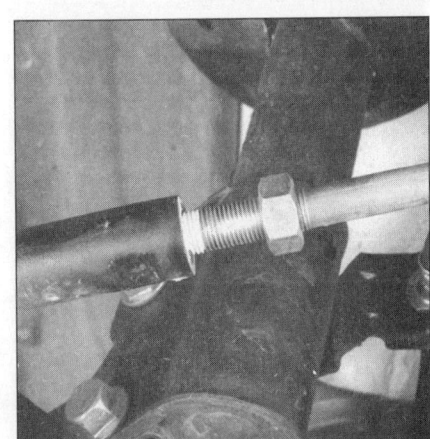

13.2b . . . then mark the position of the tie-rod end in relation to the threads

10-10 Chapter 10 Suspension and steering systems

13.4 Disconnect the tie-rod end from the steering knuckle arm with a puller (the nut has been loosened, but left on, to prevent the tie-rod end from separating violently)

14.3a The outer ends of the steering gear boots are secured by spring-type clamps

14.3b The inner ends of the steering gear boots are retained by clamps which must be cut off and discarded

steering knuckle arm with a puller **(see illustration)**. Remove the nut and separate the tie-rod.

5   Unscrew the tie-rod end from the tie-rod end.

Installation

6   Thread the tie-rod end on to the marked position and insert the tie-rod stud into the steering knuckle arm. Tighten the jam nut securely.

7   Install the castle nut on the stud and tighten it to the torque listed in this Chapter's Specifications. Install a new cotter pin.

8   Install the wheel and lug nuts. Lower the vehicle and tighten the lug nuts to the torque listed in the Chapter 1 Specifications.

9   Have the alignment checked by a dealer service department or an alignment shop.

14 Steering gear boots - replacement

Refer to illustrations 14.3a and 14.3b

1   Loosen the lug nuts, raise the vehicle

and support it securely on jackstands. Remove the wheel.

2   Remove the tie-rod end and jam nut (see Section 13).

3   Remove the outer steering gear boot clamp **(see illustration)** with a pair of pliers. Cut off the inner boot clamp **(see illustration)** with a pair of diagonal cutters. Slide the boot off.

4   Before installing the new boot, wrap the threads on the end of the steering rod with a layer of tape so the small end of the new boot isn't damaged.

5   Slide the new boot into position on the steering gear until it seats in the groove in the steering rod and install new clamps.

6   Remove the tape and install the tie-rod end (see Section 13).

7   Install the wheel and lug nuts. Lower the vehicle and tighten the lug nuts to the torque listed in the Chapter 1 Specifications.

15 Steering gear - removal and installation

Warning: *These models are equipped with airbags. The airbag is armed and can deploy*

(inflate) anytime the battery is connected. To prevent accidental deployment (and possible injury), disconnect the negative battery cable, followed by the positive battery cable, whenever working near airbag components. After the battery is disconnected, wait at least ten minutes before beginning work (the system has a back-up capacitor that must fully discharge).

Removal

Refer to illustrations 15.2, 15.3a, 15.3b, 15.5a and 15.5b

1   Loosen the front wheel lug nuts, raise the front of the vehicle and support it securely on jackstands. Apply the parking brake and remove the wheels. Remove the engine splash shields (see Chapter 1).

2   If equipped with power steering, place a drain pan under the steering gear. Detach the power steering pressure and return lines **(see illustration)** and cap the ends to prevent excessive fluid loss and contamination.

3   Remove the universal joint cover **(see illustration)**. Mark the relationship of the lower universal joint to the steering gear input shaft **(see illustration)**. Remove the lower

15.2 Disconnect the power steering line fittings (arrows)

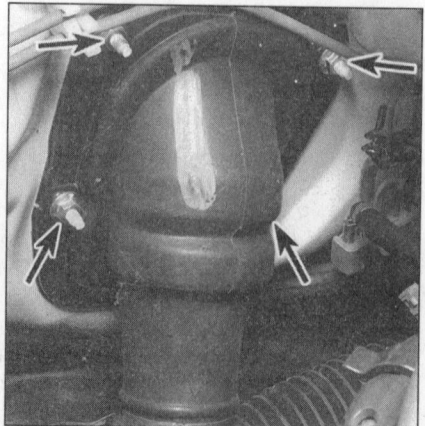

15.3a Remove the U-joint cover nuts (arrows) and peel the cover down far enough to get at the U-joint pinch bolt

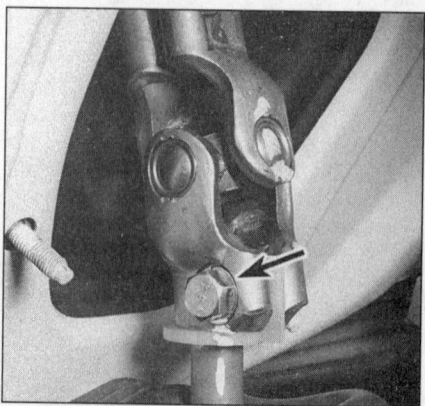

15.3b Mark the relationship of the universal joint to the steering gear input shaft, then remove the U-joint pinch bolt (arrow)

Chapter 10 Suspension and steering systems   10-11

15.5a To remove the steering gear assembly, remove the left retaining bolts (arrows) . . .

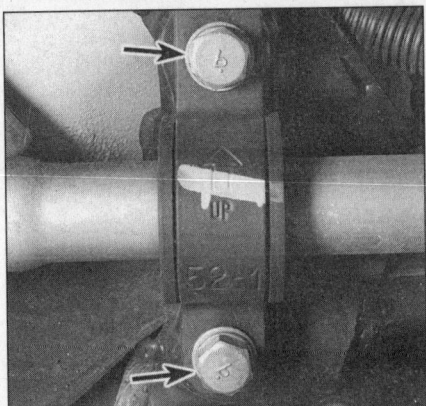

15.5b . . . and the right-side bolts (arrows)

intermediate shaft pinch bolt.
4   Separate the tie-rod ends from the steering knuckle arms (see Section 13).
5   Support the steering gear and remove the steering gear mounting bolts **(see illustrations)**. Separate the intermediate shaft from the steering gear input shaft and remove the steering gear assembly. **Warning:** *Do not turn the steering wheel while the steering gear is removed. If the steering wheel is inadvertently turned, remove the steering wheel and center the airbag spiral cable (see Section 12). To prevent the steering wheel from turning, loop the seat belt through the steering wheel and fasten it into its latch.*
6   Check the steering gear rubber mounts for excessive wear or deterioration, replacing them if necessary.

Installation

7   Raise the steering gear into position and connect the U-joint, aligning the marks.
8   Install the mounting brackets and bolts and tighten them to the torque listed in this Chapter's Specifications.
9   Connect the tie-rod ends to the steering knuckle arms (see Section 13).
10 Install the U-joint pinch bolt and tighten it to the torque listed in this Chapter's Specifications.
11 If equipped with power steering, connect the power steering pressure and return lines to the steering gear and fill the power steering pump reservoir with the recommended fluid (see Chapter 1).
12 Lower the vehicle and if equipped with power steering, bleed the steering system (see Section 17).

16 Power steering pump - removal and installation

Removal

1   Disconnect the cable from the negative battery terminal.
2   Using a large syringe or suction gun, suck as much fluid out of the power steering fluid reservoir as possible. Place a drain pan under the vehicle to catch any fluid that spills out when the hoses are disconnected.
3   Loosen the clamp and disconnect the fluid return hose from the pump.
4   Remove the pressure line-to-pump fitting or banjo bolt, then detach the line from the pump. Remove and discard the copper sealing washers. They must be replaced when installing the pump.
5   Loosen the pivot and adjuster bolt and remove the drivebelt (see Chapter 1).
6   Remove the pivot, adjuster and mounting bolts, then remove the pump from the vehicle.

Installation

7   Installation is the reverse of removal. Be sure to tighten the pressure line fitting or banjo bolt to the torque listed in this Chapter's Specifications. Adjust the drivebelt tension following the procedure described in Chapter 1.
8   Top up the fluid level in the reservoir (see Chapter 1) and bleed the system (see Section 17).

17 Power steering system - bleeding

1   Following any operation in which the power steering fluid lines have been disconnected, the power steering system must be bled to remove all air and obtain proper steering performance.
2   With the front wheels in the straight ahead position, check the power steering fluid level and, if low, add fluid until it reaches the Cold mark on the reservoir.
3   Start the engine and allow it to run at fast idle. Recheck the fluid level and add more if necessary to reach the Cold mark on the reservoir.
4   Bleed the system by turning the wheels from side to side, without hitting the stops. This will work the air out of the system. Keep the reservoir full of fluid as this is done.
5   When the air is worked out of the system, return the wheels to the straight ahead position and leave the vehicle running for several more minutes before shutting it off.
6   Road test the vehicle to be sure the steering system is functioning normally and noise free.
7   Recheck the fluid level to be sure it is up to the Hot mark on the reservoir while the engine is at normal operating temperature. Add fluid if necessary (see Chapter 1).

18 Wheels and tires - general information

Refer to illustration 18.1
1   All vehicles covered by this manual are equipped with metric-sized fiberglass or steel belted radial tires **(see illustration)**. Use of

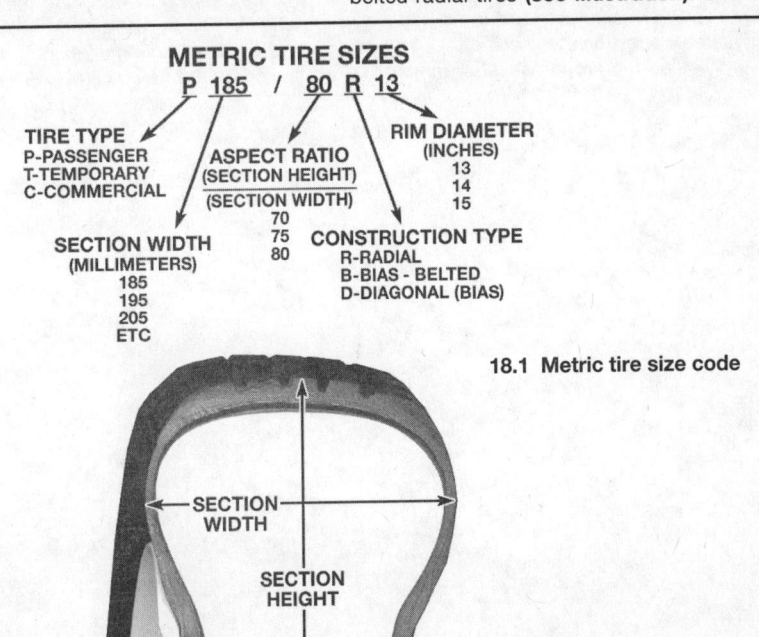

18.1 Metric tire size code

other size or type of tires may affect the ride and handling of the vehicle. Don't mix different types of tires, such as radials and bias belted, on the same vehicle as handling may be seriously affected. It's recommended that tires be replaced in pairs on the same axle, but if only one tire is being replaced, be sure it's the same size, structure and tread design as the other.

2   Because tire pressure has a substantial effect on handling and wear, the pressure on all tires should be checked at least once a month or before any extended trips (see Chapter 1).

3   Wheels must be replaced if they are bent, dented, leak air, have elongated bolt holes, are heavily rusted, out of vertical symmetry or if the lug nuts won't stay tight. Wheel repairs that use welding or peening are not recommended.

4   Tire and wheel balance is important in the overall handling, braking and performance of the vehicle. Unbalanced wheels can adversely affect handling and ride characteristics as well as tire life. Whenever a tire is installed on a wheel, the tire and wheel should be balanced by a shop with the proper equipment.

19 Wheel alignment - general information

Refer to illustration 19.1

A wheel alignment refers to the adjustments made to the wheels so they are in proper angular relationship to the suspension and the ground. Wheels that are out of proper alignment not only affect vehicle control, but also increase tire wear. The front end angles normally measured are camber, caster and toe-in **(see illustration)**. Camber and caster are preset at the factory on the vehicle covered by this manual; toe-in is the only adjustable angle on these vehicles.

Getting the proper wheel alignment is a very exacting process, one in which complicated and expensive machines are necessary to perform the job properly. Because of this, you should have a technician with the proper equipment perform these tasks. We will, however, use this space to give you a basic idea of what is involved with a wheel alignment so you can better understand the process and deal intelligently with the shop that does the work.

Toe-in is the turning in of the wheels. The purpose of a toe specification is to ensure parallel rolling of the wheels. In a vehicle with zero toe-in, the distance between the front edges of the wheels will be the same as the distance between the rear edges of the wheels. The actual amount of toe-in is normally only a fraction of an inch. Toe-in is controlled by the tie-rod end position on the tie-rod. Incorrect toe-in will cause the tires to wear improperly by making them scrub against the road surface.

Camber is the tilting of the wheels from vertical when viewed from one end of the vehicle. When the wheels tilt out at the top, the camber is said to be positive (+). When the wheels tilt in at the top the camber is negative (-). The amount of tilt is measured in degrees from vertical and this measurement is called the camber angle. This angle affects the amount of tire tread which contacts the road and compensates for changes in the suspension geometry when the vehicle is cornering or traveling over an undulating surface.

Caster is the tilting of the front steering axis from the vertical. A tilt toward the rear is positive caster and a tilt toward the front is negative caster.

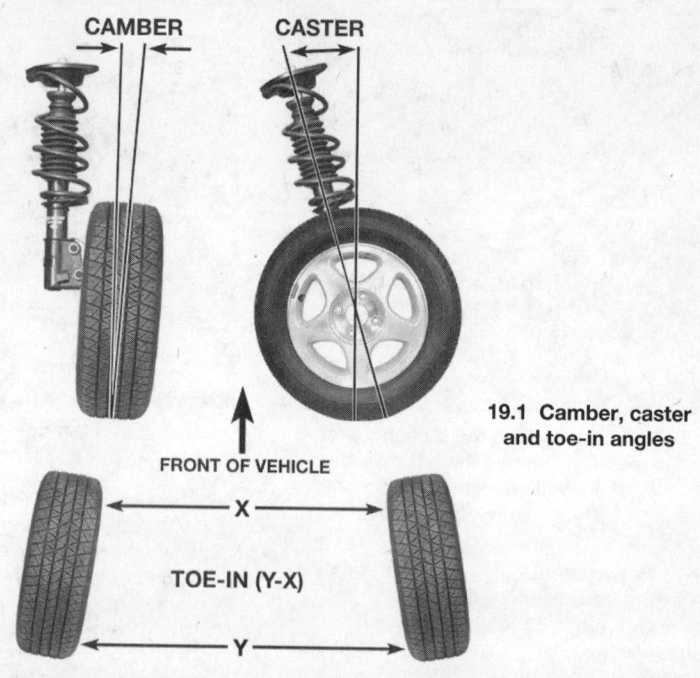

19.1 Camber, caster and toe-in angles

Chapter 11 Body

Contents

	Section
Body - maintenance	2
Body repair - major damage	6
Body repair - minor damage	5
Bumpers - removal and installation	17
Console - removal and installation	19
Dashboard trim panels - removal and installation	21
Door - removal, installation and adjustment	13
Door latch, lock cylinder and handles - removal and installation	14
Door trim panel - removal and installation	12
Door window glass - removal and installation	15
General information	1
Hinges and locks - maintenance	7
Hood - removal, installation and adjustment	9

	Section
Instrument cluster bezel - removal and installation	20
Instrument panel - removal and installation	23
Outside mirrors - removal and installation	18
Radiator grille - removal and installation	10
Seat belts - check	25
Seats - removal and installation	24
Steering column cover - removal and installation	22
Trunk lid - removal, installation and adjustment	11
Upholstery and carpets - maintenance	4
Vinyl trim - maintenance	3
Window regulator - removal and installation	16
Windshield and fixed glass - replacement	8

1 General information

These models feature a "unibody" construction, using a floor pan with front and rear frame side rails which support the body components, front and rear suspension systems and other mechanical components. Certain components are particularly vulnerable to accident damage and can be unbolted and repaired or replaced. Among these parts are the body moldings, bumpers, hood and trunk lids and all glass.

Only general body maintenance practices and body panel repair procedures within the scope of the do-it-yourselfer are included in this Chapter.

2 Body - maintenance

1 The condition of your vehicle's body is very important, because the resale value depends a great deal on it. It's much more difficult to repair a neglected or damaged body than it is to repair mechanical components. The hidden areas of the body, such as the wheel wells, the frame and the engine compartment, are equally important, although they don't require as frequent attention as the rest of the body.
2 Once a year, or every 12,000 miles, it's a good idea to have the underside of the body steam cleaned. All traces of dirt and oil will be removed and the area can then be inspected carefully for rust, damaged brake lines, frayed electrical wires, damaged cables and other problems. The front suspension components should be greased after completion of this job.
3 At the same time, clean the engine and the engine compartment with a steam cleaner or water soluble degreaser.
4 The wheel wells should be given close attention, since undercoating can peel away and stones and dirt thrown up by the tires can cause the paint to chip and flake, allowing rust to set in. If rust is found, clean down to the bare metal and apply an anti-rust paint.
5 The body should be washed about once a week. Wet the vehicle thoroughly to soften the dirt, then wash it down with a soft sponge and plenty of clean soapy water. If the surplus dirt is not washed off very carefully, it can wear down the paint.
6 Spots of tar or asphalt thrown up from the road should be removed with a cloth soaked in solvent.
7 Once every six months, wax the body and chrome trim. If a chrome cleaner is used to remove rust from any of the vehicle's plated parts, remember that the cleaner also removes part of the chrome, so use it sparingly.

3 Vinyl trim - maintenance

Don't clean vinyl trim with detergents, caustic soap or petroleum-based cleaners. Plain soap and water works just fine, with a soft brush to clean dirt that may be ingrained. Wash the vinyl as frequently as the rest of the vehicle.

After cleaning, application of a high quality rubber and vinyl protectant will help prevent oxidation and cracks. The protectant can also be applied to weatherstripping, vacuum lines and rubber hoses (which often fail as a result of chemical degradation) and to the tires.

4 Upholstery and carpets - maintenance

1 Every three months remove the carpets or mats and clean the interior of the vehicle (more frequently if necessary). Vacuum the upholstery and carpets to remove loose dirt and dust.
2 Leather upholstery requires special care. Stains should be removed with warm water and a very mild soap solution. Use a clean, damp cloth to remove the soap, then wipe again with a dry cloth. Never use alcohol, gasoline, nail polish remover or thinner to clean leather upholstery.
3 After cleaning, regularly treat leather upholstery with a leather wax. Never use car wax on leather upholstery.
4 In areas where the interior of the vehicle is subject to bright sunlight, cover leather seats with a sheet if the vehicle is to be left out for any length of time.

5 Body repair - minor damage

See photo sequence

Repair of minor scratches

1 If the scratch is superficial and does not penetrate to the metal of the body, repair is very simple. Lightly rub the scratched area with a fine rubbing compound to remove loose paint and built-up wax. Rinse the area with clean water.
2 Apply touch-up paint to the scratch, using a small brush. Continue to apply thin layers of paint until the surface of the paint in the scratch is level with the surrounding paint. Allow the new paint at least two weeks to harden, then blend it into the surrounding paint by rubbing with a very fine rubbing compound. Finally, apply a coat of wax to the scratch area.
3 If the scratch has penetrated the paint and exposed the metal of the body, causing the metal to rust, a different repair technique is required. Remove all loose rust from the bottom of the scratch with a pocket knife, then apply rust inhibiting paint to prevent the formation of rust in the future. Using a rubber or nylon applicator, coat the scratched area with glaze-type filler. If required, the filler can be mixed with thinner to provide a very thin paste, which is ideal for filling narrow scratches. Before the glaze filler in the scratch hardens, wrap a piece of smooth cotton cloth around the tip of a finger. Dip the cloth in thinner and then quickly wipe it along the surface of the scratch. This will ensure that the surface of the filler is slightly hollow. The scratch can now be painted over as described earlier in this section.

Repair of dents

4 When repairing dents, the first job is to pull the dent out until the affected area is as close as possible to its original shape. There is no point in trying to restore the original shape completely as the metal in the damaged area will have stretched on impact and cannot be restored to its original contours. It is better to bring the level of the dent up to a point which is about 1/8-inch below the level of the surrounding metal. In cases where the dent is very shallow, it is not worth trying to pull it out at all.
5 If the back side of the dent is accessible, it can be hammered out gently from behind using a soft-face hammer. While doing this, hold a block of wood firmly against the opposite side of the metal to absorb the hammer blows and prevent the metal from being stretched.
6 If the dent is in a section of the body which has double layers, or some other factor makes it inaccessible from behind, a different technique is required. Drill several small holes through the metal inside the damaged area, particularly in the deeper sections. Screw long, self-tapping screws into the holes just enough for them to get a good grip in the metal. Now the dent can be pulled out by pulling on the protruding heads of the screws with locking pliers.
7 The next stage of repair is the removal of paint from the damaged area and from an inch or so of the surrounding metal. This is done with a wire brush or sanding disk in a drill motor, although it can be done just as effectively by hand with sandpaper. To complete the preparation for filling, score the surface of the bare metal with a screwdriver or the tang of a file, or drill small holes in the affected area. This will provide a good grip for the filler material. To complete the repair, see the subsection on filling and painting later in this Section.

Repair of rust holes or gashes

8 Remove all paint from the affected area and from an inch or so of the surrounding metal using a sanding disk or wire brush mounted in a drill motor. If these are not available, a few sheets of sandpaper will do the job just as effectively.
9 With the paint removed, you will be able to determine the severity of the corrosion and decide whether to replace the whole panel, if possible, or repair the affected area. New body panels are not as expensive as most people think and it is often quicker to install a new panel than to repair large areas of rust.

10 Remove all trim pieces from the affected area except those which will act as a guide to the original shape of the damaged body, such as headlight shells, etc. Using metal snips or a hacksaw blade, remove all loose metal and any other metal that is badly affected by rust. Hammer the edges of the hole in to create a slight depression for the filler material.
11 Wire brush the affected area to remove the powdery rust from the surface of the metal. If the back of the rusted area is accessible, treat it with rust inhibiting paint.
12 Before filling is done, block the hole in some way. This can be done with sheet metal riveted or screwed into place, or by stuffing the hole with wire mesh.
13 Once the hole is blocked off, the affected area can be filled and painted. See the following subsection on filling and painting.

Filling and painting

14 Many types of body fillers are available, but generally speaking, body repair kits which contain filler paste and a tube of resin hardener are best for this type of repair work. A wide, flexible plastic or nylon applicator will be necessary for imparting a smooth and contoured finish to the surface of the filler material. Mix up a small amount of filler on a clean piece of wood or cardboard (use the hardener sparingly). Follow the manufacturer's instructions on the package, otherwise the filler will set incorrectly.
15 Using the applicator, apply the filler paste to the prepared area. Draw the applicator across the surface of the filler to achieve the desired contour and to level the filler surface. As soon as a contour that approximates the original one is achieved, stop working the paste. If you continue, the paste will begin to stick to the applicator. Continue to add thin layers of paste at 20-minute intervals until the level of the filler is just above the surrounding metal.
16 Once the filler has hardened, the excess can be removed with a body file. From then on, progressively finer grades of sandpaper should be used, starting with a 180-grit paper and finishing with 600-grit wet-or-dry paper. Always wrap the sandpaper around a flat rubber or wooden block, otherwise the surface of the filler will not be completely flat. During the sanding of the filler surface, the wet-or-dry paper should be periodically rinsed in water. This will ensure that a very smooth finish is produced in the final stage.
17 At this point, the repair area should be surrounded by a ring of bare metal, which in turn should be encircled by the finely feathered edge of good paint. Rinse the repair area with clean water until all of the dust produced by the sanding operation is gone.
18 Spray the entire area with a light coat of primer. This will reveal any imperfections in the surface of the filler. Repair the imperfections with fresh filler paste or glaze filler and once more smooth the surface with sandpaper. Repeat this spray-and-repair procedure until you are satisfied that the surface of the filler and the feathered edge of the paint are perfect. Rinse the area with clean water and allow it to dry completely.
19 The repair area is now ready for painting. Spray painting must be carried out in a warm, dry, windless and dust free atmosphere. These conditions can be created if you have access to a large indoor work area, but if you are forced to work in the open, you will have to pick the day very carefully. If you are working indoors, dousing the floor in the work area with water will help settle the dust which would otherwise be in the air. If the repair area is confined to one body panel, mask off the surrounding panels. This will help minimize the effects of a slight mismatch in paint color. Trim pieces such as chrome strips, door handles, etc., will also need to be masked off or removed. Use masking tape and several thickness of newspaper for the masking operations.
20 Before spraying, shake the paint can thoroughly, then spray a test area until the spray painting technique is mastered. Cover the repair area with a thick coat of primer. The thickness should be built up using several thin layers of primer rather than one thick one. Using 600-grit wet-or-dry sandpaper, rub down the surface of the primer until it is very smooth. While doing this, the work area should be thoroughly rinsed with water and the wet-or-dry sandpaper periodically rinsed as well. Allow the primer to dry before spraying additional coats.
21 Spray on the top coat, again building up the thickness by using several thin layers of paint. Begin spraying in the center of the repair area and then, using a circular motion, work out until the whole repair area and about two inches of the surrounding original paint is covered. Remove all masking material 10 to 15 minutes after spraying on the final coat of paint. Allow the new paint at least two weeks to harden, then use a very fine rubbing compound to blend the edges of the new paint into the existing paint. Finally, apply a coat of wax.

6 Body repair - major damage

1 Major damage must be repaired by an auto body shop specifically equipped to perform unibody repairs. These shops have the specialized equipment required to do the job properly.
2 If the damage is extensive, the body must be checked for proper alignment or the vehicle's handling characteristics may be adversely affected and other components may wear at an accelerated rate.
3 Due to the fact that all of the major body components (hood, fenders, etc.) are separate and replaceable units, any seriously damaged components should be replaced rather than repaired. Sometimes the components can be found in a wrecking yard that specializes in used vehicle components, often at considerable savings over the cost of new parts.

7 Hinges and locks - maintenance

Once every 3000 miles, or every three months, the hinges and latch assemblies on the doors, hood and trunk should be given a few drops of light oil or lock lubricant. The door latch strikers should also be lubricated with a thin coat of grease to reduce wear and ensure free movement. Lubricate the door and trunk locks with spray-on graphite lubricant.

8 Windshield and fixed glass - replacement

Replacement of the windshield and fixed glass requires the use of special fast-setting adhesive/caulk materials and some specialized tools. It is recommended that these operations be left to a dealer or a shop specializing in glass work.

9 Hood - removal, installation and adjustment

Refer to illustrations 9.1, 9.10 and 9.11
Note: *The hood is heavy and somewhat awkward to remove and install - at least two people should perform this procedure.*

Removal and installation

1 Make marks around the bolt heads to ensure proper alignment during installation **(see illustration).**
2 Use blankets or pads to cover the cowl area of the body and fenders. This will protect the body and paint as the hood is lifted off.
3 Disconnect any cables or wires that will interfere with removal.
4 Have an assistant support the hood.

9.1 Before removing the hood, make marks around the hinge plate

These photos illustrate a method of repairing simple dents. They are intended to supplement *Body repair - minor damage* in this Chapter and should not be used as the sole instructions for body repair on these vehicles.

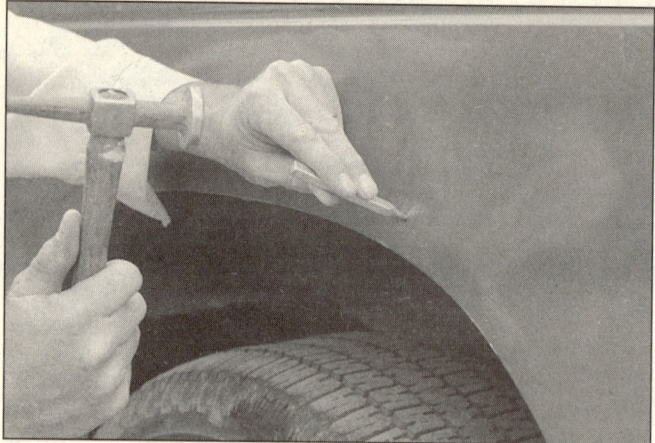

1   If you can't access the backside of the body panel to hammer out the dent, pull it out with a slide-hammer-type dent puller. In the deepest portion of the dent or along the crease line, drill or punch hole(s) at least one inch apart . . .

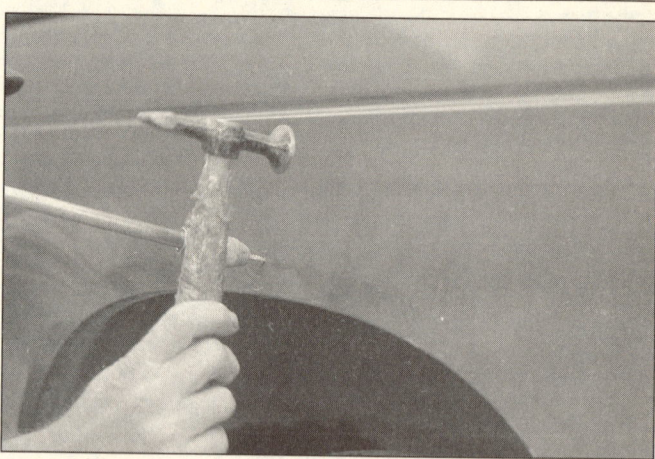

2   . . . then screw the slide-hammer into the hole and operate it. Tap with a hammer near the edge of the dent to help 'pop' the metal back to its original shape. When you're finished, the dent area should be close to its original contour and about 1/8-inch below the surface of the surrounding metal

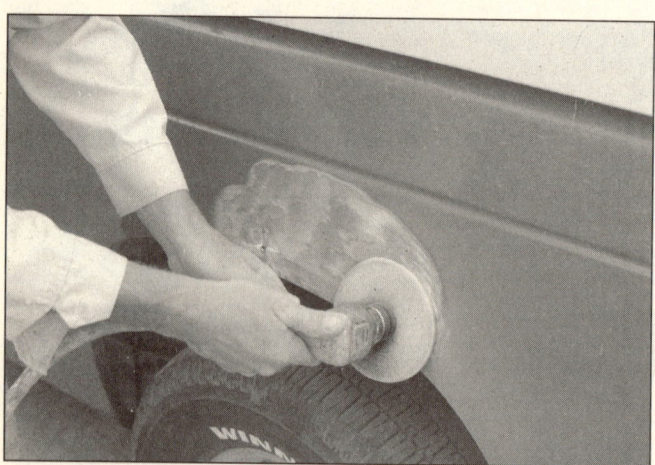

3   Using coarse-grit sandpaper, remove the paint down to the bare metal. Hand sanding works fine, but the disc sander shown here makes the job faster. Use finer (about 320-grit) sandpaper to feather-edge the paint at least one inch around the dent area

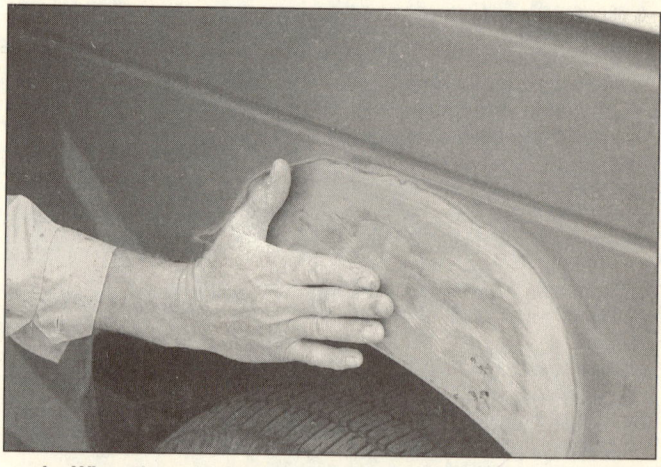

4   When the paint is removed, touch will probably be more helpful than sight for telling if the metal is straight. Hammer down the high spots or raise the low spots as necessary. Clean the repair area with wax/silicone remover

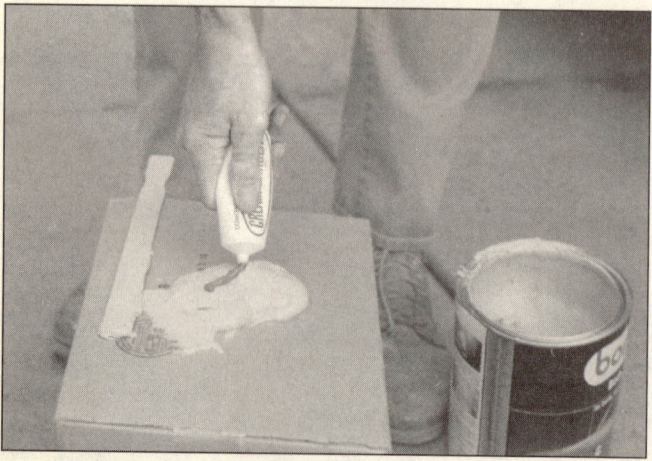

5   Following label instructions, mix up a batch of plastic filler and hardener. The ratio of filler to hardener is critical, and, if you mix it incorrectly, it will either not cure properly or cure too quickly (you won't have time to file and sand it into shape)

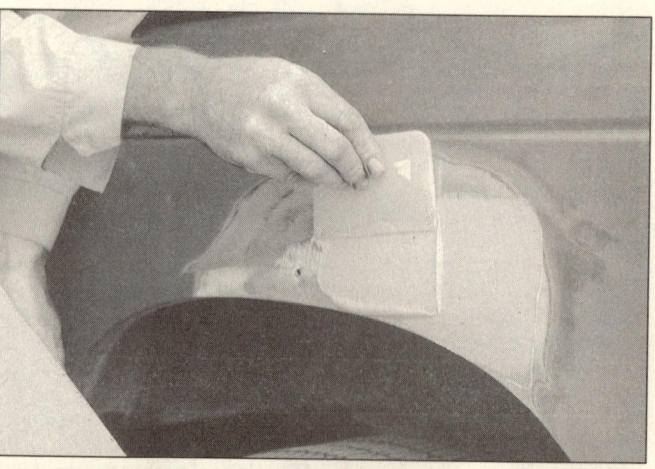

6   Working quickly so the filler doesn't harden, use a plastic applicator to press the body filler firmly into the metal, assuring it bonds completely. Work the filler until it matches the original contour and is slightly above the surrounding metal

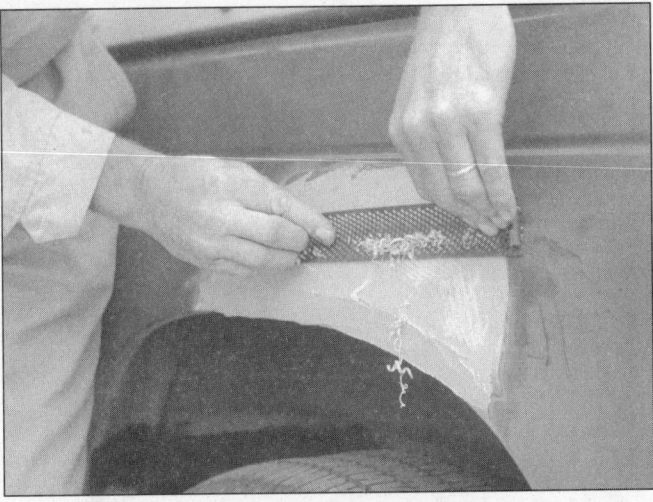

7 Let the filler harden until you can just dent it with your fingernail. Use a body file or Surform tool (shown here) to rough-shape the filler

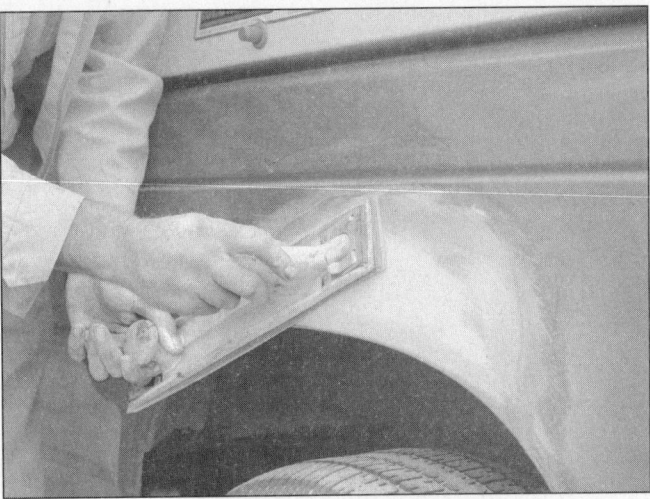

8 Use coarse-grit sandpaper and a sanding board or block to work the filler down until it's smooth and even. Work down to finer grits of sandpaper - always using a board or block - ending up with 360 or 400 grit

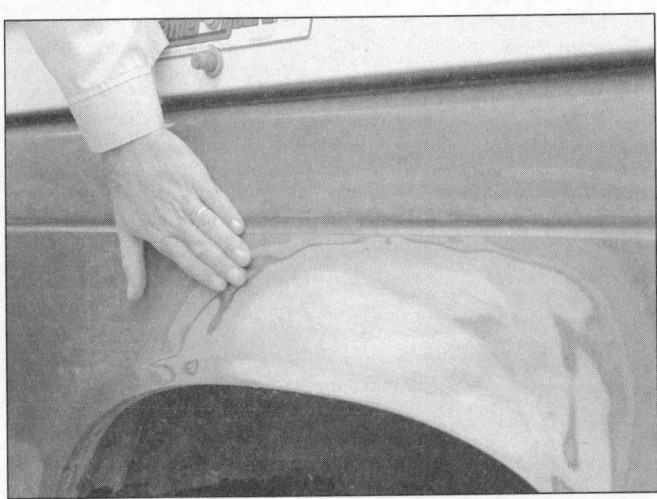

9 You shouldn't be able to feel any ridge at the transition from the filler to the bare metal or from the bare metal to the old paint. As soon as the repair is flat and uniform, remove the dust and mask off the adjacent panels or trim pieces

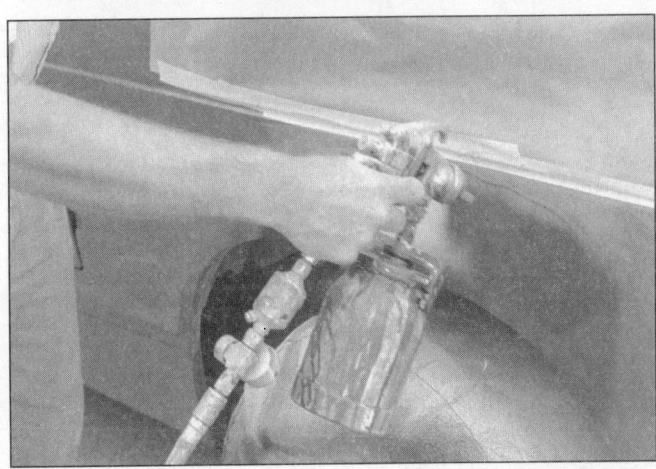

10 Apply several layers of primer to the area. Don't spray the primer on too heavy, so it sags or runs, and make sure each coat is dry before you spray on the next one. A professional-type spray gun is being used here, but aerosol spray primer is available inexpensively from auto parts stores

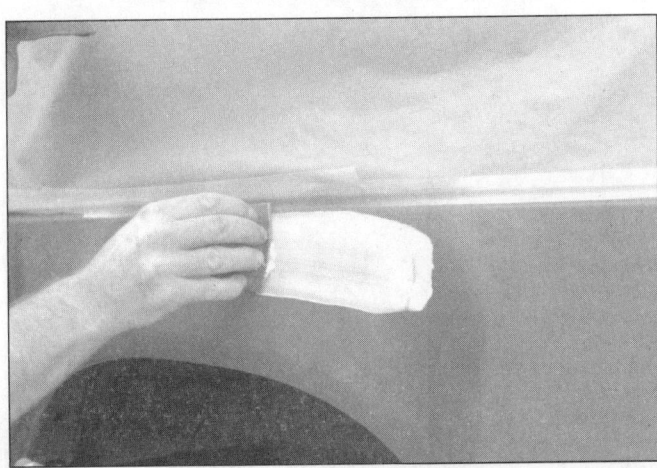

11 The primer will help reveal imperfections or scratches. Fill these with glazing compound. Follow the label instructions and sand it with 360 or 400-grit sandpaper until it's smooth. Repeat the glazing, sanding and respraying until the primer reveals a perfectly smooth surface

12 Finish sand the primer with very fine sandpaper (400 or 600-grit) to remove the primer overspray. Clean the area with water and allow it to dry. Use a tack rag to remove any dust, then apply the finish coat. Don't attempt to rub out or wax the repair area until the paint has dried completely (at least two weeks)

9.10 Loosen the hood latch bolts, move the latch and retighten bolts, then close the hood to check the fit - repeat the procedure until the hood is flush with the fenders

9.11 Adjust the hood closing height by turning the hood bumpers in or out

Remove the hinge-to-hood screws or bolts.
5   Lift off the hood.
6   Installation is the reverse of removal.

Adjustment

7   Fore-and-aft and side-to-side adjustment of the hood is done by moving the hinge plate slot after loosening the bolts or nuts.
8   Scribe a line around the entire hinge plate so you can judge the amount of movement **(see illustration 9.1)**
9   Loosen the bolts or nuts and move the hood into correct alignment. Move it only a little at a time. Tighten the hinge bolts and carefully lower the hood to check the position.
10   If necessary after installation, the entire hood latch assembly can be adjusted up-and-down as well as from side-to-side on the radiator support so the hood closes securely, flush with the fenders. To make the adjustment, scribe a line around the hood latch mounting bolts to provide a reference point, then loosen them and reposition the latch assembly, as necessary **(see illustration)**. Following adjustment, retighten the mounting bolts.
11   Finally, adjust the hood bumpers on the radiator support so the hood, when closed, is flush with the fenders **(see illustration)**.
12   The hood latch assembly, as well as the hinges, should be periodically lubricated with white, lithium-base grease to prevent binding and wear.

10   Radiator grille - removal and installation

Refer to illustration 10.1

1   Use needle-nose pliers to squeeze the retaining clips while pulling out and push them through with a screwdriver if necessary and detach the grille **(see illustration)**.

2   Place the grille in position and press it into place until the clips lock in place.

11   Trunk lid - removal, installation and adjustment

Refer to illustration 11.3

1   Open the trunk lid and cover the edges of the trunk compartment with pads or cloths to protect the painted surfaces when the lid is removed.
2   Disconnect any cables or wire harness connectors attached to the trunk lid that would interfere with removal.
3   Make alignment marks around the hinge mounting bolts with a marking pen **(see illustration)**.
4   While an assistant supports the lid, remove the lid-to-hinge bolts on both sides and lift it off.
5   Installation is the reverse of removal.

10.1 Use needle-nose pliers to squeeze the clip to release the clips while pulling out on the grille

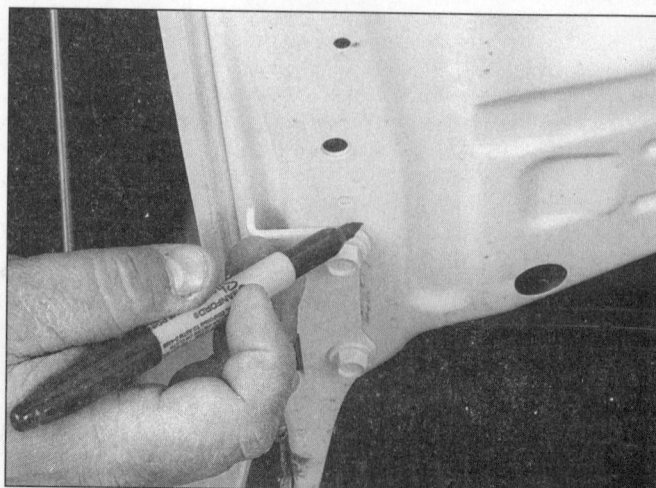

11.3 Scribe a mark around the bolt heads or hinge plate to help with lid realignment on installation

Chapter 11 Body

11-7

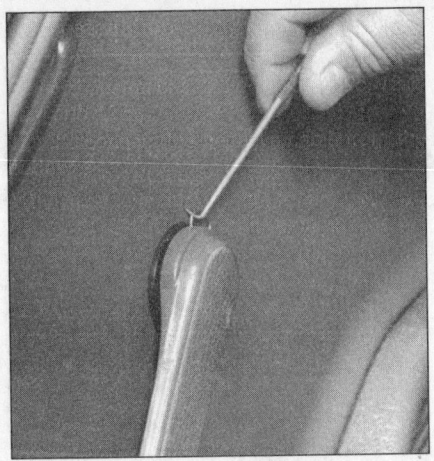

12.2a Use a hooked tool like this to remove the window regulator retaining clip

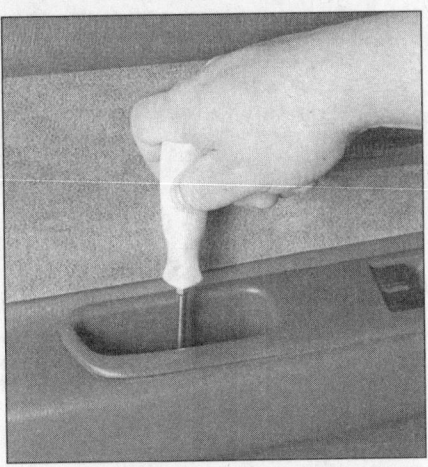

12.2b Remove the switch housing retaining screw and rotate the rear of the housing forward to remove it

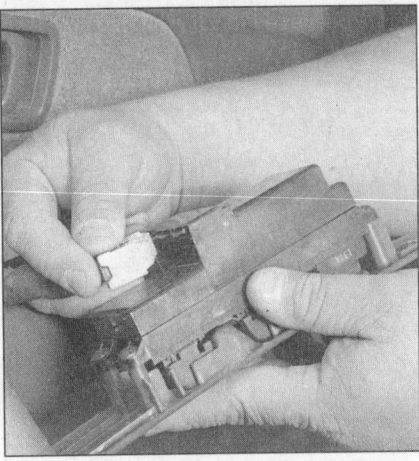

12.2c Unplug the electrical connector from the switch assembly

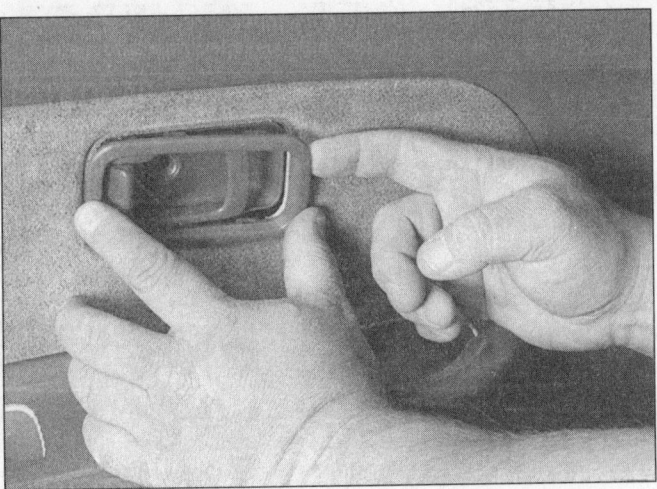

12.3 Grasp the door handle escutcheon securely and detach it from the door panel

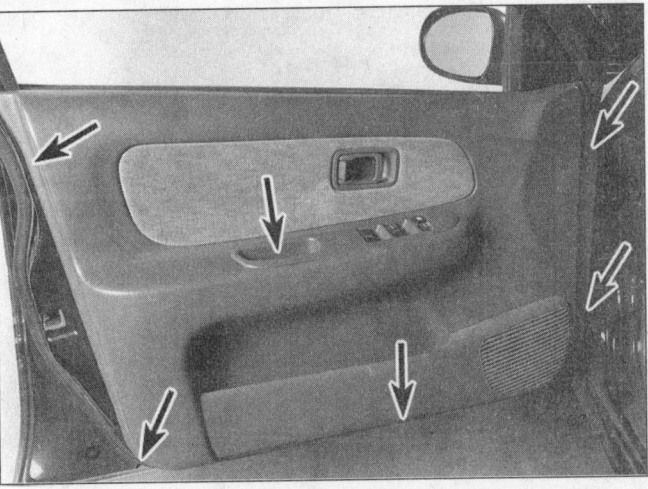

12.4 The trim panel is secured to the door with clips and screws (arrows)

Note: *When reinstalling the trunk lid, align the lid-to-hinge bolts with the marks made during removal.*

6 After installation, close the lid and make sure it's in proper alignment with the surrounding panels.

7 Forward-and-backward and side-to-side adjustments are made by loosening the hinge-to-lid bolts and gently moving the lid into correct alignment.

8 Vertical adjustments to the lid are made by adding or removing the shims used on the hinge.

9 To adjust the lid so it is flush with the body when closed, loosen the mounting bolts and move the lock and striker.

12 Door trim panel - removal and installation

Refer to illustrations 12.2a 12.2b, 12.2c, 12.3, 12.4 and 12.5

1 Disconnect the negative cable from the battery.

2 On manual window regulator equipped models, remove the window crank, using a hooked tool to remove the retainer clip **(see illustration)**. A special tool is available for this purpose, but it's not essential. With the clip removed, pull off the handle. On power window models, remove the retaining screw, pry up at the rear edge of the switch housing and rotate it forward, then unplug the electrical connector **(see illustrations)**.

3 Remove the inside door handle escutcheon **(see illustration)**.

4 Remove the retaining screws, detach any clips, unplug any electrical connectors and remove the trim panel from the vehicle by gently pulling it up and out **(see illustration)**.

5 For access to the inner door remove the plastic watershield. Peel back the watershield, taking care not to tear it **(see illustration)**. To install the trim panel, first press the watershield back into place. If necessary, add more sealant to hold it in place.

6 Prior to installation of the door panel, be sure to reinstall any clips in the panel which

12.5 If the plastic watershield is peeled off carefully, it can be reused

may have come out during the removal procedure and stayed in the door.

7 Plug in any electrical connectors and place the panel in position. Press it into place

11

11-8   Chapter 11   Body

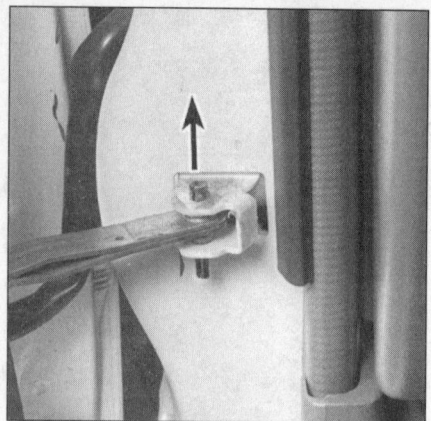

13.3 Gently tap the pin for the door stop strut in the direction shown

13.4 Before loosening or removing them, draw around the door bolt locations (arrows) with a marking pen

13.6 Adjust the door lock striker by loosening the mounting screws and gently tapping the striker in the desired direction

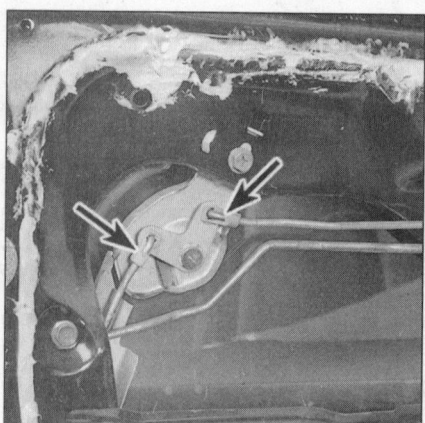

14.2 Detach the links on the two rods (arrows) leading to the latch

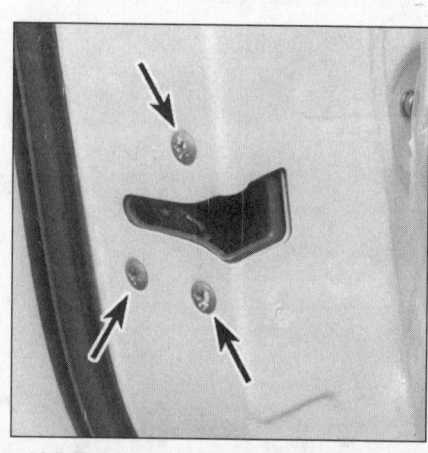

14.3 Remove the latch screws (arrows) from the end of the door

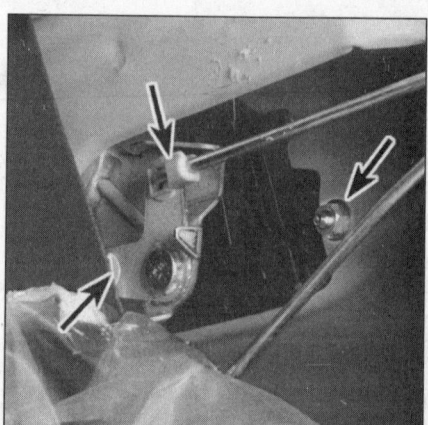

14.6 The two links and outside handle retention nuts (arrows) can be reached through the access hole in the door frame

until the clips are seated and install any retaining screws and armrest/door pulls. Install the manual regulator window crank or power switch assembly.

13 Door - removal, installation and adjustment

Removal and installation

Refer to illustrations 13.3 and 13.4

1   Remove the door trim panel (see Section 12) and disconnect any electrical connectors and push them through the door opening so they won't interfere with removal.
2   Position a jack under the door or have an assistant on hand to support the door when the hinge bolts are removed. **Note:** *If a jack or stand is used, place a rag between it and the door to protect the door's paint.*
3   Remove the door stop strut center pin **(see illustration)**.
4   Scribe or draw a line around the door hinge bolts **(see illustration)**.
5   Remove the hinge-to-door bolts and

carefully detach the door. Installation is the reverse of removal.

Adjustment

Refer to illustration 13.6

6   Following installation, make sure the door is aligned properly. Adjust it if necessary as follows:

a) Up-and-down and forward-and-backward adjustments are made by loosening the hinge-to-body bolts and moving the door, as necessary. A special offset tool may be required to reach some of the bolts.
b) In-and-out and up-and-down adjustments are made by loosening the door side hinge bolts and moving the door, as necessary. A special offset tool may be required to reach some of the bolts.
c) The door lock striker can also be adjusted both up-and-down and sideways to provide a positive engagement with the locking mechanism. This is done by loosening the screws and moving the striker, as necessary **(see illustration)**.

14 Door latch, lock cylinder and handles - removal and installation

1   Remove the door trim panel and the plastic watershield (see Section 12).

Door latch

Refer to illustrations 14.2 and 14.3

2   Reach behind inside the door panel and disconnect the control links from the latch **(see illustration)**.
3   Remove the latch retaining screws from the end of the door **(see illustration)**.
4   Detach the door latch and (if equipped) the door lock solenoid.
5   Installation is the reverse of removal.

Lock cylinder and outside handle

Refer to illustrations 14.6 and 14.7

6   Disconnect the control link rods from the outside handle, remove the outside handle retention nuts and pull the handle from

Chapter 11 Body

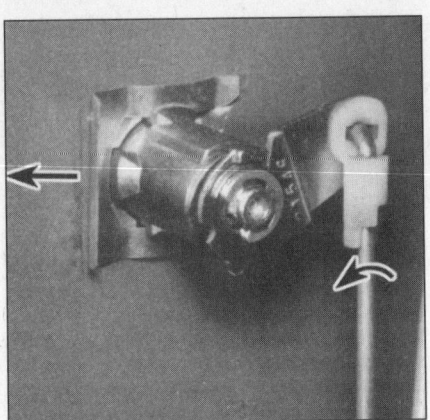

14.7 To remove the lock cylinder, detach the link and pry the retaining clip off

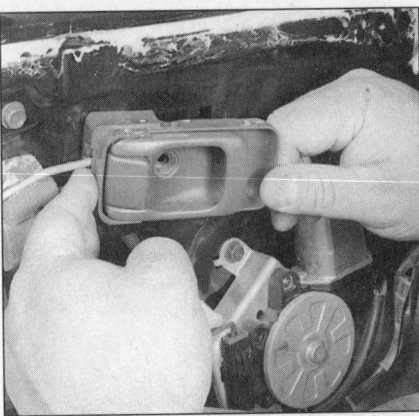

14.10 Remove the handle retaining screw and rotate the handle out, then detach the control link

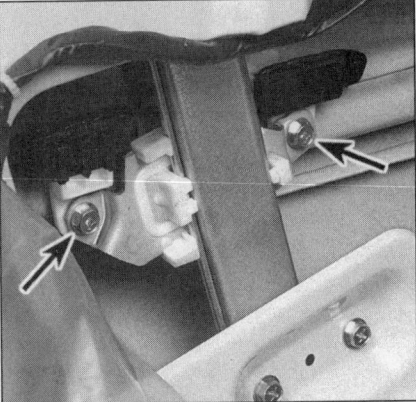

15.4 Remove the two mounting bolts (arrows), detach the glass and lift it out of the door - place a rag over the glass to help prevent scratching it

16.3 On power window models, detach the electrical connector (arrow) and unplug it

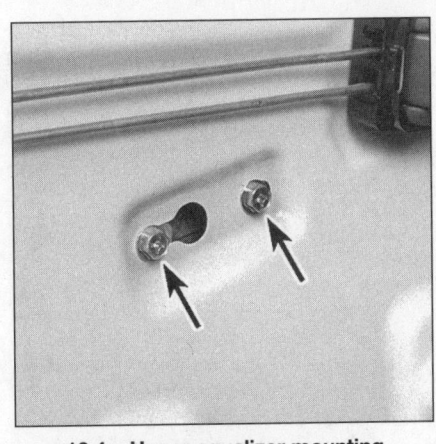

16.4a Upper equalizer mounting bolts (arrows)

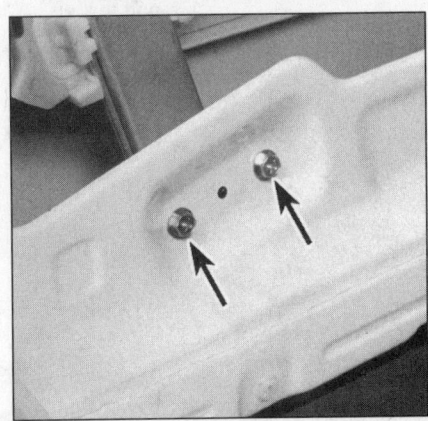

16.4b Lower equalizer mounting bolts (arrows)

the door **(see illustration)**.
7   Detach the control link from the lock cylinder, use a screwdriver to pry the retaining clip off and remove the lock cylinder from the door **(see illustration)**.
8   Installation is the reverse of removal.

Inside handle

Refer to illustration 14.10

9   Remove the door trim panel.
10 Remove the retaining screw, pull the handle free, disconnect the link(s) from the inside handle control and remove the handle from the door **(see illustration)**.
11 Installation is the reverse of removal.

15 Door window glass - removal and installation

Refer to illustration 15.4

1   Remove the door trim panel and the plastic watershield (see Section 12).
2   Lower the window glass.
3   Carefully pry the inner weatherstrip out of the door window opening.

4   Place a rag inside the door panel to help prevent scratching the glass and remove the two glass mounting bolts **(see illustration)**.
5   Remove the glass by pulling it up.
6   Installation is the reverse of the removal procedure.

16 Window regulator - removal and installation

Refer to illustrations 16.3, 16.4a, 16.4b and 16.4c

1   Remove the door trim panel and the plastic watershield (see Section 12).
2   Remove the window assembly (see Section 15).
3   If so equipped, unclip the power window electrical connector from the panel and disconnect it **(see illustration)**.
4   Remove the equalizer arm bracket and the regulator mounting bolts **(see illustrations)**.
5   Pull the regulator through the service hole to remove it.
6   Installation is the reverse of removal.

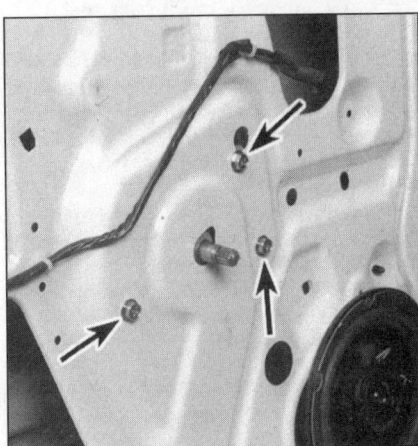

16.4c Regulator bolt locations

17 Bumpers - removal and installation

Refer to illustrations 17.3a and 17.3b

1   Apply the parking brake, raise the vehicle and support it securely on jackstands.

11-10   Chapter 11 Body

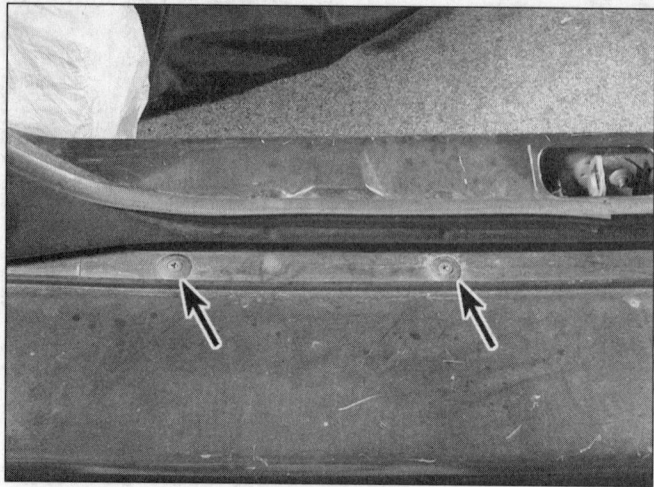

17.3a The bumper covers are held in place by plastic screws along the top . . .

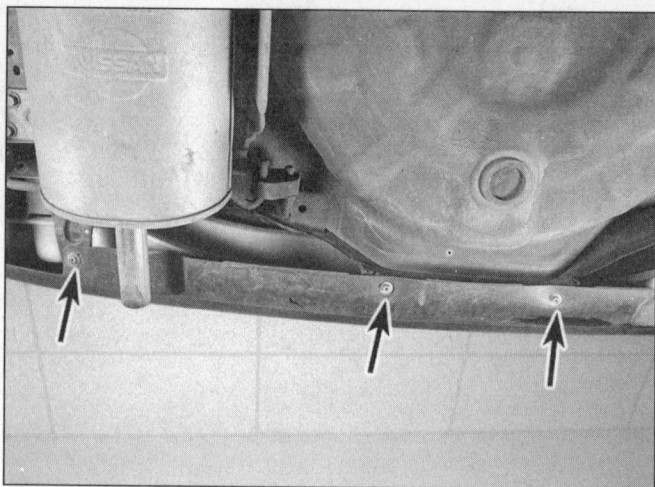

17.3b . . . and bottom edge

18.2 Remove the three screws securing the outside mirror to the door

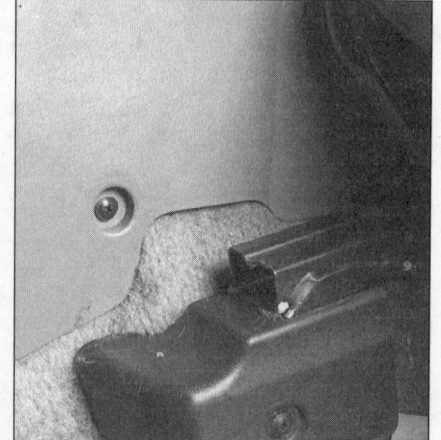

19.3a Remove the screws at each side of the rear of the console

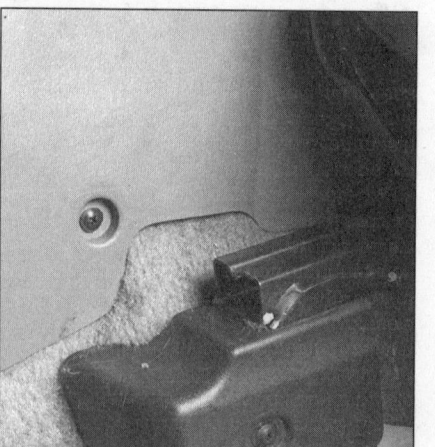

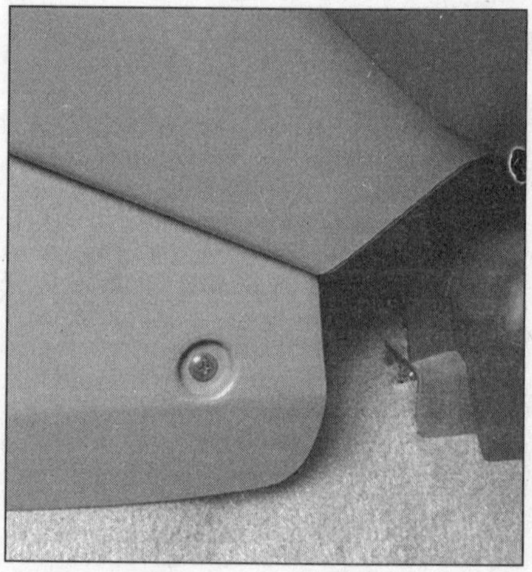

19.3b Remove the front console screws

2   Disconnect the cable from the negative battery terminal and disconnect any wiring that would interfere with bumper removal. On front bumpers, remove the radiator grille (see Section 10) and the fog light housings for access.
3   Remove the bumper cover, if equipped, taking care to avoid damaging the cover and the fender **(see illustrations)**.
4   Working under the vehicle remove the bumper retention bolts.
5   Detach the bumper assembly from the vehicle.
6   Installation is the reverse of the removal procedure.

18 Outside mirrors - removal and installation

Refer to illustration 18.2
1   On manually operated mirrors, remove the control handle. Pry off the mirror cover.
2   Remove the three retaining screws and detach the mirror **(see illustration)**.
3   Installation is the reverse of removal.

19 Console - removal and installation

Refer to illustrations 19.3a, 19.3b, 19.4 and 19.5
Warning: *These models are equipped with airbags. The airbag is armed and can deploy (inflate) anytime the battery is connected. To prevent accidental deployment (and possible injury), disconnect the battery cables whenever working near airbag components. After the battery is disconnected, wait at least ten minutes before beginning work (the system has a back-up capacitor that must fully discharge).*
1   Disconnect the cable from the negative battery terminal, followed by the positive cable. Wait ten minutes before proceeding.
2   Pull the parking brake on.
3   Remove retaining screws at the front and rear of the console **(see illustrations)**.
4   Pry off the cover with a small screwdriver to pry off the cover and remove the front console screws **(see illustration)**. Remove the shift knob.
5   Lift the console up over the shift lever,

remove the screws and separate the two halves of the console **(see illustration)**. Unplug any electrical connections and remove the console assembly from the vehicle.
6   Installation is the reverse of the removal procedure.

20 Instrument cluster bezel - removal and installation

Refer to illustrations 20.2 and 20.3
Warning: *These models are equipped with airbags. The airbag is armed and can deploy (inflate) anytime the battery is connected. To prevent accidental deployment (and possible injury), disconnect the battery cables whenever working near airbag components. After the battery is disconnected, wait at least ten minutes before beginning work (the system has a back-up capacitor that must fully discharge).*
1   Disconnect the cable from the negative battery terminal, followed by the positive cable. Wait ten minutes before proceeding.

Chapter 11 Body   11-11

2 Remove the bezel retaining screws **(see illustration)**.

3 Insert a screwdriver between the bezel and the instrument panel to release the retaining pawls, then grasp the bezel securely and pull out to detach metal clips **(see illustration)**.

4 Installation is the reverse of the removal procedure.

21 Dashboard trim panels - removal and installation

Warning: *These models are equipped with airbags. The airbag is armed and can deploy (inflate) anytime the battery is connected. To prevent accidental deployment (and possible injury), disconnect the battery cables whenever working near airbag components. After the battery is disconnected, wait at least ten minutes before beginning work (the system has a back-up capacitor that must fully discharge).*

1 Disconnect the cable from the negative battery terminal, followed by the positive cable. Wait ten minutes before proceeding.

19.4 Remove the two center screws and lift the console up

Knee protector panel
Refer to illustrations 21.3a and 21.3b

2 Remove the screws, detach the hood release handle and lower the it from the instrument panel.

3 Remove the screws, grasp the panel securely and detach the clips **(see illustration)**. For access under the dash, remove the bolts and detach the knee protector bracket **(see illustration)**.

4 Installation is the reverse of the removal procedure.

19.5 Lift the console up, remove the two screws and separate the assembly

20.2 Remove the upper bezel screws (arrows)

20.3 Pull straight out on the bezel to detach these clips (arrows)

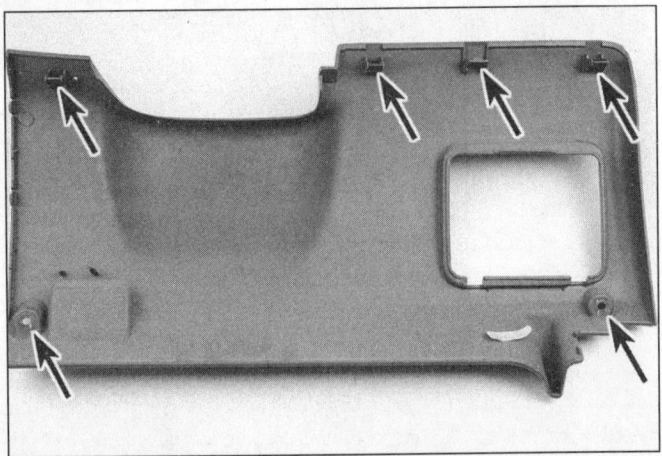

21.3a Knee protector panel screw and clip locations (arrows)

21.3b Remove the bolts (arrows) and detach the knee protector bracket

11

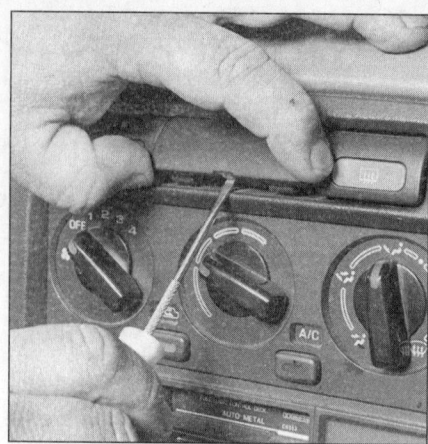
21.5a Use a small screwdriver to pry off the screw cover

21.5b Remove the retaining screw

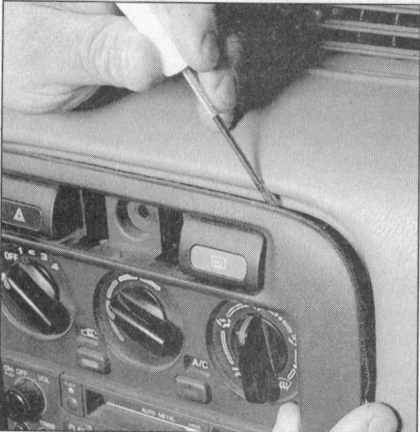
21.6 Detach the clips with a screwdriver and remove the bezel panel

Center bezel panel

Refer to illustrations 21.5a, 21.5b and 21.6

5   Use a small screwdriver to pry the screw cover out of the dash, then remove the retaining screw **(see illustrations)**.
6   Use a screwdriver to detach the retaining clips and remove the bezel from the instrument panel. **(see illustration)**.

Glove box

Refer to illustrations 21.7, 21.8a and 21.8b

7   Reach under the glove box, rotate the two plastic retainers, then withdraw remove the glove box door **(see illustration)**.
8   Remove the glove box latch lock assembly and the four screws at each corner and lower the glove box housing from the instrument panel **(see illustrations)**.
9   Rotate the two plastic retainers and withdraw them, then remove the screws and lower the glove box from the instrument panel.

22 Steering column cover - removal and installation

Refer to illustrations 22.2a and 22.2b
Warning: *These models are equipped with airbags. The airbag is armed and can deploy (inflate) anytime the battery is connected. To prevent accidental deployment (and possible injury), disconnect the battery cables whenever working near airbag components. After the battery is disconnected, wait at least ten minutes before beginning work (the system has a back-up capacitor that must fully discharge).*
1   Remove the steering column cover screws.
2   Separate the cover halves and detach them from the steering column **(see illustrations)**.
3   Disconnect any electrical connections and remove the covers.
4   Installation is the reverse of the removal procedure.

21.7 Rotate the plastic retainers, open the glove box door and rotate it out of the dash

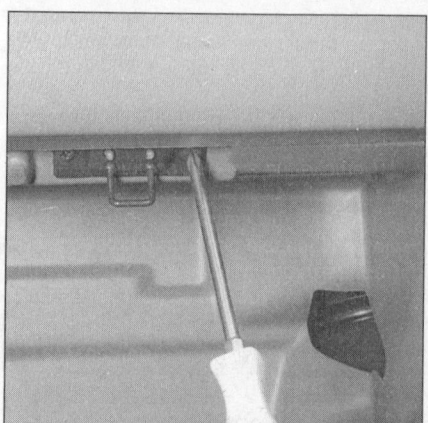
21.8a Remove the two screws and remove the latch striker

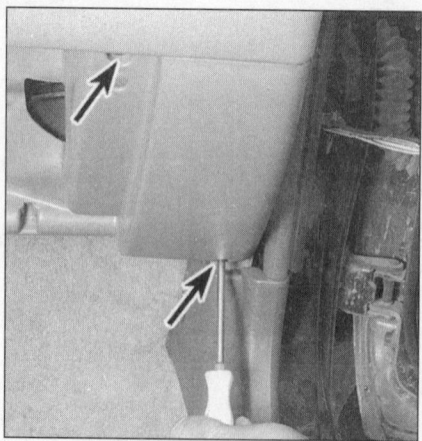
21.8b Remove the two screws at each side of the glove box door opening (arrows)

23 Instrument panel - removal and installation

Refer to illustrations 23.9a, 23.9b, 23.9c and 23.9d
Warning: *These models are equipped with airbags. The airbag is armed and can deploy*

22.2a After removing the screws, detach the lock cylinder cover (steering wheel removed for clarity)

(inflate) anytime the battery is connected. To prevent accidental deployment (and possible injury), disconnect the battery cables whenever working near airbag components. After the battery is disconnected, wait at least ten minutes before beginning work (the system has a back-up capacitor that must fully discharge).

Chapter 11 Body

11-13

22.2b Separate the steering column cover halves (steering wheel removed for clarity)

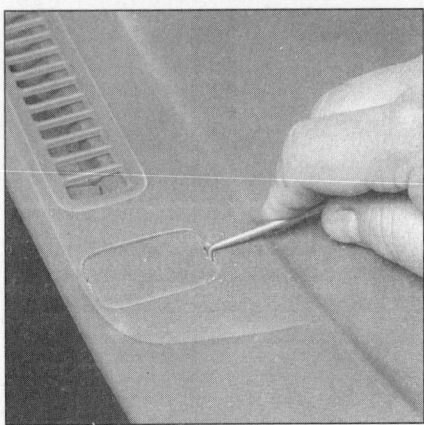

23.9a Pry the covers off for access to . . .

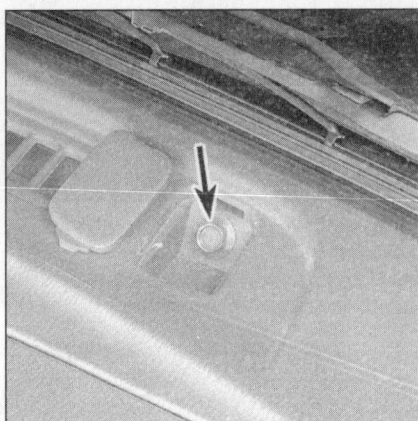

23.9b . . . the two bolts along the top edge of the instrument panel

23.9c The nuts retaining the lower part of the instrument panel are located at the right . . .

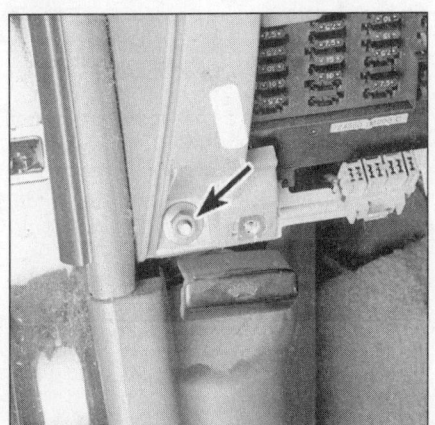

23.9d . . . and left corners (arrows)

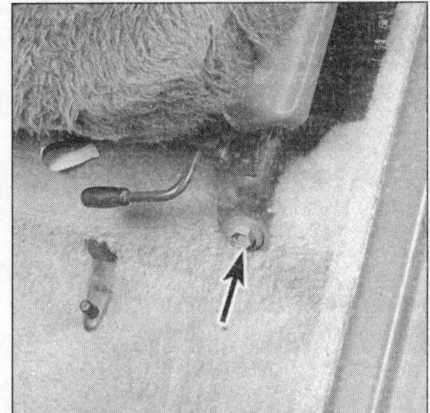

24.1a Remove the front seat bolts (arrow)

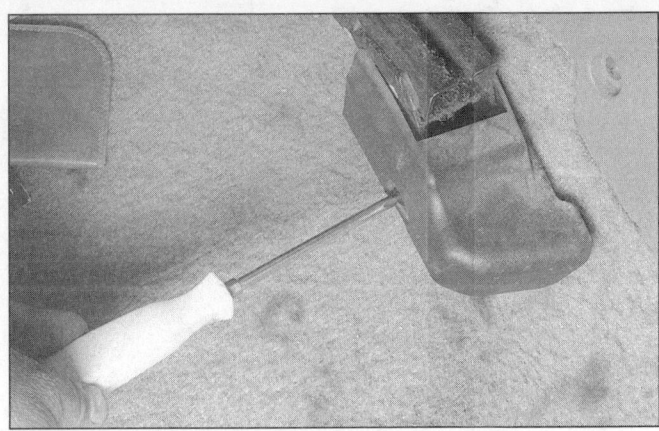

24.1b Remove the screw and detach the seat rear retaining bolt covers

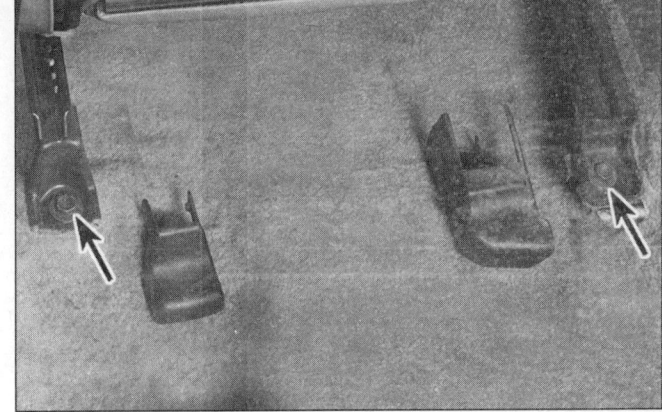

24.1c Rear seat bolt locations (arrows)

1 Disconnect the cables from the negative battery terminal, followed by the positive cable. Wait ten minutes before proceeding.
2 Remove the center console (see Section 19).
3 Remove the steering column cover (see Section 22).
4 Remove the instrument cluster bezel (see Section 20).
5 Remove the instrument cluster (see Chapter 12).

6 Remove the dashboard trim panels (see Section 21).
7 Remove the radio and heater/air conditioner controls (see Chapter 3 and Chapter 12).
8 Remove or disconnect any other components that will interfere with removal.
9 Remove the instrument panel (see illustrations).
10 Installation is the reverse of removal.

24 Seats - removal and installation

Front

Refer to illustrations 24.1a, 24.1b and 24.1c

1 Remove the retaining bolts, detach any clips, unplug any electrical connectors and lift the seat from the vehicle (see illustrations).
2 Installation is the reverse of removal.

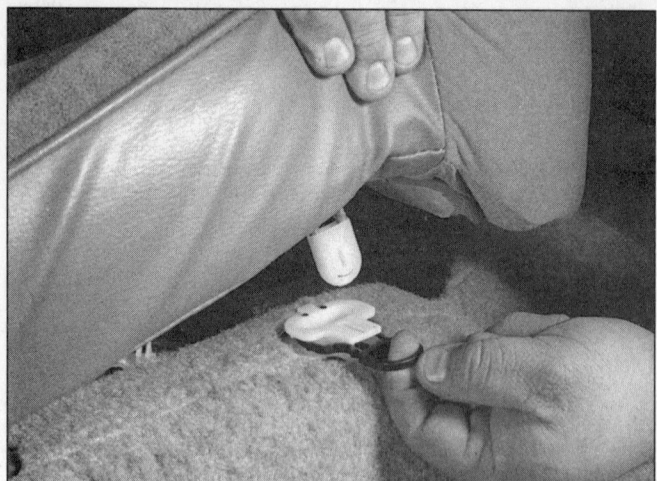

24.3a Pull out on the clip retainer ring while pulling sharply upward to detach the seat cushion

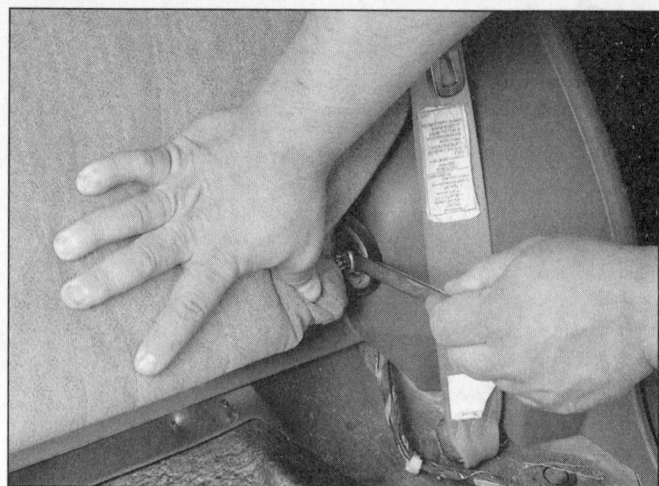

24.3b Remove the seatback bolts or screws

Rear

Refer to illustrations 24.3a and 24.3b

3   Remove the seat cushion by pulling out on the retainer, then pulling up sharply on the cushion **(see illustrations)**. After removing the cushion, remove the screws or bolts and lift the seat back out

4   Installation is the reverse of removal.

25   Seat belts - check

1   Check the seat belts, buckles, latch plates and guide loops for any obvious damage or signs of wear.

2   Make sure the seat belt reminder light comes on when the key is turned on.

3   The seat belts are designed to lock up during a sudden stop or impact, yet allow free movement during normal driving. The retractors should hold the belt against your chest while driving and rewind the belt when the buckle is unlatched.

4   If any of the above checks reveal problems with the seat belt system, replace parts as necessary.

Chapter 12
Chassis electrical system

Contents

	Section		Section
Airbag system - general information	27	Headlight bulb - replacement	15
Antenna - removal and installation	14	Horn - check and replacement	22
Bulb replacement	18	Ignition switch and key lock cylinder - removal and installation	10
Circuit breakers - general information	5	Instrument cluster - removal and installation	21
Combination switch - removal and installation	8	Power door lock system - description and check	25
Composite headlight housing - removal and installation	17	Power window system - description and check	24
Cruise control system - description and check	23	Radio and speakers - removal and installation	13
Daytime Running Lights (DRL) - general information	19	Rear window defogger - check and repair	12
Electric rear view mirrors - description and check	26	Rear window defogger switch - check and replacement	11
Electrical troubleshooting - general information	2	Relays - general information and testing	6
Fuses - general information	3	Steering column switches - check and replacement	9
Fusible links - general information	4	Turn signal/hazard flashers - check and replacement	7
General information	1	Wiper motor - removal and installation	20
Headlights - adjustment	16	Wiring diagrams - general information	28

1 General information

The electrical system is a 12-volt, negative ground type. Power for the lights and all electrical accessories is supplied by a lead/acid-type battery which is charged by the alternator.

This Chapter covers the various electrical components not associated with the engine. Information on the battery, alternator, distributor and starter motor can be found in Chapter 5.

It should be noted that when portions of the electrical system are serviced, the cable should be disconnected from the negative battery terminal to prevent electrical shorts and/or fires.

2 Electrical troubleshooting - general information

A typical electrical circuit consists of an electrical component, any switches, relays, motors, fuses, fusible links or circuit breakers related to that component and the wiring and electrical connectors that link the component to both the battery and the chassis. To help you pinpoint an electrical circuit problem, wiring diagrams are included at the end of this book.

Before tackling any troublesome electrical circuit, first study the appropriate wiring diagrams to get a complete understanding of what makes up that individual circuit. Trouble spots, for instance, can often be narrowed down by noting if other components related to the circuit are operating properly. If several components or circuits fail at one time, chances are the problem is in a fuse or ground connection, because several circuits are often routed through the same fuse and ground connections.

Electrical problems usually stem from simple causes, such as loose or corroded connections, a blown fuse, a melted fusible link or a bad relay. Visually inspect the condition of all fuses, wires and connections in a problem circuit before troubleshooting it.

If testing instruments are going to be utilized, use the diagrams to plan ahead of time where you will make the necessary connections in order to accurately pinpoint the trouble spot.

12-2   Chapter 12 Chassis electrical system

3.1a The interior fuse box is located under the left (driver's) side of the instrument panel, under a cover

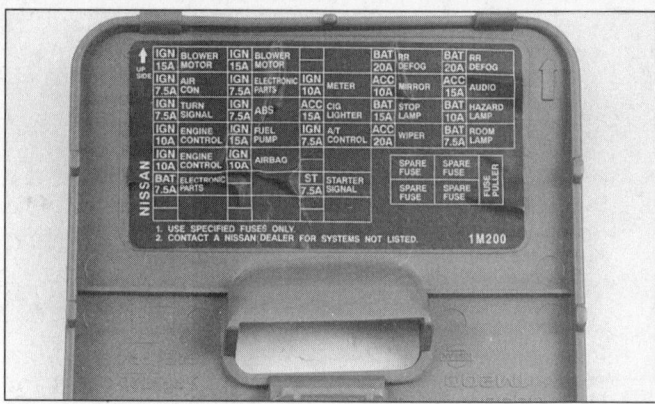

3.1b The key to the fuse location in the interior fuse box is found on the fuse box cover

The basic tools needed for electrical troubleshooting include a circuit tester or voltmeter (a 12-volt bulb with a set of test leads can also be used), a continuity tester, which includes a bulb, battery and set of test leads, and a jumper wire, preferably with a circuit breaker incorporated, which can be used to bypass electrical components. Before attempting to locate a problem with test instruments, use the wiring diagram(s) to decide where to make the connections.

Voltage checks

Voltage checks should be performed if a circuit is not functioning properly. Connect one lead of a circuit tester to either the negative battery terminal or a known good ground. Connect the other lead to a electrical connector in the circuit being tested, preferably nearest to the battery or fuse. If the bulb of the tester lights, voltage is present, which means that the part of the circuit between the electrical connector and the battery is problem free. Continue checking the rest of the circuit in the same fashion. When you reach a point at which no voltage is present, the problem lies between that point and the last test point with voltage. Most of the time the problem can be traced to a loose connection. **Note:** *Keep in mind that some circuits receive voltage only when the ignition key is in the Accessory or Run position.*

Finding a short

One method of finding shorts in a circuit is to remove the fuse and connect a test light or voltmeter in its place to the fuse terminals. There should be no voltage present in the circuit. Move the wiring harness from side to side while watching the test light. If the bulb goes on, there is a short to ground somewhere in that area, probably where the insulation has rubbed through. The same test can be performed on each component in the circuit, even a switch.

Ground check

Perform a ground test to check whether a component is properly grounded. Disconnect the battery and connect one lead of a self-powered test light, known as a continuity tester, to a known good ground. Connect the other lead to the wire or ground connection being tested. If the bulb goes on, the ground is good. If the bulb does not go on, the ground is not good.

Continuity check

A continuity check is done to determine if there are any breaks in a circuit - if it is passing electricity properly. With the circuit off (no power in the circuit), a self-powered continuity tester can be used to check the circuit. Connect the test leads to both ends of the circuit (or to the "power" end and a good ground), and if the test light comes on the circuit is passing current properly. If the light doesn't come on, there is a break somewhere in the circuit. The same procedure can be used to test a switch, by connecting the continuity tester to the power in and power out sides of the switch. With the switch turned On, the test light should come on.

Finding an open circuit

When diagnosing for possible open circuits, it is often difficult to locate them by sight because oxidation or terminal misalignment are hidden by the electrical connectors. Merely wiggling an electrical connector on a sensor or in the wiring harness may correct the open circuit condition. Remember this when an open circuit is indicated when troubleshooting a circuit. Intermittent problems may also be caused by oxidized or loose connections.

Electrical troubleshooting is simple if you keep in mind that all electrical circuits are basically electricity running from the battery, through the wires, switches, relays, fuses and fusible links to each electrical component (light bulb, motor, etc.) and to ground, from which it is passed back to the battery. Any electrical problem is an interruption in the flow of electricity to and from the battery.

3   Fuses - general information

Refer to illustrations 3.1a, 3.1b, 3.1c, 3.1d and 3.3

The electrical circuits of the vehicle are

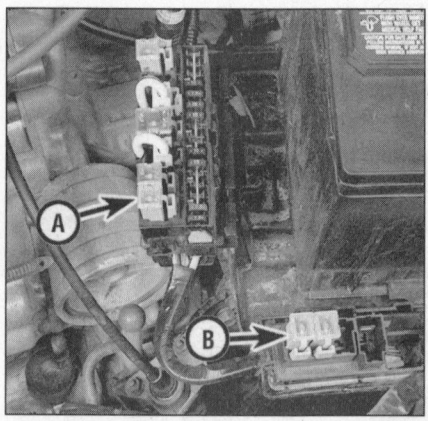

3.1c The engine compartment fuse box (A) is located next to the battery in the left front corner of the engine compartment and can contain relays as well as fuses and fusible links - a fusible link box (B) is located in front of the battery

protected by a combination of fuses, circuit breakers and fusible links. The fuse blocks are located under the instrument panel on the left side of the dashboard and in the engine compartment **(see illustrations)**.

Each of the fuses is designed to protect a specific circuit, and the various circuits are identified on the fuse panel itself.

Miniaturized fuses are employed in the fuse blocks. These compact fuses, with blade terminal design, allow fingertip removal and replacement. If an electrical component fails, always check the fuse first. The best way to check the fuses is with a test light. Check for power at the exposed terminal tips of each fuse. If power is present at one side of the fuse but not the other, the fuse is blown. A blown fuse can also be identified by visually inspecting it **(see illustration)**.

Be sure to replace blown fuses with the correct type. Fuses of different ratings are physically interchangeable, but only fuses of the proper rating should be used. Replacing a fuse with one of a higher or lower value than specified is not recommended. Each electrical circuit needs a specific amount of protection. The amperage value of each fuse is

Chapter 12 Chassis electrical system

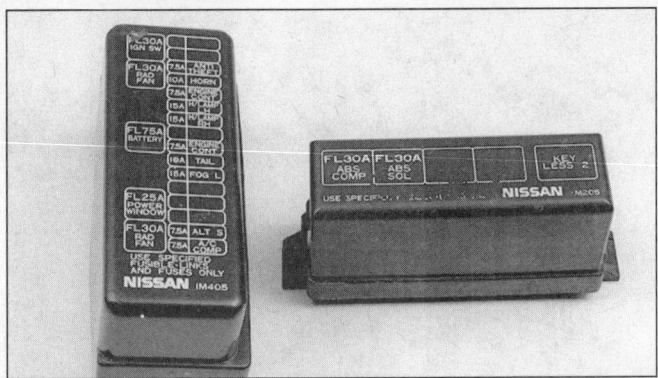

3.1d The information key is found on the covers of the engine compartment fuse and fusible link boxes

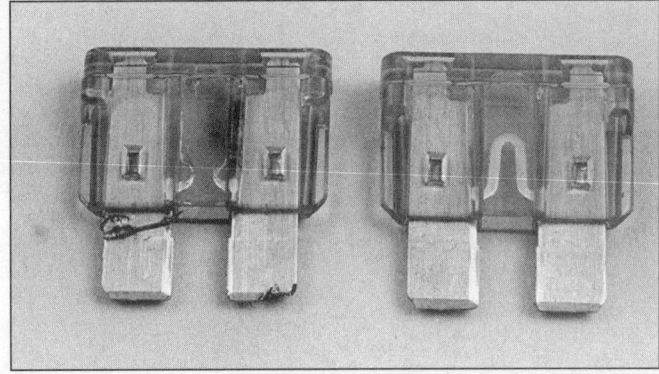

3.3 The fuses on these models can be easily checked visually to see if they are blown

molded into the fuse body.

If the replacement fuse immediately fails, don't replace it again until the cause of the problem is isolated and corrected. In most cases, this will be a short circuit in the wiring caused by a broken or deteriorated wire.

4 Fusible links - general information

Some circuits are protected by fusible links. The links are used in circuits which are not ordinarily fused, such as the ignition circuit.

The fusible links on these models are located in the engine compartment fuse block next to the battery **(see illustration 3.1c)** and are similar to fuses, but larger.

To replace a fusible link, first disconnect the negative cable from the battery. Unplug the burned-out link and replace it with a new one (available from your dealer or auto parts store). Always determine the cause for the overload which melted the fusible link before installing a new one.

5 Circuit breakers - general information

Circuit breakers protect the power windows, power door locks and power sunroof.

Because the circuit breakers reset automatically, an electrical overload in a circuit breaker protected system will cause the circuit to fail momentarily, then come back on. If the circuit does not come back on, check it immediately.

6 Relays - general information and testing

General information

Refer to illustrations 6.2a and 6.2b

1 Several electrical accessories in the vehicle use relays to transmit the electrical signal to the component. If the relay is defective, that component will not operate properly.
2 The various relays are mounted in engine compartment and several locations throughout the vehicle **(see illustrations)**.

Testing

Refer to illustration 6.5

3 It's best to refer to the wiring diagram for the circuit to determine the proper hook-ups for the relay you're testing. However, if you're not able to determine the correct hook-up from the wiring diagrams, you may be able to determine the test hook-ups from the information that follows.

6.2a A relay box is located next to the power steering reservoir

4 On most relays with four or six terminals, two of the terminals are the relay's control circuit (they connect to the relay coil which, when energized, closes the large contacts to complete the circuit). The other terminals are the power circuit (they are connected together within the relay when the control-circuit coil is energized).
5 Relays are sometimes marked as an aid to help you determine which terminals are the control circuit and which are the power circuit **(see illustration)**. As a general rule, the two thicker wires connected to the relay are the

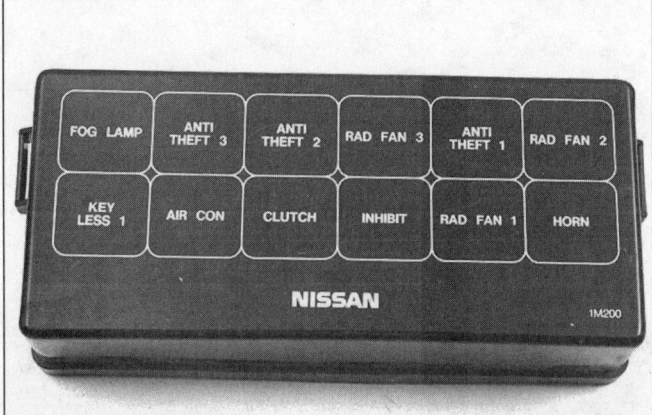

6.2b A key to the relay box contents is found on the cover

6.5 Most relays are marked on the outside to easily identify the control circuit and power circuit

12-4   Chapter 12 Chassis electrical system

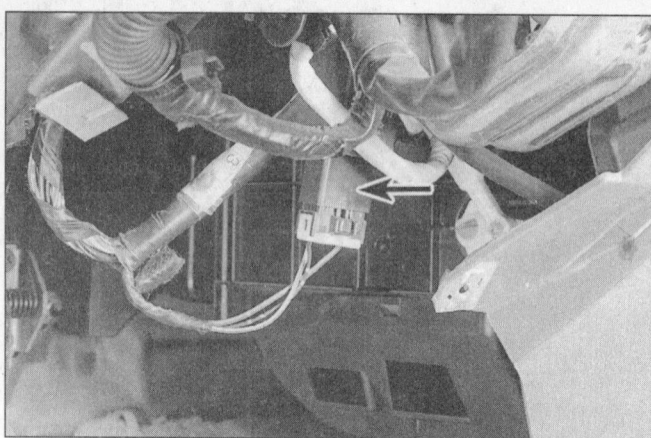

7.4 Turn signals and hazard flasher unit (arrow)

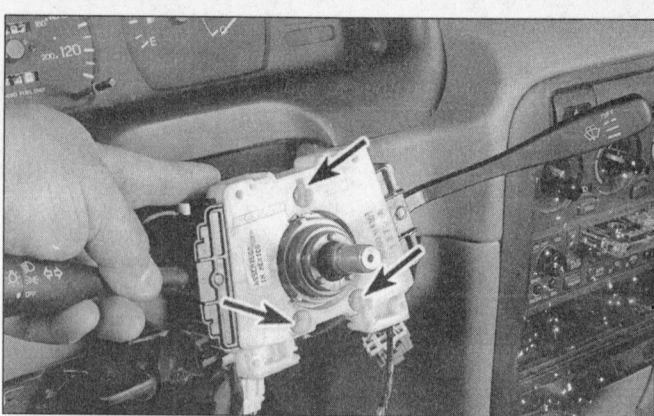

8.4 The combination switch is held in place by three screws (arrows)

power circuit; the thinner wires are the control circuit.

6   Remove the relay from the vehicle and check for continuity between the relay power circuit terminals. There should be no continuity.

7   Connect a fused jumper wire between one of the two control circuit terminals and the positive battery terminal. Connect another jumper wire between the other control circuit terminal and ground. When the connections are made, the relay should click. On some relays, polarity may be critical, so, if the relay doesn't click, try swapping the jumper wires on the control circuit terminals.

8   With the jumper wires connected, check for continuity between the power circuit terminals. Now, there should be continuity.

9   If the relay fails any of the above tests, replace it.

7 Turn signal/hazard flashers - check and replacement

Refer to illustration 7.4

Warning: *These models are equipped with airbags. To prevent accidental deployment (and possible injury) when working near airbag components, disconnect the battery cables (negative first, then positive) and wait at least ten minutes before beginning work (the system has a back-up capacitor that must fully discharge). For more information see Section 27.*

1   The turn signal and hazard flasher is a single combination unit.

2   When the flasher unit is functioning properly, an audible click can be heard during its operation. If the turn signals fail on one side or the other and the flasher unit does not make its characteristic clicking sound, a faulty turn signal bulb is indicated.

3   If both turn signals fail to blink, the problem may be due to a blown fuse, a faulty flasher unit, a broken switch or a loose or open connection. If a quick check of the fuse box indicates that the turn signal fuse has blown, check the wiring for a short before installing a new fuse.

4   To replace the flasher, disconnect the electrical connector and remove the flasher unit from its mounting bracket located under the instrument panel to the right of the steering column (see illustration).

5   Make sure that the replacement unit is identical to the original. Compare the old one to the new one before installing it.

6   Installation is the reverse of removal.

8 Combination switch - removal and installation

Refer to illustrations 8.4 and 8.5

Warning: *These models are equipped with airbags. To prevent accidental deployment (and possible injury) when working near airbag components, disconnect the battery cables (negative first, then positive) and wait at least ten minutes before beginning work (the system has a back-up capacitor that must fully discharge). For more information see Section 27.*

1   Disconnect the cables from the battery terminals.

2   Remove the steering wheel (see Chapter 10).

3   Remove the dashboard knee protector panel and steering column cover (see Chapter 11).

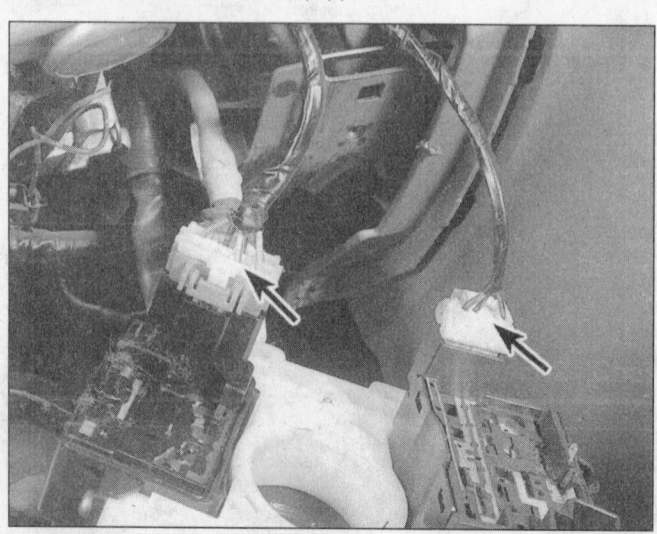

8.5 Unplug the electrical connectors (arrows) from the combination switch

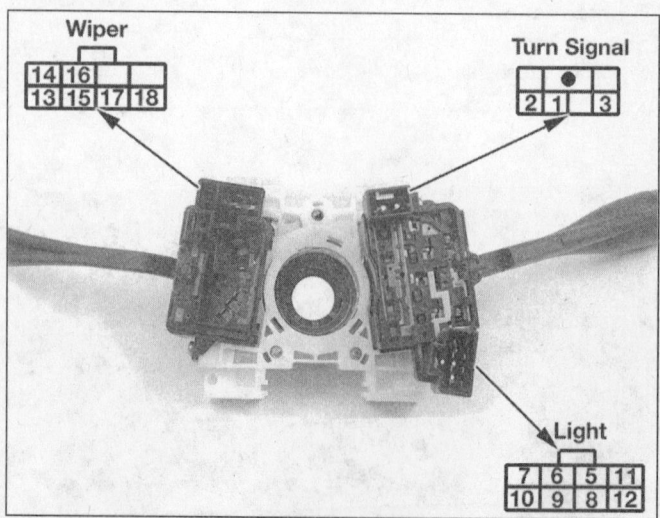

9.4a Terminal guide for the wiper/washer, turn signal and headlight switches

Chapter 12 Chassis electrical system

Headlight Switch

	off			1st.			2nd.		
	A	B	C	A	B	C	A	B	C
5		○			○			○	
6	○			○			○		
7						○			
8	○		○	○		○	○		○
9		○			○			○	
10						○			
11				○	○	○	○	○	○
12				○	○	○	○	○	○

○ Indicates continuity between terminals.

72051-12-9.4B HAYNES

9.4b Continuity table for the headlight switch

Turn signal switch

	R	N	L
1	○		○
2	○		
3			○

○ Indicates continuity between terminals.

72051-12-9.4C HAYNES

9.4c Continuity table for the turn signal switch

Wiper/Washer Switch

	off	Int	Lo	Hi	Wash
13	○	○			
14	○	○			
15		○			
16			○		
17		○	○	○		○
18					○

○ Indicates continuity between terminals.

72051-12-9.4D HAYNES

9.4d Continuity table for the windshield wiper/washer switch

---

4 Remove the combination switch retaining screws **(see illustration)**.
5 Slide the switch up off the column and unplug the connectors **(see illustration)**.
6 Installation is the reverse of removal. Refer to Chapter 10, Section 12, and center the spiral cable before installing the steering wheel.

9 Steering column switches - check and replacement

Warning: *These models are equipped with airbags. To prevent accidental deployment (and possible injury) when working near airbag components, disconnect the battery cables (negative first, then positive) and wait at least ten minutes before beginning work (the system has a back-up capacitor that must fully discharge). For more information see Section 27.*

1 Disconnect the cables from the battery terminals (negative first, then positive) and wait ten minutes before proceeding to the next Step.
2 Remove the dashboard knee protector panel and steering column cover (see Chapter 11).
3 Remove the combination switch (see Section 8).

Check

Refer to illustrations 9.4a, 9.4b, 9.4c and 9.4d
4 Using an ohmmeter, check for continuity between the indicated terminals with the various switches in each of the indicated positions **(see illustrations)**.
5 If the continuity is not as specified, replace the defective switch.

Replacement

Refer to illustration 9.7
6 Remove the combination switch (see Section 8). Remove the four screws retaining the airbag spiral cable and remove the spiral cable from the combination switch.
7 Remove the retaining screws from the switch being replaced and remove the defective switch from the switch body **(see illustration)**.
8 Insert the terminals from the new switch into the connector, pushing in until they are securely locked in place.
9 The remainder or installation is the reverse of removal. Refer to Chapter 10, Section 12 and center the spiral cable before installing the steering wheel.

10 Ignition switch and key lock cylinder - removal and installation

Refer to illustration 10.4
Warning: *These models are equipped with airbags. To prevent accidental deployment (and possible injury) when working near airbag components, disconnect the battery cables (negative first, then positive) and wait at least ten minutes before beginning work (the system has a back-up capacitor that must fully discharge). For more information see Section 27.*

1 Disconnect the cables from the battery terminals (negative first, then positive) and wait ten minutes before proceeding to the next Step.
2 Remove the steering wheel (see Chapter 10).
3 Remove the steering column cover (see Chapter 11).
4 Remove the shear-head bolts retaining the ignition switch/lock cylinder assembly and separate the bracket halves from the steering column. This can be accomplished by drilling out the bolts and unscrewing them with a screw extractor **(see illustration)**.
5 Place the new switch assembly position, install the new shear-head bolts and tighten them until the heads snap off. Be sure to center the spiral cable before installing the steering wheel (refer to Chapter 10, Section 12).

9.7 The steering column switches can be removed from the combination switch base on most models by removing the screws (arrows)

10.4 To remove the ignition/lock cylinder assembly, drill out the two retaining bolts (arrows)

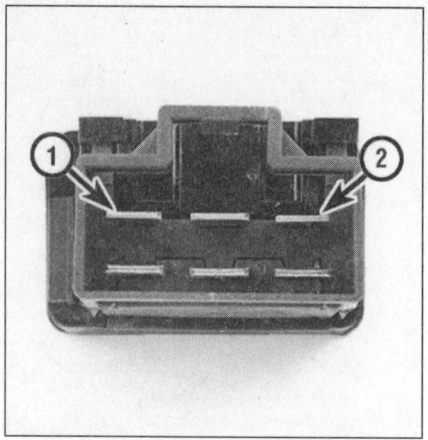

11.4 Continuity check for the rear window defogger switch - there should be continuity between terminals 1 and 2 with the switch On and no continuity with the switch Off

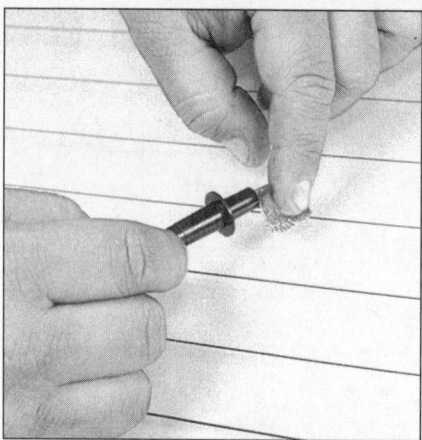

12.5 When measuring the voltage at the rear window defogger grid, wrap a piece of aluminum foil around the negative probe of the voltmeter and press the foil against the wire with your finger

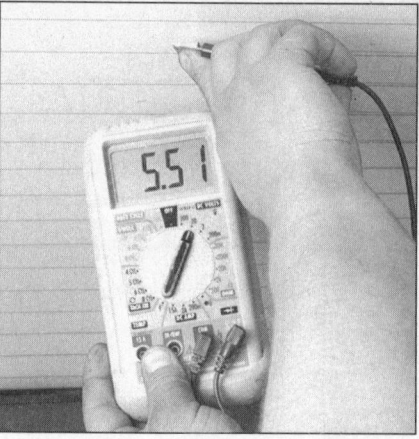

12.6 To determine if a heating element has broken, check the voltage at the center of each element - if the voltage is 6-volts, the element is unbroken

11 Rear window defogger switch - check and replacement

Refer to illustration 11.4

Warning: *These models are equipped with airbags. To prevent accidental deployment (and possible injury) when working near airbag components, disconnect the battery cables (negative first, then positive) and wait at least ten minutes before beginning work (the system has a back-up capacitor that must fully discharge). For more information see Section 27.*

1 Disconnect the cables from the battery terminals (negative first, then positive) and wait ten minutes before proceeding to the next Step.
2 Remove the center bezel panel from the dash (see Chapter 11).
3 Unplug the electrical connector and use a small screwdriver to detach the switch and pull it gently free.
4 Use an ohmmeter to check for continuity at the indicated terminals with the switch in the indicated position **(see illustration)**.
5 Replace the switch if the continuity is not as specified.

12 Rear window defogger - check and repair

1 The rear window defogger consists of a number of horizontal elements baked onto the glass surface.
2 Small breaks in the element can be repaired without removing the rear window.

Check

Refer to illustrations 12.5 and 12.6

3 Turn the ignition switch and defogger system switches to On.
4 Using a voltmeter, place the positive probe against the battery feed terminal and the negative probe against the negative (ground) bus bar. The positive terminal is located on the driver's side and the negative terminal is located on the passenger's side. If battery voltage is not indicated, check the fuse, defogger switch and related wiring.
5 When measuring voltage during the next two tests, wrap a piece of aluminum foil around the tip of the voltmeter negative probe and press the foil against the wire with your finger **(see illustration)**.
6 Place the negative lead against the negative (ground) bus bar. Check the voltage at the center of each heating element **(see illustration)**. If the voltage is 6-volts, the element is okay (there is no break). If the voltage is 10-volts or more, the element is broken between the mid-point and ground. If the voltage is 0-volts, the element is broken between the mid-point and the positive side.
7 To find the break, slide the probe toward the positive side. The point at which the voltmeter deflects from zero to several volts is the point at which the heating element is broken. **Note:** *If the heating element is not broken, the voltmeter will indicate 12-volts at the positive side and gradually decrease to 0-volts as you slide the positive probe toward the ground side.*

Repair

Refer to illustration 12.13

8 Repair the break in the element using a repair kit specifically recommended for this purpose, such as Dupont paste No. 4817 (or equivalent). Included in this kit is plastic conductive epoxy.
9 Prior to repairing a break, turn off the system and allow it to cool off for a few minutes.
10 Lightly buff the element area with fine steel wool, then clean it thoroughly with rubbing alcohol.
11 Use masking tape to mask off the area being repaired.
12 Thoroughly mix the epoxy, following the

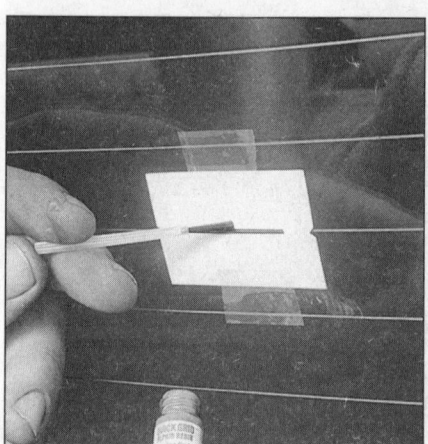

12.13 To use a defogger repair kit, apply masking tape to the inside of the window at the damaged area, then brush on the special conductive coating

instructions provided with the repair kit.
13 Apply the epoxy material to the slit in the masking tape, overlapping the undamaged area about 3/4-inch on either end **(see illustration)**.
14 Allow the repair to cure for 24 hours before removing the tape and using the system.

13 Radio and speakers - removal and installation

Warning: *These models are equipped with airbags. To prevent accidental deployment (and possible injury) when working near airbag components, disconnect the battery cables (negative first, then positive) and wait at least ten minutes before beginning work (the system has a back-up capacitor that must fully discharge). For more information see Section 27.*

1 Disconnect the cables from the battery terminals (negative first, then positive) and

Chapter 12 Chassis electrical system

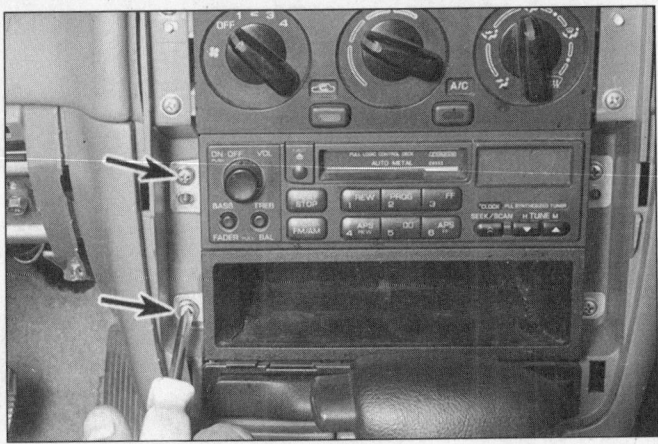

13.3a Remove the screws and pull the radio out ...

13.3b ... then unplug the connectors and remove the unit

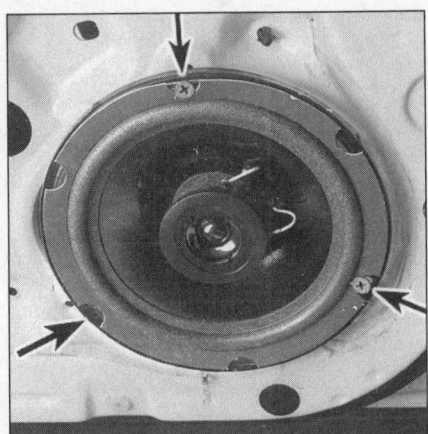

13.6 Remove the screws (arrows), pull the speaker out and unplug it

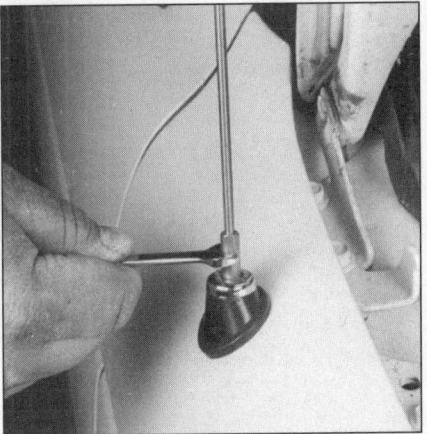

14.1 Use a small wrench to unscrew the antenna mast

15.2 Detach the wire clip and rotate it out of the way (housing removed for clarity)

15.3 Pull the bulb assembly straight out of the housing

wait ten minutes before proceeding to the next Step.

Radio

Refer to illustrations 13.3a and 13.3b

2 Remove the center bezel panel from the dash (see Chapter 11).
3 Remove the screws, pull the radio out, then unplug the electrical connector and the antenna lead and lift the radio out **(see illustrations)**.
4 Installation is the reverse of removal.

Speakers

Refer to illustration 13.6

5 Remove the front door trim panel (see Chapter 11).
6 Remove the speaker retaining screws/nuts. Unplug the electrical connector and remove the speaker **(see illustration)**.
7 Installation is the reverse of removal.

14 Antenna - removal and installation

Refer to illustration 14.1

1 Use a small wrench to unscrew the mast

(see illustration).
2 Install the new antenna mast and tighten it securely with the wrench.

15 Headlight bulb - replacement

Refer to illustrations 15.2 and 15.3

Warning: *These models are equipped with halogen gas-filled bulbs which are under pressure and may shatter if the surface is scratched or the bulb is dropped. Wear eye protection and handle the bulbs carefully, grasping only the base whenever possible. Do not touch the surface of the bulb with your fingers because the oil from your skin could cause it to overheat and fail prematurely. If you do touch the bulb surface, clean it with rubbing alcohol.*
1 Open the hood and locate the bulb assembly on the back of the headlight housing. On the right side headlight, grasp the coolant reservoir securely, lift it up and place it out of the way. Unplug the electrical connector.
2 Detach the headlight retaining clip **(see illustration)**.
3 Withdraw the bulb assembly from the headlight housing **(see illustration)**.

4 Without touching the glass with your bare fingers, insert the new bulb assembly into the headlight housing and secure it with the clip.
5 Plug in the electrical connector. Test headlight operation, then close the hood.

16.1 Use a Phillips screwdriver or small wrench to adjust the headlights - the adjuster closest to the fender (A) controls the horizontal movement and the one closest to the radiator (B), vertical movement

16 Headlights - adjustment

Refer to illustrations 16.1 and 16.2
Note: *It is important that the headlights are aimed correctly. If adjusted incorrectly they could blind the driver of an oncoming vehicle and cause a serious accident or seriously reduce your ability to see the road. The headlights should be checked for proper aim every 12 months and any time a new headlight is installed or front end body work is performed. It should be emphasized that the following procedure is only an interim step which will provide temporary adjustment until the headlights can be adjusted by a properly equipped shop.*

1 These models are equipped with composite headlights with two adjustment screws, one controlling left-and-right movement and one for up-and-down movement **(see illustration)**.
2 There are several methods of adjusting the headlights. The simplest method requires a blank wall 25 feet in front of the vehicle and a level floor **(see illustration)**.
3 Position masking tape vertically on the wall in reference to the vehicle centerline and the centerlines of both headlights.
4 Position a horizontal tape line in reference to the centerline of all the headlights.
Note: *It may be easier to position the tape on the wall with the vehicle parked only a few inches away.*
5 Adjustment should be made with the vehicle sitting level, the gas tank half-full and no unusually heavy load in the vehicle.
6 Starting with the low beam adjustment, position the high intensity zone so it is two inches below the horizontal line and two inches to the right of the headlight vertical line. Twist the adjustment screws until the desired level has been achieved.
7 With the high beams on, the high inten-

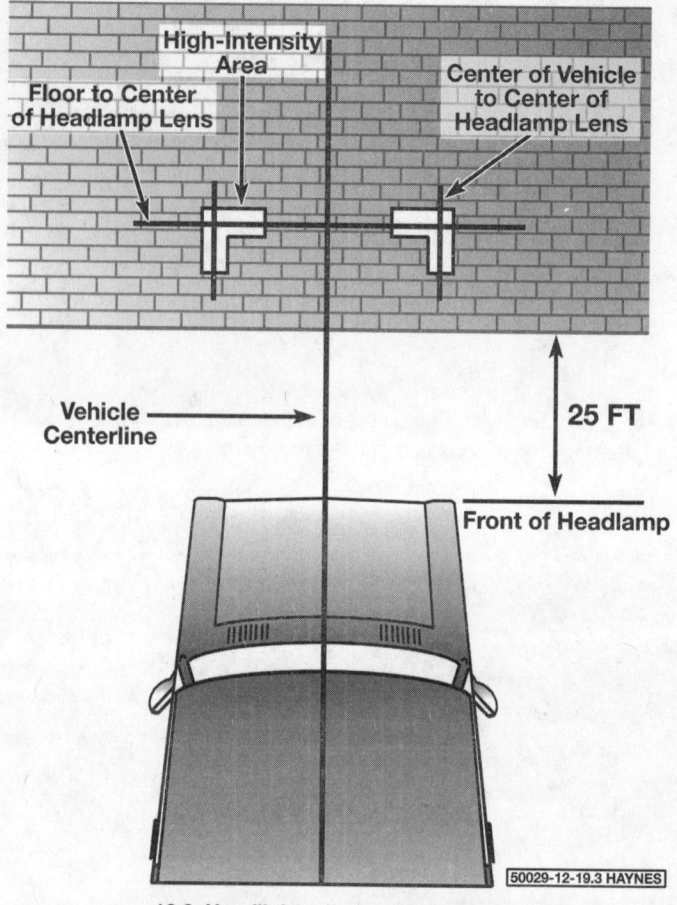

16.2 Headlight adjustment details

sity zone should be vertically centered with the exact center just below the horizontal line. **Note:** *It may not be possible to position the headlight aim exactly for both high and low beams. If a compromise must be made, keep in mind that the low beams are the most used and have the greatest effect on driver safety.*
8 Have the headlights adjusted by a dealer service department at the earliest opportunity.

17 Composite headlight housing - removal and installation

Refer to illustrations 17.5a and 17.5b
1 Disconnect the cable from the negative battery terminal.
2 Remove the headlight bulb (Section 15).
3 Remove the radiator grille (Chapter 11).
4 Remove the parking light housing (Section 18).

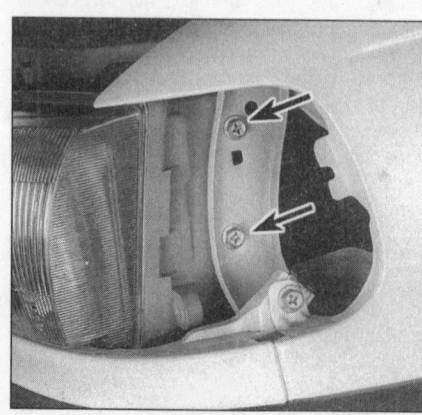

17.5a There are four headlight retaining bolts and nuts: two bolts (arrows) are on the fender brace . . .

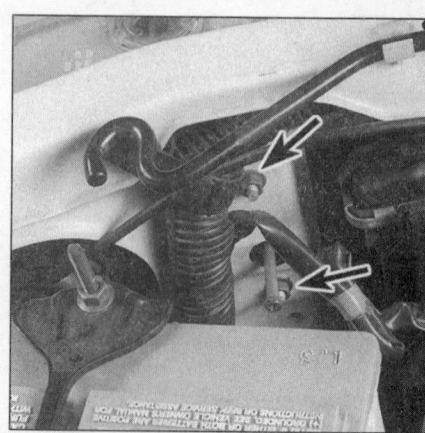

17.5b . . . and the two nuts (arrows) are accessible from behind the radiator brace

Chapter 12 Chassis electrical system

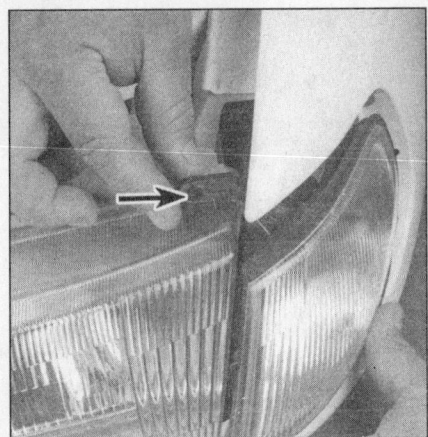

18.1 Remove the screw (arrow) and push the turn signal housing straight forward

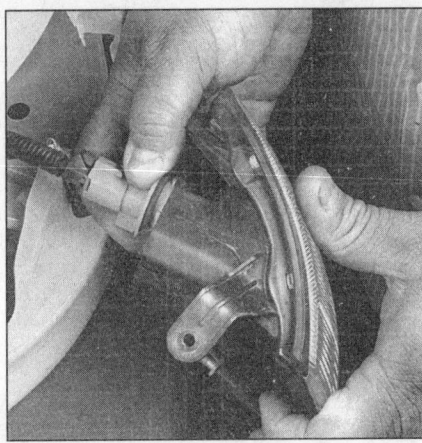

18.2 The bulb is removed by turning the holder counterclockwise, then pushing in and turning the bulb

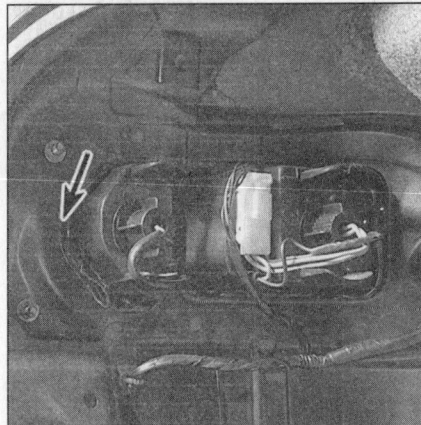

18.3 Rotate the bulb holder counterclockwise to remove it from the housing

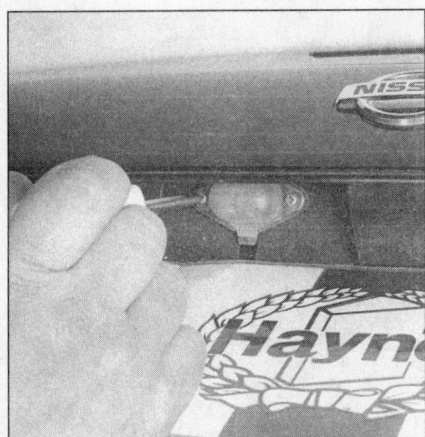

18.5 Remove the license plate light lens with a small screwdriver

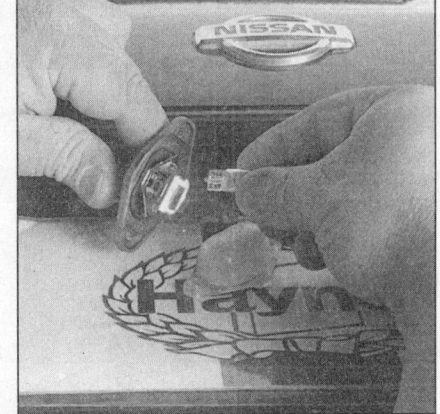

18.6 Pull the license plate bulb straight out

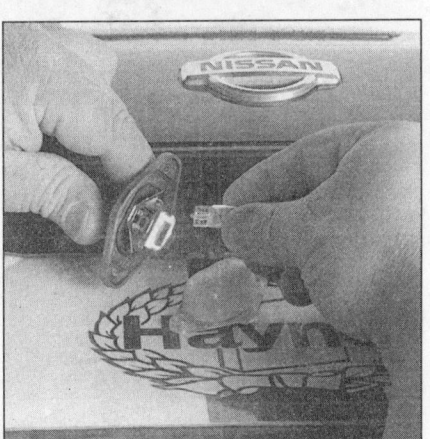

18.8 Rotate the high-mounted brake light bulb counterclockwise to remove it

5 Remove the retaining nuts and bolts, detach the housing and withdraw it from the vehicle **(see illustrations)**.
6 Installation is the reverse of removal.

18 Bulb replacement

Front parking, side marker/turn signal lights

Refer to illustrations 18.1 and 18.2

1 Remove the turn signal housing screw and push the housing straight forward to remove it **(see illustration)**.
2 Turn the bulb holder counterclockwise and remove it from the back of the housing **(see illustration)**. Push in and rotate the bulb counterclockwise to remove it from the holder.

Tail and back-up lights

Refer to illustration 18.3

3 Open the trunk and remove the trim cover for access to the housing. Rotate the holder counterclockwise to remove it **(see illustration)**.

4 Push in on the bulb and rotate it counterclockwise to remove it from the holder.

License plate light

Refer to illustrations 18.5 and 18.6

5 Remove the screws, detach the lens and pull the bulb holder down for access to the bulb **(see illustration)**.
6 Pull the bulb straight out to replace it **(see illustration)**.

High-mounted brake light

Refer to illustration 18.8

7 On parcel shelf-mounted lights, push the housing cover to the right, then pull to remove it. The bulbs pull straight out.
8 On trunk lid-mounted lights, rotate bulb holder counterclockwise to remove **(see illustration)**. Replace the bulb by pushing it in bulb and rotating it counterclockwise.

Interior lights

9 Remove dome lamp lens and replace the bulb by spreading the contacts at either end slightly.
10 Replace the trunk light bulb by detaching the lens, then pulling the bulb straight out.

Instrument cluster illumination

Refer to illustration 18.11

11 To gain access to the instrument cluster illumination lights, the instrument cluster will have to be removed (see Section 21). The bulbs can then be removed and replaced from the rear of the cluster **(see illustration)**.

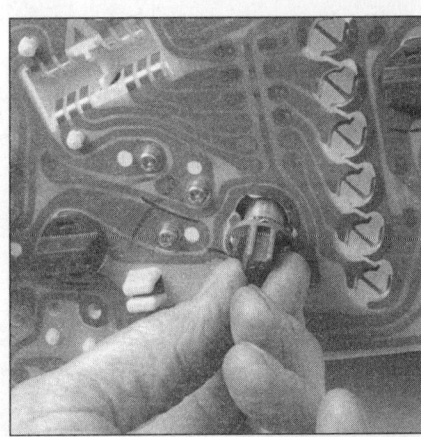

18.11 To remove an instrument cluster light bulb, depress it and turn it counterclockwise to release it

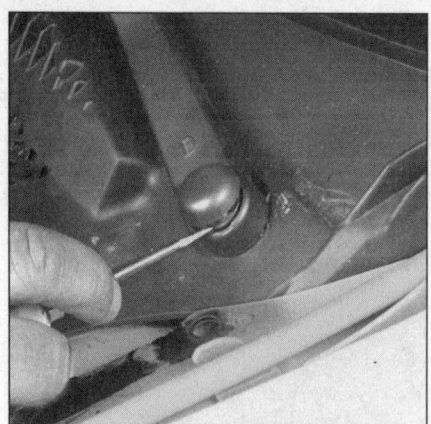

20.6a Use a small screwdriver to pry off the wiper arm nut cover, then remove the nut and pull the arm straight off its splined shaft

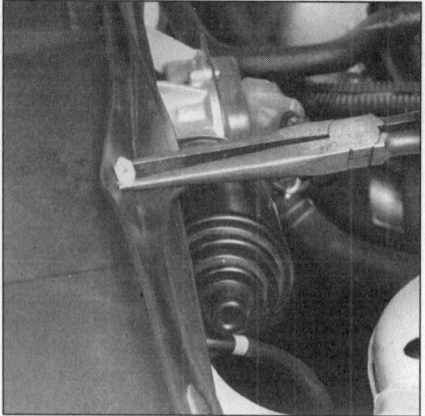

20.6b Use needle-nose pliers to remove the watershield clips (later models)

20.7 Unscrew the motor spindle nut (arrow)

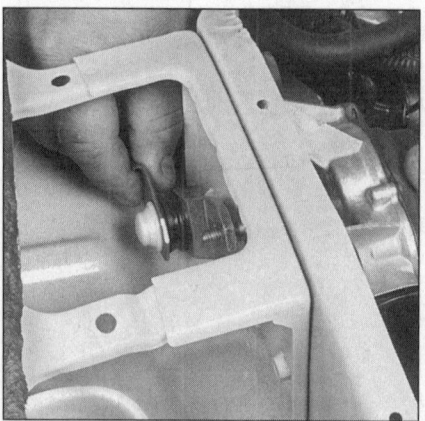

20.8 Detach the arm from the spindle and pull the motor out from the firewall

20.9 Unplug the electrical connector and remove the wiper motor

19 Daytime Running Lights (DRL) - general information

The Daytime Running Lights (DRL) system used on Canadian models turns the headlights on whenever the engine is started. The only exception is when the engine is turned on when the parking brake is engaged. Once the parking brake is released, the lights will remain on as long as the ignition switch is on, even if the parking brake is later applied.

The DRL system supplies reduced power to the headlights so they won't be too bright for daytime use while prolonging headlight life.

20 Wiper motor - removal and installation

Refer to illustrations 20.6a, 20.6b, 20.7, 20.8 and 20.9

Warning: *These models are equipped with airbags. To prevent accidental deployment (and possible injury) when working near airbag components, disconnect the battery cables (negative first, then positive) and wait at least ten minutes before beginning work (the system has a back-up capacitor that must fully discharge). For more information see Section 27.*

1   Disconnect the cables from the battery terminals (negative first, then positive) and wait ten minutes before proceeding to the next Step. The windshield wiper motor is located on the right (passenger) side of the underhood compartment.

Check

2   Refer to Section 9 for the wiper/washer switch check procedure.
3   If the continuity is not as specified, replace the switch.
4   If the motor doesn't work or doesn't park properly and the switch checks out

okay, the relay or the motor must be replaced.

Replacement

5   Disconnect the cable from the negative battery terminal.
6   Pry off the cover for the wiper arm nut, then unscrew the nut and remove the arm **(see illustration)**. Remove the screws and detach the cowl cover and wiper motor shield **(see illustration)**.
7   Remove the motor spindle nut **(see illustration)**.
8   Remove the motor bracket retaining bolts, detach the wiper arm then lower the wiper motor and bracket assembly **(see illustration)**.
9   Unplug the electrical connector and remove the motor from the vehicle **(see illustration)**.
10 Installation is the reverse of removal.

21 Instrument cluster - removal and installation

Refer to illustrations 21.3 and 21.4

Warning: *These models are equipped with airbags. To prevent accidental deployment (and possible injury) when working near airbag components, disconnect the battery cables (negative first, then positive) and wait at least ten minutes before beginning work (the system has a back-up capacitor that must fully discharge). For more information see Section 27.*

1   Disconnect the cables from the battery terminals (negative first, then positive) and wait ten minutes before proceeding to the

21.3 Remove the instrument cluster screws (arrows)

Chapter 12 Chassis electrical system

12-11

21.4 Pull the cluster out of the dash and unplug the electrical connectors from the backside

22.3 Unplug the electrical connector and remove the bolts (arrows), then detach the horn

next step.
2   Remove the instrument cluster bezel (see Chapter 11).
3   Remove the retaining screws and pull the cluster forward (see illustration).
4   Unplug the electrical connectors and remove the cluster from the vehicle (see illustration).
5   Installation is the reverse of the removal procedure.

22 Horn - check and replacement

Refer to illustration 22.3
Note: *Check the fuses before beginning electrical diagnosis.*
1   Unplug the electrical connector from the horn.
2   To test the horn, connect battery voltage to the two terminals with a pair of jumper wires. If the horn doesn't sound, replace it. If it does sound, the problem lies in the switch, relay or the wiring between the components.
3   To replace the horn, unplug the electrical connector and remove the bracket bolt (see illustration).
4   Installation is the reverse of removal.

23 Cruise control system - description and check

Refer to illustrations 23.5a and 23.5b
1   The cruise control system maintains vehicle speed with a vacuum-actuated servo motor located on the firewall in the engine compartment, which is connected to the throttle linkage by a cable. The system consists of the servo motor, brake switch, control switches, a relay and associated vacuum hoses. Some features of the system require special testers and diagnostic procedures which are beyond the scope of the home mechanic. Listed below are some general procedures that may be used to locate common problems.
2   Check the fuse (see Section 3).

23.5a The cruise control servo is located on the firewall - make sure the vacuum hose (arrow) is connected securely

3   Have an assistant operate the brake lights while you check their operation (voltage from the brake light switch deactivates the cruise control).
4   If the brake lights don't come on or don't shut off, correct the problem and retest the cruise control.
5   Visually inspect the vacuum hose connected to the servo and check the control linkage between the cruise control servo and the throttle linkage and replace as necessary (see illustrations).
6   Check the Vehicle Speed Sensor (see Chapter 6).
7   Test drive the vehicle to determine if the cruise control is now working. If it isn't, take it to a dealer service department or an automotive electrical specialist for further diagnosis and repair.

24 Power window system - description and check

1   The power window system operates the electric motors mounted in the doors which

25.5b Make sure the cruise control (A) and accelerator (B) linkage mounted on the throttle body are not damaged and that they operate smoothly together when the throttle is opened

lower and raise the windows. The system consists of the control switches, the motors (regulators), glass mechanisms and associated wiring.
2   Power windows are wired so they can be lowered and raised from the master control switch by the driver or by remote switches located at the individual windows. Each window has a separate motor which is reversible. The position of the control switch determines the polarity and therefore the direction of operation. Some systems are equipped with relays that control current flow to the motors.
3   Some vehicles are equipped with a separate circuit breaker for each motor in addition to the fuse or circuit breaker protecting the whole circuit. This prevents one stuck window from disabling the whole system.
4   The power window system will only operate when the ignition switch is ON. In addition, many models have a window lockout switch at the master control switch which, when activated, disables the switches at the rear windows and, sometimes, the

12

switch at the passenger's window also. Always check these items before troubleshooting a window problem.

5 These procedures are general in nature, so if you can't find the problem using them, take the vehicle to a dealer service department.

6 If the power windows don't work at all, check the fuse or circuit breaker.

7 If only the rear windows are inoperative, or if the windows only operate from the master control switch, check the rear window lockout switch for continuity in the unlocked position. Replace it if it doesn't have continuity.

8 Check the wiring between the switches and fuse panel for continuity. Repair the wiring, if necessary.

9 If only one window is inoperative from the master control switch, try the other control switch at the window. **Note:** *This doesn't apply to the drivers door window.*

10 If the same window works from one switch, but not the other, check the switch for continuity.

11 If the switch tests OK, check for a short or open in the wiring between the affected switch and the window motor.

12 If one window is inoperative from both switches, remove the trim panel from the affected door and check for voltage at the motor while the switch is operated.

13 If voltage is reaching the motor, disconnect the glass from the regulator (see Chapter 11). Move the window up and down by hand while checking for binding and damage. Also check for binding and damage to the regulator. If the regulator is not damaged and the window moves up and down smoothly, replace the motor. If there's binding or damage, lubricate, repair or replace parts, as necessary.

14 If voltage isn't reaching the motor, check the wiring in the circuit for continuity between the switches and motors. You'll need to consult the wiring diagram for the vehicle. Some power window circuits are equipped with relays. If equipped, check that the relays are grounded properly and receiving voltage from the switches. Also check that each relay sends voltage to the motor when the switch is turned on. If it doesn't, replace the relay.

15 Test the windows after you are done to confirm proper repairs.

25 Power door lock system - description and check

1 The power door lock system operates the door lock actuators mounted in each door. The system consists of the switches, actuators and associated wiring. Diagnosis can usually be limited to simple checks of the wiring connections and actuators for minor faults which can be easily repaired.

2 Power door lock systems are operated by bi-directional solenoids located in the doors. The lock switches have two operating positions: Lock and Unlock. These switches activate a relay which in turn connects voltage to the door lock solenoids. Depending on which way the relay is activated, it reverses polarity, allowing the two sides of the circuit to be used alternately as the feed (positive) and ground side.

3 Some vehicles may have keyless entry, electronic control modules and anti-theft systems incorporated into the power locks. If you are unable to locate the trouble using the following general steps, consult your a dealer service department. **Note:** *Some vehicles also have control switches connected to the key locks in the doors which unlock all the doors when one is unlocked.*

4 Always check the circuit protection first. Some vehicles use a combination of circuit breakers and fuses.

5 Operate the door lock switches in both directions (Lock and Unlock) with the engine off. Listen for the faint click of the relay operating.

6 If there's no click, check for voltage at the switches. If no voltage is present, check the wiring between the fuse panel and the switches for shorts and opens.

7 If voltage is present but no click is heard, test the switch for continuity. Replace it if there's not continuity in both switch positions.

8 If the switch has continuity but the relay doesn't click, check the wiring between the switch and relay for continuity. Repair the wiring if there's no continuity.

9 If the relay is receiving voltage from the switch but is not sending voltage to the solenoids, check for a bad ground at the relay case. If the relay case is grounding properly, replace the relay.

10 If all but one lock solenoids operate, remove the trim panel from the affected door (see Chapter 11). and check for voltage at the solenoid while the lock switch is operated. One of the wires should have voltage in the Lock position; the other should have voltage in the Unlock position.

11 If the inoperative solenoid is receiving voltage, replace the solenoid.

12 If the inoperative solenoid isn't receiving voltage, check for an open or short in the wire between the lock solenoid and the relay. **Note:** *It's common for wires to break in the portion of the harness between the body and door (opening and closing the door fatigues and eventually breaks the wires).*

26 Electric rear view mirrors - description and check

1 Most electric rear view mirrors use two motors to move the glass; one for up and down adjustments and one for left-right adjustments.

2 The control switch has a selector portion which sends voltage to the left or right side mirror. With the ignition ON but the engine OFF, roll down the windows and operate the mirror control switch through all functions (left-right and up-down) for both the left and right side mirrors.

3 Listen carefully for the sound of the electric motors running in the mirrors.

4 If the motors can be heard but the mirror glass doesn't move, there's probably a problem with the drive mechanism inside the mirror. Remove and disassemble the mirror to locate the problem.

5 If the mirrors don't operate and no sound comes from the mirrors, check the fuse (see Chapter 1).

6 If the fuse is OK, remove the mirror control switch from its mounting without disconnecting the wires attached to it. Turn the ignition ON and check for voltage at the switch. There should be voltage at one terminal. If there's no voltage at the switch, check for an open or short in the wiring between the fuse panel and the switch.

7 If there's voltage at the switch, disconnect it. Check the switch for continuity in all its operating positions. If the switch does not have continuity, replace it.

8 Re-connect the switch. Locate the wire going from the switch to ground. Leaving the switch connected, connect a jumper wire between this wire and ground. If the mirror works normally with this wire in place, repair the faulty ground connection.

9 If the mirror still doesn't work, remove the mirror and check the wires at the mirror for voltage. Check with ignition ON and the mirror selector switch on the appropriate side. Operate the mirror switch in all its positions. There should be voltage at one of the switch-to-mirror wires in each switch position (except the neutral "off" position).

10 If there's not voltage in each switch position, check the wiring between the mirror and control switch for opens and shorts.

11 If there's voltage, remove the mirror and test it off the vehicle with jumper wires. Replace the mirror if it fails this test.

27 Airbag system - general information

These models are equipped with a Supplemental Restraint System (SRS), more commonly known as airbags. This system is designed to protect the driver and front seat passenger from serious injury in the event of a head-on or frontal collision. It consists of airbag modules in the center of the steering wheel and the right side of the dash and a diagnostic/control unit located inside the passenger compartment.

Airbag modules

The airbag modules consist of a housing incorporating the cushion (airbag) and inflator unit. The inflator assembly is mounted on the back of the housing over a hole through which gas is expelled, inflating the bag almost instantaneously when an electrical signal is

Chapter 12 Chassis electrical system

sent from the system. The specially wound wire on the driver's side that carries this signal to the module is called a spiral cable. The spiral cable is a flat, ribbon-like electrically conductive tape which is wound many times so that it can transmit an electrical signal regardless of steering wheel position.

Diagnostic/control unit and tunnel/safing sensors

The diagnostic/control unit contains an on-board microprocessor which monitors the operation of the system, and also contains a crash sensor. It checks this system every time the vehicle is started, causing the "AIRBAG" light to go on then off, if the system is operating properly. If there is a fault in the system, the light will go on and stay on and the unit will store fault codes indicating the nature of the fault. If the AIRBAG light goes on and stays on, the vehicle should be taken to your dealer immediately for service.

Precautions

Warning: *Failure to follow these precautions could result in accidental deployment of the airbag and personal injury.*

Whenever working in the vicinity of the steering wheel, steering column or any of the other SRS system components, the system must be disarmed. To disarm the system:
a) Point the wheels straight ahead and turn the key to the Lock position.
b) Disconnect the battery cables.
c) Wait at least 3 minutes for the back-up power supply to be depleted

Whenever handling an airbag module, always keep the airbag opening pointed away from your body. Never place the airbag module on a bench of other surface with the airbag opening facing the surface. Always place the airbag module in a safe location with the airbag opening facing up.

Never measure the resistance of any SRS component. An ohmmeter has a built-in battery supply that could accidentally deploy the airbag.

Never use electrical welding equipment on a vehicle equipped with an airbag without first disconnecting the yellow airbag connectors.

Never dispose of a live airbag module. Return it to your dealer for safe deployment, using special equipment, and disposal.

28 Wiring diagrams - general information

Since it isn't possible to include all wiring diagrams for every year covered by this manual, the following diagrams are those that are typical and most commonly needed.

Prior to troubleshooting any circuits, check the fuse and circuit breakers (if equipped) to make sure they are in good condition. Make sure the battery is properly charged and has clean, tight cable connections (see Chapter 1).

When checking the wiring system, make sure that all electrical connectors are clean, with no broken or loose pins. When unplugging an electrical connector, do not pull on the wires, only on the connector housings themselves.

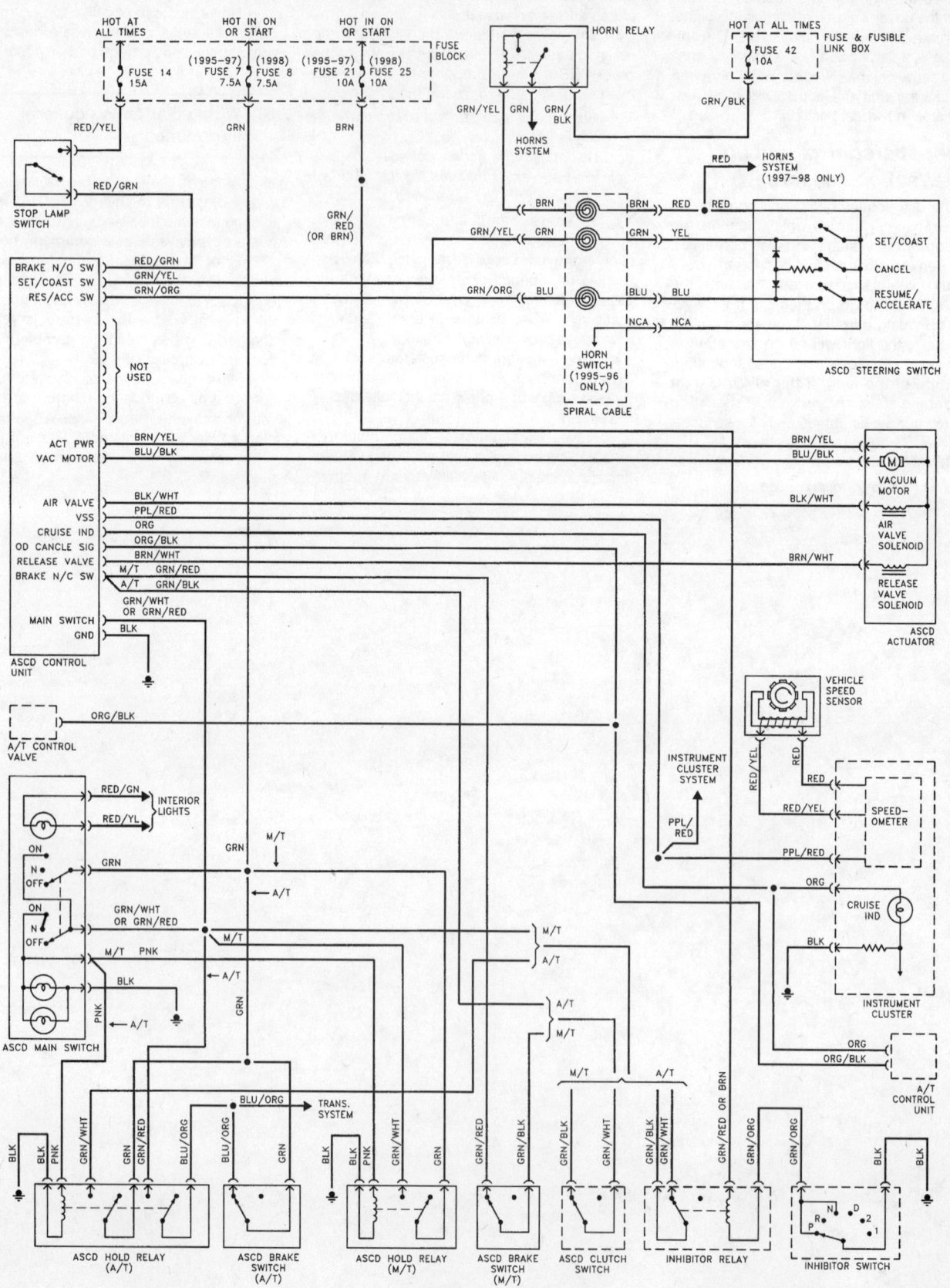

Cruise control system wiring diagram

Chapter 12 Chassis electrical system

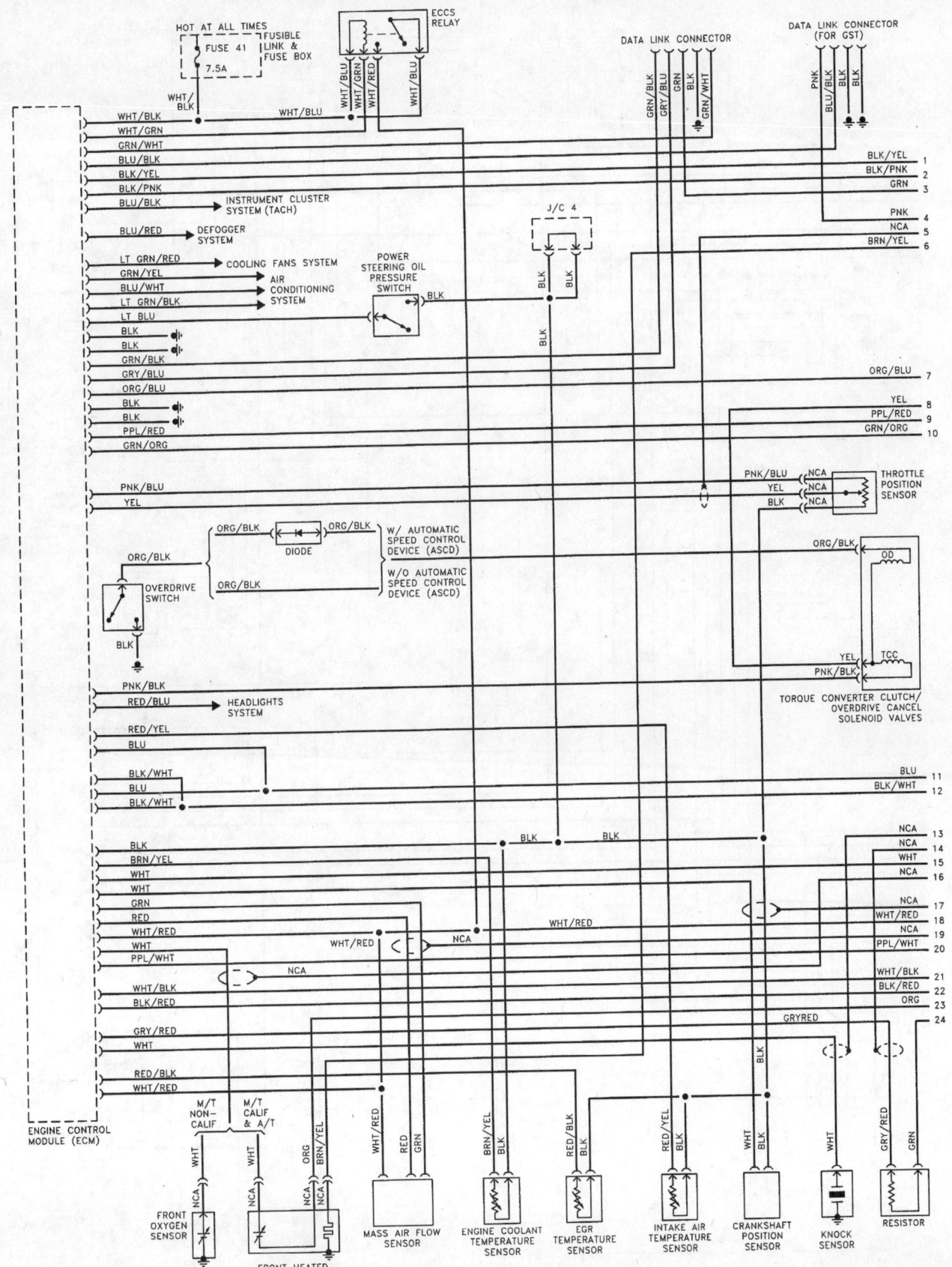

1995 and 1996 GA16DE engine control wiring diagram (1 of 2)

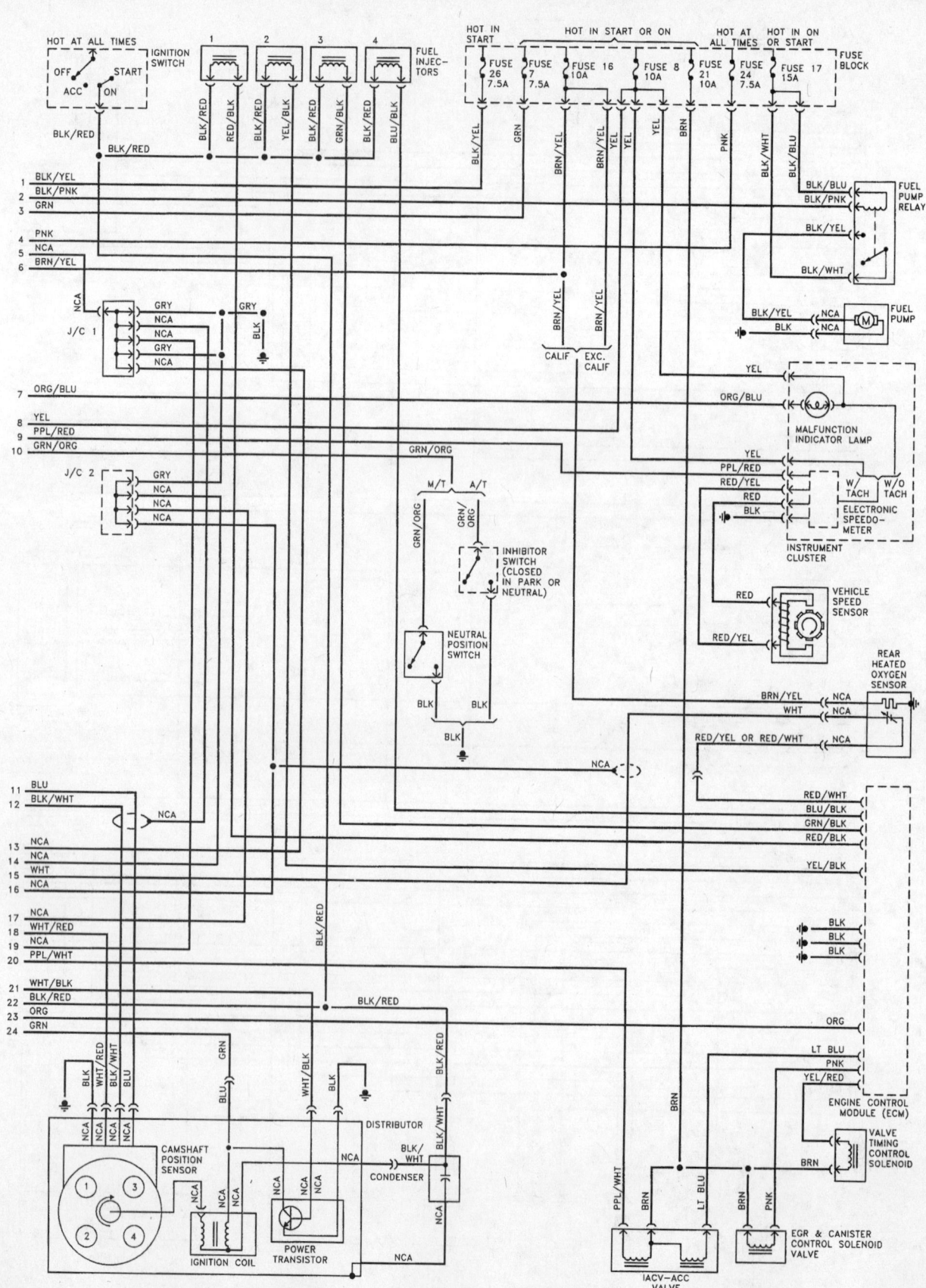

1995 and 1996 GA16DE engine control wiring diagram (2 of 2)

Chapter 12 Chassis electrical system

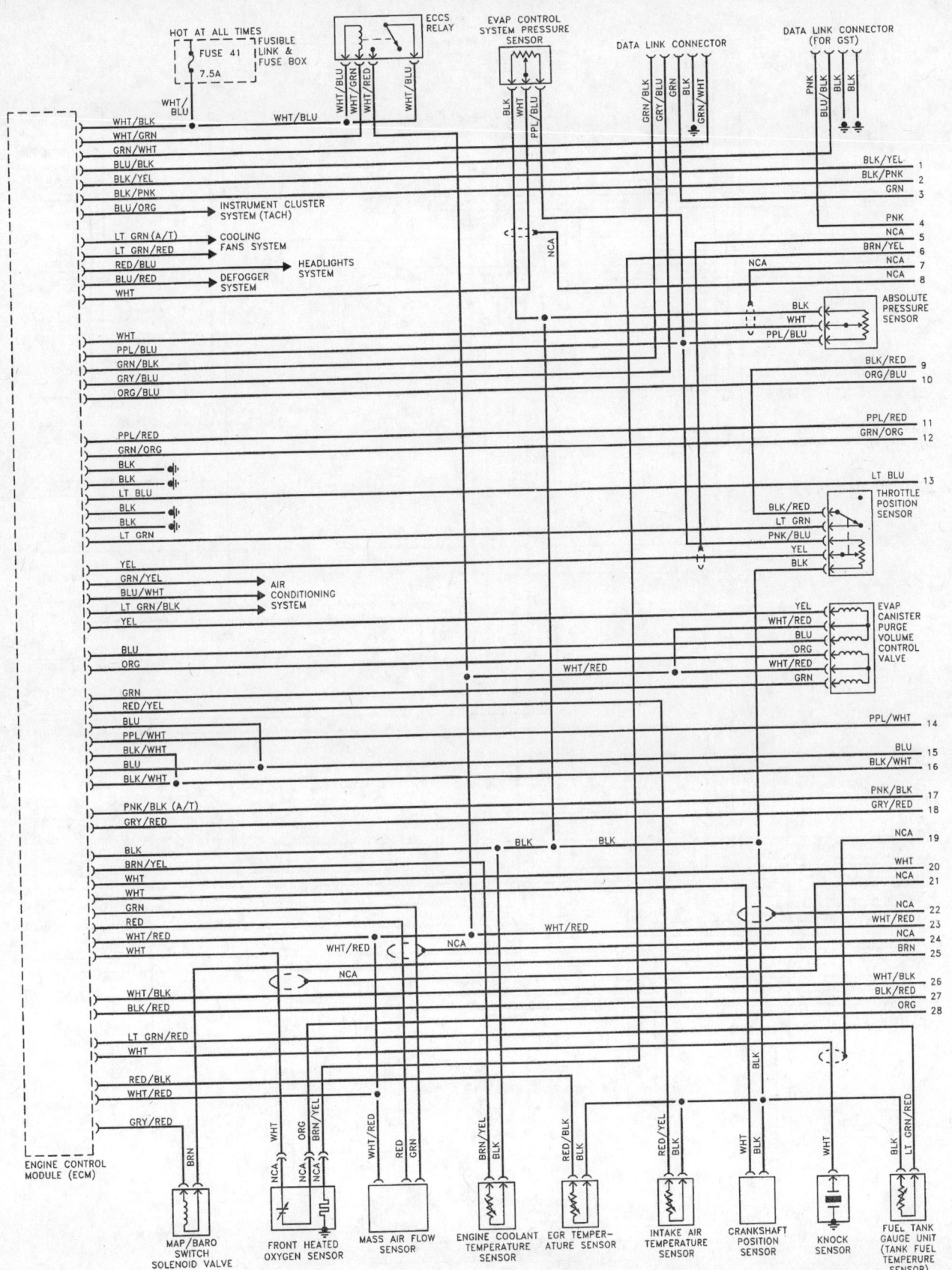

1997 and 1998 GA16DE engine control wiring diagram (1 of 3)

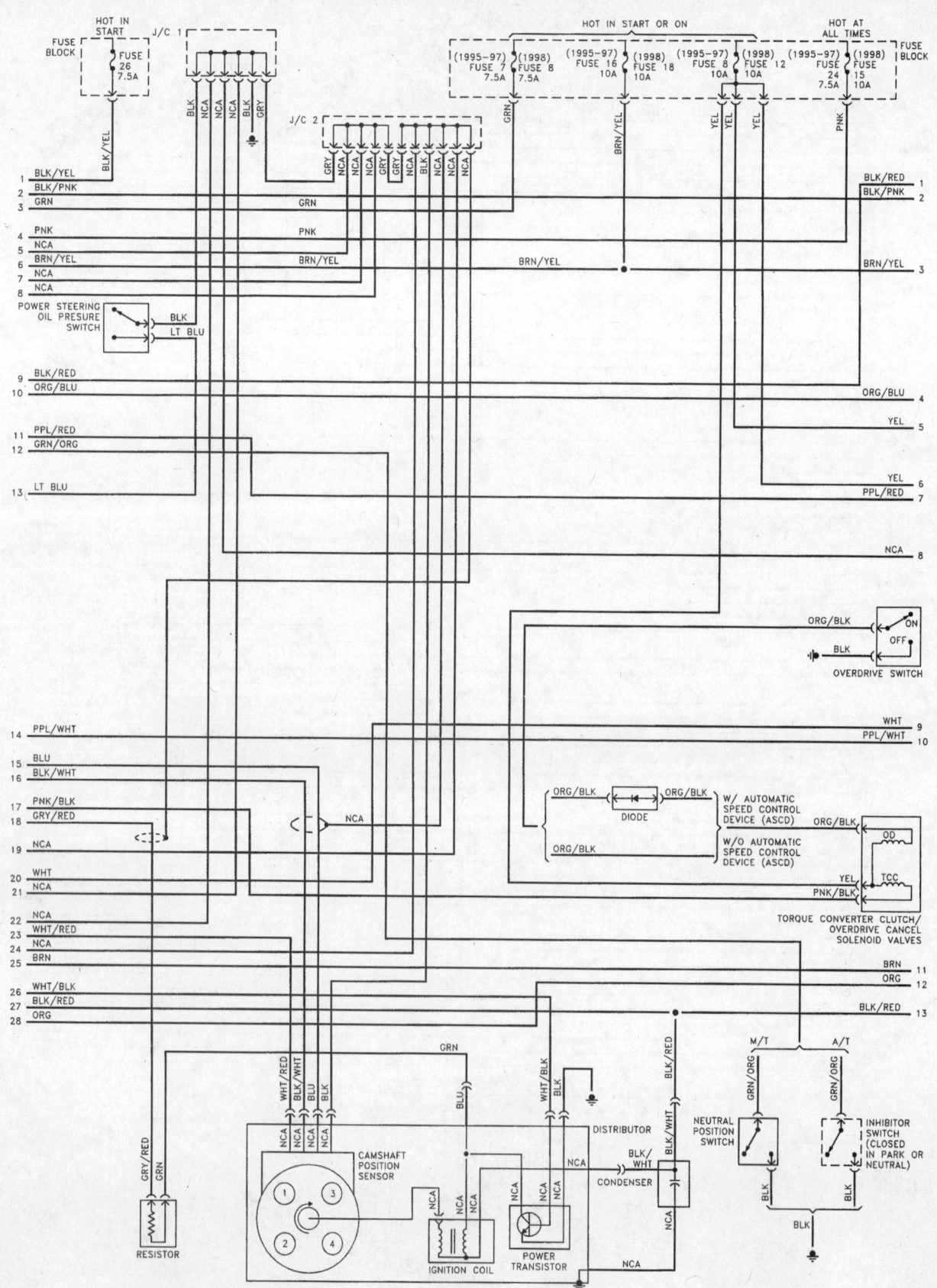

1997 and 1998 GA16DE engine control wiring diagram (2 of 3)

Chapter 12 Chassis electrical system

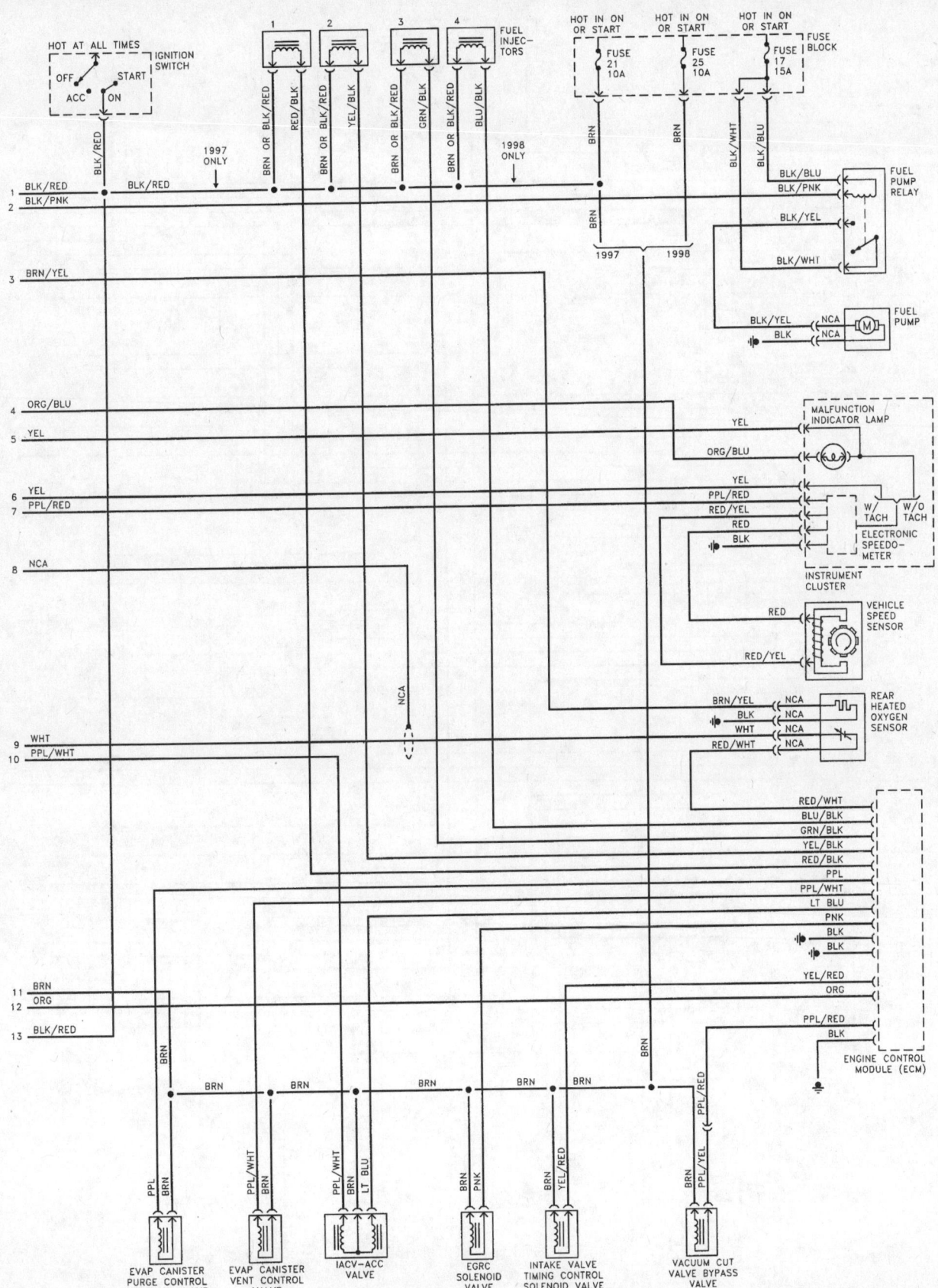

1997 and 1998 GA16DE engine control wiring diagram (3 of 3)

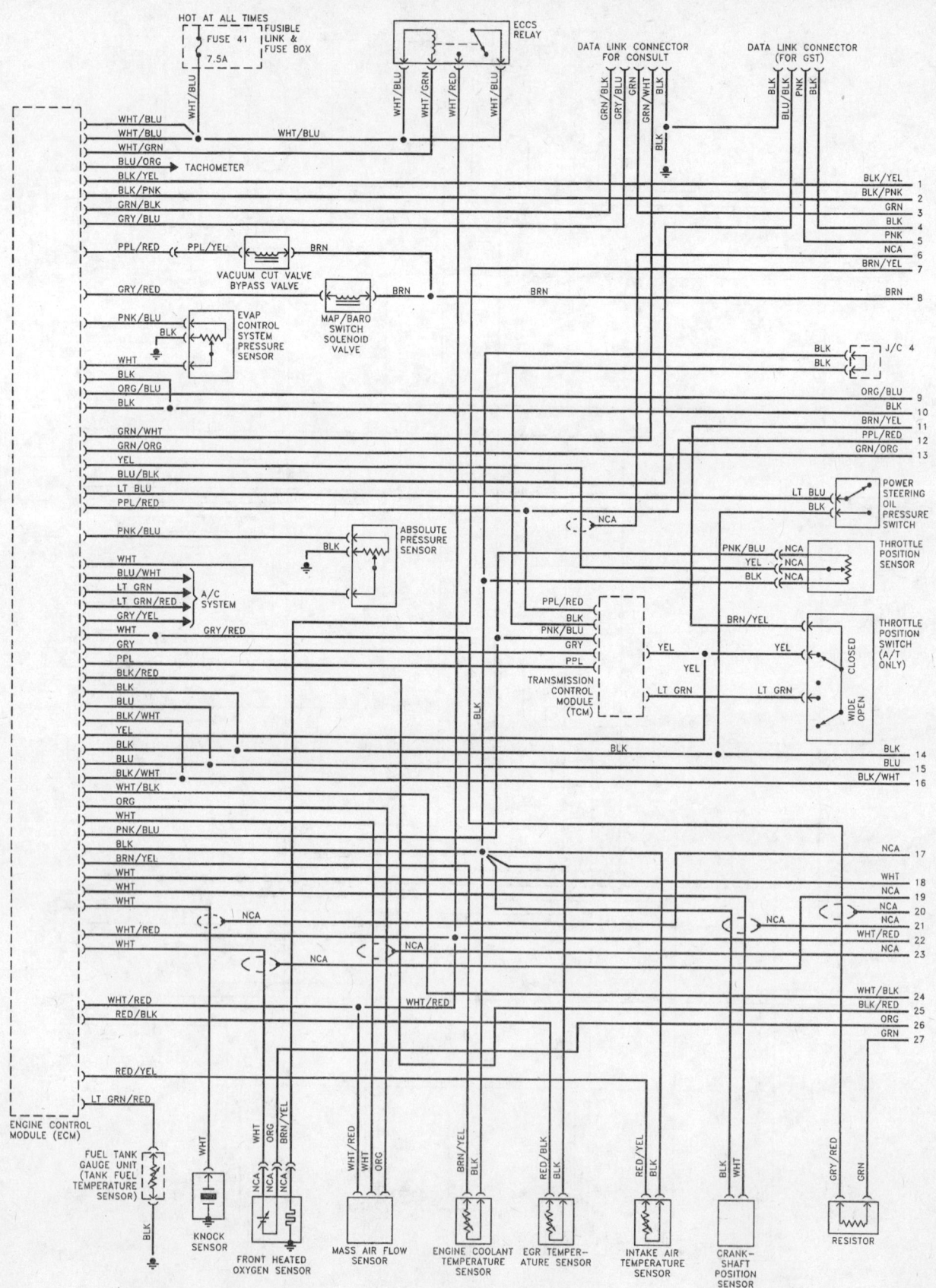

SR20DE engine control wiring diagram (1 of 2)

Chapter 12 Chassis electrical system

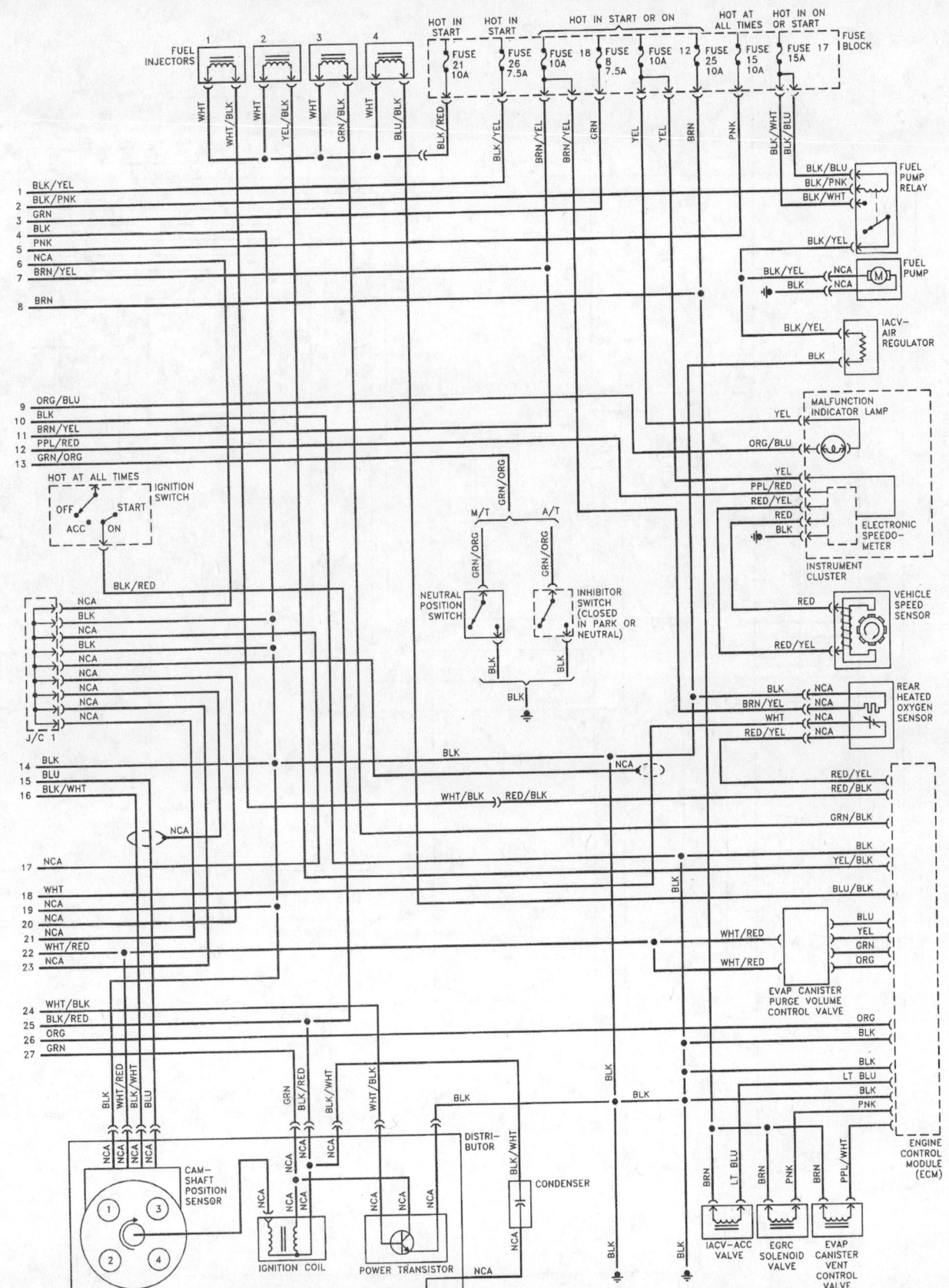

SR20DE engine control wiring diagram (2 of 2)

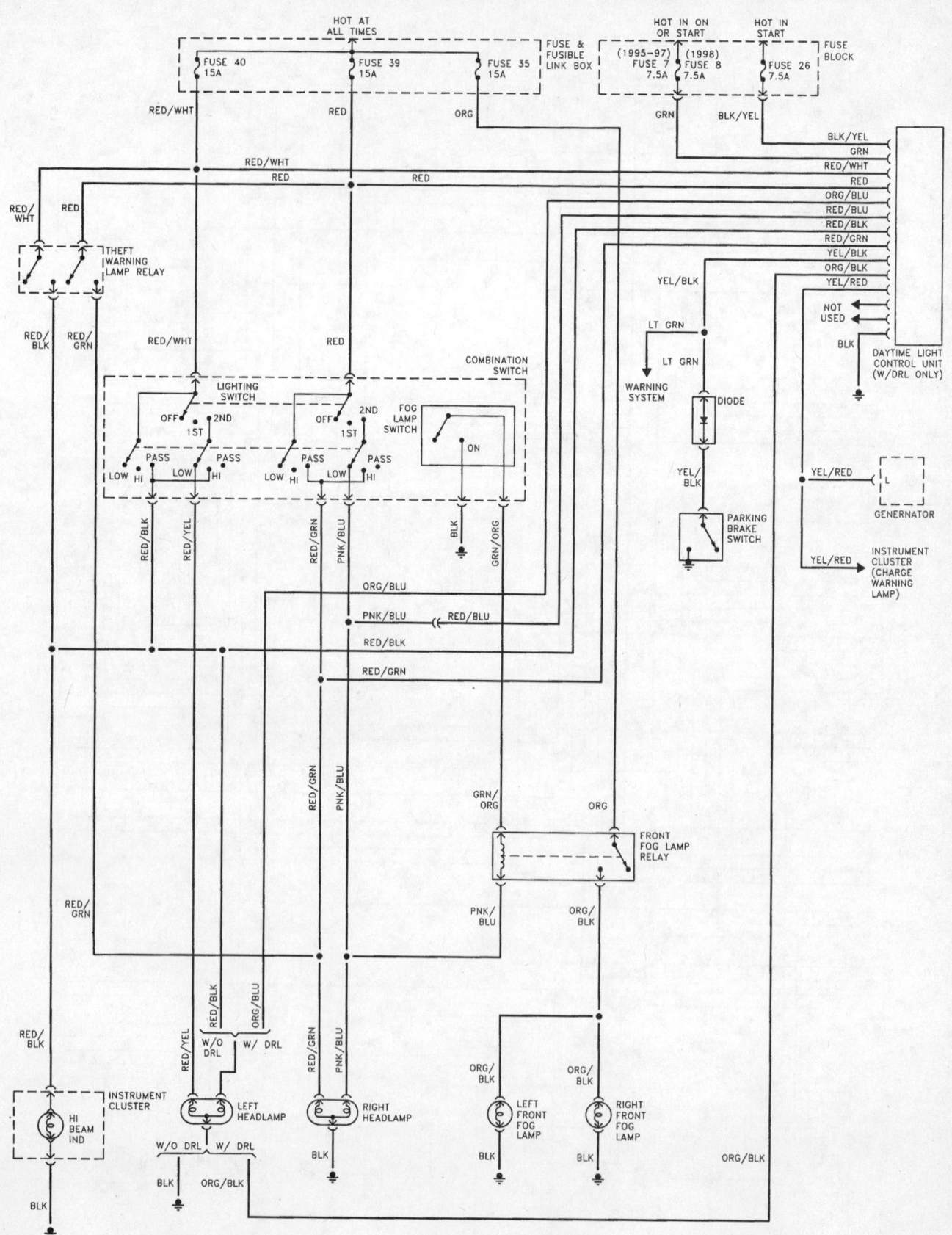

Headlight system wiring diagram

Chapter 12 Chassis electrical system

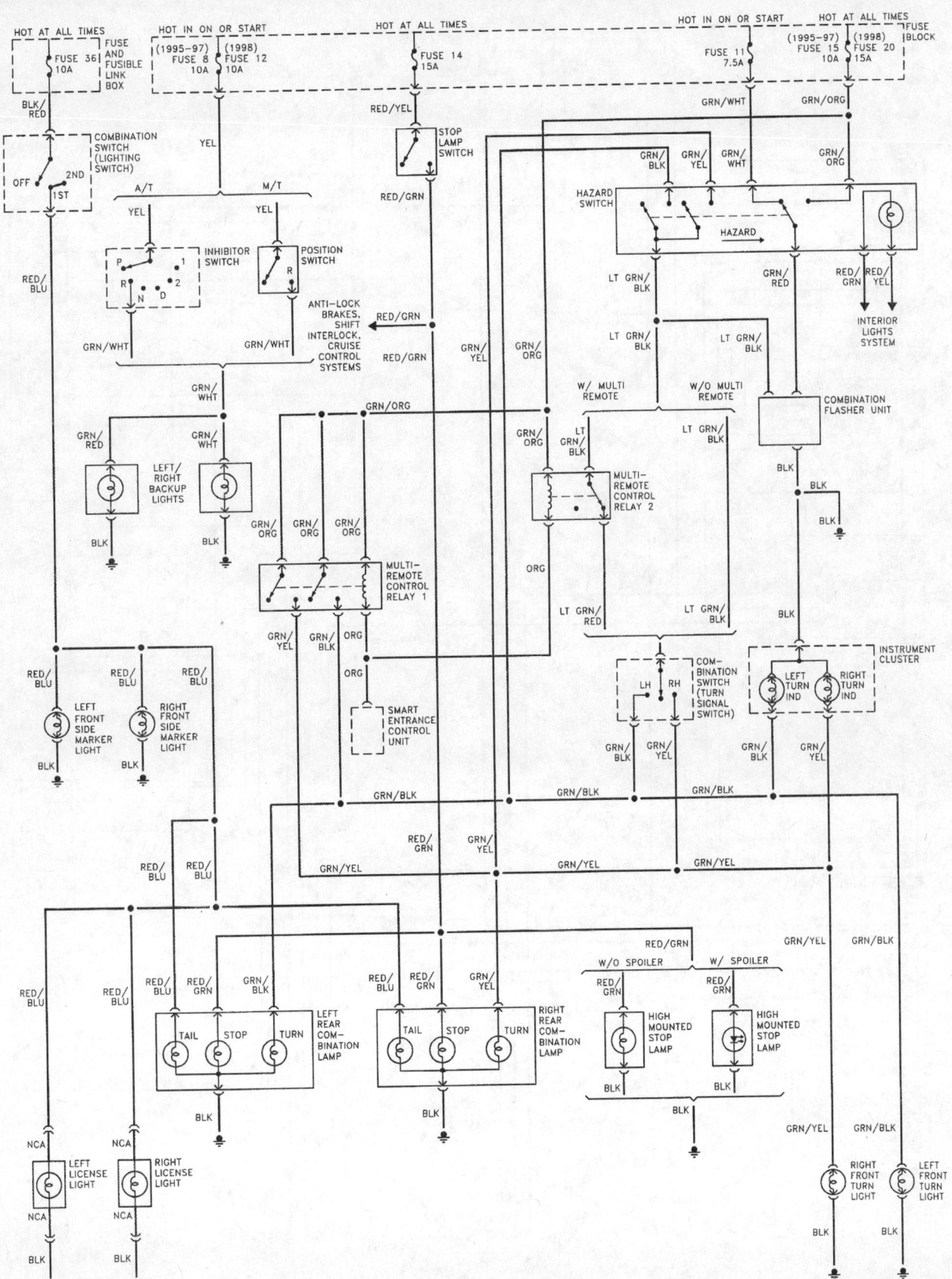

Exterior lighting system wiring diagram (except headlights)

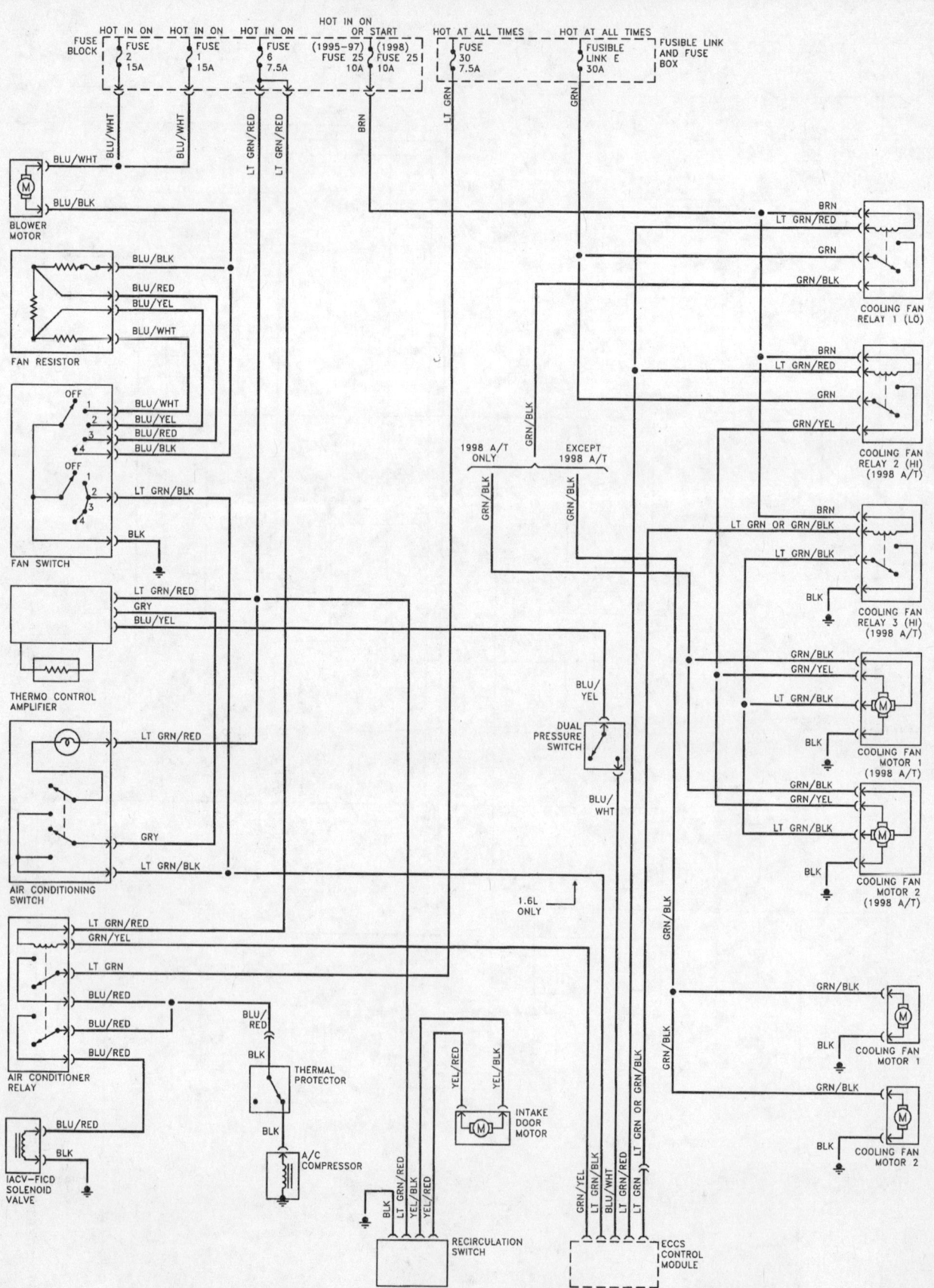

Heating, air conditioning and engine cooling fan systems wiring diagram

Chapter 12 Chassis electrical system

12-25

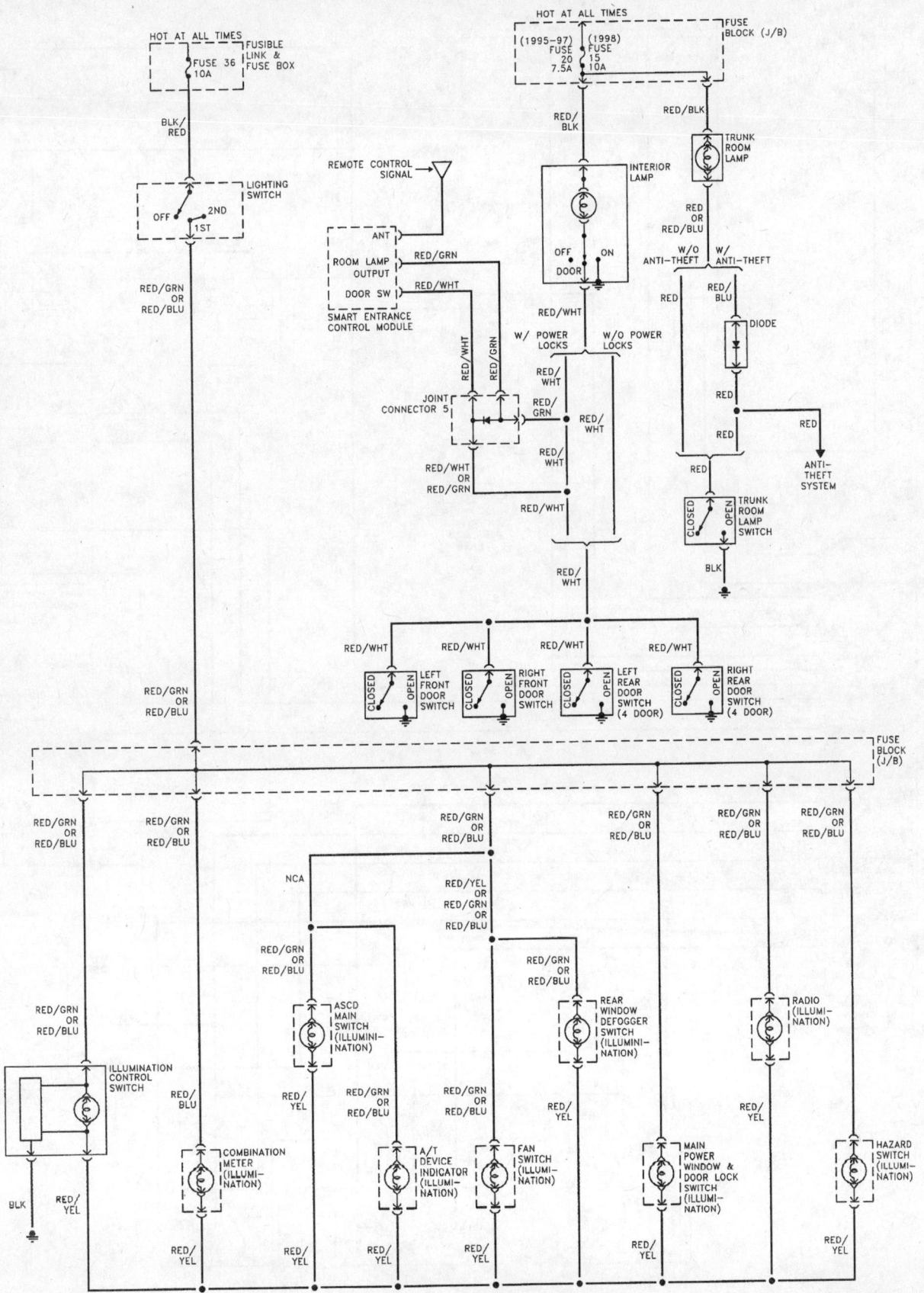

Interior lighting system wiring diagram

12-26 Chapter 12 Chassis electrical system

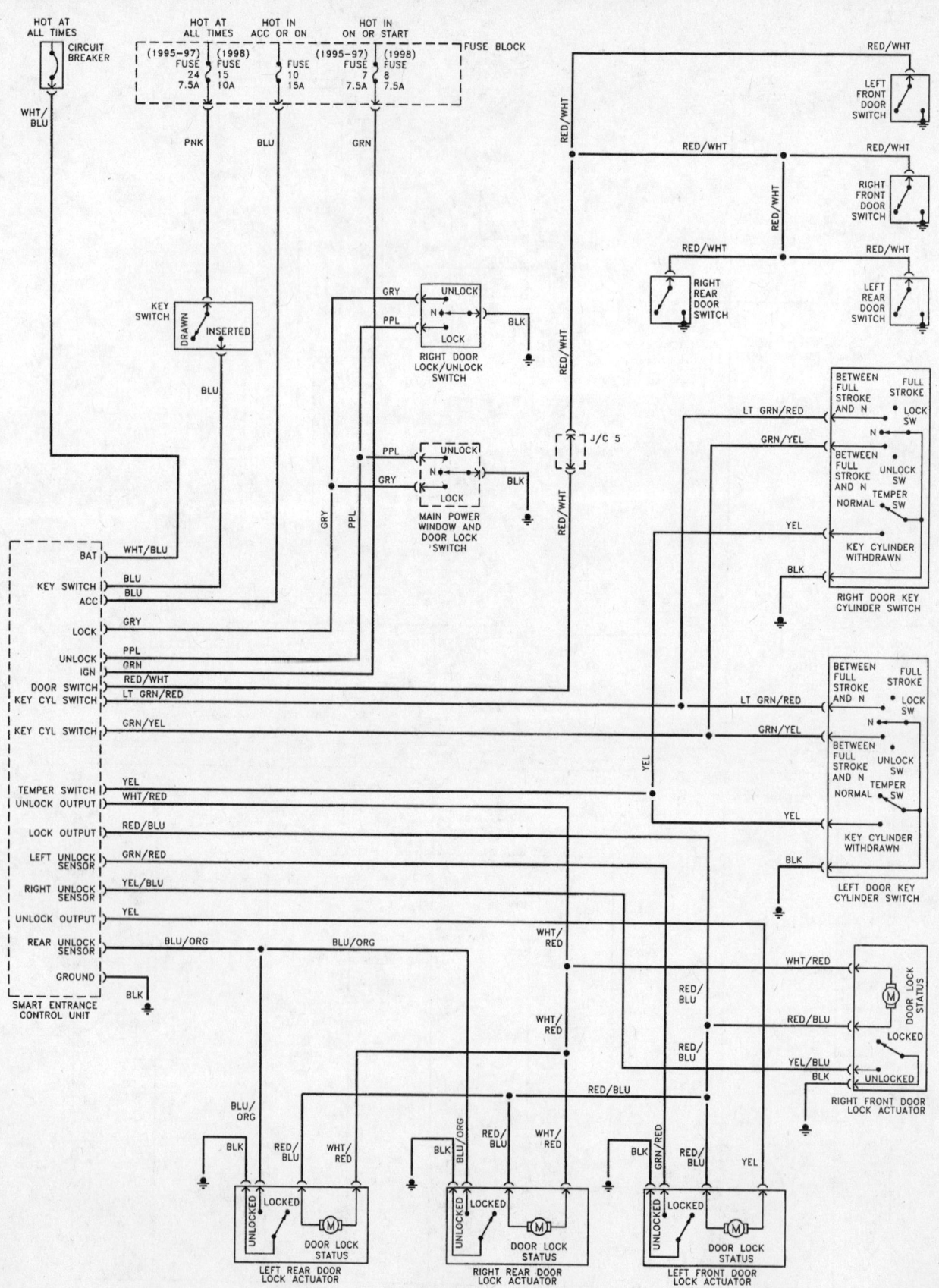

Power door lock system wiring diagram (with keyless entry)

Chapter 12 Chassis electrical system

12-27

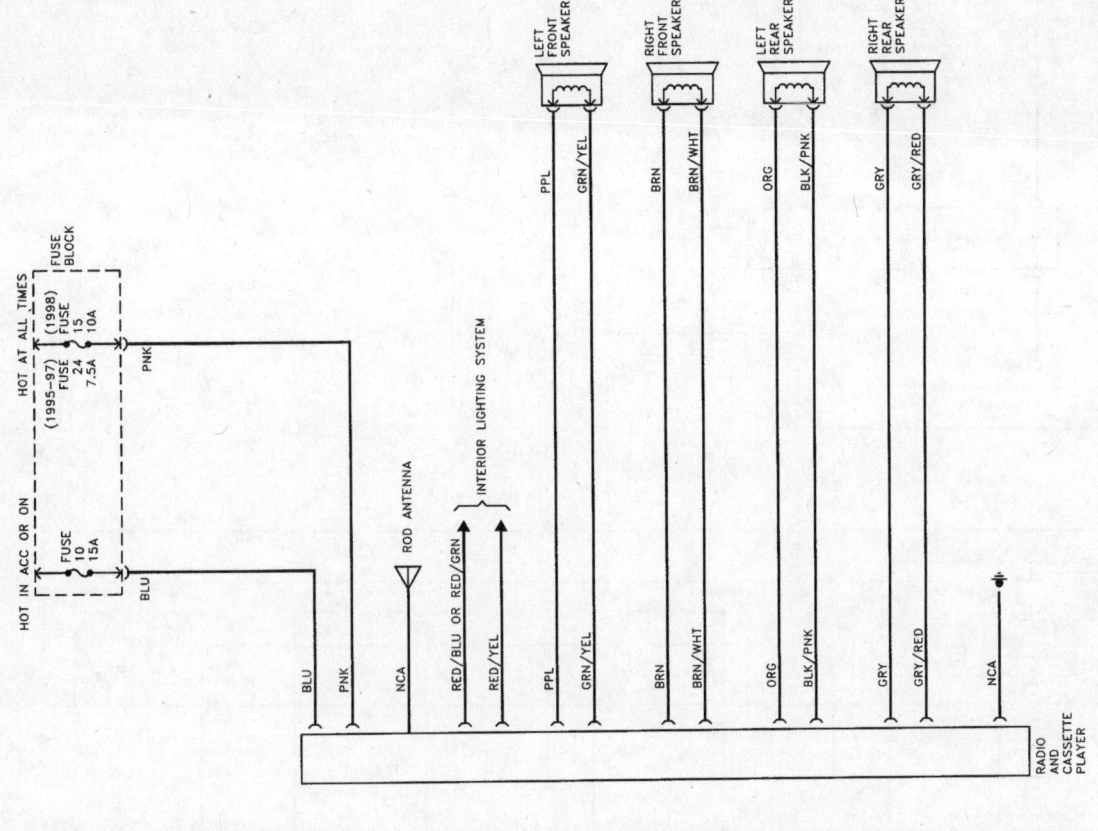

Audio system wiring diagram

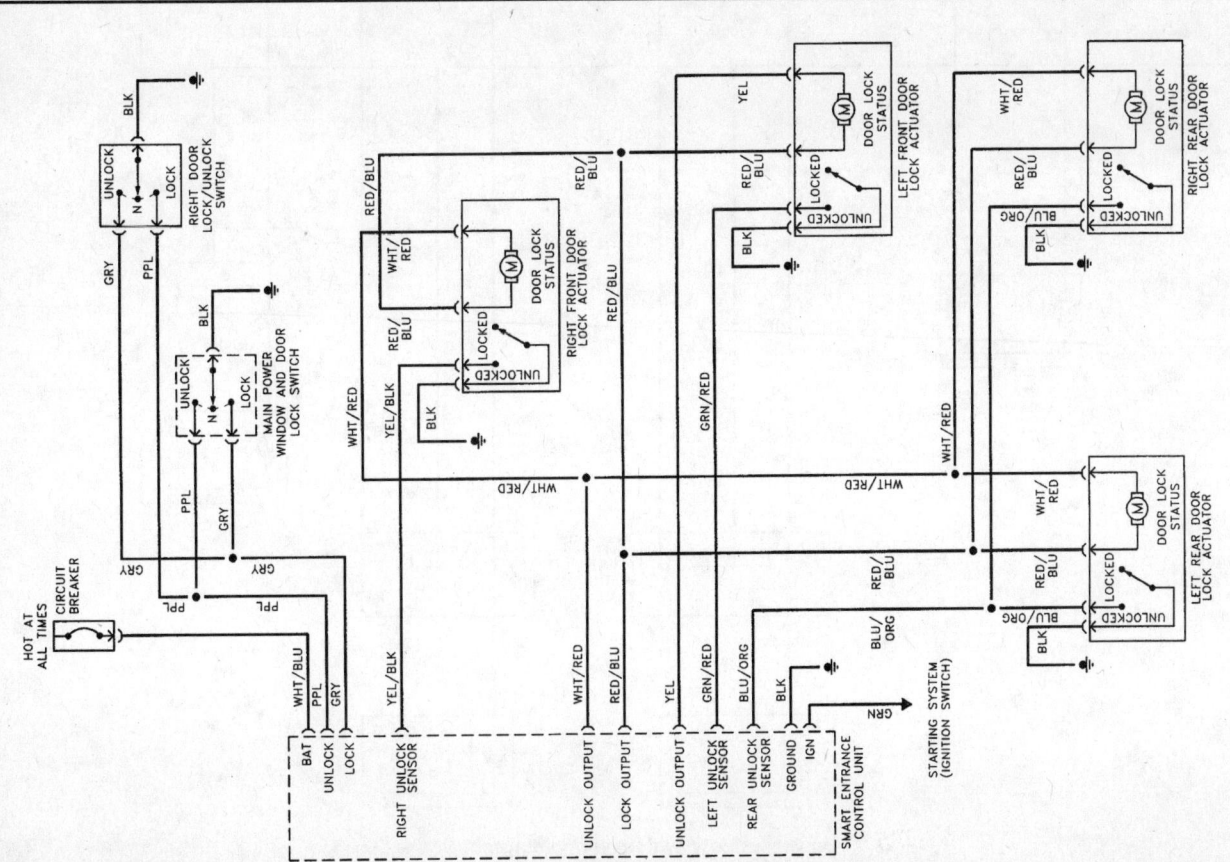

Power door lock system wiring diagram (without keyless entry)

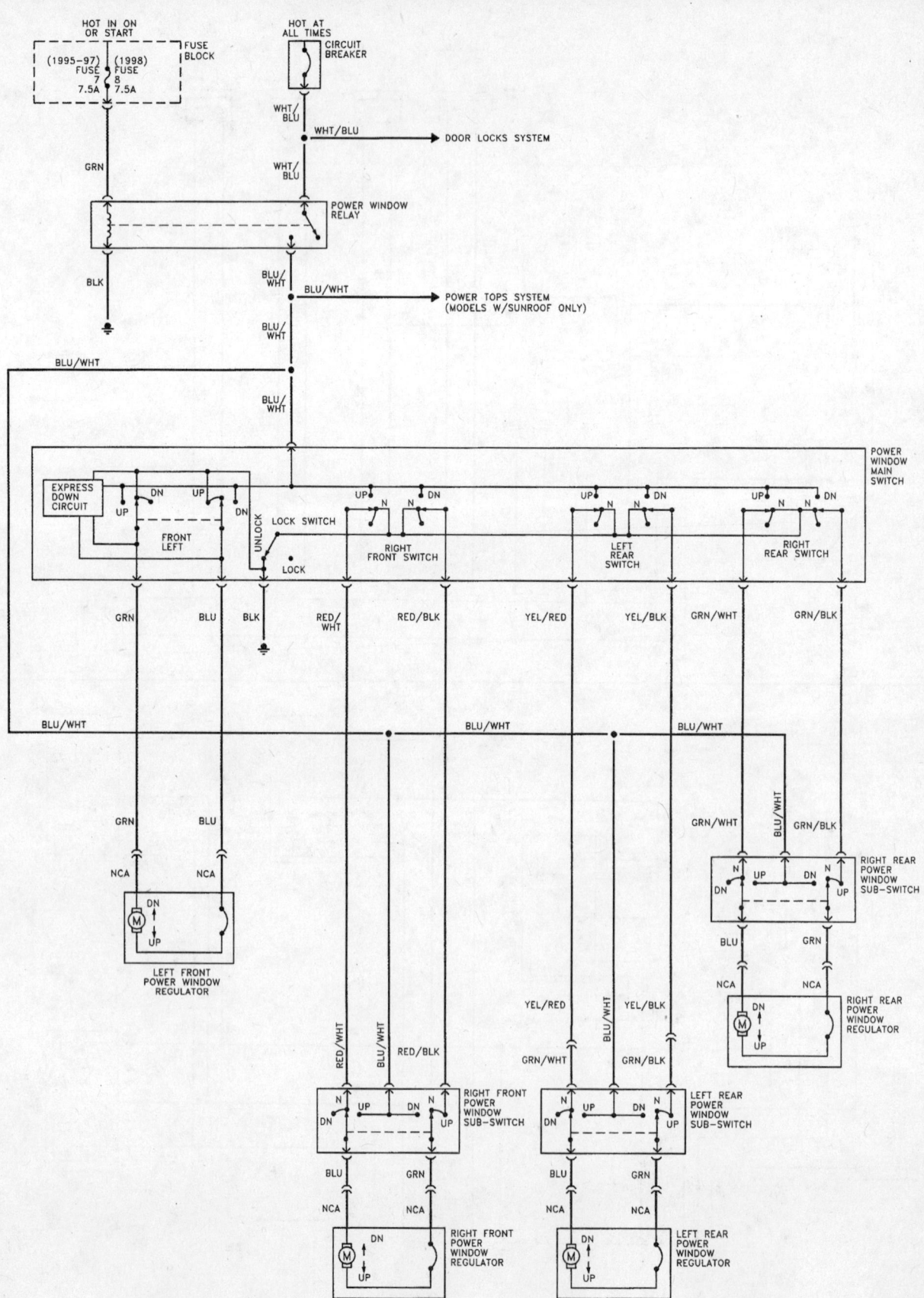

Power window system wiring diagram

Chapter 12 Chassis electrical system

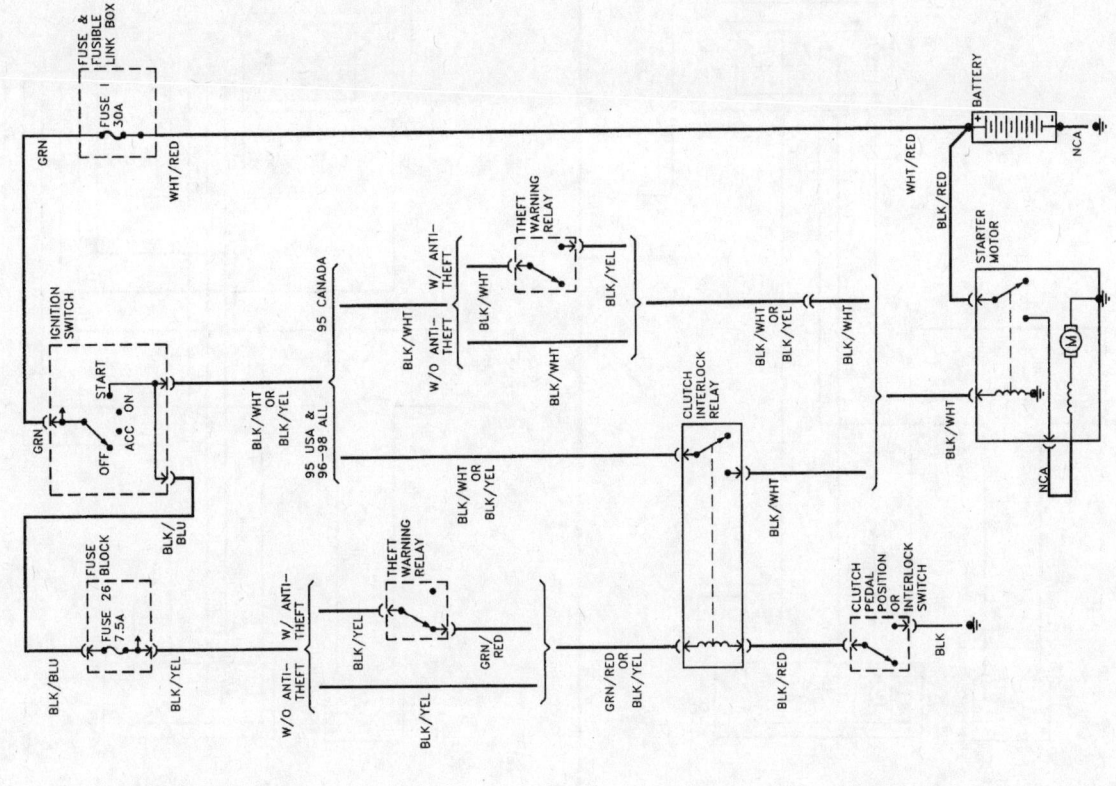

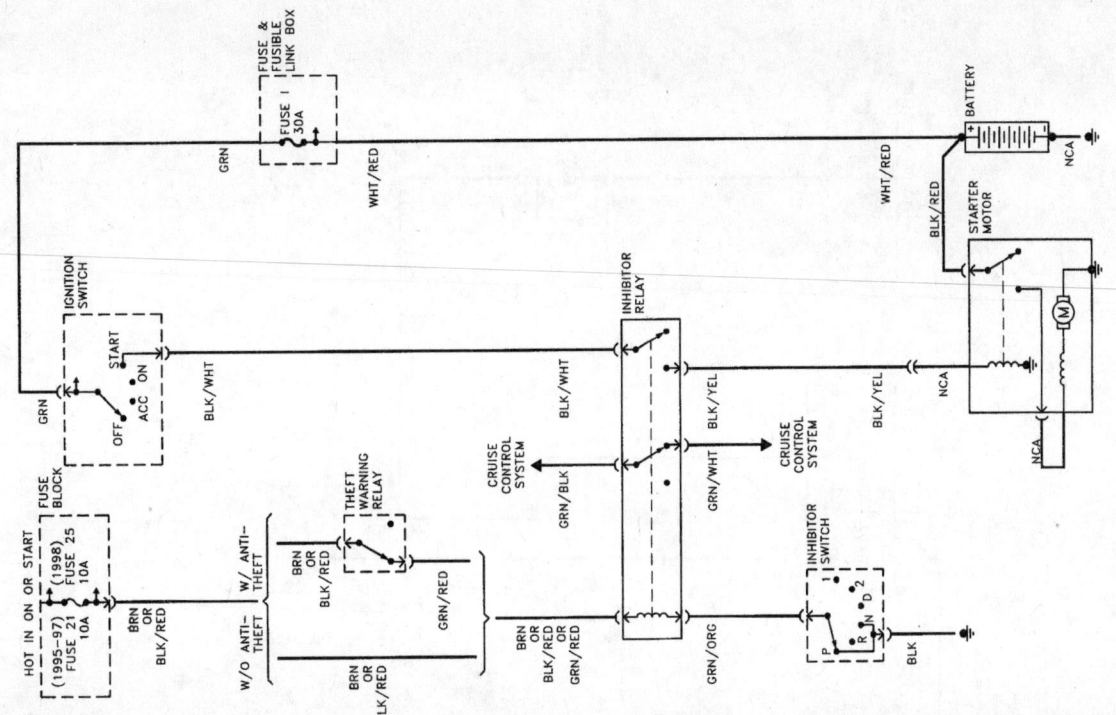

Starting system wiring diagram

12-30 Chapter 12 Chassis electrical system

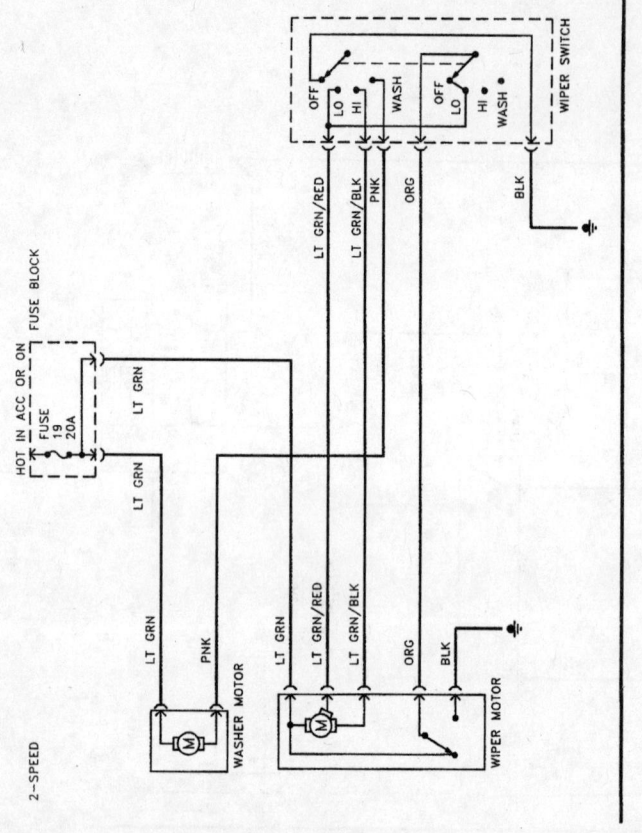
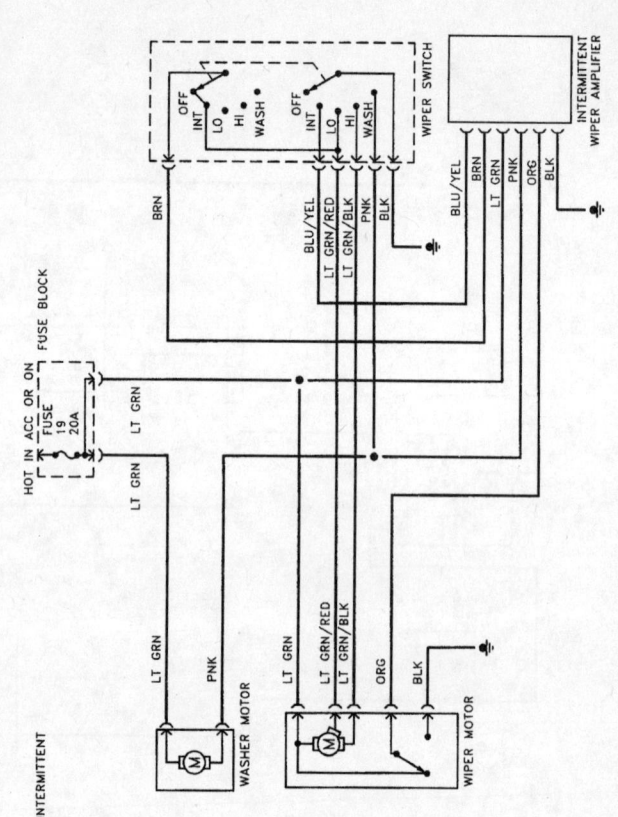

Wiper/washer system wiring diagram

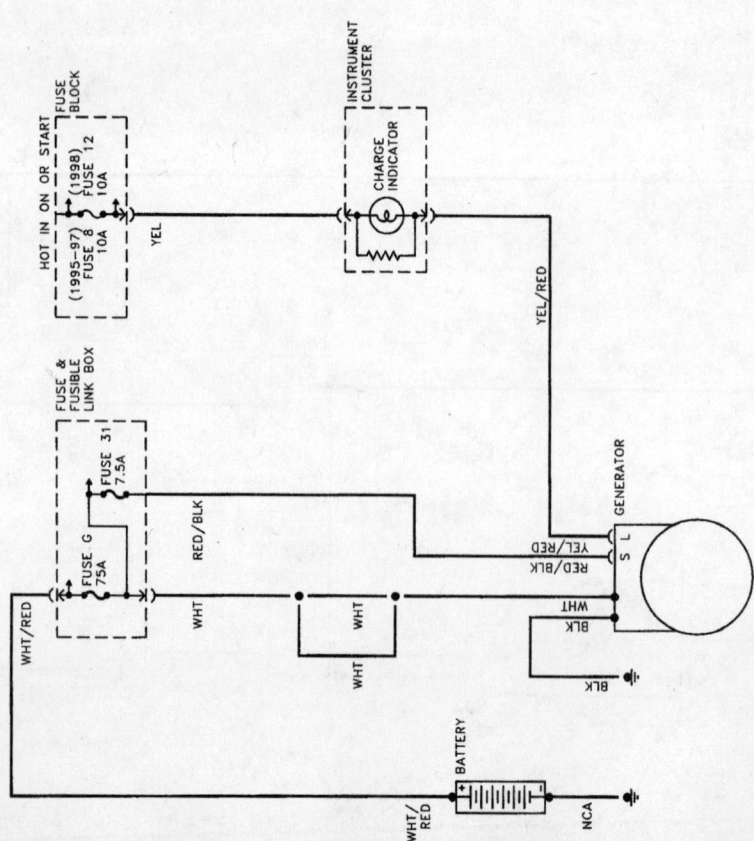

Charging system wiring diagram

Chapter 12 Chassis electrical system

12-31

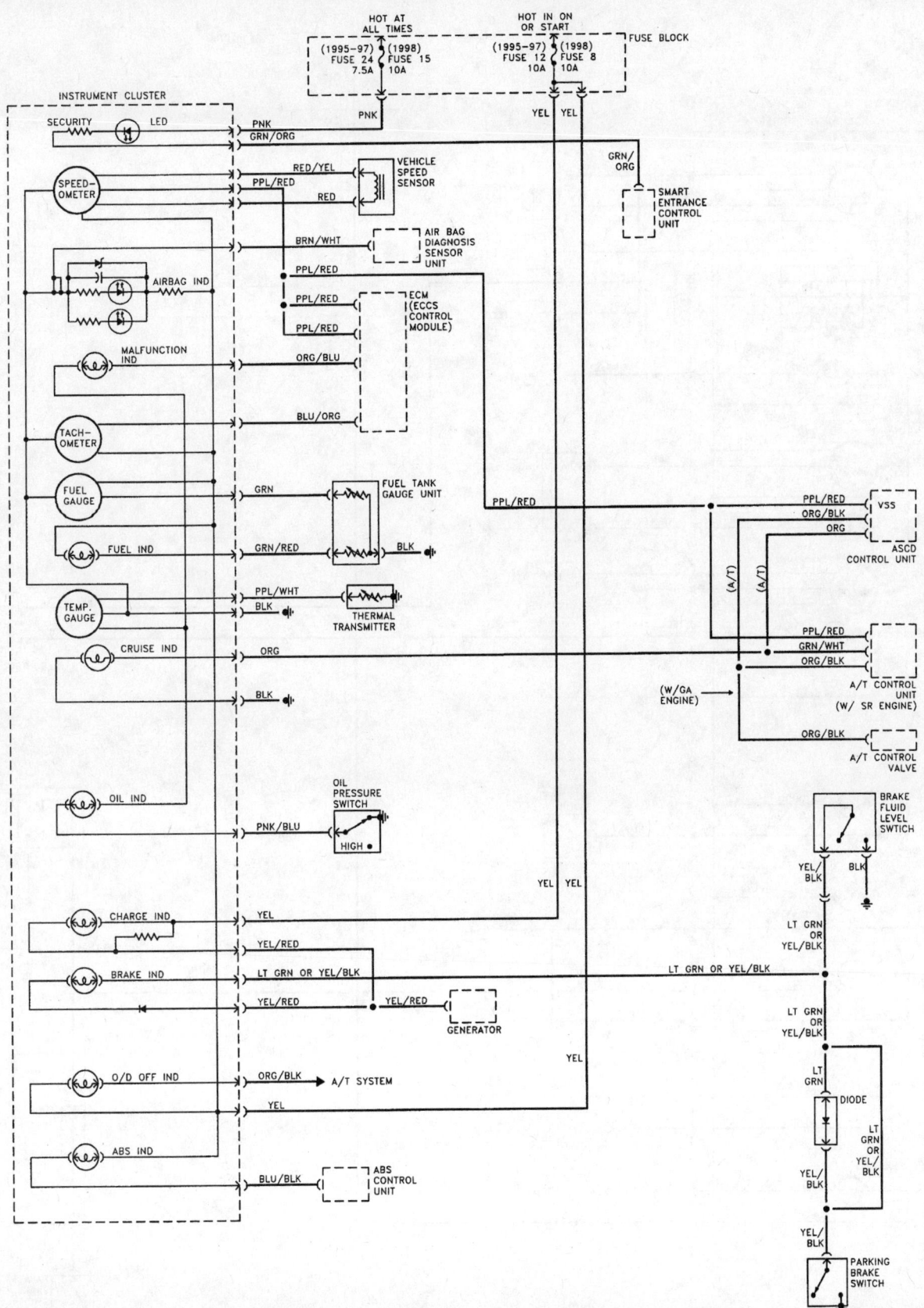

Instrument panel wiring diagram (with tachometer)

Chapter 12 Chassis electrical system

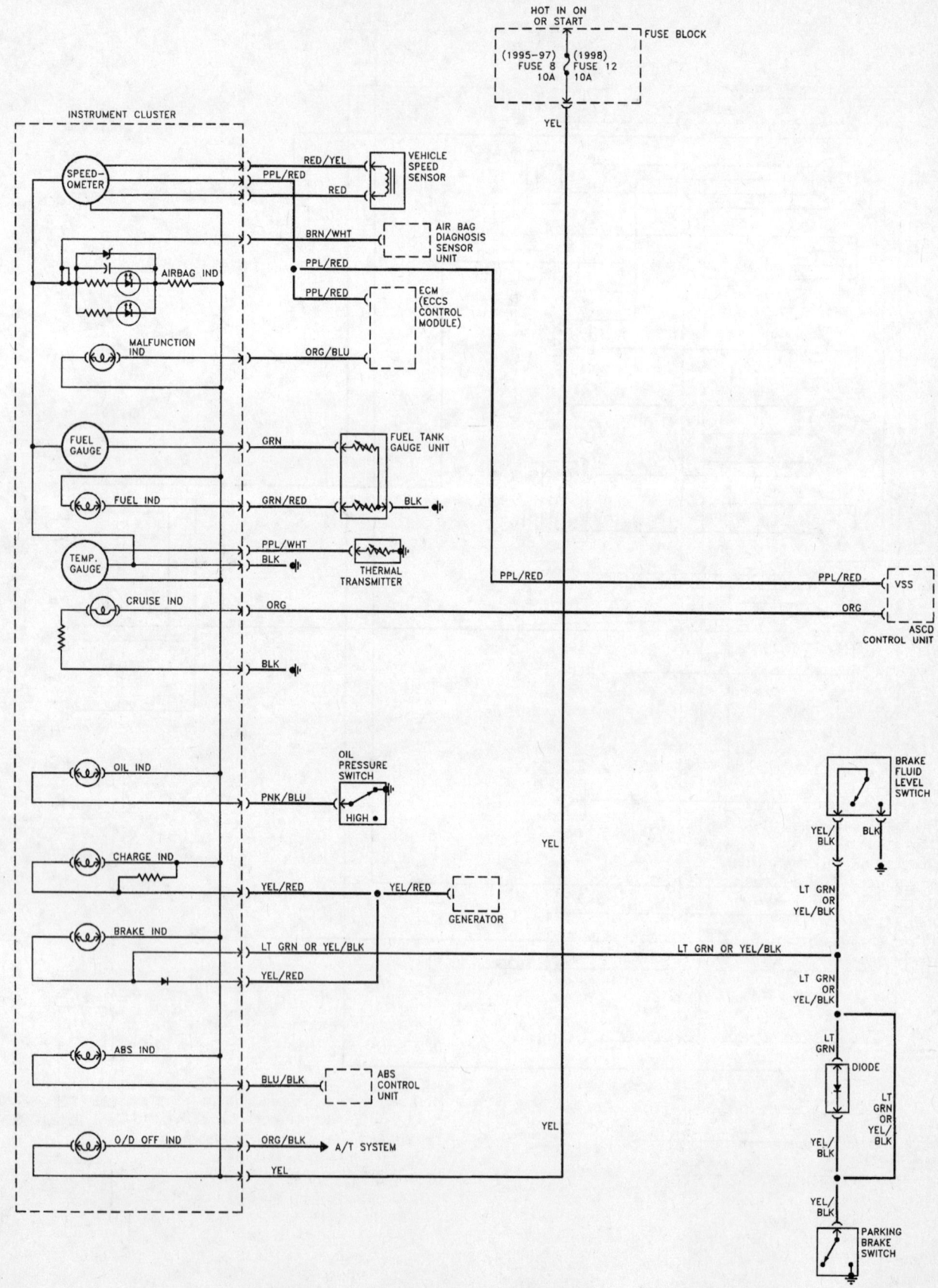

Instrument panel wiring diagram (without tachometer)

Index

A

About this manual, 0-2
Absolute Pressure Sensor, 6-13
Accelerator cable, removal, installation and adjustment, 4-6
Air conditioning
    and heater control assembly, removal and installation, 3-10
    and heating system, check and maintenance, 3-10
    compressor, removal and installation, 3-11
    condenser, removal and installation, 3-12
    evaporator and expansion valve, removal and installation, 3-12
    receiver/drier, removal and installation, 3-11
Air filter assembly, removal and installation, 4-6
Air filter replacement, 1-2
Air regulator (SR20DE engine only), 4-10
Airbag system, general information, 12-12
Alternator
    components, check and replacement, 5-5
    removal and installation, 5-5
Antifreeze, general information, 3-2
Anti-lock Brake System (ABS), general information, 9-2
Automatic transaxle, 7B-1 through 7B-6

B

Back-up light and neutral switch (position switch), check and replacement, 7A-4
Balljoints, replacement, 10-6
Battery
    cables, check and replacement, 5-2
    check, maintenance and charging, 1-1
    removal and installation, 5-2
Blower motor and circuit, check, removal and component replacement, 3-7
Body repair
    major damage, 11-3
    minor damage, 11-2
Booster battery (jump) starting, 0-12

Brakes, 9-1 through 9-22
    caliper, removal, overhaul and installation, 9-9
    check, 1-2
    disc, inspection, removal and installation, 9-10
    hoses and lines, inspection and replacement, 9-18
    hydraulic system, bleeding, 9-19
    light switch, check and replacement, 9-21
    master cylinder, removal, overhaul and installation, 9-15
    pads, disc, replacement, 9-4
    parking
        cables, replacement, 9-20
        check and adjustment, 9-20
    proportioning valve, replacement, 9-18
    shoes, drum, replacement, 9-11
    wheel cylinder, removal, overhaul and installation, 9-14
Bulb replacement, 12-9
Bumpers, removal and installation, 11-9
Buying parts, 0-6

C

Camshaft position sensor (CMPS), 6-13
Camshafts, removal, inspection and installation, 2A-13
Capacities, 1-1
Catalytic converter, 6-17
Charging system, check, general information and precautions, 5-4
Chassis electrical system, 12-1 through 12-32
Circuit breakers, general information, 12-3
Clutch and driveaxles, 8-1 through 8-12
Clutch
    cable, removal and installation, 8-2
    components, removal, inspection and installation, 8-2
    description and check, 8-2
    pedal height, freeplay and release lever freeplay check and adjustment, 1-1
    release bearing and lever, removal, inspection and installation, 8-4
    start switch, check and replacement, 8-5
Combination switch, removal and installation, 12-4

Index

Composite headlight housing, removal and installation, 12-8
Compression check, 2B-5
Connecting rod bearing oil clearance check, 2B-22
Console, removal and installation, 11-10
Conversion factors, 0-14
Coolant reservoir, removal and installation, 3-5
Coolant temperature gauge sending unit, check and replacement, 3-7
Cooling system, check and servicing, 1-2
Cooling, heating and air conditioning systems, 3-1 through 3-12
Crankshaft
   inspection, 2B-17
   installation and main bearing oil clearance check, 2B-21
   removal, 2B-13
   front oil seal, replacement, 2A-12
   position sensor (CKPS), 6-13
Cruise control system, description and check, 12-11
Cylinder head
   cleaning and inspection, 2B-9
   disassembly, 2B-8
   reassembly, 2B-11
   removal and installation, 2A-16
Cylinder honing, 2B-15

D

Dashboard trim panels, removal and installation, 11-11
Daytime Running Lights (DRL), general information, 12-10
Diagnosis, general, 7B-1
Disc brake
   caliper, removal, overhaul and installation, 9-9
   inspection, removal and installation, 9-10
   pads, replacement, 9-4
Distributor, removal and installation, 5-3
Door
   latch, lock cylinder and handles, removal and installation, 11-8
   removal, installation and adjustment, 11-8
   trim panel, removal and installation, 11-7
   window glass, removal and installation, 11-9
Driveaxle boot
   check, 1-2
   replacement and CV joint overhaul, 8-9
Driveaxles, general information, inspection, removal and installation, 8-6
Drivebelt check, adjustment and replacement, 1-1
Drum brake shoes, replacement, 9-11

E

EGR gas temperature sensor, 6-12
Electric rear view mirrors, description and check, 12-12
Electrical troubleshooting, general information, 12-1
Electronic fuel injection system
   check, 4-7
   general information, 4-6
Emissions and engine control systems, 6-1 through 6-18
Engine
   block
      cleaning, 2B-13
      inspection, 2B-14
   camshafts, removal, inspection and installation, 2A-13
   compression check, 2B-5
   coolant, 1-1
   crankshaft front oil seal, replacement, 2A-12
   crankshaft
      inspection, 2B-17
      installation and main bearing oil clearance check, 2B-21
      removal, 2B-13
   cylinder head
      cleaning and inspection, 2B-9
      disassembly, 2B-8
      reassembly, 2B-11
      removal and installation, 2A-16
   cylinder honing, 2B-15
   exhaust manifold, removal and installation, 2A-6
   flywheel/driveplate, removal and installation, 2A-22
   general overhaul procedures, 2B-1
   idle speed check and adjustment, 1-2
   initial start-up and break-in after overhaul, 2B-24
   intake manifold, removal and installation, 2A-5
   main and connecting rod bearings, inspection and selection, 2B-18
   mounts, check and replacement, 2A-23
   oil pan, removal and installation, 2A-18
   oil pump, removal, inspection and installation, 2A-20
   overhaul
      disassembly sequence, 2B-7
      general information, 2B-4
      reassembly sequence, 2B-19
   piston rings, installation, 2B-19
   pistons/connecting rods
      inspection, 2B-16
      installation and rod bearing oil clearance check, 2B-22
      removal, 2B-12
   rear main oil seal
      installation (during overhaul), 2B-22
      replacement (in-vehicle), 2A-23
   rebuilding alternatives, 2B-7
   removal, methods and precautions, 2B-5
   repair operations possible with the engine in the vehicle, 2A-3
   timing chain, removal, inspection and installation, 2A-6
   vacuum gauge diagnostic checks, 2B-4
   valve cover, removal and installation, 2A-4
   valve springs, retainers and seals, replacement, 2A-15
   valves, servicing, 2B-11
Engine Control Module (ECM), removal and installation, 6-8
Engine Coolant Temperature (ECT) sensor, 6-8
Engine cooling fan(s) and circuit(s), check and component replacement, 3-4
Engine electrical systems, 5-1 through 5-8
Engines, 2A-1 through 2A-24
Evaporative Emissions Control System (EVAP), 1-2, 6-16
Exhaust Gas Recirculation (EGR) system, 6-14
Exhaust manifold, removal and installation, 2A-6
Exhaust system, 4-1 through 4-12
   air filter assembly, removal and installation, 4-6
   check, 1-2

F

Fluid level checks, 1-1
Flywheel/driveplate, removal and installation, 2A-22
Fuel system, 4-1 through 4-12
   electronic fuel injection system
      check, 4-7
      general information, 4-6
   fuel
      filter replacement, 1-2
      injectors, 4-8
      level sending unit, check and replacement, 4-5
      lines and fittings, repair and replacement, 4-4
      pressure regulator, 4-8
      pressure relief procedure, 4-2
      pump, removal and installation, 4-3
      pump/fuel pressure, check, 4-2
      system check, 1-2
      tank
         removal and installation, 4-5
         cleaning and repair, general information, 4-5

Index

Fuses, general information, 12-2
Fusible links, general information, 12-3

G

General engine overhaul procedures, 2B-1 through 2B-24

H

Headlight bulb, replacement, 12-7
Headlights, adjustment, 12-8
Heated oxygen sensor (HO2S) (front and rear), 6-9
Heater core, removal and installation, 3-8
Hinges and locks, maintenance, 11-3
Hood, removal, installation and adjustment, 11-3
Horn, check and replacement, 12-11
Hub and bearing assembly, removal and installation
  front, 10-6
  rear, 10-7

I

Idle Air Control/Auxiliary Air Control (IAC-AAC) valve, 4-9
Idle speed, 1-1
Ignition switch and key lock cylinder, removal and installation, 12-5
Ignition system, 1-1
  check, 5-3
  general information, 5-2
Ignition timing, check and adjustment, 1-1 and 1-2
Information sensors, 6-8
  Absolute Pressure Sensor, 6-13
  Camshaft position sensor (CMPS), 6-13
  Crankshaft position sensor (CKPS), 6-13
  EGR gas temperature sensor, 6-12
  Engine Coolant Temperature (ECT) sensor, 6-8
  Heated oxygen sensor (HO2S) (front and rear), 6-9
  Intake air temperature (IAT) sensor, 6-12
  Knock sensor (KS), 6-13
  Mass Airflow (MAF) sensor, 6-11
  Park/Neutral Back-up light switch, 6-14
  Throttle Position Sensor (TPS), 6-10
  Vehicle Speed Sensor (VSS), 6-12
Initial start-up and break-in after overhaul, 2B-24
Instrument cluster
  bezel, removal and installation, 11-10
  removal and installation, 12-10
Instrument panel, removal and installation, 11-12
Intake air temperature (IAT) sensor, 6-12
Intake manifold, removal and installation, 2A-5
Introduction to the Nissan Sentra and 200SX, 0-4

J

Jacking and towing, 0-12

K

Knock sensor (KS), 6-13

M

Main and connecting rod bearings, inspection and selection, 2B-18

Maintenance schedule, 1-1
Maintenance techniques, tools and working facilities, 0-6
Manual transaxle, 7A-1 through 7A-6
  back-up light and neutral switch (position switch), check and replacement, 7A-4
  lubricant, change and level check, 1-2
  mount, check and replacement, 7A-3
  oil seal, replacement, 7A-2
  overhaul, general information, 7A-6
  removal and installation, 7A-4
Mass Airflow (MAF) sensor, 6-11
Master cylinder, removal, overhaul and installation, 9-15
Multi Port Fuel Injection (MPFI) system components, check, removal and installation, 4-7

N

Neutral start/back-up light (inhibitor) switch, check, adjustment and replacement, 7B-4

O

Oil pan, removal and installation, 2A-18
Oil pump, removal, inspection and installation, 2A-20
On Board Diagnosis (OBD) system and trouble codes, 6-2
Outside mirrors, removal and installation, 11-10

P

Park/Neutral Back-up light switch, 6-14
Parking brake cables, replacement, 9-20
Parking brake, check and adjustment, 9-20
Piston rings, installation, 2B-19
Pistons/connecting rods
  inspection, 2B-16
  installation and rod bearing oil clearance check, 2B-22
  removal, 2B-12
Positive Crankcase Ventilation (PCV) system, 6-17
Power brake booster, check, removal and installation, 9-19
Power door lock system, description and check, 12-12
Power steering
  fluid level check, 1-1
  pump, removal and installation, 10-11
  system, bleeding, 10-11
Power window system, description and check, 12-11
Proportioning valve, replacement, 9-18

R

Radiator
  grille, removal and installation, 11-6
  removal and installation, 3-4
Radio and speakers, removal and installation, 12-6
Rear axle assembly, removal and installation, 10-6
Rear main oil seal
  installation (during overhaul), 2B-22
  replacement (in-vehicle), 2A-23
Rear window defogger switch, check and replacement, 12-6
Recommended lubricants and fluids, 1-1
Relays, general information and testing, 12-3
Repair operations possible with the engine in the vehicle, 2A-3

S

Safety first!, 0-15
Seat belts, check, 11-14
Seats, removal and installation, 11-13
Shift cable, check, adjustment and replacement, 7B-2
Shift linkage, removal and installation, 7A-3
Shift lock system, description, check and component replacement, 7B-4
Shock absorber/coil spring assembly (rear), removal, inspection and installation, 10-7
Spark plug check and replacement, 1-2
Spark plug wire, distributor cap and rotor check and replacement, 1-2
Stabilizer bar (front), removal and installation, 10-5
Starter
- motor and circuit, in-vehicle check, 5-7
- motor, removal and installation, 5-8
- solenoid, replacement, 5-8

Starting system, general information and precautions, 5-6
Steering
- column cover, removal and installation, 11-12
- column switches, check and replacement, 12-5
- power steering pump, removal and installation, 10-11
- power, bleeding, 10-11

Strut assembly (front), removal, inspection and installation, 10-3
Strut/shock absorber or coil spring, replacement, 10-4
Support bearing assembly, removal, overhaul and installation, 8-8
Suspension and steering systems, 10-1 through 10-12
- check, 1-2
- control arm, removal, inspection and installation, 10-5
- gear boots, replacement, 10-10
- gear, removal and installation, 10-10
- hub and bearing assembly (front), removal and installation, 10-6
- knuckle and hub, removal and installation, 10-6
- rear axle assembly, removal and installation, 10-6
- shock absorber/coil spring assembly (rear), removal, inspection and installation, 10-7
- strut assembly (front), removal, inspection and installation, 10-3
- wheel alignment, general information, 10-12
- wheel, removal and installation, 10-8

T

Thermostat, check and replacement, 3-3
Throttle body, removal and installation, 4-8
Throttle Position Sensor (TPS), 4-8, 6-10
Tie-rod ends, removal and installation, 10-9
Timing chain, removal, inspection and installation, 2A-6
Tire
- pressure checks, 1-1
- rotation, 1-2

Tools, 0-8
Top Dead Center (TDC) for number one piston, locating, 2A-3
Trouble code chart
- 1995 models, 6-3
- 1996 and later models, 6-5

Troubleshooting, 0-16
Trunk lid, removal, installation and adjustment, 11-6
Tune-up and routine maintenance, 1-1 through 1-28
Turn signal/hazard flashers, check and replacement, 12-4

U

Underhood hose check and replacement, 1-1
Upholstery and carpets, maintenance, 11-2

V

Vacuum gauge diagnostic checks, 2B-4
Valve clearance check and adjustment (GA16DE engine only), 1-2
Valve cover, removal and installation, 2A-4
Valve springs, retainers and seals, replacement, 2A-15
Valves, servicing, 2B-11
Vehicle Identification Number (VIN), 0-5
Vehicle Speed Sensor (VSS), 6-12
Vinyl trim, maintenance, 11-2

W

Water pump, check and replacement, 3-6
Wheel alignment, general information, 10-12
Wheel cylinder, removal, overhaul and installation, 9-14
Wheels and tires, general information, 10-11
Window regulator, removal and installation, 11-9
Windshield and fixed glass, replacement, 11-3
Windshield wiper blade inspection and replacement, 1-1
Wiper motor, removal and installation, 12-10
Wiring diagrams, general information, 12-13
Working facilities, 0-11

Haynes Automotive Manuals

NOTE: New manuals are added to this list on a periodic basis. If you do not see a listing for your vehicle, consult your local Haynes dealer for the latest product information.

ACURA
- *12020 Integra '86 thru '89 & Legend '86 thru '90

AMC
- Jeep CJ - see JEEP (50020)
- 14020 Mid-size models, Concord, Hornet, Gremlin & Spirit '70 thru '83
- 14025 (Renault) Alliance & Encore '83 thru '87

AUDI
- 15020 4000 all models '80 thru '87
- 15025 5000 all models '77 thru '83
- 15026 5000 all models '84 thru '88

AUSTIN-HEALEY
- Sprite - see MG Midget (66015)

BMW
- *18020 3/5 Series not including diesel or all-wheel drive models '82 thru '92
- *18021 3 Series except 325iX models '92 thru '97
- 18025 320i all 4 cyl models '75 thru '83
- 18035 528i & 530i all models '75 thru '80
- 18050 1500 thru 2002 except Turbo '59 thru '77

BUICK
- Century (front wheel drive) - see GM (829)
- *19020 Buick, Oldsmobile & Pontiac Full-size (Front wheel drive) all models '85 thru '98
  Buick Electra, LeSabre and Park Avenue;
  Oldsmobile Delta 88 Royale, Ninety Eight and Regency; Pontiac Bonneville
- 19025 Buick Oldsmobile & Pontiac Full-size (Rear wheel drive)
  Buick Estate '70 thru '90, Electra '70 thru '84, LeSabre '70 thru '85, Limited '74 thru '79
  Oldsmobile Custom Cruiser '70 thru '90, Delta 88 '70 thru '85, Ninety-eight '70 thru '84
  Pontiac Bonneville '70 thru '81, Catalina '70 thru '81, Grandville '70 thru '75, Parisienne '83 thru '86
- 19030 Mid-size Regal & Century all rear-drive models with V6, V8 and Turbo '74 thru '87
  Regal - see GENERAL MOTORS (38010)
  Riviera - see GENERAL MOTORS (38030)
  Roadmaster - see CHEVROLET (24046)
  Skyhawk - see GENERAL MOTORS (38015)
  Skylark '80 thru '85 - see GM (38020)
  Skylark '86 on - see GM (38025)
  Somerset - see GENERAL MOTORS (38025)

CADILLAC
- *21030 Cadillac Rear Wheel Drive all gasoline models '70 thru '93
  Cimarron - see GENERAL MOTORS (38015)
  Eldorado - see GENERAL MOTORS (38030)
  Seville '80 thru '85 - see GM (38030)

CHEVROLET
- *24010 Astro & GMC Safari Mini-vans '85 thru '93
- 24015 Camaro V8 all models '70 thru '81
- 24016 Camaro all models '82 thru '92
  Cavalier - see GENERAL MOTORS (38015)
  Celebrity - see GENERAL MOTORS (38005)
- 24017 Camaro & Firebird '93 thru '97
- 24020 Chevelle, Malibu & El Camino '69 thru '87
- 24024 Chevette & Pontiac T1000 '76 thru '87
  Citation - see GENERAL MOTORS (38020)
- *24032 Corsica/Beretta all models '87 thru '96
- 24040 Corvette all V8 models '68 thru '82
- *24041 Corvette all models '84 thru '96
- 10305 Chevrolet Engine Overhaul Manual
- 24045 Full-size Sedans Caprice, Impala, Biscayne, Bel Air & Wagons '69 thru '90
- 24046 Impala SS & Caprice and Buick Roadmaster '91 thru '96
  Lumina - see GENERAL MOTORS (38010)
- 24048 Lumina & Monte Carlo '95 thru '98
  Lumina APV - see GM (38035)
- 24050 Luv Pick-up all 2WD & 4WD '72 thru '82
- *24055 Monte Carlo all models '70 thru '88
  Monte Carlo '95 thru '98 - see LUMINA (24048)
- 24059 Nova all V8 models '69 thru '79
- *24060 Nova and Geo Prizm '85 thru '92
- 24064 Pick-ups '67 thru '87 - Chevrolet & GMC, all V8 & in-line 6 cyl, 2WD & 4WD '67 thru '87; Suburbans, Blazers & Jimmys '67 thru '91
- *24065 Pick-ups '88 thru '98 - Chevrolet & GMC, all full-size pick-ups, '88 thru '98; Blazer & Jimmy '92 thru '94; Suburban '92 thru '98; Tahoe & Yukon '98
- 24070 S-10 & S-15 Pick-ups '82 thru '93, Blazer & Jimmy '83 thru '94,
- *24071 S-10 & S-15 Pick-ups '94 thru '96 Blazer & Jimmy '95 thru '96
- *24075 Sprint & Geo Metro '85 thru '94
- *24080 Vans - Chevrolet & GMC, V8 & in-line 6 cylinder models '68 thru '96

CHRYSLER
- 25015 Chrysler Cirrus, Dodge Stratus, Plymouth Breeze '95 thru '98
- 25025 Chrysler Concorde, New Yorker & LHS, Dodge Intrepid, Eagle Vision, '93 thru '97
- 10310 Chrysler Engine Overhaul Manual
- *25020 Full-size Front-Wheel Drive '88 thru '93
  K-Cars - see DODGE Aries (30008)
  Laser - see DODGE Daytona (30030)
- *25030 Chrysler & Plymouth Mid-size front wheel drive '82 thru '95
  Rear-wheel Drive - see Dodge (30050)

DATSUN
- 28005 200SX all models '80 thru '83
- 28007 B-210 all models '73 thru '78
- 28009 210 all models '79 thru '82
- 28012 240Z, 260Z & 280Z Coupe '70 thru '78
- 28014 280ZX Coupe & 2+2 '79 thru '83
  300ZX - see NISSAN (72010)
- 28016 310 all models '78 thru '82
- 28018 510 & PL521 Pick-up '68 thru '73
- 28020 510 all models '78 thru '81
- 28022 620 Series Pick-up all models '73 thru '79
  720 Series Pick-up - see NISSAN (72030)
- 28025 810/Maxima all gasoline models, '77 thru '84

DODGE
- 400 & 600 - see CHRYSLER (25030)
- *30008 Aries & Plymouth Reliant '81 thru '89
- 30010 Caravan & Plymouth Voyager Mini-Vans all models '84 thru '95
- *30011 Caravan & Plymouth Voyager Mini-Vans all models '96 thru '98
- 30012 Challenger/Plymouth Saporro '78 thru '83
- 30016 Colt & Plymouth Champ (front wheel drive) all models '78 thru '87
- *30020 Dakota Pick-ups all models '87 thru '96
- 30025 Dart, Demon, Plymouth Barracuda, Duster & Valiant 6 cyl models '67 thru '76
- *30030 Daytona & Chrysler Laser '84 thru '89
  Intrepid - see CHRYSLER (25025)
- *30034 Neon all models '95 thru '97
- *30035 Omni & Plymouth Horizon '78 thru '90
- *30040 Pick-ups all full-size models '74 thru '93
- *30041 Pick-ups all full-size models '94 thru '96
- *30045 Ram 50/D50 Pick-ups & Raider and Plymouth Arrow Pick-ups '79 thru '93
- 30050 Dodge/Plymouth/Chrysler rear wheel drive '71 thru '89
- *30055 Shadow & Plymouth Sundance '87 thru '94
- *30060 Spirit & Plymouth Acclaim '89 thru '95
- *30065 Vans - Dodge & Plymouth '71 thru '96

EAGLE
- Talon - see Mitsubishi Eclipse (68030)
- Vision - see CHRYSLER (25025)

FIAT
- 34010 124 Sport Coupe & Spider '68 thru '78
- 34025 X1/9 all models '74 thru '80

FORD
- 10355 Ford Automatic Transmission Overhaul
- *36004 Aerostar Mini-vans all models '86 thru '96
- *36006 Contour & Mercury Mystique '95 thru '98
- 36008 Courier Pick-up all models '72 thru '82
- 36012 Crown Victoria & Mercury Grand Marquis '88 thru '96
- 10320 Ford Engine Overhaul Manual
- 36016 Escort/Mercury Lynx all models '81 thru '90
- *36020 Escort/Mercury Tracer '91 thru '96
- *36024 Explorer & Mazda Navajo '91 thru '95
- 36028 Fairmont & Mercury Zephyr '78 thru '83
- 36030 Festiva & Aspire '88 thru '97
- 36032 Fiesta all models '77 thru '80
- 36036 Ford & Mercury Full-size,
  Ford LTD & Mercury Marquis ('75 thru '82);
  Ford Custom 500, Country Squire, Crown Victoria & Mercury Colony Park ('75 thru '87);
  Ford LTD Crown Victoria &
  Mercury Gran Marquis ('83 thru '87)
- 36040 Granada & Mercury Monarch '75 thru '80
- 36044 Ford & Mercury Mid-size,
  Ford Thunderbird & Mercury Cougar ('75 thru '82);
  Ford LTD & Mercury Marquis ('83 thru '86);
  Ford Torino, Gran Torino, Elite, Ranchero pick-up, LTD II, Mercury Montego, Comet, XR-7 & Lincoln Versailles ('75 thru '86)
- 36048 Mustang V8 all models '64-1/2 thru '73
- 36049 Mustang II 4 cyl, V6 & V8 models '74 thru '78
- 36050 Mustang & Mercury Capri all models Mustang, '79 thru '93; Capri, '79 thru '86
- *36051 Mustang all models '94 thru '97
- 36054 Pick-ups & Bronco '73 thru '79
- 36058 Pick-ups & Bronco '80 thru '96
- 36059 Pick-ups, Expedition & Mercury Navigator '97 thru '98
- 36062 Pinto & Mercury Bobcat '75 thru '80
- 36066 Probe all models '89 thru '92
- 36070 Ranger/Bronco II gasoline models '83 thru '92
- *36071 Ranger '93 thru '97 & Mazda Pick-ups '94 thru '97
- 36074 Taurus & Mercury Sable '86 thru '95
- *36075 Taurus & Mercury Sable '96 thru '98
- *36078 Tempo & Mercury Topaz '84 thru '94
- 36082 Thunderbird/Mercury Cougar '83 thru '88
- *36086 Thunderbird/Mercury Cougar '89 and '97
- 36090 Vans all V8 Econoline models '69 thru '91
- *36094 Vans full size '92 thru '95
- *36097 Windstar Mini-van '95-'98

GENERAL MOTORS
- *10360 GM Automatic Transmission Overhaul
- *38005 Buick Century, Chevrolet Celebrity, Oldsmobile Cutlass Ciera & Pontiac 6000 all models '82 thru '96
- *38010 Buick Regal, Chevrolet Lumina, Oldsmobile Cutlass Supreme & Pontiac Grand Prix front-wheel drive models '88 thru '95
- *38015 Buick Skyhawk, Cadillac Cimarron, Chevrolet Cavalier, Oldsmobile Firenza & Pontiac J-2000 & Sunbird '82 thru '94
- *38016 Chevrolet Cavalier & Pontiac Sunfire '95 thru '98
- 38020 Buick Skylark, Chevrolet Citation, Olds Omega, Pontiac Phoenix '80 thru '85
- 38025 Buick Skylark & Somerset, Oldsmobile Achieva & Calais and Pontiac Grand Am all models '85 thru '95
- 38030 Cadillac Eldorado '71 thru '85, Seville '80 thru '85, Oldsmobile Toronado '71 thru '85 & Buick Riviera '79 thru '85
- *38035 Chevrolet Lumina APV, Olds Silhouette & Pontiac Trans Sport all models '90 thru '95
- General Motors Full-size Rear-wheel Drive - see BUICK (19025)

(Continued on other side)

---

* Listings shown with an asterisk (*) indicate model coverage as of this printing. These titles will be periodically updated to include later model years - consult your Haynes dealer for more information.

Haynes North America, Inc., 861 Lawrence Drive, Newbury Park, CA 91320-1514 • (805) 498-6703

Haynes Automotive Manuals (continued)

NOTE: New manuals are added to this list on a periodic basis. If you do not see a listing for your vehicle, consult your local Haynes dealer for the latest product information.

GEO
- Metro - see CHEVROLET Sprint (24075)
- Prizm - '85 thru '92 see CHEVY (24060), '93 thru '96 see TOYOTA Corolla (92036)
- *40030 Storm all models '90 thru '93
- Tracker - see SUZUKI Samurai (90010)

GMC
- Safari - see CHEVROLET ASTRO (24010)
- Vans & Pick-ups - see CHEVROLET

HONDA
- 42010 Accord CVCC all models '76 thru '83
- 42011 Accord all models '84 thru '89
- 42012 Accord all models '90 thru '93
- 42013 Accord all models '94 thru '95
- 42020 Civic 1200 all models '73 thru '79
- 42021 Civic 1300 & 1500 CVCC '80 thru '83
- 42022 Civic 1500 CVCC all models '75 thru '79
- 42023 Civic all models '84 thru '91
- *42024 Civic & del Sol '92 thru '95
- *42040 Prelude CVCC all models '79 thru '89

HYUNDAI
- *43015 Excel all models '86 thru '94

ISUZU
- Hombre - see CHEVROLET S-10 (24071)
- *47017 Rodeo '91 thru '97; Amigo '89 thru '94; Honda Passport '95 thru '97
- *47020 Trooper & Pick-up, all gasoline models Pick-up, '81 thru '93; Trooper, '84 thru '91

JAGUAR
- *49010 XJ6 all 6 cyl models '68 thru '86
- *49011 XJ6 all models '88 thru '94
- *49015 XJ12 & XJS all 12 cyl models '72 thru '85

JEEP
- *50010 Cherokee, Comanche & Wagoneer Limited all models '84 thru '96
- 50020 CJ all models '49 thru '86
- *50025 Grand Cherokee all models '93 thru '98
- 50029 Grand Wagoneer & Pick-up '72 thru '91 Grand Wagoneer '84 thru '91, Cherokee & Wagoneer '72 thru '83, Pick-up '72 thru '88
- *50030 Wrangler all models '87 thru '95

LINCOLN
- Navigator - see FORD Pick-up (36059)
- 59010 Rear Wheel Drive all models '70 thru '96

MAZDA
- 61010 GLC Hatchback (rear wheel drive) '77 thru '83
- 61011 GLC (front wheel drive) '81 thru '85
- *61015 323 & Protegé '90 thru '97
- *61016 MX-5 Miata '90 thru '97
- *61020 MPV all models '89 thru '94
- Navajo - see Ford Explorer (36024)
- 61030 Pick-ups '72 thru '93 Pick-ups '94 thru '96 - see Ford Ranger (36071)
- 61035 RX-7 all models '79 thru '85
- *61036 RX-7 all models '86 thru '91
- 61040 626 (rear wheel drive) all models '79 thru '82
- *61041 626/MX-6 (front wheel drive) '83 thru '91

MERCEDES-BENZ
- 63012 123 Series Diesel '76 thru '85
- *63015 190 Series four-cyl gas models, '84 thru '88
- 63020 230/250/280 6 cyl sohc models '68 thru '72
- 63025 280 123 Series gasoline models '77 thru '81
- 63030 350 & 450 all models '71 thru '80

MERCURY
- See FORD Listing.

MG
- 66010 MGB Roadster & GT Coupe '62 thru '80
- 66015 MG Midget, Austin Healey Sprite '58 thru '80

MITSUBISHI
- *68020 Cordia, Tredia, Galant, Precis & Mirage '83 thru '93
- *68030 Eclipse, Eagle Talon & Ply. Laser '90 thru '94
- *68040 Pick-up '83 thru '96 & Montero '83 thru '93

NISSAN
- 72010 300ZX all models including Turbo '84 thru '89
- *72015 Altima all models '93 thru '97
- *72020 Maxima all models '85 thru '91
- *72030 Pick-ups '80 thru '96 Pathfinder '87 thru '95
- 72040 Pulsar all models '83 thru '86
- *72050 Sentra all models '82 thru '94
- *72051 Sentra & 200SX all models '95 thru '98
- *72060 Stanza all models '82 thru '90

OLDSMOBILE
- *73015 Cutlass V6 & V8 gas models '74 thru '88
- For other OLDSMOBILE titles, see BUICK, CHEVROLET or GENERAL MOTORS listing.

PLYMOUTH
- For PLYMOUTH titles, see DODGE listing.

PONTIAC
- 79008 Fiero all models '84 thru '88
- 79018 Firebird V8 models except Turbo '70 thru '81
- 79019 Firebird all models '82 thru '92
- For other PONTIAC titles, see BUICK, CHEVROLET or GENERAL MOTORS listing.

PORSCHE
- *80020 911 except Turbo & Carrera 4 '65 thru '89
- 80025 914 all 4 cyl models '69 thru '76
- 80030 924 all models including Turbo '76 thru '82
- *80035 944 all models including Turbo '83 thru '89

RENAULT
- Alliance & Encore - see AMC (14020)

SAAB
- *84010 900 all models including Turbo '79 thru '88

SATURN
- 87010 Saturn all models '91 thru '96

SUBARU
- 89002 1100, 1300, 1400 & 1600 '71 thru '79
- *89003 1600 & 1800 2WD & 4WD '80 thru '94

SUZUKI
- *90010 Samurai/Sidekick & Geo Tracker '86 thru '96

TOYOTA
- 92005 Camry all models '83 thru '91
- 92006 Camry all models '92 thru '96
- 92015 Celica Rear Wheel Drive '71 thru '85
- *92020 Celica Front Wheel Drive '86 thru '93
- 92025 Celica Supra all models '79 thru '92
- 92030 Corolla all models '75 thru '79
- 92032 Corolla all rear wheel drive models '80 thru '87
- 92035 Corolla all front wheel drive models '84 thru '92
- *92036 Corolla & Geo Prizm '93 thru '97
- 92040 Corolla Tercel all models '80 thru '82
- 92045 Corona all models '74 thru '82
- 92050 Cressida all models '78 thru '82
- 92055 Land Cruiser FJ40, 43, 45, 55 '68 thru '82
- 92056 Land Cruiser FJ60, 62, 80, FZJ80 '80 thru '96
- *92065 MR2 all models '85 thru '87
- 92070 Pick-up all models '69 thru '78
- *92075 Pick-up all models '79 thru '95
- *92076 Tacoma '95 thru '98, 4Runner '96 thru '98, & T100 '93 thru '98
- *92080 Previa all models '91 thru '95
- 92085 Tercel all models '87 thru '94

TRIUMPH
- 94007 Spitfire all models '62 thru '81
- 94010 TR7 all models '75 thru '81

VW
- 96008 Beetle & Karmann Ghia '54 thru '79
- 96012 Dasher all gasoline models '74 thru '81
- *96016 Rabbit, Jetta, Scirocco, & Pick-up gas models '74 thru '91 & Convertible '80 thru '92
- 96017 Golf & Jetta all models '93 thru '97
- 96020 Rabbit, Jetta & Pick-up diesel '77 thru '84
- 96030 Transporter 1600 all models '68 thru '79
- 96035 Transporter 1700, 1800 & 2000 '72 thru '79
- 96040 Type 3 1500 & 1600 all models '63 thru '73
- 96045 Vanagon all air-cooled models '80 thru '83

VOLVO
- 97010 120, 130 Series & 1800 Sports '61 thru '73
- 97015 140 Series all models '66 thru '74
- *97020 240 Series all models '76 thru '93
- 97025 260 Series all models '75 thru '82
- *97040 740 & 760 Series all models '82 thru '88

TECHBOOK MANUALS
- 10205 Automotive Computer Codes
- 10210 Automotive Emissions Control Manual
- 10215 Fuel Injection Manual, 1978 thru 1985
- 10220 Fuel Injection Manual, 1986 thru 1996
- 10225 Holley Carburetor Manual
- 10230 Rochester Carburetor Manual
- 10240 Weber/Zenith/Stromberg/SU Carburetors
- 10305 Chevrolet Engine Overhaul Manual
- 10310 Chrysler Engine Overhaul Manual
- 10320 Ford Engine Overhaul Manual
- 10330 GM and Ford Diesel Engine Repair Manual
- 10340 Small Engine Repair Manual
- 10345 Suspension, Steering & Driveline Manual
- 10355 Ford Automatic Transmission Overhaul
- 10360 GM Automatic Transmission Overhaul
- 10405 Automotive Body Repair & Painting
- 10410 Automotive Brake Manual
- 10415 Automotive Detailing Manual
- 10420 Automotive Eelectrical Manual
- 10425 Automotive Heating & Air Conditioning
- 10430 Automotive Reference Manual & Dictionary
- 10435 Automotive Tools Manual
- 10440 Used Car Buying Guide
- 10445 Welding Manual
- 10450 ATV Basics

SPANISH MANUALS
- 98903 Reparación de Carrocería & Pintura
- 98905 Códigos Automotrices de la Computadora
- 98910 Frenos Automotriz
- 98915 Inyección de Combustible 1986 al 1994
- 99040 Chevrolet & GMC Camionetas '67 al '87 Incluye Suburban, Blazer & Jimmy '67 al '91
- 99041 Chevrolet & GMC Camionetas '88 al '95 Incluye Suburban '92 al '95, Blazer & Jimmy '92 al '94, Tahoe y Yukon '95
- 99042 Chevrolet & GMC Camionetas Cerradas '68 al '95
- 99055 Dodge Caravan & Plymouth Voyager '84 al '95
- 99075 Ford Camionetas y Bronco '80 al '94
- 99077 Ford Camionetas Cerradas '69 al '91
- 99083 Ford Modelos de Tamaño Grande '75 al '87
- 99088 Ford Modelos de Tamaño Mediano '75 al '86
- 99091 Ford Taurus & Mercury Sable '86 al '95
- 99095 GM Modelos de Tamaño Grande '70 al '90
- 99100 GM Modelos de Tamaño Mediano '70 al '88
- 99110 Nissan Camionetas '80 al '96, Pathfinder '87 al '95
- 99118 Nissan Sentra '82 al '94
- 99125 Toyota Camionetas y 4Runner '79 al '95

Over 100 Haynes motorcycle manuals also available

* Listings shown with an asterisk (*) indicate model coverage as of this printing. These titles will be periodically updated to include later model years - consult your Haynes dealer for more information.

5-98

Haynes North America, Inc., 861 Lawrence Drive, Newbury Park, CA 91320-1514 • (805) 498-6703